Evolution of Water Supply Throughout the Millennia

Evolution of Water Supply Throughout the Millennia

Andreas N Angelakis, Larry W Mays, Demetris Koutsoyiannis and Nikos Mamassis

IWA Publishing
London • New York

Published by

IWA Publishing
Alliance House
12 Caxton Street
London SW1H 0QS, UK
Telephone: +44 (0)20 7654 5500
Fax: +44 (0)20 7654 5555
Email: publications@iwap.co.uk
Web: www.iwapublishing.com

First published 2012

British Library Cataloguing in Publication Data
A CIP catalogue record for this book is available from the British Library

Library of Congress Cataloging-in-Publication Data

ISBN 9781843395409 (Paperback)
ISBN 9781780401041 (eBook)

Contents

List of authors

1. Dr. Andreas N. Angelakis, Institute of Iraklion, National Foundation for Agricultural Research (N.AG. RE.F.), 71307 Iraklion, Greece, info@a-angelakis.gr.
2. Dr. Diana Birrichaga, Universidad Autónoma del Estado de México, Toluca, Estado de México, México, dianabirrichaga@gmail.com.
3. Dr. Emmanouil G. Chalkiadakis, University of Oxford, St Antony's College, Oxford, UK, Greece, manoschalkiadakis@yahoo.gr.
4. Dr. Stathis D. Chiotis, Institute of Geology and Mineral Exploration, Athens, Greece, echiotis@otenet.gr.
5. Lamprini E. Chioti, Department of Archaeology and History, University of Athens, Athens, Greece, chioti_lambrini@hotmail.com.
6. Prof. Giovanni De Feo, University of Salerno, Department of Civil Engineering, via Ponte don Melillo, 1 – 84084 Fisciano (SA), Italy, g.defeo@unisa.it.
7. Dr. Vassilios K. Despotakis, Mastracha 4, 71202 Iraklion, Greece, vdesp@tee.gr.
8. Dr. Manolis G. Dialynas, DIALYNAS SA., 71202, Iraklion, Greece, md@dialynas.com.
9. Dr. Renato Drusiani, FederUtility, Rome, Italy, acqua@federutility.it
10. Prof. Pengfei Du, School of Environment, Tsinghua University, Beijing 100084, P.R. China, dupf@tsinghua.edu.cn
11. Prof. Fatma El Gohari, National Research Center, Water Pollution Research Department El behooth Str., Dokki, Cairo, Egypt, fgohary@hotmail.com.
12. Dr. Nikos M. Gigourtakis, Department of History and Archaeology University of Crete, Rethymnon, Greece, ngigourtakis@hotmail.com.
13. Prof. Benoit Haut, Université Libre de Bruxelles, C.P. 165/67, 50 Av. F.D. Roosevelt, 1050 Bruxelles, Belgium, bhaut@ulb.ac.be.
14. Prof (Em.) Arie Issar, The Institutes for Desert Research, Ben-Gurion University of The Negev Kiryat Sde-Boker 84990, Israel, issar@bgu.ac.il.
15. Prof. Blanca Jiménez, Universidad Nacional Autónoma de México, Instituto de Ingenieria, Apdo Postal 70472, 04510 México D. F., bjimenezc@iingen.unam.mx.

16. Dr. Chrisostomos A. Kambanelllas, Hydranos Ltd, P. O. Box: 23699, CY-1685, Nicosia, Cyprus, kambanel@cytanet.com.cy.
17. Prof. Albert Koenig, Department of Civil Engineering, The University of Hong Kong, Hong Kong SAR, P.R. China, kalbert@hkucc.hku.hk.
18. Prof. Demetris Koutsoyiannis, Department of Water Resources, School of Civil Engineering, National Technical University of Athens, Heroon Polytechneiou 5, GR-15780 Zographou, Greece, dk@itia.ntua.gr.
19. Dr. Pietro Laureano, IPOGEA,Vico Conservatorio s.n., 75100 Matera, Italy, ipogea@ipogea.org.
20. Prof. Nikos Manassis, Department of Water Resources, School of Civil Engineering, National Technical University of Athens, Heroon Polytechneiou 5, GR-15780 Zographou, Greece, nikos@itia.ntua.gr.
21. Seyed Ali Mahmoudian and Shahid Abdollah Zadeh, St. Keshavarz Bulvd, National Water and Wastewater Engineering Company, Tehran, Iran, seyedali@nww.co.ir
22. Seyed Navid Mahmoudian, Regierungsbaumeister Schlegel GmbH & Co. KG • Guntherstraße 29 • D-80639 München, Germany, Navid.Mahmoudian@ib-schlegel.de
23. Prof. Pierluigi Martini, Associazione Idrotecnica Italiana, Rome, Italy, segreteria@idrotecnicaitaliana.it
24. Prof. Larry W. Mays, School of Sustainable Engineering and the Built Environment, Arizona State University, Tempe, Arizona, USA, mays@asu.edu
25. Dr. Joanna M. Pope, Dept. of Fisheries and Wildlife, Michigan State Univ., East Lansing, MI 48824, USA, maesbriallu@yahoo.com
26. Dr. Alexander Reyes-Knoche, Latin America Consultant Company, Allmendweg 5, D-68256 Ladenburg, Germany, alexander.reyes-knoche@web.de
27. Prof. Joan B. Rose, Dept. of Fisheries and Wildlife, Michigan State Univ., East Lansing, MI 48824, USA, rosejo@msu.edu
28. Prof. Miquel Salgot, Soil Science Laboratory, Facultat de Farmàcia, Universitat de Barcelona, Barcelona, Spain, salgot@ub.edu
29. Dr. Markos Skivaniotis, Municipal Enterprise for Water and Sewage of Patras, Patras, Greece, mark_sklian@freemail.gr
30. Prof. Anna I. Strataridaki, Department of History and Education, University of Crete, Rethymnon, Greece, astrat@edc.uoc.gr
31. Prof. Didier Viviers, Université Libre de Bruxelles, C.P. 130, 50 Av. F.D. Roosevelt, 1050 Bruxelles, Belgium, Marylene.Poelaert@ulb.ac.be
32. Prof. Kostas Voudouris, Lab. of Engineering Geology & Hydrogeology, School of Geology, Aristotle Univ., GR 51124, Thessaloniki, Greece, kvoudour@geo.auth.gr
33. Dr. Mark H. Weir, Dept. of Fisheries and Wildlife, Michigan State Univ., East Lansing, MI 48824, USA, weirmarkh@gmail.com
34. Nikos Zarkadoulas, Department of Water Resources, School of Civil Engineering, National Technical University of Athens, Heroon Polytechneiou 5, GR-15780 Zographou, Greece, ps03014@yahoo.gr
35. Dr. Matti Zohar, Mishlat St. 12b, POB 82548, Ma'oz Tsion 90805, Israel, megalith@zahav.net.il

List of reviewers

1. Prof (Em.) Stylianos Alexiou, Department of History and Archaeology University of Crete, Rethymnon, Greece, geohatzakis@gmail.com
2. Prof. Apostolos Alexopoulos, Department of Geology, University of Athens, Athens, Greece, aalexopoulos@geol.uoa.gr
3. Dr. Andreas N. Angelakis, Institute of Iraklion, National Foundation for Agricultural Research (N.AG. RE.F.), 713 07 Iraklion, Greece, info@a-angelakis.gr
4. Prof. Giovanni De Feo, University of Salerno, Department of Civil Engineering, via Ponte don Melillo, 1 – 84084 Fisciano (SA), Italy, g.defeo@unisa.it
5. Prof. Yuri Gorokhovich, Lehman College, City University of New York (CUNY) Department of Environmental, Geographical and Geological Sciences Gillet Hall 315, 250 Bedford Park Blvd. West, Bronx, NY 10468, USA, yuri.gorokhovich@lehman.cuny.edu.
6. Prof. Benoit Haut, Univ. Libre de Bruxelles, Chem. Eng., 1050 Bruxells, Belgium, bhaut@uib.ac.be
7. Prof. Blanca Jiménez, Universidad Nacional Autónoma de México, Instituto de Ingenieria, Apdo Postal 70472, 04510 México D. F., bjimenezc@iingen.unam.mx
8. Prof. Amada Kelly, Classics Department, Office 507, Tower II, NUI Galway, Ireland, amanda.kelly@nuigalway.ie
9. Prof. Albert Koenig, Department of Civil Engineering, The University of Hong Kong, Hong Kong SAR, P.R. China, kalbert@hkucc.hku.hk
10. Prof. Demetris Koutsoyiannis, Department of Water Resources, School of Civil Engineering, National Technical University of Athens, Heroon Polytechneiou 5, GR-15780 Zographou, Greece, dk@itia.ntua.gr.
11. Dr. Pietro Laureano, IPOGEA,Vico Conservatorio s.n., 75100 Matera, Italy, ipogea@ipogea.org
12. Prof. Yiannis Lolos, Department of History, Archaeology and Social Anthropology, University of Thessaly, Volos, Greece, ylolos@otenet.gr
13. Prof. Nikos Manassis, Department of Water Resources, School of Civil Engineering, National Technical University of Athens, Heroon Polytechneiou 5, GR-15780 Zographou, Greece, nikos@itia.ntua.gr.

14. Prof. Larry W. Mays, School of Sustainable Engineering and the Built Environment, Arizona State University, Tempe, Arizona, USA, mays@asu.edu
15. Prof. Nikos Paranychianakis, Dept. of Environmental Engineering, Technical University of Crete, Chania, Greece, niko.paranychianakis@enveng.tuc.gr
16. Prof. Kalliopi A. Roubelakis-Angelakis, Department of Biology, University of Crete, P.O. Box 2208, 71409 Iraklion, Greece, poproube@biology.uoc.gr
17. Prof. Miquel Salgot, Soil Science Laboratory, Facultat de Farmàcia, Universitat de Barcelona, Barcelona, Spain, salgot@ub.edu
18. Prof. Stylianos Spyridakis, Dept. of History, University of California Davis, Davis, Calif., USA, svspyridakis@ucdavis.edu
19. Prof. Anna I. Strataridaki, Department of History and Education, University of Crete, Rethymnon, Greece, astrat@edc.uoc.gr
20. Nikos Zarkadoulas, Department of Water Resources, School of Civil Engineering, National Technical University of Athens, Heroon Polytechneiou 5, GR-15780 Zographou, Greece, ps03014@yahoo.gr

Prolegomena
The evolution of water supply throughout the millennia

Since the dawn of humankind on Earth, the adequacy of available water has been of utmost importance for survival and prosperity. Assurance of an adequate supply of water to an ever growing and developing population of today is a world-wide problem. Water scarcity has become a major global concern, with a major part of the world facing water shortages. Rapid population growth with the resulting change in demographics has become the world's greatest problem associated with water resources. Cities around the world have been experiencing water shortages and scarcity coupled with water quality problems. This situation is expected to worsen due to the population growth and relocation to urban areas, particularly in the developing countries. The effects of climatic variability exacerbate the problem even further, particularly when water management practices are poor. Even areas with an abundance of available water are recognising water supply as a natural resource superior to all others.

There have been a great deal of unresolved problems related to the management principles, such as the decentralisation of the processes, the cost effectiveness, the durability of the water projects, and sustainability issues. In the developing parts of the world, such problems have been intensified to an unprecedented degree. Moreover, new problems have arisen such as the contamination of surface- and ground-water. The intensification of unresolved problems has led societies to revisit the past and to reinvestigate the successful past achievements. To their surprise, those who attempted this retrospect, based on archaeological, historical, and technical evidence were impressed by two things: the similarity of principles with those of the present, and the advanced level of water engineering and management practices.

Technological advancements related to water during the 20th century created a disdain for the past achievements. Many have felt the achievements of the past are not solutions for the present and the future. Past water technologies, were regarded to be far behind those of the present. However, many of those technologies developed in ancient times could be solutions for many parts of the world. Many of the technologies developed during the Bronze Age could be considered in today's development and management plans.

Many of our present water technological principles have a foundation dating back three to four thousand years ago. These achievements include technologies such as dams, wells, cisterns, aqueducts, baths, recreational structures, and even water reuse. These hydraulic works also reflect technical and scientific knowledge, which for instance allowed the construction of tunnels from two openings and the transportation of water both by open channels and closed conduits under pressure. Certainly, technological developments were driven by the necessities for efficient use of natural resources in order to make civilizations more resistant to destructive natural elements and to improve the standards of life. With respect to the latter, certain civilizations developed an advanced, comfortable and hygienic lifestyle, as manifested from public and private bathrooms and flushing toilets, which can only be compared to our modern facilities which were re-established in Europe and North America at the beginning of the last century.

The Evolution of Water Supply Throughout the Millennia examines some of the major achievements in nearly all scientific fields of water supply technologies and management by ancient civilizations. This Book provides valuable insights into the ancient water supply technologies with apparent characteristics of their durability, their adaptability to the environment, and their sustainability. A comparison of the water technological developments in several civilizations is also undertaken. These technologies are the underpinning of modern achievements in water engineering and management practices. It is the best proof that "*the past is the key for the future*." The ancient technologies and water management practices will be a useful tool for future cities' planning.

Thirty-six authors from several disciplines developed the chapters in this book. The disciplines include archaeology, water sciences, engineering, life sciences, environmental sciences, health sciences, biology and geosciences. The geographical coverage is very wide, with prominence in the Mediterranean world. However, several other civilizations from other parts of the world, such as Asia (Iran, China) and America (south-western United States, Mexico, South America) are included. The book is organized in four parts. The first four chapters are introductory and refer to general subjects. The next 11 chapters refer to different civilizations over the globe. The next five chapters deal with major cities with long histories. In the last chapter, conclusions and lessons learned are included. The themes of the chapters included are from prehistoric to medieval and even modern times.

Publication of a book of this scope and magnitude could be accomplished only with the help of many people, and our appreciation is gratefully offered to the authors of the 21 chapters. Their contribution to the quality of this book is evident. Also, the editors would like to express their gratitude to IWA Publishing and to Dr. Maggie Smith of IWA, for her patience and understanding, and her valuable managerial advice. We sincerely appreciate the work of 20 reviewers, who provided valuable assistance and authoritative guidance for each of the chapters which were reviewed by at least two reviewers.

Andreas N. Angelakis, Institute of Iraklion, National Foundation for Agricultural Research (N.AG.RE.F.), 71307 Iraklion, Greece, info@a-angelakis.gr

Larry W. Mays, Civil and Environmental Engineering Group, School of Sustainable Engineering and the Built Environment, Arizona State University, Tempe, Arizona, USA, mays@asu.edu

Demetris Koutsoyiannis, Department of Water Resources, School of Civil Engineering, National Technical University of Athens, Heroon Polytechneiou 5, GR-15780 Zographou, Greece, dk@itia.ntua.gr

Nikos Manassis, Department of Water Resources, School of Civil Engineering, National Technical University of Athens, Heroon Polytechneiou 5, GR-15780 Zographou, Greece, nikos@itia.ntua.gr

Chapter 1

Ancient gods and goddesses of water

L. W. Mays and A. N. Angelakis

1.1 INTRODUCTION: MYTHOLOGY, GODS AND GODDESSES

Water is the beginning of everything

Thales of Miletus (636–546 BC)

To ancient civilizations and cultures, water was considered as a potential destructive element, but also a necessity to life, next to food, safety and hygiene; usually some deity was associated with water. Also, water was important for purifying the body and for cleaning. Dieties associated with water or various bodies of water were important in almost all mythologies. Deities and religious beliefs prescribed its uses, collection and purification. Minoans, Babylonians, Greeks, Romans, and the American Indians all had gods and goddesses for water irrigation of crops as well as water purity. Temples water delivery systems were often built to honour these gods and certain elements were incorporated to honour the god. Also gods and/or goddesses were associated with making rain in order to face severe droughts (Haland, 2007). Water deities were usually more important among civilizations when a sea or an ocean, or a great river was important.

Gods and goddesses have had a significant influence on the human society since the ancient times. This Chapter explores some of the ancient gods and goddesses of rain, water, rivers, and sea. The gods are considered collectively as a "pantheon" of a particular mythology. Many ancient societies had a "Creator God" and many lower levels of gods and goddesses. Most ancient societies had a belief system that was "polytheistic," meaning they worshipped many gods. They believed that there was a god for every aspect (or element) of the Earth such as the sun, the moon, wind, lightning, rain, water, etc. In many ancient cultures the moon was seen as gentle and feminine in contrast to the sun which was seen as masculine. Ancient Egyptians believed that the gods and goddesses maintained the balance of chaos and order on earth. Some ancient gods and goddesses are still worshipped today, such as Chac, the Mayan god of rain. Gods and goddesses could be aspects of one another and could have shifting roles and levels of importance such as the Vedic gods in the Vedic religion in India.

Gods and goddesses have been depicted by various cultures in many different ways. For example ancient Indian gods and goddesses have been depicted with many arms showing a sign of great power, with multiple heads, and with combined human and animal characteristics. Sun gods were often depicted with attributes of wheels, while sky gods were depicted as hammers to symbolise thunder. Ancient Egyptian gods and goddesses took on many human and animal forms. The gods of the ancient Greeks were almost all conceived of as being human in form. Other gods, such as those at the founding of Rome, were *numina* (divine manifestations, faceless, and formless) but no less powerful. For the Romans everything in nature was thought to be inhabited by *numina*, explaining the large number of deities in the Roman pantheon. Gods and goddesses of the Tibetan Buddhism take on a variety of forms ranging from the fierce to the peaceful. Goddesses, in some cultures, are associated with earth, motherhood, love, and the household, while in other cultures, they rule over war, death, and destruction as well as healing.

In the following sections the major water gods and goddesses in several ancient civilizations and/or religions are presented and discussed. Emphasis is given to Mesopotamians and Egyptians, to the Bronze era, to Hellenistic and Roman civilizations, to the Celtic and Hindu mythologies and to the ancient Mesoamericans (e.g. Aztecs, Mayas, and Incas).

1.2 ANCIENT MESOPOTAMIAN MYTHOLOGY

In Mesopotamia humans were created to alleviate the gods from the hard work they had to do. Mesopotamia, "the land between the rivers," that is the Tigris and Euphrates Rivers, is in the Fertile Crescent. During the time of ancient Mesopotamia urban civilization grew, the wheel was invented, arts and sciences flourished, writing was pioneered, and humankind began to create a record of the past. All life originated with water for these people (Woolf, 2005).

Water was ever present in all aspects of Mesopotamian life, including religion, politics, law, economy, international affairs, war, etc. Enuma Elish is the Babylonian creation myth recorded on seven clay tablets in Old Babylonian. The Enuma Elish indicated that in the beginning water already existed, distinguishing between Apsu, conceived as a male god of freshwater (sweet) waters, and Tiamat, his spouse, a goddess of salt water. Also, Apsu denotes the freshwater upon which the earth floated. As underground waters, Apsu may be reached when laying the foundations of a temple, and also appears naturally in pools and marshes (Jacobsen, 1946). The Apsu is the domain of one of the most important gods in the Mesopotamian pantheon, Enki (Sumerian) and later Ea (Akkadian), who was depicted with cascade of water emanating from his shoulders, or holding a vase from where water emerges, as shown in Figure 1.1a. Enki, as master of the fresh water was a creator god, a wise god, always ready to help humans. The god Murdak (Figure 1.1b) was recognised as the son of Ea. His powers and attributes were passed to him from Ea and another god, Enlil.

Tablet I from the Atra-Hasis epic about the creation and early history of man begins: 'When the gods like mėn / bore the work and suffered the toil…'. The heavy work included digging and maintenance of canals. But the lesser gods did not tolerate this state of affairs, and they rebelled ('… they set fire to their tools, / fire to their spades they put / and flame to their hods …'). As a result, mankind was created: '… Let the birth-goddess create offspring / and let man bear the toil of the gods …'. And men 'with picks and spades they built the shrines, / they built the big canals' banks / for food of the peoples, for sustenance of the gods'. However, 'Twelve hundred years had not yet passed', the peoples multiplied and with their noise disturbed the gods who decided to exterminate mankind. At this point, Enki intervened and alerted one man, Atra-Hasis, who was instructed on how to escape from the plan of the gods and let mankind survive. Tablet II refers to one of the ways the gods attempted to eradicate mankind from the earth: '… Adad (god of weather) should withhold his rain, / and below, the flood should not come up from the abyss / …. / let the fields diminish their yields …'. Famine arose not only from the drought, but also because the soil became unsuitable for agriculture due to salinisation: '… the black fields became white, / the broad plain was choked with salt …'. Not succeeding with famine (due to Enki's intervention), the gods decided on a devastating flood that 'tear up the mooring poles' and 'make the dykes overflow'. Tablet III describes that 'for seven days and seven nights / came the deluge, the storm, the flood'.

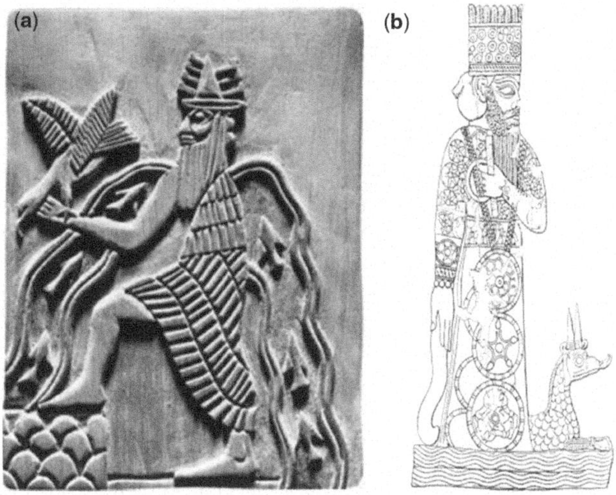

Figure 1.1 Mesopotamian gods. (a) Enki (Sumerian) or Ea (Akkadian) and (b) Marduk, considered as the son of Enki (*Source*: Wikipedia Commons, in public domain).

The Sumerian version of the flood (Lambert & Millard, 1999) explicitly says that construction and maintenance of water channels was a god's decision '... he (the god) did not stop the yearly flood, but dug the ground and brought the water, / he established the cleaning of the small canals and the irrigation ditches ...'.

1.3 ANCIENT EGYPT MYTHOLOGY

Ancient Egyptian civilization was centred about its highly complex religion (Woolf, 2005). In ancient Egypt all aspects of human life were reflected in the divine world and the range of associations held by the deities and goddesses (Oakes & Gahlin, 2003). They believed that the balance of order and chaos in the universe could only be maintained by the gods and goddesses and their representative on earth (the king or the pharaoh). Ancient Egyptian gods and goddesses took on many different human and animal different forms. For example Aker the earth god, appeared as two lions back-to-back on a tract of land with lion or human heads at either end. Anat, the goddess of war appeared as a woman with a lance, axe and shield. Many of the deities were worshipped in temples and shrines throughout Egypt. Hundreds of deities were in the Egyptian pantheon. Many of these originated as local gods who became the central focus of important cults and others borrowed from different cultures.

For thousands of years the people of Egypt have owed their very existence to a river that flowed mysteriously and inexplicably out of the greatest and most forbidding desert in the world (Hillel, 1994). The ancient Egyptians depended upon the Nile not only for their livelihoods, but they also considered the Nile to be a deific force of the universe, to be respected and honoured if they wanted it to treat them favourably. Its annual rise and fall were likened to the rise and fall of the sun, each cycle equally important to their lives, though both remaining a mystery. Since the Nile sources were unknown up until the 19th century, the ancient Egyptians believed it to be a part of the great celestial ocean, or the sea that surrounds the whole world. Shown in Figure 1.2 is Hapi – the Nile god, first shown as one god and then as two gods, portrayed with breasts to show his capacity to nurture.

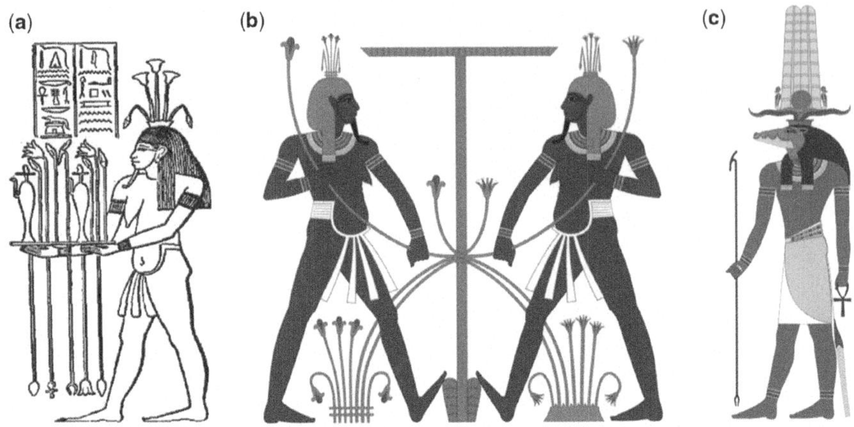

Figure 1.2 Egyptian gods associated with the Nile River: (a) Hapi shown as one god with breasts (*Source*: Illustration from the Encyclopaedia Biblica, a 1903 publication, now in the public domain), (b) Hapi shown as a pair of gods symbolically tying together upper and lower Egypt (Source: Wikipedia Commons, licensed under Creative Commons, GNU Free Documentation Licence), and (c) Sobek, god of crocodiles who created the Nile by Jeff Dahl (*Source*: Wikipedia, under GNU Free Documentation Licence, Version 1.2).

Ancient Egyptians considered the sun (the dominant factor in Egyptian life) to be a potent life force along with the annual inundation of the Nile, which was responsible for the successful harvests. Re or Ra, meaning "the Sun," was the pre-eminent god of solar and appeared as a ram or falcon head with sun disc and cobra headdress. The cult centre for Re was in Heliopolis. For further reading on ancient Egyptian gods and goddesses refer to Wilkinson (2003).

1.4 THE BRONZE ERA

The Minoan civilization (*ca.* 3200–1100 BC) was based on the island of Crete, a very fertile land. The importance of religion of the Cretans is somewhat sketchy because they built no great temples or carved statues of gods. However the Cretans did leave behind many other items such as small shrines, and representations of sacred birds, trees, bulls, and snakes. They also built lustral basins (rooms possibly set aside for cult worship), temple repositories, double-axes, and sacral horns. They also used sacred caves and shrines set aside on mountain peaks. Religious scenes are shown in Figure 1.3 and lustral basins are shown in Figure 1.4a.

Goddesses seem to have had the dominant role in Cretan religion as they played a dominant role in their society (Graham, 1987). A so-called Poppy goddess, with three incised poppy seeds on her head that was found at a small ritual shrine suggests that opium was used to induce religious ecstasy (Freeman, 2004). Referring to Figure 1.4b women had a dominant role as they had elaborate costumes indicating their importance. The colourful frescos of women found in palaces may indicate a more feminine influence than other societies of the ancient times. In fact archaeologist Jacquetta Hawkes in her book, The Dawn of the Gods (London, 1968) argued that Minoan Crete was essentially a feminine society. Also Minoan ritual included the possible sacrificing of children (Freeman, 2004).

Figure 1.3 Scenes from Hagia Trida Sarcophagus: (a) Offerings and (b) bull sacrifice (*Source*: Graham (1987).

Figure 1.4 Lustral basins: (a) North lustral basin at Knossos (Evans thought that these basins were used for purification ceremonies and that Knossos palace was a sacred place) and (b) Lustral basin at Kato Zakros (with permission of L.W. Mays).

1.5 GREEK MYTHOLOGY

Hydros (or Hydrus) was the *protogenos* of the primordial waters. In the *Orphic Theogonies* Water was the first being to emerge at creation alongside Creation (Thesis, goddess of creation) and Mud. The primordial mud solidified into Gaia (Earth) and with Hydros produced Kronos (time) and Ananke (compulsion). These two in turn caught the early cosmos in the coils, and split it apart to form the god Phanes (creator of life), and the four ordered elements of Heaven (fire), Earth, Air and Sea (water). The *Orphic Rhapsodies* later discarded the figures the Kronos and Ananke, and have Phanes instead born directly from Hydros and Gaia [*Homer, Iliad 14. 200 ff (trans. Lattimore) Greek epic C8th BC*].

In Greek mythology the Twelve Olympians were the principal gods of the Greek pantheon, residing atop Mount Olympus. There were, at various times, fourteen different gods recognised as Olympians, though never more than twelve at one time. In Greek mythology, the twelve gods and goddesses ruled the universe from atop Greece's Mount Olympus. These Olympians, all related to one another, had come to power after their leader, Zeus, overthrew his father, Kronos, leader of the Titans. Later the Romans adopted most of these Greek gods and goddesses, but with new names.

1.5.1 Olympian gods and goddesses

Achelous (Figure 1.5), named after the Achelous River (largest river in Greece), was the chief of all river deities. The name is pre-Greek and the meaning unknown. Every river had its own river spirit. Tethys was a Greek deity who oversaw the fresh water rivers of the world and the mother and grandmother of thousands of other deities.

Figure 1.5 Achelous was often reduced to a bearded mask, an inspiration for the medieval Green Man. Shown is floor mosaic of Achelous in Zeugma, Turkey (*Source*: Wikipedia Commons, in public domain).

Tethys (Figure 1.6) was a goddess who most probably was a primordial deity in Archaic Greece, but who was seen in Classical myths as the deity responsible for the fresh water rivers of the world and the progenitor of thousands of water deities. Tethys was described in classical myths as the deity responsible for the fresh water rivers of the world and the progenitor of thousands of water deities. Tethys was considered as an embodiment of the waters of the world making her also a counterpart of Thalassa, the embodiment of the sea.

Demeter (Figure 1.7), the goddess of earth, agriculture, water and fertility, was the second daughter of the major Titans Rhea and Cronus, after Hestia, the goddesses of the hearth. Demeter was a peace-loving deity and the source of all growth and life; she was the goddess who provided all nutrition on the earth and taught mortals how to cultivate the earth and ease life. Demeter was most appreciated for introducing wheat to mankind, making man different from animals.

Figure 1.6 Mosaics of Tethys: (a) Mosaic from Phillopolis that dates to the mid-fourth century (Source: Wikipedia Commons, licensed under Creative Commons, GNU Free Documentation Licence) and (b) Mosaic on the floor at the bathhouse of Garni with the image of Tethys, constructed between the first and third centuries AD (*Source*: Wikipedia Commons, in public domain).

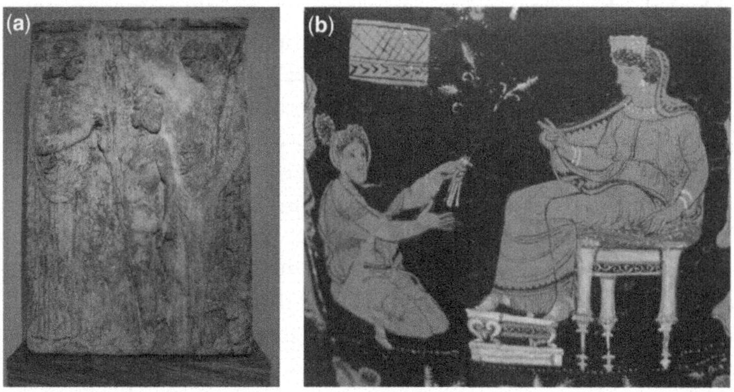

Figure 1.7 Demeter: (a) Votive relief of Demeter, on the left, during a ritual holding a sceptre in her left hand offering wheat to Triptolemos, son of the Eleusinian King Keleos, for bestowing on mankind. On the right Persephone, with a mantle and holding a torch, blesses Triptolemos with her right hand. This is the largest and most known votive relief (*ca.* 440–430 BC), which was found in Eleusis, and now resides in the Archaeological Museum of Athens, Greece. The relief was apparently famous in antiquity and was copied in the Roman period with one copy now in the Metropolitan Museum of New York. (with permission of A. N. Angelakis) and (b) Demeter and Metanira. Detail of the belly of an Apulian red-figure hydria, *ca.* 340 BC (*Source*: Wikipedia Commons, in public domain).

1.5.2 Zeus or Poseidon of rain

The ancient Greeks worshipped Zeus or Poseidon (Figure 1.8) god of rain and in colloquial speech one can say "Zeus is raining" (Haland, 2007). Children in Ancient Greece sang: "Rain, rain, o dear Zeus, on the fields of the Athenians." According to the tradition, Zeus was the god of rain (Hes. *Op*. 488).

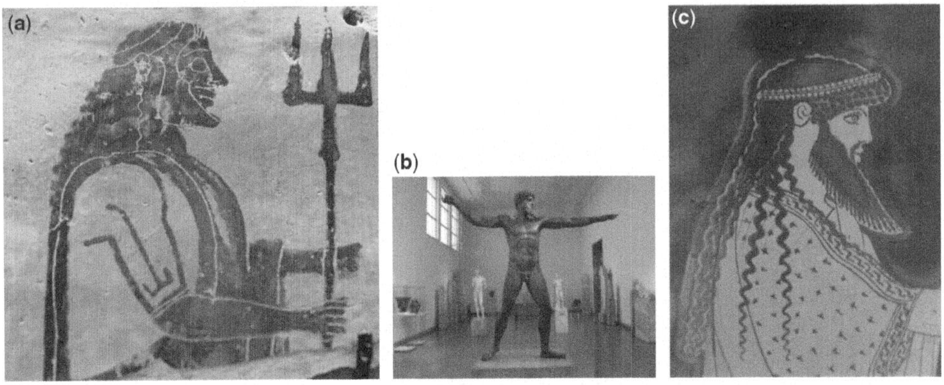

Figure 1.8 Poseidon (a) Poseidon holding a trident (Corinthian plaque, 550–525 BC) from Penteskouphia at Louvre Museum, photographed by Marie-Lan Nguyen, Jastrow (2006) (Source Wikipedia Commons, in public domain); (b) Bronze statue of Zeus or Poseidon (from Archaeological Museum of Athens). It is certainly the work of a great sculptor of the early Classical period *ca*. 460 BC (with permission of A. N. Angelakis); and (c) Head of Poseidon, photographed by Marie-Lan Nguyen, Jastrow (*Source*: Wikipedia Commons, licensed under the Creative Commons Attribution 2.5 Generic licence).

Fully-fledged rain-magic is found in the cult of Zeus Lykaios in Arkadia, where nevertheless one of the Nymphs who reared him also has something to say: if a severe drought lasts a long time the priest of Zeus will go to the spring of the Nymph Hagno, make a sacrifice, and let the blood run into the spring (Haland, 2007). Then, after prayer, he dips a branch from an oak (the sacred tree of Zeus) into the water, and forthwith a vapour will rise up from the spring like a mist, 'and a little way off the mist becomes a cloud, collects other clouds, and makes the rain drop on Arkadian land, (Haland, 2009).

Among the ancient Greeks, a king is often a magician in the service of the gods. Part of his duty is to be a weather-king. He is "making the weather", and this means that he is making rain, for example by shaking rattles or by other means trying to make thunder and lightning. In ancient Thessaly, when the land suffered from drought, they shook a bronze wagon by way of praying the god for rain, and it was said rain came. This was a traditional public ceremony for the making of rain (Haland, 2007).

According to (Haland, 2009) 'Greece had been withering under a drought: neither inside the isthmus (of Corinth) nor outside it would rain, until they sent to Delphi to discover the reasons and ask for relief. The Pythian priestess told them to placate Zeus, but, if he were to listen, it had to be Aiakos who made the ritual supplication. They sent men from every city to beseech Aiakos son of Zeus; he sacrificed and prayed to Panhellenic Zeus, and brought rain to Greece; so the Aiginetans made the portraits of the ambassadors' (Haland, 2007).

1.5.3 Nymphs

In Greek mythology a nymph was a female minor nature deity typically associated with a particular location or landform. Nymphs were generally regarded as divine spirits who animated nature, and were usually depicted as beautiful, young maidens who loved to dance and sing. Water nymphs (*Hydriades or Ephydriades*) included freshwater nymphs (*naiads or naides*) who presided over fountains, wells, springs, streams, and brooks, but not rivers. *Naiads* (see Figure 1.9) were either daughters of Poseidon or various *Oceanids*, who were the patrons of a particular spring, river, sea, lake, pond, pasture, flower or cloud. *Oceanids* were the three thousand daughters of Oceanus and Tethys. Water nymphs associated with particular springs were known all through Europe at locations with no direct connection with Greece.

Figure 1.9 A *naiad*, a water nymph approaches the sleeping Hylas by Artist John William Waterhouse, 1893. In Greek mythology, Hylas, the son of Hercules, was kidnapped by nymphs of the spring of Pegae (*Source*: Wikipedia Commons, in public domain).

Limnades or leimenides were naiads that lived in freshwater lakes, whose parents were river or lake gods. Pegaeae were naiads that lived in springs and were often considered daughters of the river gods (Potamoi). They established a mythological relationship between a river and its springs. Crinaeae were naiads associated with fountains or wells. Eleionomae were naiads associated with living in marshes and often misled travellers with their illusions that were images of a traveller's loved ones. They also lured young, virgin boys seducing them with their beauty (see Figüre 1.9).

1.6 ROMAN MYTHOLOGY

The Etruscans were a continuous population during the Mediterranean Iron Age in the second half of the first millennium BC, located in the Po Valley and some of its alpine slopes, southward along the west coast of Italy. The Estruscans were at their peak around 500 BC, having a significant maritime power with a presence in Sardinia and the Aegean Sea. They had a distinct language and culture during the period of earliest European writing and worshipped a pantheon of twelve gods. In fact they were very dedicated to religion, doing nothing without proper consultation with the gods and signs from them. These practices were later taken over by the Romans. Roman mythology definitely took on many Greek things, including the Greek gods. The Dii Consentes, a group of twelve gods were especially honoured by the Romans. The number of twelve gods came from the Etruscans; however the Dii Consentes were identified with the Greek Olympian gods, even though the original character of the Roman gods was different.

One of the twelve Roman gods was Neptunus. In Roman mythology and religion Neptune (*Neptūnus*) was the god of fresh water (from rivers and springs) and the sea (see Figure 1.10). Similarly to Poseidon, Neptune was also worshipped as a god of horses, under the name Neptunus Equester, a patron god of horse-racing. The festival of Neptunalia is celebrated on July 23. Hadrian, the emperor, ordered the construction of the Pantheon as a sanctuary dedicated to all the gods. Roman temples were the main centers of religion serving as symbolic homes for the gods, which were represented by statues.

(a) (b)

Figure 1.10 Neptunus or Neptune (a) [*Source*: Wikipedia Commons, licensed under Creative Commons, GNU Free Documentation Licence] and (b) Mosaic at the Museo archeologico regionale di Palermo, (Photo by Giovanni Dall'Orto, licensed under the Creative Commons Attribution-Share Alike 2.5 Generic licence).

Cloacina (*cloaca*: "sewer" or "drain") was the goddess of sewers who presided over the Cloaca Maxima ("Great Drain"), which was the main trunk of the system of sewers in Rome. She was originally derived from Etruscan mythology. The Romans placed a great deal of faith and trust in her for the wellbeing of Rome's sewers (and workers) which was very important to their desired way of life and good health through sanitation. The Shrine of Venus Cloacina was a small sanctuary on the Roman Forum, honouring the divinity of the *Cloaca Maxima*, the spirit of the "Great Drain" or Sewer of Rome. Two images of the "Shrine of Venus Cloacina" in the Forum are shown in Figure 1.11.

Figure 1.11 Two images of the "Shrine of Venus Cloacina" in the Forum Romanum from Hülsen, Christian (1906), *The Roman Forum – Its History and Its Monuments*, Ermanno Loescher and Co: Publishers to H. M. the Queen of Italy, p. 138.

Roman mythology also included nymphs. A couple examples of naiads include Albunea and Appias. Albunea was a prophetic nymph or Sibyl, who was a naiad who lived in the sulphuric spring near Tibur (Tivoli). Appias was one of the Crinaeae, who was a naiad who lived in the Appian Well outside the temple to Venus Genitrix in the Roman Forum.

Finally, in the left side of the crypt of the famous Cathedral in Bari, Italy, which was built between the late 12th and late 13th centuries, one sarcophagus contains the relics of Saint Columba, recently restored, who is related to the rain and fire. She was pagan and became Christian when she was 16-years old. She escaped from Spain (257–273 AD) due to Emperor Adriano's persecution (Figure 1.12). She went to Sens in France where she was kept and imprisoned anyway. The legend says that a prison officer tried to rape her but a bear from the nearby amphitheatre saved her. Then, she was sent to the stake but a strange rain saved her from fire. That's why people pray for her in the case of water being needed to extinguish a fire. She was killed later due to Emperor Aureliano's persecution.

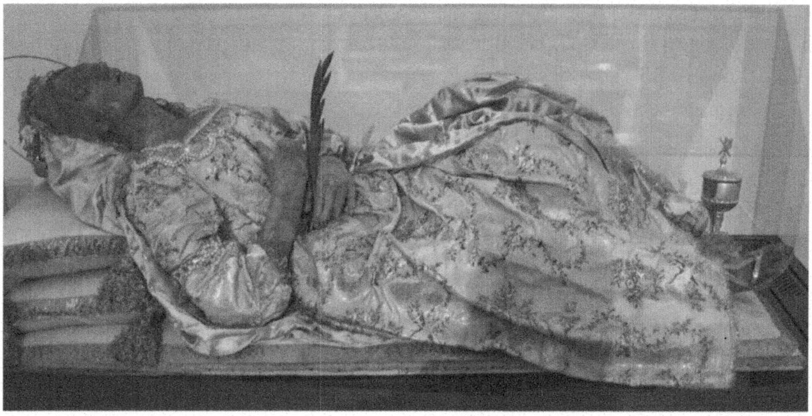

Figure 1.12 Saint Columba di Sens (Spain 257 AC–Sens 273 AC) in the crypt of the Cathedral, Bari, Italy (with permission of A. N. Angelakis).

1.7 CELTIC MYTHOLOGY

Celtic mythology is the mythology of Celtic polytheism, or the religion of the Iron Age Celts. In Europe, as early as the Bronze Age, the association between water and spiritual power was evident. The reverence for water was a hallmark of the Celtic religion, as rivers and springs were very important in Celtic myth. The ancient Celtics had a great deal of trust for the curative powers of springs. Springs were therefore sites of great sanctuaries. Sequana was the goddesses of the Seine River in France. She was often depicted in a duck-shaped boat. Her sacred place was among the springs at the headwaters of the Seine River in Burgundy. Goddess Sulis Minerva had her sanctuary at the ancient springs of Bath, England. Lakes were important sites for ritual activities.

Arausio was a local Celtic water god who gave his name to the town of Arausio (Orange) in southern Gaul. Dea Icaunis was the goddess of the Yonne River in Gaul in Gallo-Roman religion. In Celtic mythology, Condatis (waters meet) was a deity associated with the confluences of rivers (in particular the Tyne and the Tees). He was worshipped primarily in northern Britain and in Gaul. Latis was the anglo-Celtic goddess associated with water. She was originally a lake goddess who became a goddess of ale and mead. There were many other gods and goddesses related to specific rivers in the Celtic mythology.

1.8 HINDU/VEDIC MYTHOLOGY

As in other mythologies, deities associated with water or various bodies of water are known in Hindu mythology. As already mentioned, water deities were common in civilizations (e.g. Hindu) in which the sea or ocean, or a great river played an important role in their development. Thus, Varuna was the Lord of the oceans and Apam Napat was the god of fresh water, such as in rivers and lakes (Darian, 2001).

To understand the river Ganga means to understand a significant part of India. It holds a place unique in all mythologies, theologies and beliefs of the world. In no other culture was a natural feature assumed so religious and of psychological significance. The Ganga is so intertwined with the Indian imagination that even for people who never live nearby, it will always be the supreme river (Darian, 2001). In such circumstances, it is not surprising that the Ganga River is also one of the most popular goddesses in India (Figure 1.13). Jawaharlal Nehru, First Prime Minister of India stated: "The Ganga, especially, is the river of India, beloved of her people, round which are intertwined her memories, her hopes and fears, her songs of triumph, her victories and her defeats. She has been a symbol of India's age long culture and civilization, ever changing, ever flowing, and yet ever the same Ganga". Also various rivers are associated with goddesses in the Rigveda, such as Sarasvati (Sarasvati River) and Yamuna.

Figure 1.13 In the cannons of Indian art, Ganga is visualised as all other major Indian goddesses are, voluptuous and beautiful. Their ample breasts and, sturdy, child bearing hips, giving adequate testimony to their fecundating powers (with permission of K. Vipin)

1.9 ANCIENT MESOAMERICA

1.9.1 The Aztecs

Mesoamericans had a large number of fertility gods and goddesses, of which the rain god was among the most senior. The nucleus of the Aztec Empire was the Valley (or Basin) of Mexico starting in the 12th century. The ancient city of Tenochtitlan, located on a reclaimed island in Lake Texcoco, was the capital of the Aztec Empire. The Aztecs were faithful worshippers of the rain god, Tlaloc, also known as Nuhualpilli, whose cult dates back as far as the Olmec civilization (Woolf, 2005). Tlaloc (Figures 1.14 and 1.15) was honoured with sacrifices in the form of blood and other offerings. He stored rainwater in four huge jars, which he kept in the north, south, east, and west, and from the eastern jar he sent life giving rains, and from others storms and droughts.

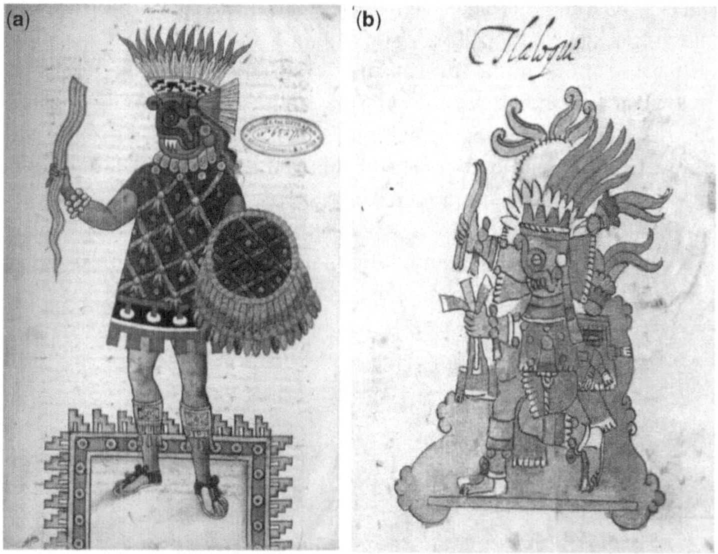

Figure 1.14 Tlaloc the rain god: (a) From the Collection of E. Eug. Goupil, Author unknown, 17th century) and (b) Tlaloc as shown in late 16th century, Codex Rios (*Source*: Wikipedia Commons, in public domain).

Figure 1.15 Fragments of a brazier depicting Tlaloc from Stage IVB of the Templo Mayor in Mexico City (*Source*: Wikipedia Commons in public domain).

Tlaloc was greatly feared among the Aztecs, who drowned children to appease him. They believed that Tlaloc was responsible for both floods and droughts, and that he had been created by the other gods. Tlaloc was depicted as a goggle-eyed blue being with fangs. Human sacrifices, usually children, were often made in his honour. Before the victims were actually sacrificed, their tears were collected in a ceremonial bowl, to serve as an offering. Tlaloc was also worshipped in pre-Aztec times, by the Teotihuacan and Toltec civilizations.

1.9.2 The Maya

The Mayan civilization survived the longest of all the great cultures of ancient America. This was also the widest spread of all these cultures. The ancient Maya lived in a vast area covering parts of present-day Guatemala, Mexico, Belize, and the western areas of Honduras and El Salvador. Mayans were fascinated by time and the heavens, charting the stars and planets, and developed the most accurate calendar in existence before modern times (Woolf, 1987).

Mayans settled in the last millennium BC and their civilization flourished until around 870 AD The Mayas settled in the lowlands of the Yucatan Peninsula and the neighbouring coastal regions. The large aquifer under this area is in an extensive, porous limestone layer (Karst terrain), which allows tropical rainfall to percolate down to the aquifer. Because of this and the fact that few rivers or streams exist in the area, surface water is scarce. One important water supply source for the Maya, particularly in the north, was the underground caves (see Figure 1.16) called cenotes (se-NO-tes), which also had religious significance (portals to the underworld where they journeyed after death to meet the gods and ancestors). In Yucatan there are over 2,200 identified and mapped cenotes.

Figure 1.16 Sacred *cenote* at Chichen Itza (which means mouth of the well of the Itzas). The word *cenote* is derived from *tz'onot*, the Maya term for the natural sinkholes. This *cenote*, which measures about 50 m form north to south and 60 m from east to west, was used for sacrifices of young men and women, warriors and even children to keep alive the prophecy that all would live again. Shown at the left are the remains of a building once used as a steam bath, or *temezcal*, to purify those to be sacrificed. Those sacrificed were tossed from a platform that jutted out over the edge of the *cenote* (with permission of L.W. Mays)

The Mayans believed in many gods and goddess, having a god or goddesses for almost everything. Chaac or Chac (Figure 1.17), the god of rain, lightning and thunder, was both adored and feared as the rain was for growing crops, but if the rain was too heavy the crops could be destroyed. The storms and lightning could also cause death and disaster. He is often depicted as blue symbolising rain. Today Chac is still worshipped by the Maya, where in the Yucatan toward the end of the dry season shrines are prepared. Small boys are tied to the corners of altars and croak like frogs to urge Chac to cause the rains.

Figure 1.17 Chac, Mayan rain god: (a) Terra cotta image of Maya Rain God Chac at San Francisco's deYoung Museum (*Source*: Wikipedia Commons, in public domain) and (b) Chac, artist unknown. Source: Francis Robicsek: The Maya Book of the Dead. The Ceramic Codex, University of Virginia Art Museum (1981) (*Source*: Wikipedia Commons, in public domain)

1.9.3 The Inca

The Inca civilization emerged from fragmented independent societies by *ca.* 1000 AD (D'Altroy, 2003). The extent of the Inca Empire included parts of modern day Columbia, Ecuador, Peru, Bolivia, Chile, and Argentina. The city-state of Cuzco became the capital. Incas developed water technologies to supply water for irrigation and domestic uses in addition to religious purposes.

Wiraqocha (or Viracocha) Pachyachachic, "Creator of All Things", emerged from the water of Lake Titicaca into an empty world of darkness. He peopled the world, fashioning the first men and women from solid stone. After creating this new world in darkness, humanity became miserable because of greed and conflict. Wiraqocha then turned some of the men and women he had created into stone and had some swallowed up by the ocean or earth. Then he caused sixty days and nights of rain so that the resulting floods washed away his first creation with the exception of two that were saved (possibly his sons). So now he created the light, calling forth the sun and moon, from the waters of Lake Titicaca to the sky, so they could take turns illuminating the Earth. Even though Wiraqocha was the creator of the cosmos he probably played a small role in the everyday functioning of the Inca universe. The everyday functioning was shared by a pantheon of gods and goddesses, the most prominent of them being Wiraqocha's son Inti. Inti married Mama-Quilla, the goddess of the moon. Inti-Ilapa was the thunder god, the feared warrior in the sky, who also obtained water for the rain from the Milky Way, thought to be a starlit river.

1.10 SUMMARY AND CONCLUSIONS

The availability of water played an important role in the selection of the location for establishing a community in most of the ancient civilizations. The water was thought to be particularly healing and purifying during the rituals and festivals. Gods and goddesses of water have been depicted by various cultures in many different ways as illustrated in this chapter. People have traditionally expressed their beliefs through rituals connected to purity and water by fetching Holy water from the caves dedicated to these divinities. Water seems to have had the dominant role in Minoan religion as it played a unique role in the society. This was probably the main reason for the great water technologies achievements during the Bronze Era.

In ancient Mesopotamia the Enuma Elish indicated that in the beginning water already existed. Enki (Sumerian), and later Ea (Akkadian), was the master of the freshwater. Ancient Egyptians believed that the gods and goddesses maintained the balance of chaos and order on earth and that the Nile was a deific force of the universe with the Nile god Hapi. In Roman mythology, Neptune (Neptūnus) was the god of freshwater (from rivers and springs) and the sea. Similar to the Greek god, Poseidon, Neptune was also worshipped as a god of horses. Tlaloc, the Aztec god was greatly feared among the Aztecs, who drowned children to appease him, believing the he was responsible for both floods and droughts, and that he had been created by the other gods. Inti-Ilapa was the Inca thunder god, the feared warrior in the sky, who obtained water for the rain from the Milky Way. Chac, the Mayan god of rain, lightning and thunder, was both adored and feared as the rain was for growing crops, but if the rain was too heavy the crops could be destroyed. Some ancient gods and goddesses are still worshipped today, such as Chac.

REFERENCES

D'Altroy T. (2003). The Incas. Blackwell Publishing, Oxford, UK.

Freeman C. (2004). Egypt, Greece, and Rome: Civilizations of the Ancient Mediterranean, 2nd edn, Oxford University Press, Oxford, UK.

Darian S. G. (2001). The Ganges in Myth and History. Motilal Ranarsidass, Delhi, India.

Graham J. W. (1987). The Palaces of Crete. Princeton University Press, Princeton, NJ, USA.

Hillel D. (1994). Rivers of Eden: the Struggle for Water and the Quest for Peace in the Middle East. Oxford University Press, Oxford, UK.

Haland E. J. (2007). Let it Rain or Rain Conceive: Rituals of Magical Rain – Making in Modern and Ancient Greece. In: European Landscapes and Lifestyles: The Mediterranean and Beyond, Z. Roca, T. Spek, T. Terkeli, T. Plieninger and F. HöChtl (eds), Edições Universitárias Lusófonas, Lisbon, Portugal, pp. 285–304.

Haland E. J. (2009). Water sources and the sacred in modern and ancient greece and beyond. *Water Hist.*, **1**, 83–108.

Harland J. P. (1960). An ancient survival in modern aigina. In: Studies in honor of B. L. Ullman presented to him on the occasion of his seventy-fifth birthday, L. B. Lawler, D. M. Robathan and W. C. Korfmacher (eds), St. Louis University Classical Bulletin, St. Louis, USA, pp. 17–18.

Jacobsen T. (1946). Sumerian mythology: A review article. *Journal of near Eastern Studies*, **5**(2), pp. 128–152.

King L. W. (1902). The Seven Tablets of Creation, Vol. I, Luzac's Semitic Text and Translation Series, Vol. xii-xiii, Luzac And Co., London, UK.

Lambert W. G. and Millard A. R. (1999). Atra-Hasis, the Babylonian Story of the Flood, Eisenbrauns, Winona Lake, Indiana, USA.

Mays L. W. (2010). Ancient Water Technologies, Springer, Dordrecht, The Netherlands.

Oakes L. and Gahlin L. (2003). Ancient Egypt, Barnes And Noble Books, New York, USA. (Previously Published as The Mysteries Of Ancient Egypt).

Platon M. (1990). New indications for the problems of purgatory cisterns and bathrooms in Minoan world. In: Proceedings of the 6th International Cretologic Congress, Literary Association Chrysostomos, Chania, Greece, A2, pp. 141–155 (in Greek).

Tamburrino A. (2010). In: Ancient Water Technologies, L. W. Mays (ed.), Springer, Dordrecht, The Netherlands.

Vipin K. (2011). Personal communication. http://www.exoticindia.Com, 13/11/2011 .

Wilkinson R. H. (2003). The Complete Gods and Goddesses of Ancient Egypt. Thames and Hudson, London, UK.

Woolf G. (General Editor) (2005). Ancient Civilizations. Duncan Baird Publishers, London, UK.

Chapter 2

Water for human consumption through history

L. W. Mays, M. Sklivaniotis and A. N. Angelakis

2.1 PROLEGOMENA

In the long history of humankind the basic, most powerful, propelling force that shaped his action was the need to secure food and water. Air was beyond his control. This is shown practically in all excavations of prehistoric human habitations. They all had one thing in common; all were located near sources of a spring, river, lake or stream.

Humans have spent most of their history as hunting and food gathering beings. Only in the last 9000 to 10,000 years did they discover how to grow crops and tame animals. Such revolution probably first took place in the hills to the north of Mesopotamia. From there the agricultural revolution spread to the Nile and Indus Valleys. During this agricultural revolution, permanent villages replaced a wandering existence. About 6000 to 7000 years ago, farming villages of the Near East and Middle East became cities. During the Neolithic age (*ca.* 5700–3200 BC), the first successful efforts to control the flow of water were driven by agricultural needs (irrigation) and were implemented in Mesopotamia and Egypt. Remains of these prehistoric irrigation canals still exist (Mays *et al.* 2007).

Urban hydraulic systems are dated at the Bronze Age (*ca.* 3200–1100 BC). There are several astonishing examples of urban water systems from about the mid-third millennium BC. Mohenjo-Daro, located in the province of Sindh, Pakistan, one of the largest settlements of the ancient Indus Valley Civilization, developed a sophisticated system for water supply and sewage. Built *ca.* 2600 BC, it was one of the world's earliest major urban settlements, existing at the same time as the civilizations of ancient Egypt, Mesopotamia, and Minoan Crete. Water came from more than 700 wells and supplied not only domestic demands but also a system of private baths and a Great Bath for public use (Jansen, 1989). The Mesopotamians were not far behind (Adams, 1981). An example is the city of Eshnunna (80 km northeast of the present Baghdad) dated in the same period, where archaeological excavations exposed sewers constructed of brick, with laterals connecting to houses (Gray, 1940). In the Sumerian city of Nippur the excavations exposed clay pipes as well as tee- and angle-joints. Rainwater harvesting and collection in cisterns for urban water supply was practised at about the same period in Jawa in north-eastern Jordan (Abdel Khaleq & Alhaj Ahmed, this issue). Another collection system, this time for groundwater, was developed in Persia, again in the same period (Hassan, 2003). This is the well-known "qanat" – a subterranean system of tunnels connecting wells and dug using vertical shafts, designed to collect and transport water, sometimes over long distances, from highlands to low-lying farming land.

Amazing characteristics of the latter technology are its application over a tremendously long period up to modern times and the durability of the systems constructed. For instance, in Iran, over millennia, 22,000 qanat units were constructed, comprising more than 270,000 kilometres of underground channels. In the 1960s, 75% of all the water used in that country for irrigation and domestic consumption was provided by such systems (Wulff, 1968). Over the centuries, the technology was transferred to all civilizations and become known with different names such as "karez" (Afghanistan and Pakistan), "kanerjing" (China), "falaj" (United Arab Emirates) and "foggara/fughara" (North Africa). A "foggara" in Algeria is shown in Figure 2.1. According to UNESCO recommendations, some of those projects are protected as monuments of world heritage (Mays *et al.* 2007).

Figure 2.1 A foggara at Adrar, Algeria: (left) shaft; (right) distribution and metering weir at the foggara outlet (Mays *et al.* 2007)

As in the Indus Valley civilization, the history of water supply engineering on Crete dates back more than ca. 5000 years. From the early Minoan period (ca. 3200–2300 BC) issues related to water supply were considered of great importance and were accordingly developed. Archaeological and other evidence indicate that during the Bronze Age advanced water management and sanitary techniques were practised in several settlements in Crete (Angelakis & Koutsoyiannis, 2003). The emergence of the palaces reveals a remarkable development of water management in the urban context. Moreover, during the Middle Minoan and the beginning of the Late Minoan periods (ca. 2000–1500 BC) a "cultural explosion" occurred on the island. A striking indication of this is manifested, inter alia, in the advanced water management techniques practised in Crete at that time (Angelakis & Spyridakis, 1996; Koutsoyiannis *et al.* 2008). These included various scientific fields of water resources, such as wells and ground-water hydrology, aqueducts and domestic water supply according to local conditions in terms of climate and geomorphology. Additionally, the constructions and use of sanitary and purgatory facilities, even the recreational uses of water, signify attitudes of life and taste (Angelakis *et al.* 2005; Antoniou & Angelakis, 2011).

Minoan hydraulic technologies were developed further in Greece during several stages of the Greek civilizations. New more advanced water technologies were also invented there, with a peak in the Hellenistic period which followed Alexander the Great, during which they spread over a geographical area from Greece to India to the east and Egypt to the south. The Romans, whose empire replaced the Greek rule in most part of this area, inherited the technologies and developed them further also changing their application scale from small to large and implementing them in almost every large city. The Greek and Roman water technologies are not only a cultural heritage but are the underpinning of modern achievements in water engineering and management. The durability of some of the constructions that operated up to modern times, as well as the support of the technologies and their scientific background by written documents, enabled these technologies to pass to modern societies despite regressions that have occurred through the centuries (e.g. in the Dark Ages).

The availability of drinking water is an essential condition for the development of human civilization. For thousands of years this factor has been critical for the choice of location for the development of the cities. The first great civilizations developed close to rivers both for the availability of water suitable for human needs as well as agricultural and transportation purposes. Initially the water purification was very limited. It is only the last 200 years that water processing has been developed aiming at the improvement of hygiene and aesthetic conditions.

Every settlement of the humankind basically depends on a sufficient water supply. This applies especially for arid and semi-arid climate conditions in the regions around the Mediterranean basin and the Near East, where water resources availability is extremely limited mainly during the summer. Meanwhile, in these regions the first cultures rose. The scope of this Chapter is to present the main achievements in selected fields of potable water management chronologically including water supply technological principles and consumption by humans, extending from the earliest to the present. It is not an exhaustive presentation of what is known today about water treatment, related technologies and their uses in water supply since the beginning of humans' quest for water supply systems. Emphasis is given to the periods of great achievements.

2.2 NEOLITHIC AGE

There is an intriguing similarity between the lives of these people in spite of the distances that separated them. Water canals for drinking purposes and irrigation of the land were very much in use by them. Incidentally, and perhaps by coincidence, they were the first to use the land and developed early forms of agriculture instead of depending on nature alone to provide food for them as was the case with primitive man. We find similar man-made water canals and other water-saving devices in ancient Egypt as early as *ca.* 4000 BC.

The Egyptians believed in life after death and spoke freely about it. They prepared themselves all through life for what they would do or what they would say when the time came to go to that unknown after-death world. Water was sacred to the Egyptians. On one of the stone slabs unearthed was inscribed the "confession" of a man who was readying himself for that journey and who was prepared to say that he had not 'held up water in its season' nor had ever 'built a dam against running water'.

During the Neolithic age (ca. 10,000–3000 BC), the first successful efforts to control the flow of water were driven by agricultural needs (irrigation). Irrigation probably began to develop at a small scale during the Neolithic age in the so-called "fertile crescent" (Figure 2.2), an arc constituting the hills of Syria – Palestine and the feet of the Tauraus and Sagros mountains, as well as the southeast of the Caspian Sea (Viollet, 2006).

Mesopotamia, named by the ancient Greeks as being the land between the Euphrates and Tigris Rivers and its tributaries, is basically the eastern side of the fertile crescent. This area roughly comprises modern day Iraq and Syria. Larger scale irrigation developed when early cultivators settled in the low plains where the Tigris and Euphrates Rivers join. The large alluvial plain between the Tigris and Euphrates Rivers has been occupied by humans since around 5000 BC, with the beginning of the Sumerian civilization. The large scale diversion of water by humans probably had its origin in ancient Mesopotamia.

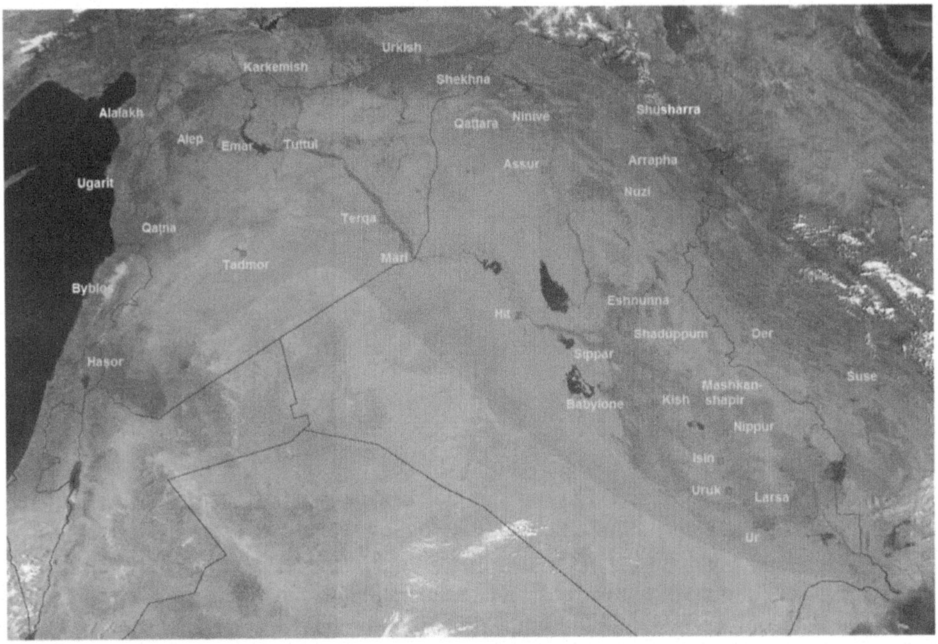

Figure 2.2 Fertile Crescent (*Source*: www.Wikipedia.org, GNU Free Documentation Licence)

2.3 BRONZE AGE

When the human mind overcame the barriers of nature and man was no longer a total slave to its uncontrolled powers, when man had risen to a position that enabled him to harness or manipulate the forces of nature, the basis for a well-conceived and fairly social type of organisation started to develop. Water could be brought from long distances by human effort by detouring streams, using ditches or channels and gradually by constructing cisterns, digging wells or building aqueducts (Phillips, 1972).

One of the early Bronze Age civilizations was Mohenjo-Daro (Mound of the Dead) one of the major urban centres of the Harappa Culture or Indus Civilization (Mays, 2010). Mohenjo-Daro, located on the right bank of the Indus about 400 km south of Karachi, Pakistan, was a deliberately planned city built around 2450 BC over a relatively short time period (Jansen, 1989). This planned city, located in a semi-arid environment, was serviced by at least 700 wells, with an average frequency of one in every third house (Jansen, 1989). The cylindrical well shafts were constructed using wedge-shaped bricks.

The following, from ancient Sanskrit writings (ca. 2000 BC) concerning medical treatment, give evidence of the first writings of water purification. From the Sus'ruta Samhita, 'impure water should be purified by being boiled over a fire, or being heated in the sun, or by dipping a heated iron into it, or it may be purified by filtration through sand and coarse gravel and then allowed to cool'. The Sanskrit Ousruta Sanghita included, 'It is good to keep water in copper vessels, to expose it to sunlight, and filter through charcoal,' according to Francis Evelyn Place (in 1905) who studied Sanskrit medical lore.

The technologies for water resources for urban areas of Mesopotamia during the Bronze Age included short canals connected to rivers in cities such as Uruk, Ur, Mari, and Babylon (all cities in the Tigris and Euphrates valleys). Other examples including gutters and cisterns for collecting and storing rainwater were used in Mari, and wells were used in Ugarit (Syria). The shaduf, a device for lifting water from a source, a well, or a river, were used in Mesopotamia, as evidenced by a shaduf represented on a cylindrical seal from Mesopotamia dated around 2200 BC.

The Minoan culture flourished during the Bronze Age in Crete. A systematic evolution of water management in ancient Greece began in Crete during the early Bronze Age, for example, the Early Minoan period (ca. 3200–2300 BC). Wells, cisterns, water distribution, fountains, and even recreational functions existed. In prehistoric Crete rivers and springs provided people with water. Starting the Early Minoan period II (ca. 3200–2300 BC), a variety of technologies such as wells, cisterns, and aqueducts were used. Also the Minoan architecture, including flat rooftops, light wells, and open courts, played an important role in water management. The rooftops and open courts acted as catch basins to collect rainwater from which it flowed to storage areas or cisterns.

Historically, drinking water has been considered the clear water. Considering the scientific knowledge of that era, this simplification was totally justified. Without the tools of chemistry and microbiology, even today, clarity (and probably taste) is the main criterion for classifying water as fit for human consumption. Therefore, the first treatment attempts were aiming at the improvement of the aesthetic conditions of water. An ancient Hindu source presents, probably, the first water standard, dated 4000 years ago. It dictates that the dirty water should be exposed to the sun and then a hot copper bar be inserted seven times in it, before filtration, cooling and storage in a clay jar.

Archaeological and other evidence indicate that during the Middle Bronze Age a "cultural explosion", unparalleled in the history of other ancient civilizations, occurred on Crete. A striking indication of this is manifested, inter alia, in the advanced urban water management techniques practised in Crete at that time. One of the salient characteristics of the Minoan civilization (ca. 3200–1100 BC) was the treatment devices used for water supply in palaces, cities, and villages from the beginning of the Bronze Age. It is truly amazing that the most common water quality modification technique for providing suitable domestic water supplies was known to Minoan engineers. Thus, according to Dafner (1921), a strange, oblong device with an opening in one of its ends, was used to treat domestic water (Figure 2.3a). The device was constructed in a similar manner and with the same material as the terracotta water pipes. Spanakis (1982) theorised this device as a hydraulic filter which was probably connected to a water supply reservoir by a rope passing through its outside holds. Its operation relied on local, high speed, turbulent conditions in order to continuously clean the porous surface thus allowing the continuous flow of filtered water to the jar. For cleaning purposes after extensive solids accumulation, it was possible to release it from the pipe end by loosening the rope in the holes. Also, an Egyptian clarifying device pictured in the tomb of Amenophis II and later in the tomb of Rameses II is shown in Figure 2.3b. This device allowed impurities to settle out of the water and then the clarified water was siphoned off and stored for later use.

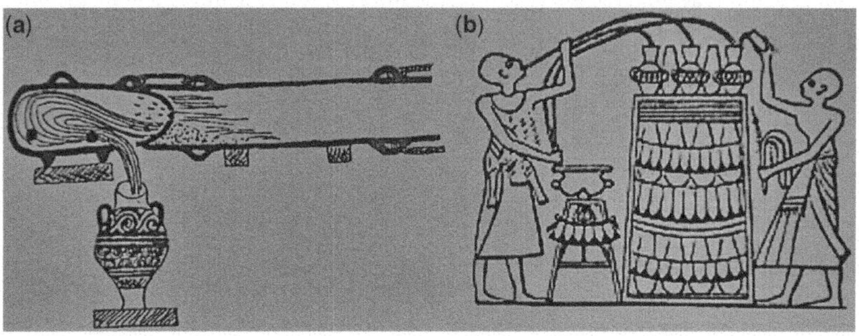

Figure 2.3 Bronze Age water purification devices. (a) Minoan water treatment device (Defner, 1921) and (b) Egyptian clarifying device (Baker & Taras, 1981)

Two other examples are:

(a) In the Tylissos houses the water was transported through an aqueduct to the three Minoan houses from the Agios Mama Spring, a distance of about 3 km. Terracotta infiltration devices were discovered in Agios Mamas, the location of the spring from where water was transported to the Tylissos village (Figure 2.4). These devices were filled with charcoal, thus playing the role of activated carbon treatment processes for removing both organic and inorganic constituents. In addition to these devices, a terracotta pipeline of 42 m length, similar to those used in Knossos, was discovered in the north-western site of House B. Other remains of the aqueduct are shown in Figure 2.5. A small cistern of stone was used for the removal of suspended solids from the water before its storage in the main cistern (Figure 2.5a). This cylindrical-shaped cistern (Figure 2.5b) was located at the northern site of House C and was considered as a part of the aqueduct (Hatzidakis, 1934).

Figure 2.4 Two terracotta conical tubes probably used as refinery devices from Agios Mamas spring in Tylissos (Archaeological Museum of Iraklion), Iraklion (with permission of M. Nikiforakis, EFIAP)

Figure 2.5 Small cistern (sedimentation basin) of stone (a) used for removal of suspended solids of water before it flowed to the main cylindrical-shaped cistern (b) used for water supply of the house (with permission A. N. Angelakis)

(b) In the palace of Phaestos, as in other cities and villages in Minoan Crete, the water supply system depended directly on precipitation: here, the rainwater was collected from the roofs and yards of buildings and stored in cisterns. In Phaestos the water supply system was dependent directly on surface runoff; there, rainfall water was collected in spatial cisterns from the roofs and yards of buildings. Special care was taken with the hygiene of water collection by: (a) cleaning the surfaces used for collecting the runoff water (Figures 2.6a and 2.6b) and (b) filtering the water in coarse sandy filters before it flowed into the cisterns in order to maintain the purity of water (Figure 2.6b) (Angelakis & Spyridakis, 1996). It was estimated that about 8 Mm^3 of rainwater were collected in an average year. That water was mainly used for washing clothes and other cleaning tasks.

Figure 2.6 Open yard (a) and special cistern (b) with sandy filter in the Phaestos palace (with permission A. N. Angelakis)

2.4 AFTER THE BRONZE AGE (IRON AGE)

In Jerusalem at times of siege (before *ca.* 1,000 BC) or when water from the Gihon spring was not accessible, the water supply was from the cisterns that could possibly supply water only for urgent drinking and cooking purposes alone at 3–5 L/inh · d for as many as 30,000 people (Shuvial, 2009). However, studies by anthropologists in developing countries show that in communities without direct access to potable water sources, the domestic water use is in the range of 10–20 L/inh · d. In such villages all water has to be carried in pots and buckets some distance from the water source to the home. For a family of six that would mean 60–120 L of water each day weighing 60–120 kg. It would hardly be feasible to carry much more water than that every day. Shuvial (2009) is assuming that after *ca.* 1000 BC in ancient Jerusalem, the water requirement for drinking, cooking, washing and minimal other domestic uses was about 20 L/inh · d and that half of the water of the minimum flow Gihon spring (650 m^3/d) was for such domestic purposes. Based on this assumption there would be more than enough water to support a population of 20,000–30,000.

Since the prehistoric era the availability of water sources was the primary criterion for the selection of small site settlements. All other criteria such as the natural protection of the site, the good soil and the ease of approach, were considered as secondary.

Gradually, as the small settlements grew into cities, during the archaic and classic period, the sources of fresh water within the boundaries of the city were not sufficient for the needs of the larger population. So they had to be enhanced with water quantities brought from sources outside the boundaries of the city. These sources had to be as close as possible in order to make the transportation of water easy and safe. The Mycenaean cities had constructed tunnels under the walls that provided safe underground passages to water sources, in order to secure adequate supply of water during seizures. During the same period important water projects were accomplished such as the draining of Lake Kopaes and the dam near Myticas in west Greece.

2.5 ARCHAIC AND CLASSICAL GREEK PERIODS

In the archaic (750–480 BC) and the classical (480–323 BC) periods of the Greek civilization, aqueducts, cisterns, and wells were similar to those built by the Minoans and Mycenaeans. However, the scientific and engineering progress during those stages enabled the construction of more sophisticated structures. An example is the Peisistratean aqueduct, constructed in Athens during the time of the tyrant Peisistratos and descendents *ca.* 510 BC. This aqueduct carried water from the foothills of the Hymettos mountain (probably from east of the present Holargos suburb for a distance of 7.5 km to the centre of the city near the Acropolis (Tassios, 2002). The pipe segments of the Peisistratean aqueduct are shown in Figure 2.7. For safety reasons, aqueducts were always subterranean, either tunnels or trenches. At the entrance of the city, aqueducts would branch into the city and would feed cisterns and public fountains in central locations. Along the bottom of trenches or tunnels of aqueducts, pipes usually made of terracotta, were laid, allowing for protection. A single pipe, or two or more pipes in parallel, would be used depending upon the flow to be conveyed.

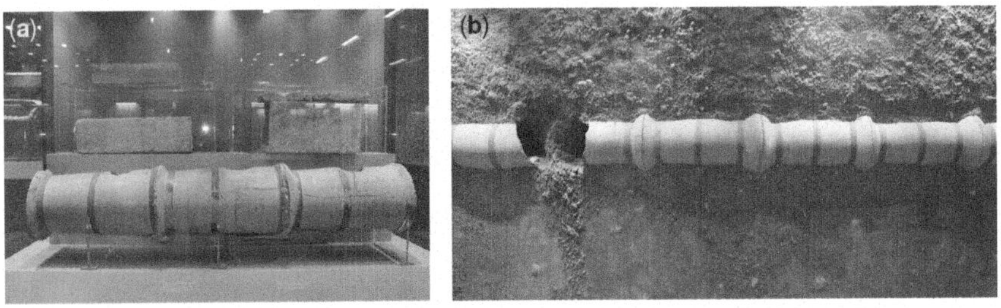

Figure 2.7 Peisistratean aqueduct terracotta pipe segments that were laid in a channel. Pipe joints were made of lead and the elliptical pipe openings were for cleaning purpose (with permission of A. N. Angelakis)

Hippocrates (460–377 BC), the Greek physician, wrote the first work on public hygiene (Air, Water, and Places) in which he noted that water varied in quality (taste and weight) and that it could be purified. His purification process, using what was later called "Hippocrates Sleeve", was to use a cloth bag through which he poured the water after it was boiled to trap sediments that caused bad taste and smell. This device was used for his patients.

2.6 HELLENISTIC GREEK PERIOD

Later, during the Hellenistic period [323 BC (death of Alexander the Great) – 146 BC (conquest of Greece by the Roman Empire)], further developments were made by the Greeks in hydraulics, such as the construction and operation of aqueducts, cisterns, wells, harbours, water supply systems, baths, toilets, and sewerage and drainage systems. During that period the political and economic situation changed leading to much more architectural development and urban beautification, of which aqueducts played a major role. Progress in science during the Hellenistic period provided a new technical expertise. Hellenistic aqueducts usually used pipes, compared to the late Roman masonry conduit. Furthermore, following the classical Greek tradition, the aqueducts continued to be subterranean for security reasons (not to be exposed to aliens, e.g. in case of war) but also for the safety of the construction during earthquakes which are frequent in the area. Again the Hellenistic technology contrasts with the later Roman technology, whose apparent characteristic was the use of arches and aqueduct bridges.

The water supply system to Pergamon, discussed further in Chapter 12, was built by the Greeks and the Romans (Ozis, 1996). The Madra Dag aqueduct to the acropolis was 42 km long (built by Eumenes II (197–159 BC). The aqueduct was constructed of triple terracotta pipelines of approximately 0.18 m in diameter and a wall thickness of 3.5 cm. The last stretch of 3.2 km of the siphon was under a maximum pressure of 190 m. No sign of lead pipes has survived, however soil samples close to the pipeline location proved to have 50 times higher lead-content than soil samples further away (Fahlbush, 2010).

During the historic years, after the eighth century BC, the settlements and cities obtained drinking water from springs, wells, rivers and precipitating water collected from roofs and stored in reservoirs. The use of remote sources required water transportation. The Hellenes (Greeks is a name imposed later by the Romans) used mostly clay pipes. Since every City-State had a limited ruling domain the water sources had to be as close as possible to the walls so the protection of the water system could be effective. The most impressive remote water transportation system was constructed in the island of Samos by the Tyrant Polycrates who constructed a 1000 m underground tunnel through the mountain based on the design of the famous engineer Eupalenos. This particular tunnel is an engineering marvel, since the construction started simultaneously from the two opposite ends and at the meeting point the deviation of the two axes was just 1.8 m. Later the King of Pergamos, Eumenes the 2nd (197–159 BC) constructed the first complete Hellenic water transportation and storage system based on branching pipes. This aqueduct is proved to have been operated for several hundred years. It is looked upon as the peak of Hellenistic water supply technology (Fahlbusch, 2010).

Despite the fact that during the classic era Hellenes had both the financial and technical capacity (the technology of arch construction was known since Mycenaean years) for large scale constructions, they did not build long, above ground, open water transportation systems for a number of reasons. First of all, for political reasons the organisation of relatively small City-States resulted in frequent conflict. Open and long water transportation systems were vulnerable. Underground pipe systems are not visible and not easy to reach and inflict damage upon. City-States are relatively small. They did not need huge amounts of water, therefore, relatively small clay pipes of manageable size for that level of technology, would suffice. Closed pipe systems are also known to offer the best protection against natural pollution. Hippocrates probably influenced the choice towards closed systems. He had identified the significance of good water quality for good health. He indicated that it is preferable to use water from a good source than attempting to treat water from an inferior source. This approach has also been adopted by Article 7 of the European Water Framework Directive established in the year 2000 AD (EU, 2000). Moreover, he is credited with the development of the "Hippocrates sleeve" made of cloth and used to filter rainwater.

The small diameter closed water transportation systems had two particular problems, besides the usual structural and leakage problems: calcium carbonate deposits and lead. The predominance of limestone formations in the Balkan Peninsula as well as in many places across Europe, result in high concentrations of calcium carbonate in surface and underground water as well as relatively alkaline pH values. These conditions cause high rates of calcium salts depositions on the surface of pipes, which in practice cannot be removed. So, the pipes gradually had their diameter reduced and the amount of water flow considerably diminished. This problem has been noticed in many excavations of ancient water pipes. Lead was used as a sealing medium in clay pipes of larger diameters as well as to form pipes of small diameters. As far as we know, ancient civilizations ignored the epidemiology of lead, but the problems arising from lead use would be the same as they are today. For the construction of lead pipes specialised workshops were set up. Their remains have been uncovered in a number of excavations.

2.7 ROMAN PERIOD

Springs were the most common sources of water for the Romans. Water sources included not only springs, percolation wells, and weirs on streams, but also reservoirs that were developed by building dams which is discussed further in Chapter 12. To properly discuss Roman water supply it is important to be aware of the treatises of Marcus Vitruvius Pollio, a Roman architect and engineer who lived in the first century BC. His "Ten Books on Architecture" or "De Architectura" (Morgan, 1914), in Book VIII (Water Supply) discussed the various elements of water supply. In Chapters 5 and 6 of Book VIII, Vitruvius addressed the quality of water.

Chapter III (Various Properties of Different Waters):

(a) 20. Then there are springs in which wine seems to be mingled, like the one in Paphlagonia, the water of which intoxicates those who drink of the spring alone without wine. …
(b) 21. In Arcadia, is the well-known city of Clitor, in whose territory is a cave flowing with running water, which makes people who drink of it become abstemious. …
(c) 22. In the island of Zeas is a spring of which those who thoughtlessly drink lose their understanding, and an epigram is cut there to the effect that a draught from the spring is delightful, but that he who drinks will become dull as a stone. …
(d) 23. At Susa, the capital of the Persian kingdom, there is a little spring, those who drink of which lose their teeth.
(e) 24. There are also in some places springs which have the peculiarity of giving fine singing voices to the natives, as at Tarsus in Magnesia, and in other countries of that kind. …

Chapter IV (Tests of Good Water):

(a) Springs should be tested and proved in advance in the flowing ways. …
(b) And if green vegetables cook quickly when put into a vessel of such water and set over a fire, it will be proof that the water is good and wholesome. …

Chapter V (Aqueducts, Wells, and Cisterns):

(a) There are three methods of conducting water, in channels through masonry conduits, or in lead pipes, or in pipes of baked clay. …
(b) 10. Clay pipes for conducting water have the following advantages. … Secondly, water from clay pipes is much more wholesome than that which is conducted through lead pipes, because lead is found to be harmful for the reason that white lead is derived from it, and this is said to be hurtful to the human system. …

2.7.1 Roman water supply systems

Various components of the Roman water supply system are illustrated in Figure 2.8. Aqueducts were used to transport water from the source to the locations where the water was needed, either for irrigation or for urban water supplies. Rome was an empire, the internal safety was not at danger and therefore the open aqueducts were not a problem for water safety. Roman aqueducts were built to promote quality of life and in general were built to serve existing urban centres, in many cases where prosperous life had existed for centuries before the aqueducts had been built. The Romans built aqueducts largely to supply baths and were an expression of civic pride. Many of the Roman inhabitants obtained their drinking water from wells prior to the construction of the aqueducts. The location and delivery point of the aqueducts were determined by geographical, economic, and social factors. Secondary lines *(vamus)* were built at some locations along the aqueduct to supply additional water. Also subsidiary or branch lines *(ramus)* were used. At distribution points water was delivered through pipes *(fistulae)* made of tile or lead.

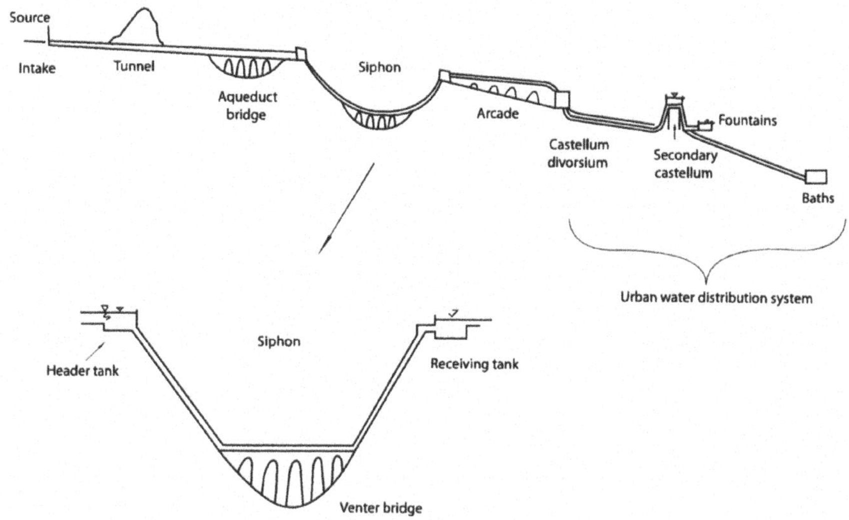

Figure 2.8 Components of the Roman water supply system (copyright with L. W. Mays)

Roman aqueducts included open channels and pipes. The most common was the open channel constructed of masonry conduit. Also open channels were made of cut stone and an open clay-lined leet. Aqueduct bridges were probably the most spectacular feature of the aqueduct systems. The Romans used siphons throughout the empire. The siphons included a header tank for transitioning the open channel flow of the aqueduct into one or more pipes, the bends called geniculus, the venter bridge to support the pipes in the valley, and the transition of pipe flow to open channel flow using a receiving tank.

There were many components in the Roman urban water distribution system. Water flowed by gravity through enclosed conduits (specus or rivus), which were typically underground, from the source to a terminus or distribution tank (castellum). Above ground aqueducts were built on a raised embankment (substructio) or on an arcade or bridge. Settling tanks (piscinae) were located along the aqueducts to remove sediments and foreign matter. These pipes were connected to the castellum by a fitting or nozzle (calix). These pipes were placed below ground level along major streets. Two approaches to laying out the pipe network were followed: (a) using a main pipe from the secondary castellum with smaller branch pipes attached to serve individual customers, and (b) not using main pipes but using individual pipes laid from the secondary castellum to the individual customer, which was the normal Roman practice (Hodge, 2002).

The Romans made extensive use of cisterns. Cisterns were used extensively for storing water from rainfall collection and from aqueducts. The households and public buildings both had very interesting systems to collect and store rainwater. Buildings with peaked roofs had gutters along the eaves to collect the rainwater and downspouts to carry the water to the cisterns located under the building. Downspouts were made of terracotta pipes and were often set inside the wall.

As the cities and states grew bigger and more confident, especially during the Roman period, they shifted their technology to longer and bigger water transportation systems, namely the open aqueducts. Typically brick was used for the aqueduct construction. The remains of Patras aqueduct in western Greece are shown in Figure 2.9. The Carthage water transportation system was 132 km long above ground. The first Roman aqueduct supported on consecutive arches, known as Aqua Marcia, was constructed in 144 BC by Marcius Rex. It was 90 km long and the last 9 km was supported on arches. When the Romans concluded the occupation of the Hellenic states, after the era of emperor August, a great deal of long aqueducts were constructed in the Hellenic region as well as all over Europe. However, the vulnerability of these open systems never ceased to be a problem. In 537 AD the aqueduct of Rome was destroyed by the Goths and was used as a weak point to enter the city.

Figure 2.9 Sections of the aqueduct of Patras in western Greece (Sklivaniotis & Angelakis, 2006)

The extensive use of open aqueducts was aided by the technology of the strong bonding mixture called pozzolana for bricks, which made the construction of aqueducts both inexpensive and fast to execute.

The distribution of water inside the cities was achieved through pipes: clay and lead for smaller diameters, brickwork sealed with specialised hydraulic plaster for larger diameters (Figures 2.10 and 2.11). These distribution systems were operating under pressure which allowed the construction of recreational jet fountains. It is known that the water system in ancient Pergamos, 45 km long, reached inside the city with a pressure of 20 atm. Ceramic pipes were usually of cylindrical shape.

Figure 2.10 Lead water pipe of the Roman period, Greece (Sklivaniotis & Angelakis, 2006)

Figure 2.11 Ceramic water pipes from an excavation in Aigio in western Greece (Sklivaniotis & Angelakis, 2006)

Despite the availability of a public water supply, individual houses maintained their own water sources. Where there was available underground water, a well was constructed. The internal surface of the well was covered with built stones or clay rings, which at regular intervals had holes to facilitate access to the bottom of the well for maintenance (Figure 2.12). They also utilised the rainwater which was collected from the roofs of the houses and stored in cisterns, usually underground. Such a cistern used to collect rainwater is shown in Figure 2.13. This water was mostly used for cleaning the house if running water was available for human consumption.

Figure 2.12 Cylindrical terracotta rings, for coating the walls of a Roman well (Sklivaniotis & Angelakis, 2006)

During the Classic period and mostly during the Roman period, water was extensively used for human personal hygiene. Baths were both private and public. Public baths were more than places for hygiene. They were places for social conduct and recreation. The "complete" Roman bath included a cold bath called the frigidarium, followed by a warm bath called the tepidariuma, followed by a hot bath called the caldarium. The bath houses were heated by hot air passing through clay pipes inside the walls and in the empty space under the false floor. Remains of Roman thermae and a public bath in western Greece are shown in Figures 2.14a and b, respectively.

Figure 2.13 House cistern used to collect rainwater in the Atrium of a Roman house in Patras in western Greece (Sklivaniotis & Angelakis, 2006)

Figure 2.14 Roman baths: (a) Baths Hypocaust columns of Roman bath at Lucentum, near Alicante, Spain (with permission of L. W. Mays) and (b) An early Christian bath in Patras in western Greece (Sklivaniotis & Angelakis, 2006)

2.7.2 Water consumption

As for other ancient periods one unresolved problem, during the Roman period is the quantity of water provided by the aqueducts. In general, data provided by archaeological evidence and written accounts fall short of that required for the computation of either per capita supply or per capita use. The cross-sectional areas of the aqueducts are known, but one cannot be sure that the conduits ever ran at their maximum capacity. Sizing of the channels was probably determined as much by the need for work space as by the volume needed for water (Hansen, 1983). It was certainly a rare occasion when the aqueducts were all working at the same time. Another uncertainty is the population of Imperial Rome. No accurate estimate exists for any particular period. It is generally assumed that during Frontinus era the population of Rome was approximately one million (Parker, 1967).

Hansen (1983) reported that the amount of water delivered by the aqueducts varied from a low of 322,000 to a high of 1,010,623 m^3/d. There is also little consensus among more recent estimates. Bruun (1991) prefers a water rate of 600,000 m^3/d. This latter figure computes a very high water supply value of

$0.6\ m^3/d \cdot inh$. However, he also makes an estimate of the per capita usage of the middle and lower classes. According to Frontinus, the aqueducts served 591 lacus (or major delivery points), and each, on average, delivered $60\ m^3/d$. He estimates that each delivery point served 900 individuals on an average basis. This computes to a water use of $67\ L/d \cdot inh$. This figure is considerably lower than all previous estimates, but it is not too low when it is considered that Romans could also recycle and reuse wastewater from baths, toilets, and other establishments. In addition, they could probably store and use rainwater and/or water directly from Tiber River. A water consumption of 67L/d, only includes the water consumed from the lacus. It would seem that for the ordinary Roman, this quantity or water was satisfactory, but not lavish. Contemporary residential customers in the USA require approximately 350 $L/inh \cdot d$ (Hansen, 1983).

Roman water supply in comparison with modern times' water usage rates appears to be problematic. It is commonly felt that water supply in Roman cities was delivered at a constant flow rate instead of on-demand (as with modern systems). For example, water flowed through a latrine in one continual stream, not just when flushed as is the case with the modern toilet. According to Hansen (1983) fountains probably ran night and day. Thus, to compare water use (on $L/inh \cdot d$ basis) in the Roman period with that in 20th century is problematic.

Bruun (1991) disagrees somewhat with the idea that Rome's water supply was constant and continuous. Arguably the Romans considered water flowing day and night as a particular sign of richness and a high standard of living. But this does not necessarily mean that they paid little attention to the storage of water, or to regularising the water flow. In fact there exists sundry evidence that measures for saving water had been adopted. It is often maintained that the Romans did not make use of taps and similar devices in this distribution net; for example Frontinus never even once refers to such objects. But as a matter of fact archaeological discoveries have brought forward a large amount of evidence for their use (Hansen, 1983).

The Romans also diverted water into storage tanks. Archaeologists have uncovered large cisterns in Rome, many received water from the aqueducts. Bruun (1991) argues that Romans were better water managers than we commonly give them credit for. It is safe to assume that Rome received an impressive supply of water, and that the rich and influential received a disproportionate amount. But the water supply for the common Roman was still sufficient. By historic standards, the Romans' water supply was a very impressive accomplishment (Hansen, 1983).

2.8 BYZANTINE PERIOD

The Byzantine Empire and Eastern Roman Empire are names used to describe the Roman Empire during the Middle Ages, with the capital in Constantinople (Istanbul). During the thousand-year existence of the empire the influence spread widely into North Africa and the near East during the Middle Ages. Several water byzantine cisterns are known in various places in Greece (e.g. Athens, Amorgos Island, Mistra, Leontari Arkadias, Monemvasia, and Iraklion Crete). A byzantine cistern at the base of Acropolis in Athens is shown in Figure 2.15a. Another one at the castle of Leontari village in Peloponnesos (Arkadia), known as the cistern of "Goula" is shown in Figure 2.15b. This village was the basis of the brother of Constantine Paleologos (Antoniou, 2011).

The Halkai water conveyance system, dating from the early Byzantine period, was the first of three main systems supplying Constantinople (Ozis, 1987). The most interesting part of the water system of the Byzantine period is the cisterns in Constantinople. During the Byzantine times in Constantinople at least 70 cisterns were constructed (Cinis, 2003). The Basilica Cistern or in Turkish the "Yerebatan Sarayi", is one of the largest ($140 \times 70\ m^2$ and with a capacity of $80{,}000\ m^3$) known covered cisterns, probably completed in the early sixth century. This cistern, shown in Figure 2.16a, is an underground cistern, having 336 marble columns each 9 m high. The columns are arranged in 12 rows each with 38 columns, spaced 4.9 m apart. The cistern was built underneath the Stoa Basilica which is a large public square on the First Hill of Constantinople during the reign of Emperor Justinianos I in the sixth century. Water was supplied to the cistern by an aqueduct from springs in Marmara, which is west of the city.

Figure 2.15 Byzantine cisterns in Greece: (a) at the base of the acropolis in Athens and (b) at the castle of Leontari village in Peloponnesos (with permission of L. W. Mays & G. Antoniou, respectively)

Another very well known cistern in Constantinople is the Philoxenous cistern located on the opposite side of the present Courthouse close to the Sultanahmet square (Figure 2.16b). The name comes from the Roman Senator Philoxenous who built the Philoxenous palace during the period of Great Constantinus (324–337 AD). Another theory suggests that the cistern was built in the middle of the sixth century by the Emperor Justinianos I based on the style and materials used. The cistern has a rectangular shape of dimensions 64.0 x 56.4 m^2. It incorporates 16 rows consisting of 14 pillars each (total of 224 pillars) spaced 3.8 m apart. (Cinis, 2003). The pillars are connected via arches and carry diagonal beams and simple pyramid shaped headings. This cistern is also known as the "1001 post" cistern because of the many posts it incorporates. As a result of drying out in subsequent years, the Ottomans used the cistern as a thread-manufacturing workshop until the beginning of the 20th century. Other underground cisterns in Constantinople are those of St. Sophia (five; two of which are below the church), the Grand Palace cisterns (various), the Theodosius cistern, the Forty Martyrs cistern, the Botniates cistern, the Myrelaion cistern, the Fatih cistern, the Studios cistern, the Zeyrek-Pantokrator cisterns (two), and the Sishane cistern.

In addition to the underground cisterns in Constantinople there were at least four open-air cisterns (above ground). The Elephant Dome or Fildami cistern is one of the four biggest reservoirs of antiquity in Byzantium, Constantinople (Figure 2.16c). Different than most of the other cisterns in Constantinople, the Fildami cistern is not situated in the historical peninsula which is covered by Land Walls constructed by Theodosius. It is an extraordinary monument, compared to the others, which have survived until now.

The Fildami cistern is an open-air reservoir of the early Byzantine Empire, 1.5 km away from the coast of the Marmara sea on the west side of the Cirpici stream. It is constructed next to the Magnaura and Jucundianae palaces (Crow *et al.* 2008). The dimensions of the cistern are 127 x 76 m and its depth is 11 m, indicating a total volume of more than 105,000 m^3. The walls are visible to a height of about 10 m. Therefore it is estimated that its original shape was deeper than its form today. In total area it is somewhat smaller than the other open reservoirs within the Land Walls. It is only 1000 m^2 larger in area than the Basilica Cistern. Other known open-air cisterns are Aspar, St. Mocius, and Aetius (Cinis, 2003).

Figure 2.16 Byzantine cisterns in Constantinople: (a) inner view of basilica cistern, (b) an old engraining of Philoxenus cistern and (c) The Elephant Dome or Fildami cistern (with permission of M. Nikiforakis, EFIAP)

The total capacity of the underground cisterns was 200,000 m^3 and that of open-air cisterns was 800,000 m^3 (Cinis, 2003). With the ancient population of Constantinople estimated to be 50,000 inhabitants, these cisterns were sufficient to provide fresh water for a period of more than one year. Thus, during the Byzantine period an average water consumption of less than 55 L/inh.d is assumed.

2.9 OTTOMAN PERIOD (*CA.* 1669–1898 AD)

Water was connected to Islam, so that during the Ottoman period there was a water tap in all mosques. Hammams, which are presently also referred to as Turkish bathes, played an important role in the Ottoman culture and served as places of social gathering, ritual cleansing, and as architectural structures, institutes etc. The cleansing of the body symbolises the cleansing of the soul, according to the Koran. The hammam is a very old Ottoman institution and was established in all the regions of the Ottoman Empire. Following the very old Moslem tradition, water supply to hammams and fountains was the major hydraulic works developed during the Ottoman period. Such a domed fountain is shown in Figure 2.17a.

Figure 2.17 Ottoman cistern and fountain: (a) Fountain in Iraklion, Greece and (b) Domed circular cistern in southwest Anatolia (with permission of A. N. Angelakis & L. W. Mays, respectively)

The most remarkable water systems of the Ottoman period served Constantinople (Istanbul). There are also cisterns from this period that were circular-shaped, domed or other (Figure 2.17b) located in several Mediterranean countries. These cisterns were constructed primarily in the 16th century (Ozis, 1987) in various places occupied by Ottommans. Some of these are still in use for livestock water supply. They originally served the Ottoman army at military logistic points.

2.10 MODERN TIMES

As in every other field of art and science, during the Middle Ages there was no progress in the area of water treatment except for limited references in the Arab world. During the eighth century AD the Arab alchemist Gever distilled water to render it free from evil spirits and in the 11th century AD the Persian physician Avicenna advised travellers to boil the water or at least filter it through a cloth.

2.10.1 17th century

The British philosopher and scientist Francis Bacon, who managed to transform nature probing to scientific conclusions, devoted considerable effort to studying water purification techniques. In 1627 he published experimental results on percolation, filtration, distillation and coagulation.

In 1680 Dutch naturalist Antony van Leeuwenhoek discovered the microscope and in 1684 gave a first account of bacteria observed in water which he named "animacules". These findings were dismissed by the scientific community as unimportant curiosities. It took another 200 years for the importance of the van Leeuwenhoek's findings for public health to be understood.

In 1685 the Italian Lu Antonio Porzio designed the first multi-pass filter containing a straining section and two sand filtration passes (see Figure 2.18).

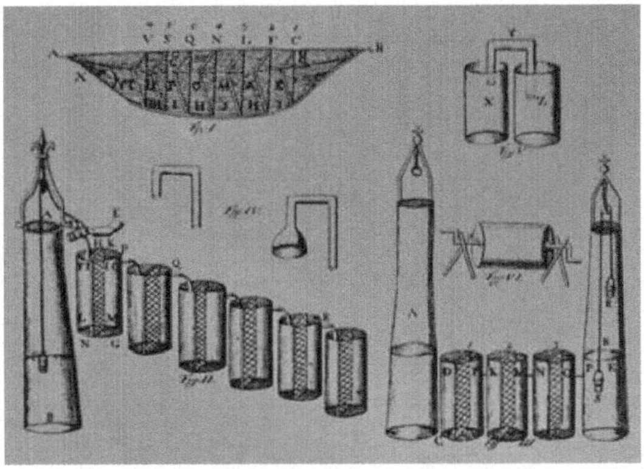

Figure 2.18 Porzio's multiple filter system. Showing the three filters: one floating, and two for land used for well water. (from Militis in Castro Sanitate Tuenda, 1685, translated into English as the Soldier's Vade Mecuum, 1747)

2.10.2 18th century

In the 18th century France took the lead. In 1703 the scientist Phillippe La Hire presented a home filter suitable for treating rainwater consisting of a sand filter and a storage tank. He also noticed that the water which came from underground aquifers was rarely polluted. In 1746 Joseph Amy received the first patent for a water filter that in 1750 he put on the market. Figures 2.19 and 2.20 illustrate Amy's filters. The filter was made of sponge, wool and sand. However, it was the British architect James Peacock that managed to claim a patent for a sand filter with backwashing.

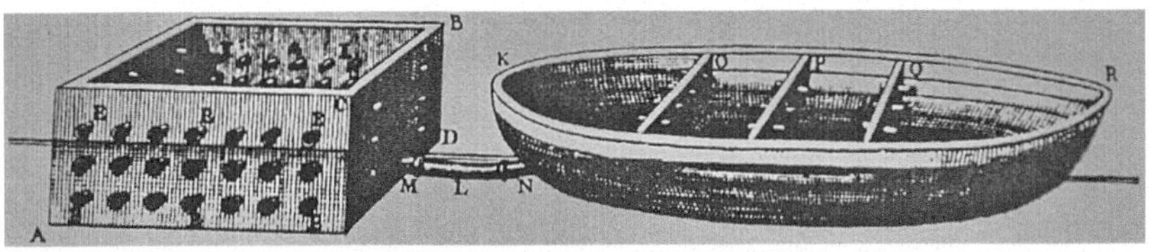

Figure 2.19 Amy's machine to purify water. This filter was approved by the French Academy of Sciences in 1745. The floating filter box and clear-water trailer are shown. Sponges were inserted in the sides of the filter box and if required were placed in the partitions of the trailer. (Source: *Machines et Inventions Approuvees par L'Academie Royale des Sciences*)

2.10.3 19th century

In 1804, Paisley, Scotland, became the site of the first filter facility to deliver water to an entire town. This filter is illustrated in Figure 2.21.

In 1806, a large water treatment plant opened in Paris, using the River Seine as a source. Water was settled for 12 hours prior to filtration then run through sponge pre-filters that were renewed every hour. The main filters consisted of coarse river sand, clean sand, pounded charcoal, and clean Fountainebleau sand. The filters were renewed every six hours. A simple form of aeration was also part of the process, and pumps were driven by horses working in three shifts (steam power was too expensive). This plant operated for 50 years. In 1807 in Glasgow, Scotland, filtered water was piped directly to customers.

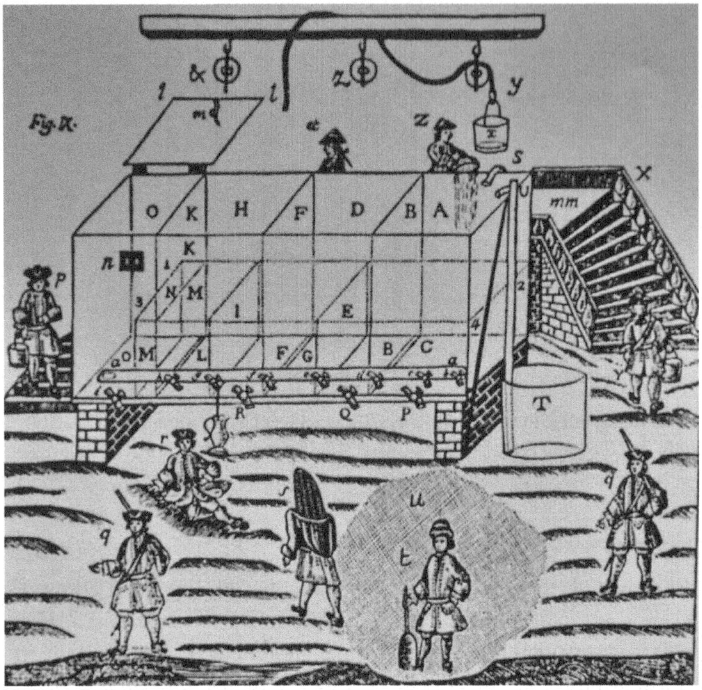

Figure 2.20 Amy's sextifold filter box for army garrisons. This filter system consisted of three pairs of down-up filters. Water flowed through 54 cm of sand. The tank was either lead-lined wood or masonry, 54 x 9 cm^2 in plan and 18 cm deep. The design closely resembled one described by Porizio in his 1685 book on military camp sanitation (Source: Amy's 1754 book)

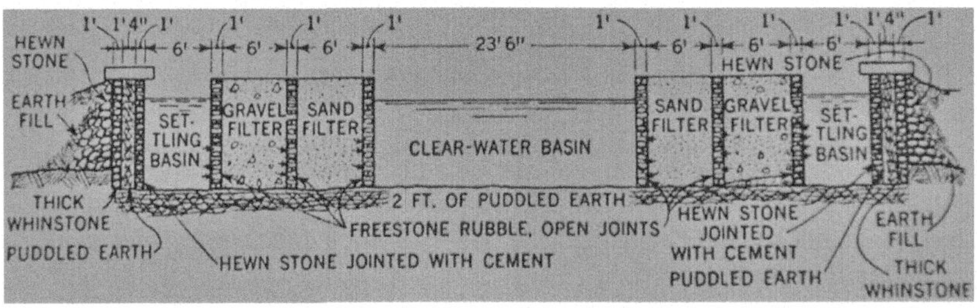

Figure 2.21 Filter system for Paisley, Scotland completed in 1804 by John Gibb. This is the first known filter to supply an entire city according to Baker (1948). Water flowed through a stone-filled trench to a ring-shaped settling chamber followed by flow through two lateral-flow filters to a central clear-water chamber. Water was then delivered to a tank on the hillside from where it was carted to consumers. (From description in Sinclair's Code of Health and Longevity, London, 1807; as presented in Baker (1948)

In 1832 the first slow sand filtration plant in the United States was built in Richmond. In 1833, the plant had 295 water subscribers. A small upward-flow filter of gravel and sand in the water works proved to be a failure for the turbid water of the James River. This was replaced by a downward flow approach which also failed. The filters were too small to handle the highly turbid water. The next US plant to open was in Elizabeth, New Jersey, in 1855. Slow sand filters were introduced in Massachusetts in the mid-1870s. Sand filters and other treatments were primarily designed to improve the aesthetic quality of water.

The evolution of rapid filters (Figure 2.22) in the USA began with developments by Patrick Clark and John W. Hyatt. Clark suspended a shallow filter in a river and provided surface-jet wash from a perforated revolving arm, with the loosened dirt being swept downstream (Baker, 1948). This filter was installed at Rahway, New Jersey waterworks around 1880. Hyatt superimposed several Clark filters in a closed tank and serviced them with a common pipe system (Baker, 1948). This method was first applied at Somerville & Raritan Water Co., New Jersey in 1882.

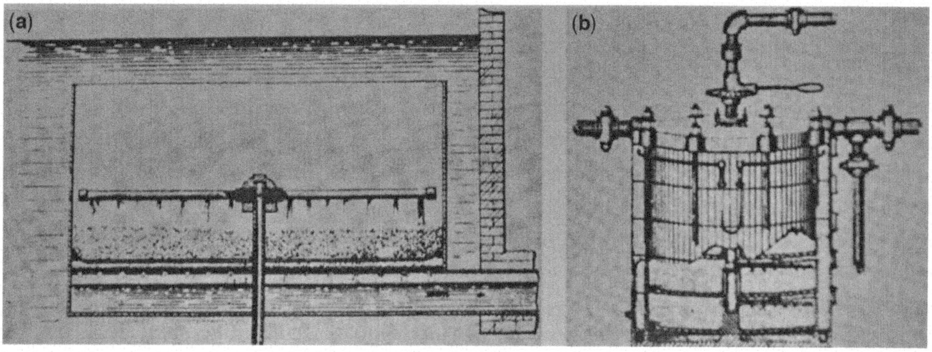

Figure 2.22 Rapid filtration advances by Clark and Hyatt. (a) Clark suspended shallow filter (from US patent drawing, June 21, 1881) (b) Hyatt superimposed several Clark filters in closed tank (from "The multifold water filter," Engineering News, January 1, 1882)

During the decades 1880 and 1890 the rapid sand technology matured and it became evident that in addition to the suspended solids removal, extensive bacteria removal is achieved. The first attempts to use chlorine and ozone for disinfection are made.

Regarding water availability and consumption, a water shortage at the end of 1861 in Athens, Greece was reported by Lambrou (1999). Total available water supplies were 600 *dramia* water in September and October of 1861 whereas the deficit to meet the water requirements of the city during that period was 450 *dramia* of water (one *drami* is equal to about one m^3). Lambrou (1999) considered 1.5 *dramia*/family · yr as an indicative water supply for the Athens Municipality. Similar water consumptions were reported by Soulis (1884).

2.10.4 20th century

In 1906 ozone was used for the first time for disinfection in Nice (France) and became very popular in Europe. In the USA, chlorine was mostly used for disinfection due to the complexity and the cost of the ozone equipment. The Europeans were particularly negative against chlorine due to its use in chemical warfare during the 1st world war. In 1908 sodium hypochloride was used in Jersey (USA) for disinfection and in 1917 chloramines were used first time for disinfection in Ottawa (Canada) and Denver (USA). The first serious efforts in water desalination technology were made during World War II for the supply of units which encountered difficulties in securing drinking water.

2.11 EPILOGUE

Water for human consumption has had a long and very interesting history. It is closely related to the development of civilization and thus a great deal of related events occurred around the world. All great civilizations had a large chapter written for water transportation, treatment and management of potable water. In this Chapter the evolution of urban water management in ancient civilizations is considered. The important steps in water supply history relevant to water treatment and hygienic technologies are chronologically presented and discussed. During the oligarchic periods the emphasis was on the construction of large-scale hydraulic projects, whereas in democratic periods the focus of water management was on sustainable small scale, safe and cost efficient management practices and institutional arrangements related to both the private and the public sectors. Such practices and institutions are relevant even today, as the water related problems of modern societies are not very different from those in antiquity.

As previously discussed several technologies for improving the water supply of settlements were accomplished very early in recorded history. Few basic principles have been added up to the present. However, the technological skill has been improved steadily such that world records in some fields have been achieved in antiquity for many, many centuries. The water supply of some ancient cities amounted to several hundred litres per day per inhabitant, a number which today is provided for cities only in a few highly industrialised countries. The fact that the water supply systems for many big cities in Central Europe were constructed in the last century, following the Greek and Roman examples, may indicate the high standard, which was achieved in several ancient civilizations (Fahlbusch, 2008). Thus, scientists of the 21st century can admire these achievements and hope that this standard will soon be gained worldwide.

Relationships based on exchange are known between the Minoans and continental Greece, Egypt, and the Levant. There is no doubt that some hydraulic practices and water management had been transferred from Mesopotamia and Syria to Crete by that time since such several works concerning the management of water had already been applied in these areas. However, the Minoans applied this skill and developed it, especially in urban hydraulics, in the palaces, cities and villages, up to a degree which had never been reached before (Lirintzis & Angelakis, 2006). Thus, the first indication for the development of some technology relevant to the treatment of urban water and hygienic water supply lies in Minoan civilization. It seems likely that these technologies were transferred to the Mycenaeans in continental Greece (Angelakis & Spyridakis, 1996). Classic period Hellenic states and Romans capitalised on this knowledge and made great advances in potable water management (Petropoulos, 2006; Fahlbusch, 2010). The unfortunate Middle Ages, a period of stagnation, was followed by the revival of letters and technology. Slow in the beginning, but faster and faster as time was passing, progress in drinking water transportation, treatment and management was made. The fast progress has created huge problems and the ease of communication has made the problems global. As usual the problems are emerging from politics in the broad sense. Looking to the future we wonder: What is going to grow faster, the problems or the technology than can solve them?

REFERENCES

Adams R. (1981). Heartland of Cities: Surveys of Ancient Settlements and Land Use on the Central Floodplain of the Euphrates. University of Chicago Press, Chicago, USA.

Amy J. (1754). Suite du Livre Intitule' "Nouvelles Fountaines, Filtrantes Approuvees par l'Academie Royale des Sciences." Antoine Boudet, Paris, France.

Angelakis A. N. and Spyridakis S. V. (1996). The status of water resources in Minoan times – A preliminary study. In: Diachronic Climatic Impacts on Water Resources with Emphasis on Mediterranean Region, A. Angelakis and A. Issar (eds). Springer–Verlag, Heidelberg, Germany, pp. 161–191.

Angelakis A. N. and Koutsoyiannis D. (2003). Urban Water Resources Management in Ancient Greek Times. In: The Encycl. of Water Sci., Markel Dekker Inc., B. A. Stewart and T. Howell (Eds), Madison Ave. New York, N.Y., USA, pp. 999–1008.

Angelakis A. N., Koutsoyiannis D. and Tchobanoglous G. (2005). Urban Wastewater and Stormwater Technologies in the Ancient Greece. *Water Res.*, **39**(1), 210–220.

Antoniou G. (2011). Study Rehabilitation buildings of Castle in Leontari Arkadias, Greece. Personal communication (in Greek).

Antoniou A. and Angelakis A. N. (2011). Evolution of Sanitary and Purgatory Structures in Ancient Greece. *Journal of Early Medicine Issue on Latrines, Hygiene and Health* (in press).

Baker M. N. (1948). The Quest for Pure Water: The History of Water Purification from the Earliest Records to the Twentieth Century. American Water Works Association (AWWA), Denver, USA.

Baker M. N. and Taras M. J. (1981). The Quest for Pure Water: The History of the Twentieth Century. Volume 1 and 2. American Water Works Association (AWWA), Denver, USA.

Bruun C. (1991). The Water Supply of Ancient Rome: A Study of Roman Imperial Administration. Societas Scientiarum Fennica, Helsinki, Filand.

Cinis N. (2003). Yerebatan Cistern and Other Cisterns in Istanbul. Duru Basim Yayin Remcilik ve Gida San, Tic, Ltd., Sti, Istanbul, Turkey.

Crow J., Rardill J. and Bayliss R. (2008). The Water Supply of Constantinople. *Journal of Roman Studies Monograph no 11*. Roman Society, London, UK.

Defner M. (1921). Late Minoan Water Treatment Device. Archaeological Newspaper no 78, Iraklion, Greece (in Greek).

EU (European Communities) (2000). Framework for Community Action in the Field of Water Policy. The Official Journal of the European Communities L 327/1, 2000/60/EC. Brussels, Belgium.

Fahlbusch H. (2008). Municipal water supply in antiquity. http://www.romanaqueducts.info/webteksten/waterinantiquity.htm (15 accessed August 2011).

Fahlbusch H. (2010). Water engineering and management in classic civilizations. In: Water Engineering and Management through Time – Learning from History, E. Cabrera and F. Arregui (eds), Taylor and Francis Group, London, UK.

Gray H. F. (1940). Sewerage in Ancient and Medieval Times. *Sewage Works Journal*, **12**(5), 939–946.

Hansen R. D. (1983). Water and Wastewater Systems in Imperial Rome. *Journal of the American Water Resources Association*, **19**(2), 263–269, www.waterhistory.org

Hassan F. A. (2003). Water management and early civilizations: From cooperation to conflict. In: History and Future of Shared Water Resources, F. A. Hassan, M. Reuss, J. Trottier, C. Bernhardt, A. T. Wolf, J. Mohamed-Katerere and P. van derZaag (eds.), (PCCP Series No. 6). UNESCO, Paris, France.

Hatzidakis J. (1934). Les Villas Minoennesde Tylissos. Version Franqaise de Chapouthier el R. Joly, Etudes Cretoises III, Paris, France.

Hodge A. T. (2002) Roman Aqueducts and Water Supply. Gerald Duckworth & Co. Ltd., London, UK.

Jansen M. (1989). Water Supply and Sewage Disposal at Mohenjo-Daro. *World Archaeology, The Archaeology of Public Health*, **21**(21), 177–192.

Koutsoyiannis D., Zarkadoulas N., Angelakis A. N. and Tchobanoglous G. (2008). Urban Water management in Ancient Greece: Legacies and Lessons. *ASCE, Journal of Water Resources Planning & Manag.*, **134**(1), 45–54.

Lambrou Y. L. (1999). Fresh Water: A Gift of Gods to Athens. Desmos, Kato Petralona, Athens, Greece (in Greek).

Lyrintzis A. and Angelakis A. N. (2006). Is the Labyrinth a Water Catchment Technique? People and water management in Minoan Crete. In: Proc. of 1st IWA International Symposium on Water and Wastewater Technologies in Ancient Civilizations, 28–30 October 2006, Iraklion, Greece, pp. 163–174.

Mays L. W. (2002). Urban water infrastructure: A historical perspective. In: Urban Water Supply Handbook, L. W. Mays (ed.). McGraw–Hill, New York, USA, pp. 1.3–1.66.

Mays L. W., Koutsoyiannis D. and Angelakis A. N. (2007) A brief history of urban water supply in antiquity. *Water Science & Technology, Water Supply*, **7**(1), 1–12

Mays L. W. (2008). A very brief history of hydraulic technology during antiquity. *Environmental Fluid Mechanics*, **8**(5), 471–484.

Mays L. W. (ed.) (2010). Ancient Water Technologies. Springer, Dordrecht Heidelberg, London, UK and New York, USA.

Morgan M. H. (transl.) (1914). Vitruvius: The Ten Books on Architecture. Cambridge, Mass. USA.

Ozis U. (1987) Ancient Water Works in Anatolia. Water Resources Development, **3**(1), Butterworth & Co. Ltd, Belfast, Ireland.

Ozis U. (1996). Historical Water Schemes in Turkey. Water Resources Development, **12**(3), Butterworth & Co. Ltd, Belfast, Ireland.

Parker J. E. (1967). Housing and population in imperial Ostia and Rome. *Journal of Roman Studies*, **57**, 82–91.

Petropoulos M. (2006). The significance of water in ancient Patra. In: The Union of Local Authorities of the Prefecture of Achaia "Waterways in Achaia Prefecture". Patra, Greece, (pp. 22–39).

Phillips H. (1972). History of the Water Supply of the Borough of Beaver. Published jointly by the Beaver Area Heritage Foundation and the Beaver County Tourist Promotion Agency. Pennsylvania, USA.

Shuvial H. (2009). The ancient water supply of Jerusalem in the biblical times – some controversial issues. In: Proc. of the 2nd IWA International Symposium on Water and Wastewater Technologies in Ancient Civilizations, May 28–30, 2009, Bari, Italy.

Sklivaniotis M. and Angelakis A. N. (2006). Historical development of the treatment of potable water. In. Proc. of 1st IWA International Symposium on Water and Wastewater Technologies in Ancient Civilizations, 28–30 October 2006, Iraklion, Greece, pp. 659–666.

Soulis A. (1884). Hydraulics. Andrea Koromila press, Athens, Greece (in Greek).

Spanakis S. G. (1981). Water Supply of the City of Iraklion 828-1939. Technical Chamber of Greece, Dept. of Eastern Crete, Iraklion, Greece (in Greek).

Tassios T. P. (2002) From "Peisistratean" to Evinos. Kathimerini news paper, Athens, Greece (in Greek).

Viollet P. L. (2010). Water management in the early Bronze age civilizations. In: Water Engineering and Management through Time: Learning from History, F. Arregui (ed.), Taylor and Francis Group, London, UK, pp. 29–54.

Chapter 3

History of water and health

Joanna M. Pope, Mark H. Weir and Joan B. Rose

3.1 INTRODUCTION

Our relationship with water and its influence on human health continues to evolve as new illnesses emerge and new technological advancements are introduced to counter agents of disease. Water-related issues adversely affect human health, with contributing factors such as poverty, civil unrest, natural disasters and failures in water treatment processes exacerbating the effects of poor quality source water. These problems often occur in regions of the world where vulnerable populations cannot avoid drinking tainted water, although this is not exclusively the case. This Chapter discusses waterborne disease through history, introducing the origins of disease terminology and outlining the influence of civilization on water and sanitation. It continues with accounts of our susceptibility to waterborne illness with the analysis of mummified Egyptian tissue samples, historical records from the plague of Athens, and medical accounts from the US Civil War, which is a reminder of the extent to which water-related disease can influence the outcome of conflict. The ongoing effects of using groundwater as a drinking water source in the Indian sub-continent is examined, where, in an attempt to improve the water quality from a microbial perspective, millions of people have been exposed to elevated levels of arsenic. These examples are discussed, accommodating evidence of outbreaks and epidemics, the influence of population density, sanitation and water treatment, and medical and legal advancements. Bringing us into contemporary times, events such as the 1993 cryptosporidiosis outbreak in Milwaukee, the 2005 hurricane in New Orleans and the 2011 Japanese tsunami explore how natural disasters, engineering failures and other incidents influence water quality. Finally we offer some thoughts on how the risks from water-associated health effects can be quantified and overcome to protect health, and how our understanding of historical events can improve our responses to future water quality issues.

3.2 THOUGHTS ON WATERBORNE DISEASE THROUGHOUT HISTORY

The history of the world is full of descriptions of people and major events (particularly wars) but there has always been a fascination around diseases, epidemics and outbreaks. Perhaps in the modern world with the emergence of new diseases and their rapid spread such as SARS and H1N1 influenza, this seems as if this is something unusual, that modern medicine, our public health structure and vaccines should be enough to address these illnesses. (Note: The World Health Organization reported that during the SARS epidemic about 29 countries reported about 6903 probable SARS cases, including 495 deaths. According to flucount.org by 7 December, 2009, the H1N1 had racked up 1,438,880 cases and 14,337 deaths in 195 countries.) Yet it must be recognised that transmission is an important component of disease and the pathway for pathogen movement from host to host is influenced by people's behaviours, our societal structures and physical infrastructures. We are facing unprecedented global population, displacement of people into temporary shelters due to conflict, wars and natural disasters including droughts (influencing food security), earthquakes, hurricanes and flooding. These global events are now highlighted quickly through our communications system, yet it seems we are powerless to address the on-going risk of disease and have failed to learn from our history.

Clean and safe water is the most important consideration of a healthy population, community, and economy, and is important for food and overall security. Without access to water life ceases to exist but without access to safe water the death toll can be unimaginable. The current numbers of waterborne pathogens which still plague the globe, including *Vibrio cholera* and poliovirus, and water-based diseases such as malaria should be of great concern and deserve much more attention. While only these few are reported at the global level, there are hundreds of other waterborne and water-based pathogens, such as *Salmonella typhi*, *E. coli* 0157:H7, *Campylobacter*, *Leptospira*, *Helicobacter*, *Arcobacter*, *Cryptosporidium*, *Giardia*, and hundreds of viruses, including coxsackievirus, foot and mouth enterovirus, hepatitis and norovirus (Table 3.1).

In 2010 cholera cases were estimated at 317,534 with 7543 deaths in 49 countries (WHO, 2011a) and currently in Haiti the estimates as of 31 July, 2011 were 419,511 cases, of those 222,359 hospitalisations, and 5968 deaths due to cholera (http://www.who.int/hac/crises/hti/en/). Polio was reported at 966 cases from 21 countries in the world statistics and Malaria dwarfed all of these with an estimated 81,735,305 cases with about 100 countries reporting (WHO, 2011b).

Table 3.1 Select waterborne pathogens and their symptoms/diseases.

Pathogen	Disease/symptoms
Viruses	
Adenovirus	Respiratory infections; gastroenteritis; febrile disease; conjunctivitis
Astrovirus	Diarrhoea and vomiting; skin rashes
Coxsackievirus	Myocarditis; pericarditis; aseptic meningitis
Echovirus	Myocarditis, aseptic meningitis
Enterovirus	Gastroenteritis; meningitis; respiratory disease; encephalitis
Hepatitis virus	Gastroenteritis; hepatitis
Norovirus/Calicivirus	Diarrhoea with vomiting
Rotavirus	Diarrhoea
Bacteria	
Arcobacter	Diarrhoea
Aeromonas hydrophila	Gastroenteritis
Campylobacter	Acute gastroenteritis; reactive arthritis
Helicobacter pylori	Gastroduodenal disease; ulcers; gastric cancer
Legionella spp.	Legionnaires disease; Pontiac fever

(*Continued*)

Table 3.1 Select waterborne pathogens and their symptoms/diseases (*Continued*).

Pathogen	Disease/symptoms
Mycobacterium avium	Pulmonary disease
Escherichia coli	Haemorrhagic colitis; haemolytic uremic syndrome
Pseudomonas aeruginosa	Nosocomial infections in immunocompromised patients; urinary tract and wound infections
Yersinia enterocolitica	Gastrointestinal infections
Parasites	
Cryptosporidium parvum	Self-limiting diarrhoea; severe diarrhoea in infants and immunocompromised persons
Cyclospora cayetanensis	Diarrhoea; abdominal cramps; low grade fever
Giardia lamblia	Diarrhoea; can be chronic, severe diarrhoea in infants and immunocompromised persons
Microsporidia	Diarrhoea; cholangiopathy in immunocompromised people
Toxoplasma gondii	Flu-like illness; swollen glands in neck, armpits or groins
Other	
Cyanobacteria toxins	Poisoning; gastrointestinal disease

Source: NRC, 2004

3.2.1 The faecal-oral cycle

Much of the global burden of disease is due to the faecal pollution of water and the faecal-oral cycle that influences not only water but the food supply (Figure 3.1). Any pathogen shed in faeces should be considered a waterborne pathogen. Thus water safety and security, sanitation, food safety and disease risks cannot be considered in isolation of one another. Water quantity and human control of water resources are well studied, known and documented through history, but water quality is not well understood even today and little attention has been paid to the waste side of the equation.

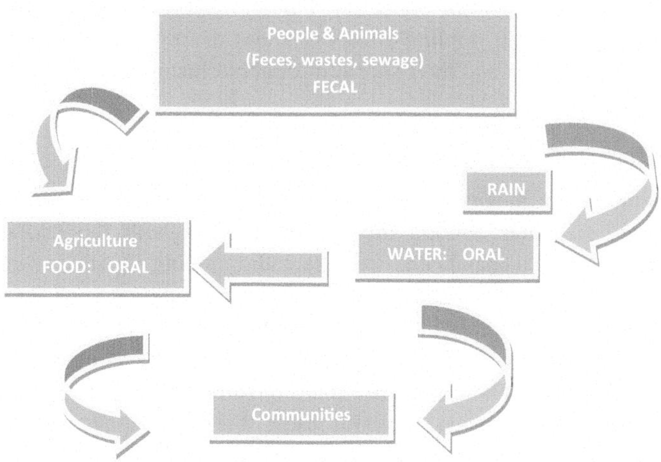

Figure 3.1 Waterborne disease cycle

Figure 3.2 Water system at Machu Picchu (Photograph by Tom Rose)

All life on earth needs water, early humans found their way to freshwater systems that supported their drinking water, food (fish as well as game that also needed water) and eventually transportation. We suggest that as humans moved from hunters and gathers to farmers and developed communities and cities, the pollution of water with both human waste and animal waste led to the widespread introduction of disease into a community which could then easily be transmitted via contaminated hands, fomites, and food.

As disease began to reverberate throughout communities and as humans began to write down their experiences and history, the language and story of infectious diseases also emerged. We now have a language of disease that has evolved but the picture of water pollution and waterborne disease remains murky. Some of the most important records are probably found during wars as the events and activities were documented. In addition, early in the history of the human race, symptomatology (description of the condition of the patient) was used to describe the diseases. While some may have been distinctive, we now know that more than one pathogen can often cause the same symptoms, and infected individuals who become better, or in some cases have no symptoms at all are "contagious" and spread the disease to the environment and to others. Scientific tools and inquiry have now begun to elucidate waterborne disease pathways and the subsequent rippling effects. The use of molecular tools in particular has begun to shed some light on the pathogens and their spread throughout history.

3.2.2 A brief history of disease and water

Table 3.2 describes some of the history of the world's various civilizations and challenges us to think about our past in regard to water. During the Palaeolithic era with the evolution of a variety of human species, groups ran in small bands and were nomadic moving to where there was access to both water and food. Faecal excretion and urination were likely done in the open on the ground or perhaps in rivers. Thus disease spread was probably via contaminated hands. These groups were often isolated and new members entering from elsewhere bringing in disease may have been rare, however zoonotic pathogens (those spread from animals to humans) could have entered via the consumption of raw meat. As *Homo sapiens* emerged and controlled fire was used, this would have decreased the potential for foodborne and zoonotic diseases but there is no evidence of boiling water for purification until much later in history.

Table 3.2 Thoughts on the condition of civilization, water and sanitation.

Time Frame	Condition	Transmission considerations	Water	Sanitation
Palaeolithic era 500,000 years ago	Hunters and gatherers, small bands may have had base camps.	Contaminated hands, and wild meat potential for zoonotic disease.	Nomadic moving to where there was water and food.	Most likely used open defecation and rivers.
400,000 to 200,000 years ago 130,000 years ago	Use of controlled fire. Modern *Homo sapiens*	Potentially decreased foodborne disease.	Water was probably not boiled as vessels would not have been available.	No evidence of impact, however with warmth provided by fire it is possible that pits may have been used near settlements.
Neolithic Age, New Stone Age, 10,700 to 9500 BC	Use of wild and domestic crops and use of domesticated animals.	Communities forming, zoonotic diseases.	Possible that shallow dug wells were used.	May have begun to use animal waste and human waste as fertiliser.
Egypt 6000 BC to 1500 BC Persians Ancient civilizations spanned over 2800 years.	Began to develop waterworks Irrigation/canals.	Contaminated water and food linked. Early diseases were alluded to as "epidemic fevers" which was written in a papyrus *ca.* 1500 BC, discovered in a tomb in Thebes, Egypt.	Some evidence of heating water, use of copper vessels and charcoal filtration.	Use of vessels, dug pits, open defecation and use of rivers. 2500 BC: In Mohenjo-daro, there existed a highly developed drainage system where waste water from each house flowed into the main drain.
Greek 2000 to 300 BC (end of Persian empire) Hellenistic kingdoms 250 BC on.	Age of Socrates (469–399 BC) Plato (427–327 BC) Aristotle (384–322 BC)	Plagues documented particularly as part of war campaigns. Alexander dies possibly	Pressurised water systems appeared in the Old World at least as long ago as 1400 BC:	1000 BC: In the Bahrain Island in the Persian Gulf, flush type toilet was discovered.

(*Continued*)

Table 3.2 Thoughts on the condition of civilization, water and sanitation (*Continued*).

Time Frame	Condition	Transmission considerations	Water	Sanitation
	Hippocrates, father of Medicine (460–354 BC)	of typhoid (323 BC) Bad air, water associated with ill health.	The remains of such a system found in a Minoan palace in Crete. Siphons used; boiled water; filtration	Drainage systems yet use outdoor toilets.
Roman Empire 44 BC to 1453 AD	95% of the population in the lower strata lived in high rises. An apartment building may have had a latrine and fountain on the ground floor for its many residents.	Dysentery epidemics (enteric fevers) were often described. Contagion well established. Malaria (*Plasmodium falciparium*, the most dangerous form) tuberculosis, typhoid fever.	Large waterworks, pipes, cisterns, fountains. Individual homes could have sinks and discharges to underground drainage.	Drainage systems developed, discharges directly through these systems to environment. Public sewers. Most were still cess pits. 69 AD: Vespasianus (Otto Empire) for the first time levied Tax on toilets. Communal baths used for health.
Mayan: 250–900 AD Aztecs: 14th, 15th and 16th centuries Incas: arose from the highlands of Peru sometime in the early 13th century. Major activity 1438 to 1533.	Civilizations in the Americas were estimated to have population densities that may have ranged from 400 to 600 people per square mile in the rural areas, and 1800 to 2600 in cities.	Well known for Spaniards bringing in new diseases which then devastated immunologically naive people.	Waterworks are well studied. Had reservoirs, piping, used irrigation systems for agriculture.	Use of pots, open defecation, perhaps faecal waste used as a fertiliser. Classic Maya (250–600 AD) constructed pressurised water system for supplying water to urban areas.
Machu Picchu (1438–1472)	140 stone buildings and some 200 skeletons discovered	No specific findings, area supported malaria.	Original water supply from the mountain still in use today, brought in by series of pipes and fountains (Figure 3.2)	Not really known, daily trips to the river were common. May have used faecal material as fertiliser.

Source: Baker (1930); Beck (2000); Cunha (2004); Crouch (1993); Clutton-Brock (1981); French & Duffy (2010); Gigante (undated); Gensini *et al.* (2004); Marlowe (2005); Price (undated); Zohary & Hopf (2000); Wrangham & Conklin-Brittain (2003).

The use of shallow hand dug wells for water supply appears to date back to *ca.* 8100–7500 BC (before the common era, Pre-Pottery Neolithic B settlement of Atlit Yam in Israel). Water quality for drinking purposes at that time could only be judged by clarity, smell and taste. Since water treatment was not practised, groundwater then, like today was probably deemed cleaner, better tasting and safer. It could be imagined that springs were great finds and often could have been the core of the structure and security of the community.

As societies grew in their sophistication and technology, both agriculture and the domestication of animals were possible due to the development of some type of waterworks. This often was simply a series of canals, but reservoirs, cisterns for storage and finally various types of water treatment for drinking water were developed and discovered dating from about 2000 BC onward including sunlight, boiling, use of copper pots, siphoning (after sedimentation), and filtration (Baker, 1930). This ability to have a secure supply and running water nearby seems that it would lend itself to the use of any drainage system for urination and defecation. Animals were also excreting faeces, thus both the contamination of the water and a closer link between water and foodborne diseases emerged. It is no accident that most of the foodborne diseases can also be waterborne and that zoonotic pathogens once again may have found a pathway to move readily from animal hosts to human hosts.

One can speculate about whether Alexander the Great died of a waterborne disease in 323 BC in Babylon. Cunha (2004) and Oldach *et al.* (1998) suggested that the symptoms described during his last days were congruent with typhoid. While West Nile Virus had been previously suspected, phylogenetic studies of the flaviviruses have indicated that the virus may not have existed at that time. In the description of Alexander's last days, it is the fever which has been used as the definitive symptom to assign the etiological agent. From investigation of the timeline it appears that the incubation time was about nine days from the onset of the fever to death (Oldach *et al.* 1998). This is very similar to the incubation described by Thucydides (see section below) who found that with the peak of the fever many people died but if they survived went on to exhibit severe diarrhoea. Alexander was in Babylon (Ancient Iraq, south of Baghdad) at the time, where the Euphrates River supplied water directly to the city. Elaborate waterworks and irrigation systems were scattered along the river and brought water into an otherwise very dry region, very rough estimates suggest about 60,000 people lived in the area. The month was June and the river may have been slowing in its flow as the rain fed (and snow melt) system is generally at peak flows in April and May. The water was probably polluted but it is really not known how much. It was said that Alexander bathed every day thus exposure to contaminated water could have occurred through drinking as well as bathing. Typhoid is a well-known waterborne disease and in countries with sewage and drinking water treatment the disease has been eradicated (with only imported cases reported). Risk estimates show how likely it is to see very low levels of infections in a community with lower doses of exposure and with the very poor documentation of disease, "outbreak" conditions would not be noted (less than 10% infection).

There is now strong evidence that urbanisation (probably due to a number of factors, including population density, new infections from new people entering the community and subsequent spread via hands, fomites, water and food) has led to an increase in infectious disease (Barnes *et al.* 2011). The study of populations ranging from 6000 BC to 1919 AD in Africa, Asia (Japan and India) and the Middle East contrasted the proxy urbanisation dates and the frequency of a genetic marker associated with exposure to infectious disease (SLC11A1 allele frequencies). The use of molecular studies on ancient remains has provided information on population exposures to pathogens and the subsequent impacts based on genes which then confer resistance. The authors surmise that one of the key pathogens was tuberculosis and do not reject the ideas of zoonotic disease transmission.

3.2.3 The language of infectious disease

Ancient medicine addressed diagnosis of illness via the description of symptoms. Early in the history of medicine it was proposed that bad air, putrid waters, and crowding were all associated with disease and it was recognised that these maladies were contagious. One of the quotes from Hippocrates, the father of medicine, demonstrates that water was considered an important issue for the populace's health. In *The Quest for Pure Water*, M. N. Baker reported in 1948 as follows: 'whoever wishes to investigate medicine properly should consider the seasons of the year, the winds and the waters in relation to health and disease.' As the 'qualities of the waters differ from one another in taste and weight as they differ much in their [other] qualities, one should consider the waters which the inhabitants use whether they be marsh and soft, to hard and running from the elevated and rocky situations and then if saltish and unfit for cooking … for water contributes much to health.' But the concept of pathogens and disease was not understood at this time, nor was the idea of "waterborne" disease affecting communities broadly.

"Plagues" were described and in particular associated with the decimation of the Greek Army near the end of the Trojan War (*ca.* 1190 BC) and with massive epidemics in Roman history in 790, 710 and 640 BC (Sherman, 2006). The understanding of the ability of pathogens to be transmitted from person to person "contagion" was said to arise during the Plague of Athens (See below).

Douglas Harper has put together a wonderful collection of the history of words and a description of what words meant hundreds and thousands of years ago (http://www.etymonline.com/abbr.php). Table 3.3 shows how our language has evolved from some of the early ideas on disease. Today this language has led us toward description of specific pathogens and exposure pathways and finally the use of quantitative microbial risk assessment to study, understand and ultimately prevent waterborne disease.

Table 3.3 The language of infectious disease. (From the online etymology Douglas Harper, http://www.etymonline.com/abbr.php).

Word	Date	Etymology
Contagion	c. 1400, old French	Contagion, from Latin contagionem (nom. contagio) a "touching, contact, contagion", related to contingere "touch closely"
Disease	Early 14c	"discomfort, inconvenience," from Old French desaise "lack, want; discomfort, distress; trouble, misfortune; disease, sickness," from des- "without, away" (see dis-) + aise "ease" (see ease). Sense of "sickness, illness" in English first recorded late 14c.; the word still sometimes was used in its literal sense early 17c.
Infection		Late 14c., "infectious disease; contaminated condition;" from Old French: infeccion "contamination, poisoning" (13c.) and dir. from Late Latin:(spoken and written c. 300–c. 700) infectionem (nom. infectio), noun of action from pp. stem of Latin: inficere (see infect). Meaning "communication of disease by agency of air or water" (distinguished from contagion, which is body-to-body communication), is from 1540s.
Epidemic	c.1600	From French: épidémique, from épidemie "an epidemic disease," from Medieval Latin (spoken c. 700–c. 1500): epidemia, from Greek: epidemia "prevalence of an epidemic disease" (especially the plague), from epi "among, upon" (see epi-) + demos "people, district" (see demotic). As a noun, from 1757; earlier epideme (see epidemy).

(Continued)

Table 3.3 The language of infectious disease. (From the online etymology Douglas Harper, http://www.etymonline.com/abbr.php) (*Continued*).

Word	Date	Etymology
Outbreak	c.1300	First used as a verb in Middle English to erupt (see below)
	c.1600	Used as a noun "eruption" (of disease or hostilities)
	c. 21st	Modern use is mostly associated with water or food: More than one person ill from a common exposure; Also used when associated with a venue Day care, nursing home, cruise ships US Centers for Disease Control suggest that: a sudden increase in disease over a short amount of time usually due to a common exposure. Gastroenteritis Outbreaks in Long Term Care Facilities B. An outbreak of gastroenteritis is defined as three or more residents from a single ward or unit, or 3% or more of the entire facility. Cruise ships participating in the Vessel Sanitation Program are required to report the total number of gastrointestinal (GI) illness cases (including zero cases) evaluated by the medical staff before the ship arrives at a US port, when sailing from a foreign port. A separate notification is required when the GI illness count exceeds 2%. For day cares: Three or more cases of GI (in a child or staff) within a program group in a 3 day period. OR One or more case(s) of a reportable enteric disease. OR Greater than 10% absenteeism above baseline due to GI illness within the program on one day.
Plague	Late 14c	"Affliction, calamity, evil, scourge," also "malignant disease," from Middle French: plague, from Late Latin (spoken and written c. 300–c. 700) plaga, used in Vulgate for "pestilence," from Latin: plaga "stroke, wound," probably from root of plangere "to strike, lament (by beating the breast)," from or cognate with Greek: (in the classical period, *ca.* 8c. BC–4c. CE) (Doric) plaga "blow," from *Proto-Indo-European** (5500 years ago) plak- "hit" (cf. O.E. flocan "to strike, beat," Goth. flokan "to bewail," Ger. fluchen, O.Fris. floka "to curse").
Pestilence	c.1300	From Old French pestilence, from Latin: pestilentia "plague," noun of action from pestilentem (nom. pestilens) "infected, unwholesome, noxious," from pestis "deadly disease, plague."
Quarantine	14c	Quarantina giorni, lit. "space of forty days," from quaranta "forty," from L. quadraginta. So called from the Venetian custom of keeping ships from plague-stricken countries waiting off its port for 40 days (first enforced at Ragusa late 14c).
Pathogen	1852	"producing disease," from Fr. pathogénique, from Greek pathos "disease" (see pathos) + Fr. génique "producing" (see-gen). Earlier pathogenetic, Related: Pathogenicity
Sicken	c.1200	"to become sick," originally the verb was simply sick (mid-12c.), from sick (adj.) + -en (1). Transf. sense of "to make sick" is recorded from 1690s. Related: Sickened. Sickening "causing revulsion" is first recorded 1789.
Transmission	1610	"conveyance from one place to another," from Latin: (Classical Latin, language of ancient Rome until about 4c.) transmissionem (nom. transmissio) "a sending over or across, passage, pass on," from transmissus, pp. of transmittere"

(*Continued*)

Table 3.3 The language of infectious disease. (From the online etymology Douglas Harper, http://www.etymonline.com/abbr.php) (*Continued*).

Word	Date	Etymology
Risk	1660s	Risque, from French: risque, from It. risco, riscio (modern rischio), from riscare "run into danger," of uncertain origin. The Anglicised spelling first recorded 1728. Spanish: riesgo and German: Risiko are Italian loan-words. The verb is from 1680s. Related: Risked; risking. Risk aversion is recorded from 1964; risk factor from 1971; risk management from 1963; risk taker from 1944.

3.3 EVIDENCE OF WATER-RELATED DISEASE IN EGYPTIAN MUMMIES

3.3.1 The influence of The Nile

Life in Egypt historically and currently centres on the Nile, the second longest river in the world. It is composed of two major tributaries, the White Nile and the Blue Nile, and its watershed covers approximately 3.25 million square kilometres, which is about 10% of the area of Africa. The Greek historian Herodotus wrote that "Egypt was the gift of the Nile" (Marozzi, 2008). As a desert nation, water is understandably a valuable commodity, and although our ideas of modern water treatment systems were not conceivable in Ancient Egypt, it would not be unreasonable to assume that the instinct for consuming clean, tasteless and odourless water would supersede stagnant, turbid drinking water. Urbanisation compromises the integrity of natural water sources, and Egypt was one of the first urbanised areas in Europe and Africa (Vuorinen *et al.* 2007). Sanitation in ancient Egypt probably did not keep human waste out of drinking or irrigation water, the spread of water-related disease may have become a significant threat to urban populations. Construction and irrigation would have facilitated the spread of waterborne and water-related vector diseases, with practices such as flooding of fields a common means of irrigation (Abdel-Wahab, 1982). Water would have been channelled onto the fields and retained for a number of weeks to allow the silt to settle and the water to infiltrate into the soil. Partially-drained fields would be replanted, and the pools of slow-moving water in and around the fields would provide enticing breeding areas for insects and molluscs. Given limited pest control measures in Ancient Egypt, these creatures would probably affect human health, and evidence discovered in mummified remains has shown that one specific water-based vector-borne disease has a long history of impacting the health of the residents of the Nile.

One such parasite associated with disease in ancient Egypt is *Schistosoma* which causes intestinal, liver and spleen infections and is associated with abdominal pain, bloody diarrhoea and fatigue. A faecal-water cycle is part of the life cycle of the parasite in which once the ova are excreted by humans and enter a snail, they can enter the next host via penetration of the skin. This can occur while working in an agricultural setting, bathing or washing.

3.3.2 Life of the *Schistosoma*

Today, Schistosomiasis affects at least 200 million people worldwide, and more than 700 million people live in endemic areas (WHO, 2010a). The majority (55%) of cases are found in Africa (WHO, 2010b), and it is currently considered a Neglected Tropical Disease by the World Health Organization as it affects the world's poorest populations in low-profile communities and areas. The manifestation of the disease is dependent on the species of trematode (flatworm) parasite causing the infection. Five predominant species of *Schistosoma* have been identified as causing disease in humans, and the most common species currently prevalent in Egypt are *S. mansoni*, *S. japonicum* and *S. haematobium* (Lambert-Zazulak *et al.* 2003). *S. mansoni* and *S. japonicum* cause intestinal and hepatosplenic schistosomiasis, manifested as abdominal pain, bloody diarrhoea and fatigue. *S. haematobium* infects the veins draining the bladder, and can cause urinary tract disease, lesions and haematuria.

The life cycle of *Schistosoma* is defined by three factors, namely a vertebrate host, a molluscan host and water. The sexual or adult schistosome resides in the vertebrate host, the asexual stage occurs in the molluscan host, and the pathway between the two stages is water. *S. mansoni* and *S. japonicum* ova are released from the vertebrate host in faeces, whereas ova from *S. haematobium* are released in urine. Upon entering environmental waters, the ova hatch releasing miracidia, or free-living larvae. These larvae penetrate the molluscan host, and in the Nile valley these freshwater snails have been identified as *Bulinus* spp., *Biomphalaria* spp., and *Oncomelania* spp. The larvae infect the snails' hepatopancreas, developing into the next larval stage as cercariae. Huge numbers of the cercariae are released from the snails back into the water ready to penetrate the skin of a vertebrate host. Exposure can occur through prolonged exposure to infected water, such as from working barefoot in contaminated irrigation water. As this occurs, the cercariae lose their tail to become schistosomules and enter the lymphatic system and bloodstream towards the liver. It is here that the male and female trematodes mature, and travel to the bowel or bladder to reproduce, thus completing the cycle.

Symptoms of schistosomiasis depend on the level and duration of exposure. Early onset symptoms include cercarial dermatitis or "swimmer's itch", a water-related condition that independently has an interesting history of affecting humans. From 1928 to 1955, Dr. William Walter Cort of the Department of Helminthology at the Johns Hopkins School of Hygiene and Public Health published a series of articles describing, analysing and summarising the distribution and manifestation of swimmer's itch throughout the world (Cort, 1950). Cort documented the global nature of the affliction that is reflected in the variety of names given to the lesions, such as "kabure" and "koganbyo" (Japan), "sawah" (Malaysia), "El Caribe" (Cuba) and "bathers' rash" (UK). In the Great Lakes Basin in the USA, schistosome dermatitis was at one time such a public health concern, that in 1939 the State of Michigan established the Division of Water Itch Control to quantify the extent of the problem and develop experimental chemical treatments for the eradication of schistosome-infected snails at recreational beaches.

Symptoms of chronic *Schistosoma* exposure can include intestinal lesions, blood in the urine or stools, anaemia, calcification of the bladder, pulmonary vascular lesions, and hepatosplenomegaly (simultaneous enlargement of both the liver and the spleen). Although schistosomiasis is believed to currently have a low mortality rate, overwhelming infection with schistosomes can prove fatal. Primary contributing factors in schistosomiasis-related death include dysentery, anaemia, circulatory failure and pellagra (Abdel-Wahab, 1982).

3.3.3 A history of mummification

Mummification in Ancient Egypt was developed in response to a gradual change in the burial preferences of its deceased. The artificial preservation of bodies, both human and animal was practised in Egypt from circa 2686 BC until the beginning of the Christian era (David & Tapp, 1984). The earliest Egyptians were buried in the sand, typically in the foetal position to reduce the size of the hole that would need to be excavated, and the hot, dry climate would desiccate the body (Fleming *et al.* 1980). The physical features of the body would be retained, and this lifelike appearance of the corpse may have supported the belief of an afterlife. As burial practices became more sophisticated, with the construction of elaborate tombs and monuments to the dead, the bodies of the deceased were no longer buried in the desert sand. However, as belief in the afterlife and rebirth was fundamental to Egyptian burial practices, mummification was developed to artificially preserve the body in readiness for the journey to the underworld and to be judged by Osiris (David, 1978).

Mummification practices were based primarily on status, and thus closely related to the cost of the procedure. "First-class" mummification involved the removal of the brain through the nasal cavity, evisceration of all internal organs (minus the heart and the kidneys), and desiccation of the body. The body would be anointed with oils, resins or other unguents and wrapped in strips of cloth. The internal organs would be treated, wrapped and returned to the corpse. Resin, plant gum, waxes or bitumen would be included within the layers of cloth to further preserve the body and maintain the integrity of the wrappings. "Second-class" mummification involved evisceration of the internal organs by injecting oil into the body and allowing the fluidised organs to be drained away, and "third-class" mummification was to purge the internal organs, prior to desiccating the body (Lambert-Zazulak *et al.* 2003).

The examination of Egyptian mummies has helped develop the field of paleoepidemiology, because minimal changes in Egypt's population dynamics in the last 5000 years have facilitated the study of disease evolution (David, 1997). Technological advancements in medical diagnostic procedures have also shown how diseases such as schistosomiasis are manifested in mummified tissue. To prevent the destruction of fragile mummified tissue, efforts have focused on minimally invasive diagnostic tools, as opposed to the historically typical unwrapping and clinical observation of mummies. As previously mentioned, calcified ova have been observed in histological kidney specimens by rehydrating, fixing, and selective staining of mummified tissue. Schistosome eggs can also be identified using UV light inspection and fluorescent microscopy to take advantage of the auto-fluorescent eggs, X-ray examination of potential calcification in the liver, kidney and bladder, enzyme-linked immunosorbent assay (ELISA) targeting schistosome antigens, and polymerase chain reaction-based methods targeting schistosome DNA (Contis & David, 1996). DNA and antibody analyses lend themselves particularly well to the diagnosis of infectious disease, but they can be controversial as they require the removal of small amounts of tissue, in the order of milligrams. Contamination with sand, mummification products and other foreign material can also compromise the analyses (Dunand & Lichtenberg, 2006).

3.3.4 Examining the mummies

The first microscopic examination of mummified tissue was in 1911 by Sir Marc Armand Ruffer, President of the Sanitary, Maritime and Quarantine Council of Egypt, and Professor of Pathology at Cairo University. Ruffer identified the calcified ova of the parasitic flatworm *Schistosoma*, also known as *Bilharzia*, in the kidneys of two twentieth dynasty mummies, circa 1187 to 1064 BC. (David & Tapp, 1984). Unfortunately the social class or occupations of these mummies are not described in Ruffer's account, but his description of the discovery is worth quoting:

> 'At the present time there is perhaps no disease more important to Egypt than that caused by the *Bilharzia haematobia*. So far no evidence has been produced to show how long it has existed in this country, although medical papyri contain prescriptions against one of its most prominent symptoms-namely, haematuria. The lesions of this disease are best seen in the bladder and rectum, but unfortunately these are just the two mummified organs which I have not been able to obtain so far. Nevertheless, in the kidneys of two mummies of the twentieth dynasty I have demonstrated in microscopic sections a large number of calcified eggs of *Bilharzia haematobia*, situated, for the most part, among the straight tubules. Although calcified, these eggs are easily recognisable and cannot be mistaken for anything else' (Ruffer, 1910).

It is the longevity of the ova and calcified lesions in the internal organs of mummies that has made schistosomiasis a particularly worthy target for infectious disease historians. The study of one particular mummy from 1200 BC has provided evidence of schistosomiasis in ancient Egyptians. Nakht (ROM 1) is known as such because hieroglyphics on his coffin signify his Egyptian name as Nakht, whilst ROM is the abbreviation for the Royal Ontario Museum, where the medical examination took place. In addition to giving his name, the hieroglyphics also described Nakht as a weaver. At the time of death, Nakht was a teenager, and the mummification method and wrappings suggest that he was not from a wealthy family. It was possible to examine the internal organs of Nakht (ROM 1) *in situ* as his body had not undergone evisceration. A multi-disciplinary team from Toronto, Detroit, Philadelphia and Cardiff identified calcified ova of *Schistosoma* sp. in the liver and kidneys; and calcified ova of both *Schistosoma* and *Taenia* spp. (tapeworm) in the large and small intestines (Hart *et al.* 1977; Ziskind, 2009).

The examination methods used on Nakht (ROM 1) included a general study of his anatomy, radiology, paleohistology, blood group testing, and chemical and microbiological analyses. The radiological examination suggested that Nakht was between 14 and 18 years and his teeth were in good condition. Anatomical observation showed hair, eyelashes and internal organs (heart, lungs, liver, gall bladder, spleen, intestines, urinary bladder, and prostate) still present, and no evidence of artificial mummification agents (e.g. natron). The histological investigations provided evidence of a number of parasites in the internal organs. Tissue samples were removed from the mummy, rehydrated and stained for microscopic examination. Within the intestinal lumens, liver and kidneys, calcified parasitic ova were numerous. No *Schistosoma* were visible in the urinary bladder, but red blood cells on the mucosal surface suggested that Nakht suffered from blood in the urine. Given the numerous locations of *Schistosoma* in the body, coupled with a lack of clear evidence suggesting the cause of death, it has been suggested that this otherwise healthy Egyptian teenager died from multiple parasitism (Cockburn *et al.* 1998).

As the examination of Egyptian mummies continues, it is fair to wonder what is the value of paleohistology and paleoepidemiology in the pursuit of knowledge. The International Ancient Egyptian Mummy Tissue Bank at Manchester Museum in England is a key example of how the development of techniques to analyse ancient human remains can influence modern science. This repository was established to improve understanding of schistosomiasis in ancient Egypt and compare the data with epidemiological research in modern communities, such as the distribution of disease by age, sex and so on (e.g. Miller *et al.* 1992). Although the disease prevalence in modern Egypt has decreased, thanks to improved sanitation, snail control, diagnostic techniques and chemotherapy, both *S. haematobium* and *S. mansoni* are endemic in the population, affecting up to six million people (WHO, 1993). These data have not changed significantly since the early 1990s because although control programs are implemented, increasing population numbers in endemic areas help to sustain the disease. The identification and genetic analysis of ancient schistosome eggs can help scientists understand how schistosomiasis manifested itself in people without access to modern medicine, how historical schistosome species differed genetically from modern species, and perhaps most importantly the distribution of modern drug-resistant species compared to their drug-susceptible counterparts. Research on waterborne and water related disease in history allows data from modern public health issues to be compared with palaeopathological evidence. In turn, this enables the history of disease in ancient civilizations to be viewed within a wider context, providing a unique historical perspective to current water quality and health considerations.

3.4 THE PLAGUE OF ATHENS: A WATERBORNE DISEASE STORY

'The greatest war in the past was the Persian War: yet in this war the decision was reached quickly as a result of two naval battles and two battles on land. The Peloponnesian War, on the other hand not only lasted for a long time but throughout its course brought with it unprecedented suffering for Hellas. Never before had so many cities been captured and then devastated, whether by foreign armies or by the Hellenic powers themselves (some of these cities after capture, were resettled with new in habitants); never had there been so many exiles; never such loss of life – both in actual warfare and in internal revolutions. Old stories of past prodigies, which had not found much confirmation in recent experience, now became credible. Wide areas for instance were affected by violent earthquakes; there were more frequent eclipses of the sun than had ever been recorded before; in various parts of the country there were extensive droughts followed by famine; and there was the plague which did more harm and destroyed more life than almost any other single factor. All of these calamities fell together upon the Hellenes after the outbreak of war.'

(Translated 1954 Rex Warner; Thucydides, History of the Peloponnesian War p. 48.)

One of the best described plagues occurred in Athens in 430 BC. The epidemic which devastated Athens as reported by the Greek historian Thucydides is often referred to as a medical mystery (http://www.perseus.tufts.edu/GreekScinence/Thuc.+2.47-55.html). In 430 to 426 BC a massive epidemic was recorded, killing about 30,000 people. Illnesses were identified and described by the symptoms. The "plague of Athens" description included fever, inflammation, blisters on the skin, open sores, nervousness, and severe ulceration and watery diarrhoea followed by death. It was noted that it was contagious and that there was some immunity as those that recovered were not attacked twice. Hypotheses have been developed on the aetiology including influenza, smallpox, bubonic plague, typhus and *Staphylococcus* (Langmuir, 1985). Some have suggested that this is an ancient disease that has since died out (Langmuir, 1985). Typhus was suggested yet it was not until 19th century that typhus and typhoid were distinguishable (1856, Corfield, 1902).

Many of the symptoms do follow those now known and ascribed to the disease typhoid. Symptoms generally associated with typhoid resemble Thucydides' description, including: a high fever from 39°C to 40°C (103°F to 104°F) that rises slowly, chills, bradycardia (slow heart rate), weakness, diarrhoea, headaches, myalgia (muscle pain), lack of appetite, constipation, stomach pains, in some cases, a rash of flat, rose-coloured spots called "rose spots", and extreme symptoms such as intestinal perforation or haemorrhage, delusions and confusion are also possible.

Rose and Masago (2007) evaluated the data and suggested that multiple pathogens could have been involved with this epidemic but that typhoid was the principal disease associated with sewage contamination of the water supply. Recent molecular evidence from bodies exhumed from a mass grave (buried during the epidemic) has supported typhoid fever as a probable cause (Papagrigorakis *et al.* 2006). DNA was extracted from the teeth from an ancient Greek burial pit. The study led by Papagrigorakis of the University of Athens, found DNA sequences similar to those of the organism that causes typhoid fever. Exposure to contaminated water could have also spread viral pathogens associated with respiratory disease and *Staphylococcus*, which has been suggested as the cause of the gangrene (Langmuir, 1985).

Water in Athens was supplied by a series of public fountains, wells and cisterns (Crouch, 1993). Springs as well as engineered and piped water (rock cut tunnels and aqueducts) fed these systems which included rainwater (in the case of the Klepsydra, rainwater drained from the Acropolis). Drainage from storm water and sewage was flushed through drains in the alley or on the side streets between houses. Channels may have run outside the community to areas where the water was reused for crops, having recognised nutrient value as well (Crouch, 1993). Cesspools were also discovered and it is uncertain when these were no longer used in favour of drainage and flushing away from the home. There is evidence that water pipes were laid and set near or within the drainage channels, or channels with freshwater were adjacent to those carrying wastewater. Combined sewer overflows where both storm water and sewer were mixed and captured were common. As the population increased to about 300,000, along with previous droughts a focus on enhanced water management including approaches used for water storage, collection of rain and runoff, transport, reuse and drainage in the karst terrain and limestone created a situation in which one could envision sewage cross-connections, and transmission of waterborne disease via drinking, bathing and through the irrigated food supply.

Translation of the original text suggested that the disease was seen first in Ethiopia. 'The plague originated, so they say in Ethiopia in Upper Egypt, and spread from there into Egypt itself and Libya and much of the territory of the King of Persia. In the city of Athens it appeared suddenly, and the first cases were among the population of Piraeus, where there were no wells at that time, so that it was supposed by them that the Peloponnesians had poisoned the reservoirs.'

The description of the plague by Thucydides suggested that there were many dead bodies which were lying unburied and that because of the fever and thirst many of the sick plunged into the water-tanks. It appeared that the incubation times ranged up to eight days prior to its height and subsequently the diarrhoea appeared. A description of the attack rates was found in one of the campaigns to capture Potidaea lead by Commander Hagnon. The plague and disease were still widespread and as Hagnon set up a siege of the city disease broke out in the troops. He returned with his ships to Athens and 1050 hoplites of 4000 troops died (an attack rate of 26%) in a span of 40 days (Translated, 1954; Rex Warner; Thucydides, History of the Peloponnesian War p. 158).

3.5 WATER, DISEASE AND DEATH IN THE AMERICAN CIVIL WAR: 1861–1866

'The manpower loss in internecine warfare by micro-organisms within an army was far greater than that lost in battle between enemy armies' (Steiner, 1968). The following section is taken primarily from the fascinating book *Disease in the Civil War, Natural Biological Warfare in 1861–1865*, written by Paul E. Steiner, a professor and M.D. at the University of Pennsylvania in 1968.

By 1860, about 125 cities in the USA had piped water and underground sewers, while in the country, wells and outdoor latrines/cess pits were used but chlorination was not being practised for drinking water. While advances were being made in medicine with smallpox vaccine development and better infectious disease documentation, it had only been a few years earlier that definitive waterborne disease transmission of cholera was documented in London (John Snow) (Beck, 2000). Thus while clean water was known to be important, the role of contaminated water in disease transmission was not understood and there were few options for treating water.

Sanitation and hygiene were the order of the day due to aesthetic reasons (bad odours) which were still associated with ill health. But even then the role of contaminated hands, the cross-contamination of food or spread of infections during surgery were not clear. It was not until after 1860 that the physician and surgeon, Joseph Lister in studying Pasteur's work on micro-organisms found through experimentation that use of carbolic acid (phenol) for hands and instruments would vastly reduce infections.

The US population was 31 million by 1860 census (http://www.thelatinlibrary.com/chron/civilwar.html) and the Civil War was associated with disease estimates of 6,029,560 cases (this is after subtracting wounds, accidents and injuries (Steiner, 1968)). The troop force during the war was estimated to be about 1.5 million for the north and 1 million for the south, but according to Steiner this did not account for thousands of short term soldiers.

Steiner summarized data from Medical and Surgical History of the War of the Rebellion, Vol. I, Part I., listed by diagnosis of the time. Table 3.4 shows the list of those possibly associated with contaminated water which equaled 2,121,541 cases with an average of 4% mortality. In addition, the water based diseases, malaria, described as "Intermittent and remittent fevers" were responsible for 1,315,955 cases and 10,063 deaths and yellow fever for 1371 cases and 436 deaths.

Table 3.4 Possible waterborne infectious diseases in the military during the US Civil War.

Diagnosis	Cases	Deaths*
Diarrhoea and dysentery, acute and chronic (Bacteria such as *Shigella* and viruses and parasites (*Entamoeba, Giardia*) may have been the culprits)	1,739,135	44,558 (2.6%)
Cholera morbus	26,366	305 (1.1%)
Typhoid	148,631	34,833 (23.4%)
Jaundice, endemic, epidemic (likely caused by hepatitis A virus)	77,236	414 (0.5%)
Acute inflammation of the liver (parasite *Entamoeba histolytica* and bacteria including Bacteroides, Enterococcus, *Escherichia coli, Klebsiella, Staphylococcus and Streptococcus I)*	12,395	327 (2.6%)
Inflammation of brain, meninges and spinal cord (caused by a variety of bacteria and viruses)	3,999	2660 (66.5%)
"Other miasmatic diseases" not classified (associated with hygiene unclear what were the symptoms as Typhus associated with filth was on the list)	94,997	2363 (2.5%)
Debility in miasmatic diseases	18,782	153 (0.8%)
TOTAL	2,121,541	85,613 (4%)

*Mortality rates
Source: Steiner (1968).

Interestingly, listed among the non-infectious diseases were many chronic conditions probably associated with the previous infections experienced by the soldiers. This included acute and chronic rheumatism (286,863 cases and 710 deaths) which is a chronic condition due to many different types of bacterial infections, neuralgia (58,774 cases, 18 deaths) a possible complication of measles and other viruses and dyspepsia (37,514 cases and 31 deaths). Dyspepsia is Latin for "bad digestion" and is characterised by chronic pain in the abdomen, it has been suggested that *Helicobacter pylori* infection, which can result in ulcers could be responsible for symptoms of dyspepsia.

It is unclear what diseases and pathogens were really a part of the category of miasmatic origin. Some described these as being associated with intermittent or continued eruptive fevers that could not fit into other categories, but it may have been a catch all category referring to the unsanitary conditions and the "bad air and environment". In 1864, W. D. Husband presented his thoughts on infant mortality at the York meeting of the United Kingdom National Association for the Promotion of Social Science (Hastings, 1864). He wrote in the transactions:

'Let us then take a hasty glance at those different causes. First with regard to miasmatic and infectious diseases, we have the authority of those who have carefully attended to sanitary matters that all who die of miasmatic diseases die of causes that may be easily prevented. It may be the want of sanitary regulations which takes away health and strength which enable a person to resist infectious or contagious disease; for remember this, that there must be a condition of public health favourable to the diffusion of disease before it spreads to any great extent and it depends much on the condition of the population and sanitary regulation that prevail, whether that disease will or will not sweep away thousands from the town or locality in which it springs up.'

Key campaigns were found to be particularly devastating and the role of water was highlighted throughout via Steiner's descriptions, summarised in Table 3.5 (Steiner, 1968).

Table 3.5 The role of disease, water and sanitation in US Civil War battles. Source: Steiner, 1968.

Dates	Campaign	Location	Major Diseases	Notes
July to Sept 1861	Confederate Western Virginia	Fought along the Kanawha River 50 miles N of Charleston, Carnifex Ferry major engagement in Sept. Hilly terrain, drainage of surface runoff into water supplies (springs).	Diarrhoea, 50% attack rates. Dysentery, typhoid. Epidemic of jaundice followed 30 to 60 days after the typhoid curve.	Rainy in July Surgeon Brown described faecal contamination.
1861 to 1862	South Carolina Coast Federal Naval campaign	Union establishes bases on Sea islands and on Hilton Head	Typhoid rate was 72 cases/1000. Outbreak of jaundice summer of 1862, near Beaufort.	Those using wells had less disease than those using surface water. In Hilton Head, sinks and latrines built over tidal waters.
Dec 1861 to Mar 1862	Eastern Kentucky Federal troops to operate against the confederates Kentucky and Tennessee gain railroad.	Valley of the Big Sandy River of the Cumberland Mountains near the Virginia border. Troops to supply their needs along the Ohio river.	30% attack rates for typhoid fever.	February heavy rain falls and flooding, increased illness rates.
Mar 1862 to Aug 1862	The Union Peninsular objective to capture Richmond	Troop campsites in Meridian Hill Kalorama Heights in D.C. area. Anchored at Potomac prior to proceeding. Army over 100,000 men would be transported by sea to the Peninsula between the James and York Rivers, to the east of the Confederate capitol of Richmond.	Typhoid highest in Brooks Brigade. Attack rates of 29.75%.	Primitive latrines, drinking water from surface waters. Used the Potomac for drinking.

There are some examples here and there of the devastation that followed the problems associated with water, sanitation and hygiene. The Meridian Hill area in DC was the campsite for the Union troops with a hospital in the area that was notorious for the spread of enteric infections (Steiner, 1964). The city's water supply (1850) had spring or well water piped in along the streets of Pennsylvania Avenue. http://www.dcwater.com/wastewater_collection/history.cfm). The sewer had been put in starting in 1810. The population was increasing during the war straining the city's water supply. The Army Corps of Engineers constructed a new aqueduct that brought in 10,000 gallons of untreated fresh water to the city each day from the Potomac. During the Peninsular campaign the illness rates in April, May, June July and August were 23.4%, 26.1%, 31.4%, 40.5% and 30%, respectively contributing to 124,027 cases of disease in a troop strength of 397,917. The primary illnesses were acute, chronic dysentery and diarrhoea (Steiner, 1964). An epidemic at Harrison's Landing resulted in 2805 cases of typhoid and 1161 cases of jaundice. Dr George Miller Sternberg was a prominent physician who served as the Surgeon General of the US from 1893 to 1902. He was appointed an assistant surgeon for the United States Army on May 28, 1861, and he participated in the Peninsular campaign where he contracted typhoid fever while at Harrison's Landing (http://www.arlingtoncemetery.net/gmsternb.htm).

Good sanitary measures for these troops included some key elements such as: (a) wells dug for water, (b) soil added to the latrine trenches once per day, (c) fresh vegetables used, (d) food prepared at the higher level by company and not by squads, (e) baths taken once per week, (f) camps located away from the woods and swamps, (g) tents raised and camps moved to new ground once per week, and (h) refuse from stables and dead animal burned or deeply buried. Yet in reality most of these conditions could not be met. There seemed to be consistent problems with new troops coming in and disease spreading quickly through the camp. Airborne person-to-person pathogens such as measles came first followed by gastrointestinal diseases such as typhoid, and in some cases followed by jaundice. These were related to routes of exposures and varying incubation times.

Multiple pathogens were found and are likely to have been associated with factors including poor nutrition. One such interaction has been reported with measles and diarrheal disease, where co-morbidity was found in communities in Africa and relationships indicated similar associations between the various shared risk factors and each of the two co-occurring diseases (Fenn *et al.* 2005).

3.6 ARSENIC EXPOSURE IN BANGLADESH: "SAFE" DRINKING WATER SOLUTIONS

3.6.1 History and geography

Bangladesh is one of the most densely-populated countries in the world, with approximately 160 million citizens living in an area of 148,000 km^2. It is also a country where efforts to provide its citizens with clean drinking water have left a legacy of chronic illness, due to a contaminant that is unseen, tasteless and odourless. Cholera had been and remains today a chronic problem in Bangladesh, affecting millions via the faecal-oral route. The rainy season is consistently associated with the disease and the bacterium *Vibrio* is now known to reside in the marine environment, interacting with zooplankton and reseeding many populations (Giebultowicz *et al.* 2011). Sewerage, wastewater treatment and drinking-water treatment were not adequate and thus international efforts to supply microbiologically-clean water to Bangladeshis, shifted efforts to providing groundwater as has been done throughout history. These actions inadvertently and catastrophically exposed the majority of the population to elevated levels of arsenic (As). The transition from consuming surface water contaminated with pathogens and faecal bacteria to drinking water free of microbial pollutants but rich in As is an example of a well-meaning but poorly-executed attempt to improve water-related morbidity and mortality. In addition, without disinfection ground water supplies are not immune to faecal contamination. Efforts to remediate the situation are on-going but as more and more communities have their water tested, the scale of the problem increases, stretching the ability of the government and international organisations to rectify the situation. The situation in Bangladesh is an important reminder of how actions to deliver drinking water assumed to be safe may not be as worthy as they first appear.

Located on the Ganges-Brahmaputra Delta, Bangladesh's subsurface is fertile alluvial soil washed into the delta by the dozens of rivers that crisscross the country. Surface water is plentiful in the tropical climate, with the monsoon season from May to August providing the majority of the annual rainfall, although the lack of management compromises its quality. In addition, 93% of the surface water flowing in the rivers originates from the neighbouring countries of India and Myanmar, meaning that pollutants from Bangladesh's neighbours enter the water en route. Predominant geological features of the subsurface include the Rangpur coal fields, Madhupur clay and Jaypurhat sediments, with naturally-occurring As concentrations of 8 to 35 mg/kg (Alam *et al.* 2002). Population demographics have shown significant changes since 1971, with a decrease in the population growth rate since the 1980s, and although the metro population of the megacity of Dhaka is approximately 13 million, Bangladesh remains a rural country with in excess of 70% of the population living outside of urban areas. The country is divided into seven administrative divisions (Barisal, Chittagong, Dhaka, Khulna, Rajshahi, Rangpur and Sylhet), named after the main city in each area, and each are divided into districts (zila) and sub-districts (upazila).

In the early 1970s, when Bangladesh was a newly-independent country, infant mortality accounted for approximately 140 deaths per thousand live births (Caldwell *et al.* 2003). Diarrheal disease accounted for many of these deaths, and the food and surface water consumed by many Bangladeshis was shown to be of poor microbial quality. Relying on surface water for consumption, either directly as drinking water, or for washing or cooking food items, exposed the majority of the population not on mains water supplies to pathogenic micro-organisms. However the underlying geology of aquifers was understood to be a reliable source of safe drinking water, not susceptible to climatic changes or anthropogenic pollution. Community treated water supplies in urban areas have historically been drawn from groundwater, and rural areas are punctuated with deep tube wells or shallow dug wells. Data from 1940 indicates at that time 40,000 tube wells were used throughout the region that was to become Bangladesh (Mukherjee & Bhattacharya, 2001). In the mid-1990s there were approximately 600,000 tube wells and by 2000 the number of wells had increased to an estimated 2 million (Anwar, 2000). Although the number of government-run and regulated wells has increased since the 1940s, the proportion of wells dug privately has increased far in excess of the government wells. Identifying and monitoring private wells for water quality issues is particularly problematic for improving water quality standards for the predominantly rural population.

3.6.2 Technological developments

In the 1970s health and policy experts from many countries, development institutions and charitable organisations sought to address the high incidence of diarrhoeal disease and associated mortality in Bangladesh. Efforts to improve the health of Bangladeshis focused on reducing a reliance on surface water sources by increasing the exploitation of groundwater. The predominantly rural communities rely on agriculture and coupled with limited clean water and sanitation infrastructure serving these people, contamination of surface waters with faecal bacteria is inevitable. Prior to this international attention, rural households with traditional shallow tube wells were outnumbered by those relying on surface water sources for domestic use. It is understood that reliance on untreated surface water for everyday domestic use means that consumers are more likely to be exposed to waterborne pathogens, hygiene standards are negatively impacted, consistent quantity cannot be assured and the labour burden for carrying water to the home is increased (Caldwell *et al.* 2003). This last factor is particularly true for women and children who are typically the primary providers of water to the household. More time spent carrying water means less time spent in education, in employment or as a caregiver. As such, the appeal of having an on-site water source close to the home extends far beyond the promise of a product that will not make the consumer ill.

Technological development programs initiated the construction of government-run tube wells. These wells were classified into three types, namely hand pumped (HPTW), shallow (STW) or deep (DTW), as shown in Table 3.6. As the name implies HPTWs rely on the labour of the user to draw water from the subsurface, whereas the STW and DTW alternatives are combined with engine-driven pumps to extract the water.

Table 3.6 Summary of the operational parameters for hand pumped, shallow and deep tube wells.

Type	Depth	Bore size	Discharge
HPTW	$<$14 m	$<$5 cm	$<$0.001 m^3/sec
STW	24–42 m	5–10 cm	0.001 m^3/sec
DTW	$>$50 m	$>$10 cm	$>$0.05 m^3/sec

3.6.3 International involvement in water quality issues

The organisations behind the effort to shift source water from the surface to underground were motivated by the fundamental assumption that water drawn from aquifers was of a better quality. However, the pace of the construction of tube wells increased exponentially, partly driven by the low cost and low technology requirements of the excavations. By the early 1990s, it was estimated that 95% of rural Bangladeshis were drinking groundwater obtained from 2.5 million tube wells (Caldwell *et al.* 2003). Data on infant mortality rates in Bangladesh do indeed show a rapid decline during the period of tube well construction. In 1980, infant mortality was approximately 129 deaths per thousand live births, and by 2001 this figure had dropped to 51 per thousand births (The World Bank, 2005). However during the same period, an increase in the use of oral rehydration therapy, improved hygiene and access to health care were also instrumental to better health standards in Bangladesh and throughout South Asia (Caldwell *et al.* 2003).

In September 2001, the UK Observer newspaper published an article titled "Scientists sued over 'poisoning': Britons face world's largest action over arsenic in Bangladesh water" (Spiller, 2001). It described the legal representation of hundreds of Bangladeshi villagers, and the accusation against the British Geological Survey (BGS) was of a failure to adequately test groundwater for elemental contaminants. The case revolved around the omission of tests specifically to determine the concentration of As in groundwater pumped to serve as drinking water in rural communities. The article continued with a discussion on other aid agencies implicated in the case, particularly the United Nations Children's Fund (Unicef). The maximum acceptable As concentration in drinking water, as defined in 1993 by WHO is 0.01 mg/L, while the Bangladesh government's standard is 0.05 mg/L. However, had the BGS (or any other agency) tested the tube well water, they could have found drinking water with As concentrations up to and in excess of 2.5 mg/L (Nordstrom, 2002). In their haste to provide microbiologically clean water, the development organisations omitted to test for a contaminant that they did not expect to find. The legal action against the BGS was dismissed as it was determined that the agency had not been obligated to test for As and did not vouch for the safety of the water for drinking purposes. Regardless of any litigation, estimates of the number of Bangladeshis affected by excessive As consumption range from 20 million to 85 million, and this figure is likely to increase as efforts to reduce the number of people relying on tube wells for potable water is offset by an increasing population and the slow adoption of alternative water sources and As-removal methods.

3.6.4 Arsenic abundance and utilisation

Arsenic is ubiquitous in the environment, and is considered an essential micro-trace element and has been around since the beginning of time. As-bearing minerals vary widely by geography, but are predominantly found to be As sulphide, As tri-sulphide and arsenopyrite. As-rich sediments include shales and clays (up to 490 mg/kg) and coals (up to 80 mg/kg) (Mukherjee & Bhattacharya, 2001). Environmental As is either organic or inorganic, with the organic species considerably less toxic to humans than the inorganic species. Inorganic As has two main oxidation states, namely trivalent arsenite and pentavalent arsenate. Used in a multitude of industrial and agricultural applications, even the ingestion of As was once encouraged. Advertisements for beauty aids such as "Dr Rose's French Arsenic Complexion Wafers" in the 1902 Sears, Roebuck & Co. catalogue were designed to develop 'a transparency and pellucid clearness of complexion, shapely contour of form, brilliant eyes, soft and smooth skin, where by nature the reverse exists'. Fowler's solution was another mechanism of ingesting arsenic, whereby a tonic of potassium arsenite was used as a treatment for a variety of skin conditions in the mid-20th century (Chappell *et al.* 1994). With changing times, the intentional consumption of As is discouraged, except in certain situations such as the administration of drugs, for example the use of arsenic trioxide as anti-cancer chemotherapy. Occurrences of elevated As levels in the environment can be attributed to either manmade or naturally-occurring deposits. Arsenic is used in a variety of applications including pesticides, herbicides, wood preservers, feed additives, drugs and poisons. Mining activities for the extraction of As for use in these applications can significantly increase exposure to people living or working in a naturally As-rich area.

Ingestion, inhalation and dermal exposure can all contribute to the body burden of As. For the people of Bangladesh, the vast majority of As exposure has been via ingestion, specifically from drinking water but also from the consumption of crops grown in As-rich soil such as rice or consumption of fish from As-rich water. Rice accounts for 80% of arable land in Bangladesh and relies on huge volumes of water for a successful harvest (Hossain, 2006). A study of trace elements in crops grown in soil considered uncontaminated with arsenic showed that rice would typically contain 110 to 200 μg/kg As based on dry weight (Kabata-Pendias & Pendias, 1992). However As concentrations in rice grown in Bangladesh have been shown to reach 46 μg/g, which is over 200 times greater than the reported "typical" values (Meharg & Rahman, 2003). In comparison the 2008 United States Food and Drug Administration Total Diet Study reports a mean arsenic concentration of 0.071 μg/g in cooked, white US rice. It is also important to note that since rice grown in contaminated soil is likely to be cooked in contaminated water from the same area, an additional As burden is placed upon the consumer. This was described in a small study published in the Lancet, where residents were asked to cook locally purchased rice in tube well water. The mean values of five experiments conducted by the authors showed that 1,700 g of rice absorbed on average 1,200 g of water, and the As concentration in the cooked rice increased by approximately 84% (Bae *et al.* 2002).

When arsenic was first identified in the water coming from the tube wells in the mid-1980s, various theories were offered with regard to possible sources. Despite widespread use of As in agricultural and industrial applications the distribution of As-tainted wells interspersed with well producing water containing lower concentrations did not universally support the theory of anthropogenic contamination. In addition, prior to the 1990s the illnesses associated with acute As toxicity were not widespread in Bangladesh. Analysis of the geology and geomorphology of the sub-surface began to reveal correlations between alluvial deposits and the most contaminated aquifers. However there is little predictive power when trying to determine how many wells in a community might be affected. This requires each well to be individually tested, and communities to be visited in turn to quantify the number of people suffering from As-related health issues.

Ingestion of organic arsenic rarely causes health issues because it is readily excreted from the human body. However inorganic arsenic is not easily metabolised and thus, it gets deposited in tissues. The acute minimal lethal dose of arsenic in adults is estimated to be 70 to 200 mg or 1 mg/kg/day, but acute arsenic toxicity below the lethal dose typically manifests itself as gastrointestinal irritation and garlic-smelling breath (Chappell *et al.* 1994). Chronic inorganic arsenic poisoning is much more insidious, and the symptomatology can be highly variable, depending on the exposure route, dose and duration of exposure. Symptoms include dermal diseases, cancers at multiple internal sites, and peripheral vascular disease, and are collectively known as arsenicosis. As-related skin disorders are often the first indication of chronic exposure, as they are readily visible and have the shortest latency period, suggested to be anything from 2 years to 10 years from initial exposure to contaminated drinking water (Rahman *et al.* 2001; Smith *et al.* 2000). Hyperkeratosis is a thickening of the skin, and this can be accompanied by hyperpigmentation and skin cancers. Internal cancers are most commonly diagnosed in the bladder kidneys and lungs, and take 20 years or more to become apparent. Peripheral vascular disease, particularly gangrene is also widely observed in persons suffering from acute As toxicity. It is possible that the elevated levels of arsenic in the diet coupled with generally poor levels of nutrition uniquely combine to cause a very specific type of gangrene, also known as black foot disease. This is a colloquial name specifically associated with arsenic-related gangrene found in Taiwan.

3.6.5 The scale of the arsenic problem

In Bangladesh, the number of people suffering from arsenicosis is reported in widely varying numbers, typically quoted to be 21 million, 40 million, or 85 million (Caldwell *et al.* 2003; Alam *et al.* 2002; Hossain, 2006). The discovery of dangerously high levels of As in Bangladesh's groundwater was a chance observation in neighbouring districts of India. Routine surveys of tube well water in West Bengal in 1983 found As concentrations in excess of the WHO permissible limit at that time (0.05 mg/L). Preliminary analyses of well water in Bangladesh funded by the World Health Organization was undertaken in 1993 but a concerted effort to identify the scale of the problem was not really undertaken until 1995, when the Bangladesh Department of Public Health Engineering worked in collaboration with the British Geological Survey to survey approximately 3 to 4 million wells (Hossain, 2006). This was three years after the BGS first surveyed tube well water, and nearly found itself the subject of the aforementioned class action lawsuit. At this point, scientific interest and public health concern in the quality of water from the millions of tube wells increased exponentially. It was discovered that the hand pumped tube wells extracting water from less than 12 metres below the surface were not typically contaminated with As, whereas the shallow and deep tube wells extending 21 to 60 metres below the surface did contain elevated As levels (Alam *et al.* 2002). This was an important finding from two perspectives: (a) had development organisations not encouraged the digging of deeper wells to move away from the microbially-contaminated surface water, the As-rich groundwater would not have become the predominant drinking water source, and (b) the fact that As contamination increased with depth suggested that this was a geological as opposed to anthropogenic problem, and any treatment solution would not be easily initiated in a predominantly poor, rural country.

3.6.6 Remediation strategies

Solving the problem of drinking water quality and quantity continues, and the number of international aid agencies, government organisations and scientific teams continues to increase as the scale of the problem has become apparent. From 1998 to 2006, a notable effort called the Arsenic Mitigation Water Supply Project was funded by the World Bank for a total amount US $44.4 million, and aimed to evaluate and mitigate arsenic contamination with the Bangladesh Ministry of Local Government, Rural Development and Co-operatives, in collaboration with the British Geological Survey and the Swiss Agency for Development and Cooperation. The focus of this project was to implement on-site mitigation, improve understanding of arsenic and health, and strengthen technical assistance and development (The World Bank, 2007). Arsenic concentrations were screened in 190 of the 482 upazilas (in excess of 3 million wells) and one outcome of this project was the identification of wells that were safe, and thus painted green, in comparison to those that were found to be unsafe, and painted red. A second achievement was the provision of arsenic safe drinking water to 1808 villages, although the original target had been 4000 villages. Mitigation methods included rainwater harvesting, sand filters and dug wells, and a lesson learned from the intervention was the difficulty in implementing As-reduction methods that were sustainable, low-maintenance and cost-effective.

Returning to a reliance on surface water is not a desirable solution, as the microbiological quality has not shown an improvement and the frequency of droughts and floods adversely affects supply. Therefore attention turns to remediation of the groundwater supply, to deliver a product that is fit to drink. Table 3.7 outlines the frequently tested methods for arsenic-removal.

Table 3.7 Commonly-used remediation technologies for arsenic-contaminated drinking water.

Treatment name	Treatment description	Arsenic removal effectiveness	Reference
Precipitation & coagulation	Ferric chloride + oxidiser (hypochlorite, permanganate, hydrogen peroxide)	92% removal of As reported in water containing 158 μg/L	Meng *et al.* 2001
Sorptive techniques	Activated carbon	84% removal of As reported in water containing 1000 μg/L	Huang & Fu, 1984
	Ion exchange	>98% removal of As reported in water containing 100 μg/L	Greenleaf *et al.* 2006
Filtration	Iron filings + sand	>90% removal of As reported in water containing 500 μg/L	Leupin *et al.* 2005
	"3-pitcher" method (iron chips, sand, charcoal)	>99% removal of As reported in water containing 1100 μg/L	Khan *et al.* 2000
Flocculation & disinfection	Commercial point-of-use water treatment	99.5% removal of As reported in water containing 430 μg/L	Souter *et al.* 2003

These methods are typically cost-effective, in terms of the technology and raw materials required, which does make their adoption more practical. For example, the "3-pitcher" filter is based on an indigenous method of filtration. Traditionally, two local clay pitchers (called kolshi) are used to filter water, where the top pitcher is partially filled with sand and charcoal, with a small hole in the bottom. A piece of cloth over the hole prevents sand from being lost from the system. Suspended matter is removed from surface water as it passes through the top pitcher, and the filtered water is stored in the bottom pitcher. A contemporary version of this system uses buckets, and the third pitcher added to the traditional two pitcher arrangement is placed above the sand/charcoal pitcher, and filled with iron filings to provide a source of iron oxide that adsorbs more As.

3.6.7 Diagnosis and treatment of arsenicosis

Treating the people who are suffering from the effects of consuming the arsenic-rich groundwater has been a priority for health professionals, aid agencies and development organisations. Assessing the severity and magnitude of the situation required the implementation of diagnostic tools and medical expertise in areas that would usually be beyond the reach of such techniques. Prolonged exposure to arsenic results in an accumulation in ectodermic tissue. Concentrations in exposed populations are typically measured from samples of urine, blood and hair. Normal arsenic levels in hair are in the range of 0.08 to 0.25 mg/kg, in fingernails 0.43 to 1.08 mg/kg and in urine 0.005 to 0.040 mg/kg (Alam *et al.* 2002). In contrast, epidemiological studies of tissues samples collected from Bangladeshis in areas of high arsenic contamination have reported As-concentrations of 1.1 to 19.84 mg/kg in hair, 1.3 to 33.98 mg/kg in nails and 0.05 to 9.42 mg/L in urine (Karim, 2000). When results such as these are discovered, intervention strategies to improve the health of the affected individuals and prevent further poisoning are needed. The first line of attack is typically the withdrawal of arsenic-contaminated water, and this is where the importance of adopting suitable mitigation strategies as described in Table 3.2 comes into play. Improving nutrition and increasing consumption of essential vitamins can help the body excrete excess arsenic. Chelation therapy using agents such as *d*-penicillamines, DMSA (dimercaptosuccinic acid) and DMPS (dimercapto-propane sulphonate) has been developed to reduce arsenic body burden. Developed in response to chemical warfare in World War One, chelating agents can be delivered intravenously, intramuscularly, or orally, and bind to the heavy metals in the body, preventing them from being deposited in the tissues. The agent-metal complex is then excreted from the body in the urine. Considerations for this method include the concurrent excretion of essential vitamins, which need to be supplemented and possible side-effects of the therapy need to be managed, such as vomiting, diarrhoea and more serious effects such as cardiac arrest and hypotension.

The current situation for the people drinking arsenic-rich waters in Bangladesh is the continued identification of affected areas, the adoption of remediation strategies to improve the environmental situation, and the treatment of individuals identified as sufferers of acute As toxicity. Agencies such as WaterAid, WHO, Unicef, and The World Bank are working with the Government of the People's Republic of Bangladesh to increase the availability of safe drinking water, improve the infrastructure for water and sanitation requirements of people in both rural and urban areas, and de-stigmatise arsenicosis to encourage people to seek treatment. Decentralising the treatment of drinking water and encouraging in-kind co-operation of citizens to build and maintain a better water infrastructure has been shown to improve the utility and maintenance of new technologies. Teaching citizens that even drinking water that appears to be clean can be the cause of chronic illness and mortality is central to the education efforts. On-going testing of existing and new water sources is needed to correctly identify susceptible populations that may need assistance with mitigation strategies and medical intervention. Finally as a central component of the Millennium Development Goal to halve, by 2015, the proportion of people without sustainable access to safe drinking-water and basic sanitation, there should not be a trade-off between microbiologically-clean drinking water and water contaminated with elemental or chemical toxins. Eliminating disparities in drinking water quality and sanitation in Bangladesh is still work in progress, and overcoming technical and logistical difficulties is key to helping some of the most vulnerable people in the world.

3.7 CONTEMPORARY OUTBREAKS OF WATER-RELATED DISEASE

Advancements in water treatment technologies and engineering have made outbreaks of water-related disease more of a rarity for those people fortunate to have sewage treatment and access to drinking water that is consistently clean and safe. However due to a combination of factors, illness and death due to diseases such as cholera, typhoid and cryptosporidiosis continue to affect millions of people throughout the world. Although serious outbreaks of cholera and typhoid are observed in developing countries, other pathogens including parasites and viruses are present and documented waterborne disease occurs throughout the economic strata. Poverty is an important issue in water supply, however evidence from recent disease outbreaks shows that it is not the only driving force that needs to be addressed to reduce the impact of water-related disease.

3.7.1 Conflict and disease

Civil unrest can lead to the destruction of infrastructure and the displacement of populations into camps with inadequate sanitation and water supply. Coupled with the vulnerable health status of displaced persons, this enhances the probability of widespread and devastating outbreaks of water-related disease. In times of conflict the ability of healthcare services to monitor the health status of a country's citizens and respond to a disease outbreak is compromised, further exacerbating the effect of illness. In recent periods of unrest, these factors have unfortunately been evident in many countries, affecting millions of people. Wars in Africa and the Middle East have resulted in some of the most devastating outbreaks of water-related disease. The Rwandan Civil War in the early 1990s resulted in the deaths of approximately 800,000 Tutsi and pro-peace Hutu civilians in a genocide for which the region has become infamous. In addition to these victims, thousands more refugees died of cholera and dysentery in camps throughout the country, and the problem was further compounded by the development of drug resistant strains of *Shigella dysenteriae*. In one Rwandan camp of 20,000 refugees, the attack rate of *S. dysenteriae* was 32% and the case fatality rate was 4% (Gayer *et al.* 2007). Unrest in Iraq has also been punctuated by outbreaks of cholera, most notably in 2003, 2007, and 2008. Documentation from the WHO's integrated global alert and response system for public health emergencies shows that cholera outbreaks were predicted based on the conflict situation and lack of water and sanitation infrastructure. From August to October 2007, the largest outbreak affected approximately 30,000 Iraqis throughout the country, with a case fatality rate of 0.52%.

3.7.2 Natural disasters

The long-term effect of natural disasters extends far beyond the initial occurrence. Earthquakes, floods and other extreme weather events can devastate vital infrastructure and if the natural disaster occurs in an already fragile region, there is potential for water-related disease to cause widespread morbidity and mortality amongst the affected population. The January 2010 earthquake that struck Haiti caused approximately 250,000 deaths, left 300,000 injured and displaced in excess of 1.3 million people (Walton & Ivers, 2011). Haiti's history is chequered with slavery, debt and hostile occupation, and, coupled with long periods of political upheaval, this has resulted in an extremely vulnerable populace. Not typically affected by cholera, Haiti suffered its first outbreak in more than 100 years in October and November 2010. According to the Pan American Health Organization's Cholera Haiti Health Cluster Bulletin dated August 16, 2011, the outbreak has caused 419,511 cases to date, of which 222,359 have been hospitalised and 5968 died, with a case fatality rate of 1.4%. Not only was the outbreak devastating in terms of the health impacts, but identification of the cholera serotype (*Vibrio cholerae* O1, serotype Ogawa, biotype El Tor) showed that the infection is likely to have originated from a human host from South Asia (Walton & Ivers, 2011). This increased social and political friction between the Haitians and aid agencies, has hindered the rebuilding efforts of the international volunteers.

The effect of natural disasters on water and health is not limited to countries with few financial resources. Two recent events show that climatological and seismic events can critically affect water quality in countries that would not typically be seen as having vulnerable populations. When Hurricane Katrina made landfall in Louisiana, USA in August 2005 its effects on water quality and infrastructure were unprecedented. Contaminants such as volatile and semi-volatile organic compounds, arsenic and faecal coliform bacteria were found to be elevated in New Orleans floodwaters (Pardue *et al.* 2005), increasing their ability to intrude into damaged sanitary infrastructure. Evacuation centres immediately began reporting illnesses as the displacement of people from New Orleans escalated. Chronic illness (e.g. diabetes, asthma, emphysema, and cardiovascular disease) accounted for 33% (4786) of the 14,531 doctor visits reported, and gastrointestinal illness, the second most commonly reported illness, accounted for 27% (3892) of total visits (September 1–22, 2001) (MMWR, 2006). Meanwhile, approximately 13,000 evacuees were taken to the Reliant Astrodome in Houston, where a norovirus outbreak resulted in 4.6 doctor visits per 1000 persons per day among the evacuees (Yee *et al.* 2007).

More recently, in the aftermath of the 2011 Japanese tsunami, the possibility of water-related disease outbreaks was of concern, due to the complex nature of the sludge that remained after the tsunami had receded (Kelley, 2011). In addition to the biological and chemical contaminants that posed a human health threat, the demographics of the affected area compounded the possibility of widespread illness. As the disaster hit a country with an aging population, certain additional considerations needed to be taken into account to prevent a predominantly vulnerable population from succumbing to disease from contaminated water. Yamamoto *et al.* (2011) describe the efforts in tsunami-affected communities to minimise outbreaks of infectious disease:

> 'A possible infectious disease outbreak was also a concern. Tap water and sewage systems were destroyed, and evacuees were advised to wrap their stools in newspaper and place them in a plastic bag. But when patients with acute gastroenteritis suggestive of norovirus infection were found, we facilitated improvement of hygiene measures, introduced chlorine-based disinfectants, and promoted accurate knowledge of virus transmission.'

The outcome in Japan was very unlike that which occurred during the Tsunami of 2004, in the case of Aceh Province. Most communities reported wide spread diarrhoea as the main cause of morbidity (85% of the cases were in children under five years of age), however no increases in mortality were reported and no outbreaks of cholera were reported (Brennen & Rimba, 2005).

3.7.3 Poverty

By far one of the most influential factors indicative of the likely spread of water-related disease is poverty. According to the most recent (2005) data from The World Bank, more than 40% of the people of sub-Saharan Africa and South Asia live on less than $1.25 per day, as compared to 17% in East Asia and the Pacific, and less than 10% in Europe and Central Asia, Latin America and the Caribbean, and the Middle East and North Africa. It is no coincidence that the 2005 data for the percentage of people with access to improved sanitation facilities mirror the poverty statistics. For sub-Saharan Africa and South Asia, 30% and 33% of people had adequate access to excreta disposal facilities, respectively. This compares to 61% in East Asia and the Pacific, and greater than 78% in Europe and Central Asia, Latin America and the Caribbean, and the Middle East and North Africa. In the developing countries, a combination of factors can accentuate the detrimental effects of poverty on the health status of vulnerable people. For example, rapid urbanisation puts stress on existing water and sanitary infrastructure, and changes in weather patterns cause flooding that spreads waterborne contaminants (Ashbolt, 2004). As such, deprivation in tandem with other factors can have devastating consequences on the ability of people to access safe water.

In 2008 and 2009, Zimbabwe experienced the largest cholera epidemic seen in Africa to date. The most recent WHO report on the outbreak dated 30 May 2009 identified 98,424 suspected cases with 4276 deaths, signifying a case fatality rate of 4.3%. Social, medical and infrastructure weaknesses combined with widespread poverty to create a "perfect storm" of waterborne disease (Mason, 2009). Hospital closures, a lack of clinical diagnostic services and a shortage of medical staff prevented the outbreak from being identified quickly and responded to in a timely and effective manner. Coupled with a lack of running water supply and subsequent dependence on water extracted from shallow wells, the poverty stricken Zimbabweans could not avoid succumbing to the cholera outbreak and were not able to access treatment. Additionally, rather than dispensing oral rehydration salts, the government recommended homemade rehydration solutions. This was a futile effort as many people could not afford to buy the sugar and salt needed to make up the solution, and lacked access to potable water to prevent reinfection (Koenig, 2009). Prior to 2008, cholera outbreaks in Zimbabwe were infrequent, small and reasonably well-contained. However in a country with an unemployment rate of 95%, hyperinflation of its currency and widespread endemic illnesses such as tuberculosis and HIV/AIDS, it is likely that water-related diseases such as cholera will continue to affect a very large number of people unable to adopt measures to protect themselves.

3.7.4 Engineering/treatment failure

Some outbreaks of water-associated illness can be attributed to a singular event, or chain of events, directly related to the treatment or distribution system. In these circumstances, it may be impossible for the consumer to avoid the adverse effects of a breakdown in the procedure designed to ensure the delivery of safe potable water. In the United States, since 1996, problems with the distribution system have been the cause of 45% of all outbreaks reported in community water systems, and most of these incidents were attributed to contamination from cross-connections and back siphonage (Craun & Calderon, 2001). Therefore, it is important to acknowledge that even in countries with economic and technical advantages, water-related adverse health outcomes continue to be an issue that should be preventable, and in less developed countries engineering failures can have even greater consequences for human health (Lee & Schwab, 2005).

Failures in water distribution systems can lead to the intrusion of contaminants into water that has already been through treatment processes and is on its way to the consumer. Pumping, piping and storage networks need to be adequately maintained to ensure that breakages, loss of hydraulic integrity or the contamination of storage tanks does not occur (Reynolds *et al.* 2008). In addition, failures in the disinfection efficacy can affect the residual disinfectant intended to maintain a pathogenically clean water supply.

Perhaps one of the most infamous outbreaks of water-related disease linked to treatment failure is the cryptosporidiosis outbreak in Milwaukee in the spring of 1993. Two water treatment plants supplied the city of Milwaukee, and from the end of March into the beginning of April, elevated turbidity levels were noted in the Lake Michigan source water. Widespread gastrointestinal illness led to an investigation and the water company issued a boil water notice and temporarily shut down the affected treatment plant. The treatment processes at the water plants comprised of coagulation using polyaluminium chloride, rapid mixing, mechanical flocculation, sedimentation and rapid sand filtration (Mac Kenzie *et al.* 1994). However, *Cryptosporidium* oocysts in the turbid lake water were not adequately removed by the treatment process, specifically during coagulation and filtration, and this allowed for widespread infection of residents and visitors to the city. The Milwaukee cryptosporidiosis outbreak was the largest outbreak of waterborne disease ever reported in the United States. It is estimated that 403,000 people suffered watery diarrhoea, cramping, nausea and fever, of which an estimated 354,600 persons (~88%) did not seek medical attention; 44,000 persons (~11%) were seen as outpatients; and 4400 persons (~1%) were hospitalised (Mac Kenzie *et al.* 1994; Corso *et al.* 2003). Fifty-four deaths were attributed to the outbreak, and of these fatalities, 85% had HIV/AIDS and 7% had coccidiosis, an intestinal disease mediated by parasites such as *Toxoplasma gondii* (Hoxie *et al.* 1997).

The financial cost of this massive outbreak per case of cryptosporidiosis was estimated to be US $79 for medical costs and US $160 for losses to productivity, totalling US $31.7 million and US $64.6 million respectively (Corso *et al.* 2003). The cost to the confidence of consumers was also substantial, particularly as the cause of the treatment failure was unclear. A combination of factors including the high turbidity of the source water, the recycling of filter backwash and an inadequate response to deterioration in the raw water quality allowed the oocysts to enter the distribution system (Mac Kenzie *et al.* 1994). Nonetheless, lessons learned from the Milwaukee cryptosporidiosis outbreak have helped prevent further comparable incidents in countries served by water treatment systems usually considered adequate, in terms of scale and likely cause.

3.8 LESSONS LEARNED OR NOT

It is the interplay between the building of knowledge in science, medicine and engineering that has lead us into the 21st century where safe water sits within its current definition. We can test water for almost any contaminant known, we have new tools such as genomics that provides us with insight into the biological nature of a "healthy" system and we have toxogenomics and physiological models which help us to understand better the human system as it is exposed to contaminants. The World Health Organization, the European Union and the United States Environmental Protection Agency have all lead the way to provide criteria and standards to protect, restore and treat water for ensuring safety. All are supporting quantitative microbial risk assessment (QMRA) (Haas *et al.* 1999) as a framework to integrate science, medicine and engineering toward meeting the goals of safe water. The latest WHO Guidelines for Drinking Water Quality recommend QMRA for health based target setting (WHO, 2011c). This has also been used for the establishment of water regulations, within the US Safe Drinking Water Act (SDWA), via institutions like the USEPA, and it has been shown that it is possible to address water safety (Public Law, 1974) using science-based approaches. New issues will always emerge which will need to be addressed such as the emergence of *Cryptosporidium* and other important issues such as the removal of lead from plumbing which in the US was enacted in the 1986 amendments, and solidified in the lead and copper rule (McGill, 1993). A substantial investment is needed in the field of risk analysis for water to lead to appropriate decision analysis tools (Kammen & Hassenzah, 1999). Overall the quality of drinking water in developed countries has been improved through the various regulations and standards evidenced through the reduction of drinking water outbreaks of disease in countries with strong regulations. Since the focus is to reduce or eliminate water-related disease, energy and effort are focused to continuously improve water systems to attain these goals. A major focus of late is the ability to predict potential outbreaks and prevent them through engineering systems. QMRA is a multidisciplinary tool which gives decision makers and engineers a means of assessing current or potential health impacts from contaminated drinking water. This field of risk assessment is an iterative set of modelling frameworks developing a continually evolving field. The power of decision support tools is that they allow for a perspective into how hazards can affect the health of populations of various sizes.

Exploring the past demonstrates that there are several issues which remain the key to understanding the risk.

(a) Potency of the organism (or chemical): The dose-response function provides the yard stick that helps us to measure the level of risk associated with the level of the exposure (level of the contamination).
(b) The consequences of the outcome (type of morbidity and mortality) which are influenced by the genetic characteristics of the pathogen, the host's specific make up in regard to susceptibility and immune status.
(c) The excretion of the pathogen into the environment and its survival. And
(d) The environmental conditions, particularly the climate which influences the transport and ultimately the temporal and spatial nature of the exposure.

The historical evidence on waterborne disease shows that schistosomiasis, typhoid, dysentery and cholera which have a faecal-water pathway caused widespread and serious disease, dramatic attack rates and high mortality. In a modern day society these diseases can be eradicated. Yet these diseases have global staying power as the science, medical and engineering knowledge has not been used. Sanitation has never received as much attention as drinking water and has been more difficult to tackle. Even with sophisticated water systems in ancient civilizations, waste was still handled with cess pits or with pots until sewers were brought into play. Even then wastewater treatment has not advanced to address nutrient recovery, energy recovery and safe water recovery globally and has only recently been getting the attention deserved.

We live in a world where in an instant we can watch and see what is happening around the globe. We see droughts, floods, famine, disasters and wars. The one thing all of these have in common is the need to provide water and sanitation to the people. The massive cholera outbreak in Haiti need not have happened, if the lessons from the past were heeded.

Bill Gates in 2011 (http://www.nytimes.com/2011/02/01/health/01polio.html) restated his continued support for eradication of polio via vaccination programs (funded largely by the Bill and Melinda Gates Foundation). The disease caused by poliovirus is transmitted via faecally-contaminated hands, food and water. Poliovirus belongs to the *Picornaviridae* family which includes many enteric viruses spread by the faecal-oral route and can be classified as a waterborne disease causing agent. While the vaccination program for this disease has been successful globally, at the same time the developed world was investing in water treatment for both sewage and drinking water. It is no accident that poliovirus remains a threat in areas where there is no sanitation and inadequate water treatment. Waterborne disease caused by faecal pollution will continue to have an important impact on communities until water treatment is in place which would not only help to eradicate polio but hundreds of other diseases.

It should be remembered that cholera and typhoid, two important faecal-oral waterborne pathogens, were eradicated in the developed world not through vaccinations but through water treatment. These ancient diseases still plague developing countries due to polluted water. Water quality diagnostics and targeted programs for corrective measures are needed.

Those of us who are microbiologists, environmental engineers and public health epidemiologists have received an integrated education. This broad overview is needed to address global health problems such as the fight against waterborne diseases. Only then will strategies and investments be maximised so that the precious dollars spent really do decrease the global burden of disease.

REFERENCES

Abdel-Wahab M .F. (1982). Schistosomiasis in Egypt. CRC Press, Inc., Boca Raton. USA.

Alam M. G. M., Allinson G., Stagnatti F., Tanaka A. and Westbrooke M. (2002). Arsenic contamination in Bangladesh groundwater: A major environmental and social disaster. *International Journal of Environmental Health Research*, **12**(3), 236–253.

Anwar J. (2000). Arsenic poisoning in Bangladesh: End of a civilization? Palash Media and Pub., Dhaka, Bangladesh.

Ashbolt N. J. (2004). Microbial contamination of drinking water and disease outcomes in developing regions, *Toxicology*, **198**(1–3), 229–238.

Bae M., Watanabe C., Inaoka T., Sekiyama M., Sudo N., Bokul M. H. and Ohtsuka R. (2002). Arsenic in cooked rice in Bangladesh. *Lancet*, **360**(9348), 1839–1840.

Baker M. N. (1948). The Quest for Pure Water. American Water Works Association, NY, USA.

Barnes I., Duda A., Pybus O. G. and Thomas M. G. (2010). Ancient Urbanization Predicts Genetic Resistance To Tuberculosis. *The Society for the Study of Evolution*, **65**(3), 842–848.

Beck R. W. (2000). A Chronology of Microbiology. ASM Press, Washington DC. USA.

British Columbia Communicable Disease Control http://www.bccdc.ca/NR/rdonlyres/E6320AA6-FEC7-42AC-840B-3721A106FE80/0/CCFOutbreakguidelines.pdf (accessed 6 September 2011).

Brennan R. J. and Rimba K. (2005). Rapid health assessment in Aceh Jaya District, Indonesia, following the December 26 tsunami. *Emergency Medicine Australasia*, **17**(4), 341–350.

Caldwell B. K., Caldwell J. C., Mitra S. N. and Smith W. (2003). Searching for an optimum solution to the Bangladesh arsenic crisis. *Social Science & Medicine*, **56**(10), 2089–2096.

Centers for Disease Control and Prevention, http://www.cdc.gov/nceh/vsp/surv/gilist.htm (accessed 6 September 2011).

Chappell W. R., Abernathy C. O. and Cothern C. R. (eds). (1994). Arsenic Exposure and Health. Science and Technology Letters, Northwood.

Clutton-Brock J. (1981). Domesticated Animals from Early Times. Univ. Texas Press, USA, Austin.

Cockburn A., Cockburn E. and Reyman T. A. (eds) (1998). Mummies, Disease and Ancient Cultures. University Press, Cambridge, UK.

Corfield W. H. (1902). The Etiology of Typhoid Fever and its Prevention, Being the Milroy Lectures Delivered at the Royal College of Physicians in 1902. HK Lewis, London.

Contis G. and David A. R. (1996). The epidemiology of bilharzia in Ancient Egypt: 5000 years of schistosomiasis. *Parasitology Today*, **12**(7), 253–255.

Corso P. S., Kramer M. H., Blair K. A., Addiss D. G., Davis J. P. and Haddix A. C. (2003). Cost of illness in the 1993 waterborne Cryptosporidium outbreak, Milwaukee, Wisconsin. *Emerging Infectious Diseases*, **9**(4), 426–431.

Cort W. W. (1950). Studies on schistosome dermatitis, XI: status of knowledge after more than twenty years. *American Journal of Hygiene*, **52**(3), 251–307.

Craun G. F. and Calderon R. L. (2001). Waterborne disease outbreaks caused by distribution system deficiencies. *Journal – American Water Works Association*, **93**(9), 64–75.

Crouch D. P. (1993). Water Management in Ancient Greek Cities. Oxford University Press, NY, USA.

Cunha B. A. (2004). Alexander the Great and West Nile Virus, Encephalitis. *Emerging Infectious Diseases*, **10**(7), 1328–1333.

David R. (ed.) (1978). Mysteries of the Mummies: The Story of the Manchester University Investigation. Cassell Ltd., London, UK.

David A. R. (1997). Disease in egyptian mummies: the contribution of new technologies. *Lancet*, **349**(9067), 1760–1763.

David R. and Tapp E. (eds) (1984). Evidence Embalmed: Modern Medicine and the Mummies of Ancient Egypt, Manchester University Press, Manchester, UK.

Dunand F. and Lichtenberg R. (2006). Mummies and Death in Egypt. Cornell University Press, Ithaca, USA.

Fenn B., Morris S. S. and Black R. E. (2005). Comorbidity in childhood in northern Ghana: magnitude, associated factors, and impact on mortality. *International Journal of Epidemiology*, **34**(2), 368–375.

Fleming S., Fishman B., O'Connor D. and Silverman D. (1980). The Egyptian Mummy: Secrets and Science. The University of Pennsylvania, Philadelphia, USA.

French K. D. and Duffy C. J. (2010). Prehispanic Water Pressure: A New World First. *Journal of Archaeological Science*, **37**, 1027–1032.

Gayer M., Legros D., Formenty P. and Connolly M. A. (2007). Conflict and emerging infectious diseases. Emerging Infectious Disease, http:www.cdc.govEIDcontent13111625.htm (accessed 17 August 2011).

Gensini G. F., Yacoub M. H. and Conti A. A. (2004). The concept of quarantine in history: from plague to SARS. *Journal of Infection*, **49**, 257–261.

Giebultowicz S., Ali M., Yunus M. and Emch M. (2011). Review Essay: A comparison of spatial and social clustering of cholera in Matlab, Bangladesh. *Health and Place*, **17**, 490–497.

Gigante L. (undated) www.innominatesociety.com/Articles/Death%20and%20Disease%20in%20Ancient%20Rome.htm (accessed September 8 2011).

Greenleaf J. E., Lin J.-C. and Sengupta A. K. (2006). Two novel applications of ion exchange fibers: arsenic removal and chemical-free softening of hard water. *Environmental Progress*, **25**(4), 300–311.

Haas C. N., Rose J. B. and Gerba C. P. (eds) (1999). Quantitative Microbial Risk Assessment, 1st edn. John Wiley & Sons, Inc., New York, USA.

Hart G. D., Cockburn A., Millet N. B. and Scott J. W. (1977). Editorial: autopsy of an Egyptian mummy. *Canadian Medical Association Journal*, **117**, 461.

Harper D. (2010). http://www.etymonline.com/abbr.php (accessed September 8 2011).

Hossain M. F. (2006). Arsenic contamination in Bangladesh—an overview. *Agriculture, Ecosystems & Environment*, **113**(1–4), 1–16.

Hoxie N. J., Davis J. P., Vergeront J. M., Nashold R. D. and Blair K. A. (1997). Cryptosporidiosis-associated mortality following a massive waterborne outbreak in Milwaukee Wisconsin. *American Journal of Public Health*, **87**(12), 2032–2035.

Huang C. P. and Fu P. (1984). Treatment of As(V) containing water by activated carbon. *Journal of the Water Pollution Control Federation*, **56**(3), 232–242.

Husband W. D. (1865). Mortality in Infancy. In: National Association for the Promotion of Social Science G. W. Hastings (ed.) p. 503. http:books.google.combooks?id=XZcSAAAAIAAJ&printsec=frontcover&source=gbs_ge_summary_r&cad=0#v=onepage&q&f=false (accessed 6 September 2011).

Kabata-Pendias A. and Pendias H. (1992). Trace Elements in Soils and Plants. CRC. Press, Inc. Boca Ratón, USA.

Karim M. M. (2000). Arsenic in groundwater and health problems in Bangladesh. *Water Research*, **34**(1), 304–310.

Kammen D. M. and Hassenzah D. M. (1999). Should We Risk It? Princeton University Press. Princeton, New Jersey USA.

Kelley T. (2011). Environmental health insights into the 2011 Tōhoku Japan earth quake disaster. *Environmental Health Insights*, **5**, 21–22.

Khan A. H., Rasul S. B., Munir A. K. M., Habibuddowla M., Alauddin M., Newaz S. S. and Hussam A. (2000). Appraisal of a simple arsenic removal method for groundwater of Bangladesh. *Journal of Environmental Science and Health*, **35**(7), 1021–1041.

Koenig R. (2009). Public health: international groups battle cholera in Zimbabwe. *Science*, **323**(5916), 860–861.

Lambert-Zazulak P. I., Rutherford P. and David A. R. (2003). The international ancient egyptian mummy tissue bank at the manchester museum as a resource for the palaeoepidemiological study of schistosomiasis. *World Archaeology*, **35**(2), 223–240.

Langmuir A. D., Worthen T. D., Solomon J., Ray C. G. and Petersen E. (1985). The thucydides syndrome: a new hypothesis for the cause of the plague of Athens. *New England Journal of Medicine*, **313**(16), 1027–1030.

Lee E. J. and Schwab K. J. (2005). Deficiencies in drinking water distribution systems in developing countries. *Journal of Water and Health*, **3**(2), 109–127.

Leupin O. X., Hug S. J. and Badruzzaman A. B. M. (2005). Arsenic removal from Bangladesh tube well water with filter columns containing zerovalent iron filings and sand. *Environmental Science & Technology*, **39**(20), 8032–8037.

Mac Kenzie W. R., Hoxie N. J., Proctor M. E., Gradus M. S., Blair K. A., Peterson D. E., Kazmierczak J. J., Addiss D. G., Fox K. R., Rose J. B. and Davis J. P. (1994). A massive outbreak in Milwaukee of Cryptosporidium infection transmitted through the public water supply. *New England Journal of Medicine*, **331**(3), 161–167.

Marlowe F. W. (2005). "Hunter-gatherers and human evolution" (PDF). *Evolutionary Anthropology*, **14**(2), 15294. http://www.anthro.fsu.edu/people/faculty/marlowe_pubs/hunter-gatherers%20and%20human%20evolution.pdf (accessed 6 September 2011).

Marozzi J. (2008). The Way of Herodotus: Travels with the Man Who Invented History. Da Capo Press, Philadelphia, USA.

Mason P. R. (2009). Zimbabwe experiences the worst epidemic of cholera in Africa. *Journal of Infection in Developing Countries*, **3**(2), 148–151.

Mcgill R. L. (1993). The influence of an educational fact sheet on small system water supplier attitudes toward the lead and copper rule. *Journal of Environmental Health*, **56**(44), 11–14.

Meharg A. A. and Rahman M. M. (2003). Arsenic contamination of Bangladesh paddy field soils: implications for rice contribution to arsenic consumption. *Environmental Science & Technology*, **37**(2), 229–234.

Meng X., Korfiatis G. P., Christodoulatos C. and Bang S. (2001). Treatment of arsenic in Bangladesh well water using a household co-precipitation and filtration system. *Water Research*, **35**(12), 2805–2810.

Miller R. L., Armelagos G. J., Ikram S., De Jonge N., Krijger F. W. and Deelder A. M. (1992). Palaeoepidemiology of schistosoma infection in mummies. *British Medical Journal*, **304**(6826), 555–556.

MMWR, (2006). Morbidity Surveillance After Hurricane Katrina — Arkansas, Louisiana, Mississippi, and Texas, September 2005; **55**(26), 727–731.

Mukherjee A. B. and Bhattacharya P. (2001). Arsenic in groundwater in the Bengal delta plain: slow poisoning in Bangladesh. *Environmental Reviews*, **9**(3), 189–220.

National Research Council (2004). Indicators for Waterborne Pathogens. National Academies Press Washington, DC, USA.

Nordstrom D. K. (2002). Public health. Worldwide occurrences of arsenic in ground water. *Science*, **296**(5576), 2143–2145.

Oldach D. W., Richard R. E., Borza E. N. and Benitz R. M. (1998). A mysterious death. *New England Journal of Medicine*, **338**, 1764–1769.

Pan American Health Organization. (2011). Cholera and Post-Earthquake Response. In: Haiti – Tuesday, August 16, 2011 – #27. http://new.paho.org/disasters/index.php?option=com_content&task=view&id=1423&Itemid=1 (accessed 18 August 2011).

Papagrigorakis M. J., Yapijakis C., Synodinos P. N. and Baziotopoulou-Valavani E. (2006). DNA examination of ancient dental pulp incriminates typhoid fever as a probable cause of the Plague of Athens. *International Journal of Infectious Disease*, **10**(3), 206–214.

Pardue J. H., Moe W. M., McInnis D., Thibodeaux L. J., Valsaraj K. T., Maciasz E., van Heerden I., Korevec N. and Yuan Q. Z. (2005). Chemical and microbiological parameters in New Orleans floodwater following Hurricane Katrina. *Environmental Science and Technology*, **39**(22), 8591–8599.

Price David. Energy and Human Evolution. http://www.dieoff.org/page137.htm (accessed 6 September 2011).

Public Law 93-523; 42 U.S.C. § 300 f 1974-12-16, http://www.law.cornell.edu/uscode/42/300f.html (accessed 6 September 2011).

Rahman M. M., Chowdhury U. K., Mukherjee S. C., Mondal B. K., Paul K., Lodh D., Biswas B. K., Chanda C. R., Basu G. K., Saha K. C., Roy S., Das R., Palit S. K., Quamruzzaman Q. and Chakraborti D. (2001). Chronic arsenic toxicity in Bangladesh and West Bengal, India—a review and commentary. *Journal of Toxicology – Clinical Toxicology*, **39**(7), 683–700.

Reynolds K. A., Mena K. D. and Gerba C. P. (2008). Risk of waterborne illness via drinking water in the United States. *Reviews of Environmental Contamination and Toxicology*, **192**, 117–158.

Rose J. B. and Masago Y. (2007). A toast to our health: our journey toward safe water. *Water Science & Technology: Water Supply*, **7**(1), 41–48.

Ruffer M. A. (1910). Note on the Presence of `Bilharzia Haematobia' in Egyptian Mummies of the XXth Dynasty [1250-1000 BC]. *British Medical Journal*, **1**, 16.

Sherman I. W. (2006). The Power of Plagues. ASM Press, Washington DC, USA.

Smith A. H., Lingas E. O. and Rahman M. (2000). Contamination of drinking-water by arsenic in Bangladesh: a public health emergency. *Bulletin – World Health Organization*, **78**(9), 1093–1103.

Souter P. F., Cruickshank G. D., Tankerville M. Z., Keswick B. H., Ellis B. D., Langworthy D. E., Metz K. A., Appleby M. R., Hamilton N., Jones A. L. and Perry J. D. (2003). Evaluation of a new water treatment for point-of-use household applications to remove microorganisms and arsenic from drinking water. *Journal of Water and Health*, **1**(2), 73–84.

Spiller S. "Scientists sued over "poisoning" The Observer. 30 Sept. 2001. http://www.guardian.co.uk/world/2001/sep/30/physicalsciences.humanities (accessed 17 August 2011).

Steiner P. E. (ed). (1968). Disease in the Civil War, Natural Biological Warfare in 1861–1865, Charles C. Thomas Publisher, Springfield IL, USA.

Thucydides, History of the Peloponnesian War (Translated in 1954 Rex Warner; Penguin Book, NY, NY, USA).

Vuorinen H. S., Juuti P. S. and Katko T. S. (2007). History of water and health from ancient civilizations to modern times. *Water Science and Technology: Water Supply*, **7**(1), 49–57.

Walton D. A. and Ivers L. C. (2011). Responding to cholera in post-earthquake Haiti. *New England Journal of Medicine*, **364**(1), 3–5.

The World Bank. (2005). Environment and Social Unit – South Asia Region. Towards a More Effective Operational Response. Arsenic Contamination of Groundwater in South and East Asian Countries Volume II Technical Report. Environmental Study. Report No. 31303.

The World Bank. (2007). Implementation Completion and Results Report (IDA-31240 SWTZ-21082) on a Credit in the Amount of SDR 24.2 Million (US$ 44.4 Million Equivalent) to Bangladesh for Arsenic Mitigation Water Supply. Document of The World Bank Report No: ICR000028: Sustainable Development Department Environment and Water Resources Unit South Asia Region.

World Health Organization. Global Alert and Response (GAR) http://www.who.int/csr/don/archive/disease/cholera/en/ (accessed 18 August 2011).

World Health Organization. (1993). The Control of Schistosomiasis Report of a WHO Expert Committee World Health Organization Technical Report Series 830, Geneva, Switzerland.

World Health Organization. (2010a). http://www.who.int/schistosomiasis/en/index.html (accessed 17 August 2011).

World Health Organization. (2010b). Weekly epidemiological record, **18**(85), 157–164.

World Health Organization. (2011a). Cholera, 2010. Weekly epidemiological record (Relevé épidémiologique hebdomadaire) 29 July 2011, 86th year/No. 31, 86, 325–340. http://www.who.int/wer (accessed 6 September 2011).

World Health Organization. (2011b). Health cluster bulletin, cholera and post-earthquake response. In: Haiti – Tuesday, August 16, 2011 – #27. http://www.who.int/hac/crises/hti/sitreps/haiti_health_cluster_bulletin_ 16aug2011.pdf. (accessed 6 September 2011).

World Health Organization. (2011c). WHO, Guidelines for Drinking-Water Quality, 4th edn. World Health Organization, Geneva, Switzerland.

Wrangham R. and Conklin-Brittain N. (2003). Cooking as a biological trait. *Comparative Biochemistry and Physiology Part A: Molecular & Integrative Physiology*, **136**(1), 35–46.

Yamamoto T., Kato M. and Shirabe S. (2011). Life, health, and community in a tsunami-affected town. *Lancet*, **378** (9788), 318.

Yee E. L., Palacio H., Atmar R. L., Shah U., Kilborn C., Faul M., Gavagan T. E., Feigin R .D., Versalovic J., Neill F. H., Panlilio A. L., Miller M., Spahr J. and Glass R. I. (2007). Widespread Outbreak of Norovirus Gastroenteritis among Evacuees of Hurricane Katrina Residing in a Large "Megashelter" in Houston, Texas: Lessons Learned for Prevention. *Clinical Infectious Diseases*, **44**, 1032–1039.

Ziskind B. (2009). La bilharziose urinaire en ancienne Égypte (Urinary schistosomiasis in ancient Egypt). *Néphrologie et Thérapeutique*, **5**(7), 658–661.

Zohary D. and Hopf M. (2000). Domestication of Plants in the Old World. Oxford University Press, UK, Oxford.

Chapter 4

Diachronic evolution of water supply in the Eastern Mediterranean

K. Voudouris

4.1 INTRODUCTION

The problem of water supply for domestic, irrigation and industrial use is a crucial issue for humanity. It is pointed out that, the socioeconomic development of a region is usually associated with the availability of water resources, in terms of amount and adequate quality. In Mediterranean countries, it is not only the water quantity but also quality deterioration that endangers the future use of water. The main pressures on water supply are: the population growth, the intensification of agriculture (80% of the total water consumption), overexploitation, urbanization-changes in land uses, disposal of untreated waste effluent in rivers and abandoned wells and quarries, deforestation, development of lignite mines and so on (EEA, 2007; Chartzoulakis *et al.* 2001; Voudouris, 2009).

About 7000 years ago the general climatic conditions in the Eastern Mediterranean became similar to the present day, with less rainfall and warmer temperatures thereafter (Bar-Matthews & Ayalon, 1997). The most recent glacial period ended 12,000 years ago followed by a warm period up to the present (Holocene). This period is characterised by the development and spread of agriculture and increased water demands.

In the Eastern Mediterranean, Near and Middle East brilliant civilizations of antiquity flourished. At the earliest stage, *ca.* 8,000 years ago people performed all their activities close to the rivers, since they could not live without water. There was great dependence on surface water from rainfall and this was the driving force for ancient people to explore the groundwater (Zhou *et al.* 2011). In the dry regions of the Mediterranean, the scarcity of water and the dominance of agriculture resulted in the early development of the art of constructing galleries, dug wells, aqueducts and other hydraulic works for water supply. The network of channels constructed by Hammurabi in the plain between Tigris and Euphrates (1760 BC) is remarkable. Accounts of well construction occur frequently in ancient literature and are especially well known in the Biblical reference of Genesis (Davis & DeWiest, 1964). The springs were the earliest water resource used by ancient people.

The Roman period extended from the 1st century BC to the end of the 4th century AD, while the Byzantine period extended from the 5th century AD to the 12th century AD in the Middle East or the 14th century AD in Greece, when the Ottomans occupied this region. The Romans had constructed water supply systems to transfer water for domestic use in the cities, as well as for irrigation purposes. The aqueducts that carried water to the Roman cities from a long distance (100 km) are admirable; the surviving remains of many aqueducts demonstrate their strong technique (Haut & Viviers, 2007).

In the Byzantine period, the construction of aqueducts was abandoned and emphasis was put on the construction of tanks or cisterns. The tanks and cisterns were constructed in order to collect rain water. It is pointed out that the modern percussion methods of well or boreholes were developed in Western Europe. One of the first wells was dug in 1126 AD by monks near to Lillers (Davis & DeWiest, 1964). The methods of drilling for water have improved rapidly during the past 150 years.

Mediterranean countries are characterised by 1) a long coastline that favours hydraulic communication between coastal aquifers and seawater, 2) non-homogeneous distribution of rainfalls and water resources, 3) non-homogeneous water demands and 4) semi-arid climate conditions. Water resources are characterised by high water requirements for agriculture and tourism during the dry period (late April to early October) when water availability is low. Water needs are mainly covered by groundwater abstracted from the aquifers via numerous wells and boreholes. As a result, a negative water balance is established in the coastal aquifer systems triggering seawater intrusion which has negative consequences in the socioeconomic development of these areas. Many aquifer systems are reported to be affected by quality deterioration (salinisation and nitrate pollution) due to irrational management (Daskalaki & Voudouris, 2007).

To cope with the growing water demands, the study of ancient hydraulic works will help solve the current water problem in many arid areas. This chapter describes the diachronic evaluation of water supply in the Eastern Mediterranean. Firstly, the availability of water, especially in Near and Middle East, including the most remarkable hydraulic technologies for water supply, is examined. Secondly, characteristic examples of water supply practices from ancient Greece are presented.

4.2 AVAILABILITY OF WATER

The water availability is correlated with the climate conditions of a region, as well as with the management of water resources (surface water, groundwater and springs). Water scarcity is endemic in many regions of the Mediterranean, making these countries particularly vulnerable to any reduction in supplies (Karas, 1997). Water needs are mainly met by exploitation of aquifer systems and surface water. The overexploitation of limited water resources in the Mediterranean region, in conjunction with future climatic changes, has led to a concern over their sustainability (Fantechi *et al.* 1995; Suppan *et al.* 2008).

Climatic changes will have a significant impact on the availability of water, as well as leading to an intensification of the hydrological cycle, resulting in high risks of more extreme and frequent floods and drought. Climate changes have been hypothesised as the cause of, or at least a major factor contributing to, the deterioration of early civilizations. The collapse of many ancient civilisations is related to extreme climate changes, for example, the Akkadian empire, centred in what is now Syria. Under the rule of Sargon of Akkad, the first empire was established between about 4300 and 4200 years before present (BP) on the broad, flat alluvial plain between the Tigris and Euphrates Rivers. The Akkadian empire collapsed abruptly at 4170 ± 150 years BP. It has been recently proposed (De Menocal, 2001) that the collapse of this civilization was due to a prolonged period of intense aridity starting at 4025 ± 150 years.

The ancient civilizations were dependent upon sophisticated systems of water supply. Water was one of the most critical issues for the survival of mankind and played an important role on the creation of the settlements. Human effort was first made to utilise surface waters and springs and therefore the settlements were close to them (Tigris, Euphrates, Nile). Groundwater has been a source of water supply since the dawn of human history and agricultural activity. A spring is the natural emergence of groundwater, so many springs can be directly used without any digging (Zhou *et al.* 2011).

The ancient Greeks, as opposed to other people, avoided living near rivers, probably for protection from floods and water related diseases, for example, malaria (Koutsoyiannis *et al.* 2008). The first settlement of many of their towns was on a hill and the water demands were met by springs. When the water demands were increased in connection with irrigated agriculture, the groundwater exploitation was expanded with the construction of qanats and dug wells. Groundwater development dates from ancient times.

Well construction in the Near East was accomplished by man and animal and was aided by hoists and primitive hand tools (Davis & DeWiest, 1964). Egyptians had developed drilling systems in rocks as early as 3000 BC. Ancient Chinese also developed a drilling tool for water wells which, in principle, is similar to modern machines. Based on the C-14 dating of the well wood, it was concluded that the oldest well in China was built in 3710 ± 125 BC (Zhou *et al.* 2011). There is evidence that, during the Minoan period (3500–1200 BC) in the island of Crete, wells (with a depth of 10–20 m and diameter less than 5 m) and springs were used for water supply. Advanced hydraulic technology, including small dams, channels, cisterns and so on was developed in the Aegean islands during the Cycladic period (3100–1600 BC), as well as on mainland Greece during the Mycenaean period (1600–1100 BC) (Papademos, 1975; Koutsoyiannis & Angelakis, 2004; De Feo *et al.* 2011).

Securing the necessary quantity of water for a city by transferring it more than 50 km requires knowledge and large technical works. Many impressive hydraulic engineering works have been found in regions of the Middle East (Mesopotamia, Arabian Peninsula, Persia, Greece and so on), as deduced from the references of Herodotus (Avgoloupis & Katsifarakis, 2009). The groundwater exploitation in Cyprus started 7000 years ago and it is connected with the oldest dug wells of the world. The oldest dug well dates from *ca.* 6500 years ago near to Paphos with a depth of 8.5 m. It is noted that in this region there are neither springs nor rivers. It is noted that in this area there are no springs, rivers or torrents. Some cities of the island developed beyond the capacity of local water supply and for their sustainability they had to transfer water from big springs up to 40 km away (Konstantinou & Konstantinou, 1999). During the Hellenistic and Roman period the water supply of cities relied on the springs and rivers. In the case of extended dry periods, the water supply was assisted by the exploitation of groundwater from shallow aquifers. These works are located in areas which are faced with the problem of water shortage today.

Ancient peoples used mainly horizontal works for water supply (tunnel, galleries, aqueduct, and qanats) as opposed to using modern vertical works (deep boreholes). The aqueduct consisted of channels, tunnels and water bridges. The study of how to meet the needs in ancient times will help solve the current problem. Below, three representative examples of hydraulic works from ancient Greece are presented.

4.3 HYDRAULIC WORKS

4.3.1 Ancient Korinthos-Hadrian's aqueduct

The issue of the water supply in Korinthos has always been, and continues to be, the major problem for the town. The ancient town is located in the southern semi-mountainous area at a distance of 5.5 km from the modern town and flourished during the 5th century BC. The water supply of the ancient town was based on springs within the walls (Figures 4.1 and 4.2), but mainly on the Hadrian aqueduct; admirable work for their era.

Pausanias, a traveller and geographer of 2nd century AD, notes in Korinthiaka '…κρήναι δέ πολλαί μέν ανά την πόλιν πεποίηνται πάσαν άτε αφθόνου ρέοντός σφισιν ύδατος και ό δή βασιλεύς Ανδριανός εσήγαγεν έκ Στυμφήλου… ' (there are many fountains in the city of Korinthos, plenty of water transferred from Stymfalia by Emperor Hadrian).

Based on investigations of Lolos (1997, 2010) who has traced the entire course of the aqueduct and described the surviving remains in detail, the Hadrian aqueduct with a length of 84–85 km was constructed by the Emperor Hadrian during the period 117–138 AD, in order to bring water from Stymfalia basin to Korinthos. In Figure 4.3, the course followed by the Hadrian aqueduct is shown. The aqueduct is still visible at various points between the two sites. It is noted that, after 1900 years the new city of Korinthos meets its water needs for domestic use (3.5×10^6 m^3 per year) from the Stymfalia area via a transfer pipe following the same course (Pana & Voudouris, 2008).

Figure 4.1 Fountain Peirini in Ancient Korinthos. (with permission of K. Voudouris)

Figure 4.2 Fountain Glafki in Ancient Korinthos. (with permission of K. Voudouris)

Figure 4.3 The course of the Hadrian aqueduct (left) and underground gallery (right) (Lolos, 1997, with modifications)

Similar aqueducts have been found in others places in Greece, for example, Athens, Argos, Megara, Samos, Knossos, Nikopolis (Crouch, 1993; Longfellow, 2009) (Figure 4.4). The aqueduct of Athens was constructed by Hadrian (134–140 AD). The Hadrian's aqueduct of Athens consisted of an underground tunnel with length 25 km and was built manually through solid rocks by slaves using simple tools. The aqueduct transferred water by gravity from the mountain of Parnitha (Tatoi area) to a final tank at the hill of Lycabettus and collected water from various sources (shallow wells or springs) along the course.

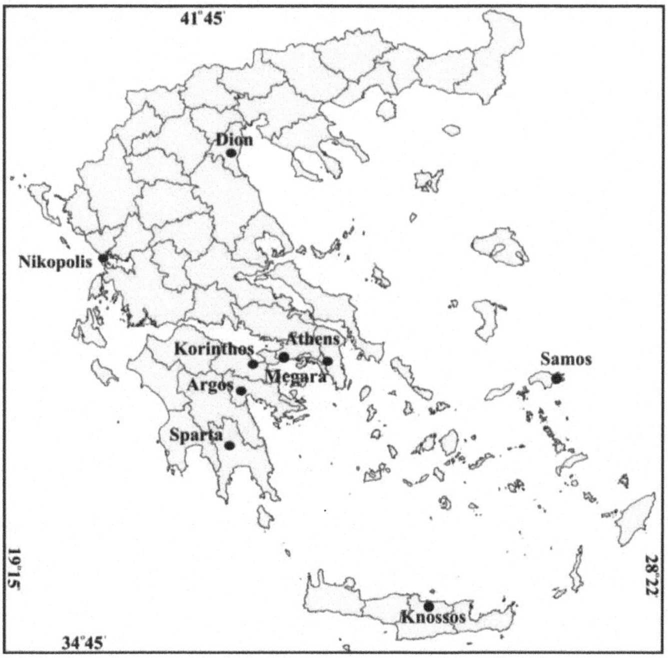

Figure 4.4 Sites of aqueducts in ancient Greece

Another example is the case of Dion, South Greece (Karadedos, 2000). In order to meet the needs for domestic use, water from a distance of 6 km was transferred by an open conduit. The main tank had a capacity 100 m^3 and consisted of three water distribution conduits (earthen) for different uses (public baths, private houses and public fountains). From the aforementioned examples, it is concluded that ancient people had developed a technology (including aqueducts, tunnels, galleries, dug wells, tanks, small dams and so on), for transporting water long distances in order to supply water to cities.

4.3.2 Qanats in Greece

The greatest achievement in groundwater exploitation by ancient peoples was in the construction of long galleries or qanats, which collected water from alluvial deposits and soft sedimentary rocks. Qanats or Kariz, which means chain of wells, are a most remarkable technology of water supply (Figure 4.5). They are gently sloping, artificially constructed underground galleries, which bring groundwater from the mountainous area to the lowlands where water is needed, sometimes many kilometres away (Weingartner, 2007). Many qanats are still in use stretching from China in the east to Morocco in the west, and even to the Americas (Keshtkar *et al.* 2005).

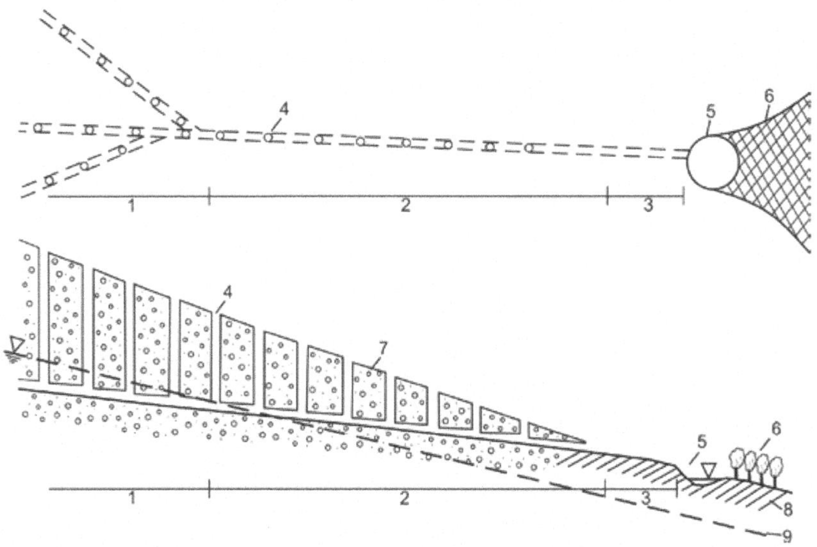

Figure 4.5 Simplified cross-section of a qanat. *Explanation*: 1 = infiltration part of the tunnel, 2 = water conveyance part, 3 = open channel, 4 = vertical shafts, 5 = small storage pond, 6 = irrigation area, 7 = sand and gravel, 8 = layers of soil, 9 = groundwater surface (Adapted from Woulff, 1968 with modifications)

At the highest point of the mountain the initial (mother) well is constructed in order to ascertain the presence of groundwater. A windlass is set up at the ground surface and the excavated soil is hauled up in buckets (Figure 4.6). Every 10–30 m, vertical wells (shafts) are dug for the removal of soil and ventilation of the tunnel (Vavliakis, 1989). Furthermore, the shafts enable access for repair-works. Then, a tunnel is constructed downstream with a height of 1.2–2 m and a width of 0.8–1.5 m. Qanat tunnels were hand-dug, just large enough to fit the person doing the digging. In unstable soils, reinforcing rings are installed in the tunnel to prevent cave-ins (Keshtkar & Salajegheh, 2005). The water moves along the bottom of the tunnel. The maximum length is approximately 40–50 km. The first qanats were constructed in Persia (ancient Iran) and then spread towards Arabian Peninsula and Egypt (Lightfood, 2000). One extensive qanat built about 500 BC in Egypt is said to have irrigated 4700 km^2 of fertile land west of the Nile (Davis & DeWiest, 1964). In Iran there are still about 20,000 qanats in operation. East of Iran the qanats expanded towards Pakistan, Afghanistan and along the oases of Silk Road to China (Weingartner, 2007). The longest qanat near Zarand, Iran, is 29 km long with a mother well depth of 96 m and with 966 shafts along its length (Beaumont, 1971).

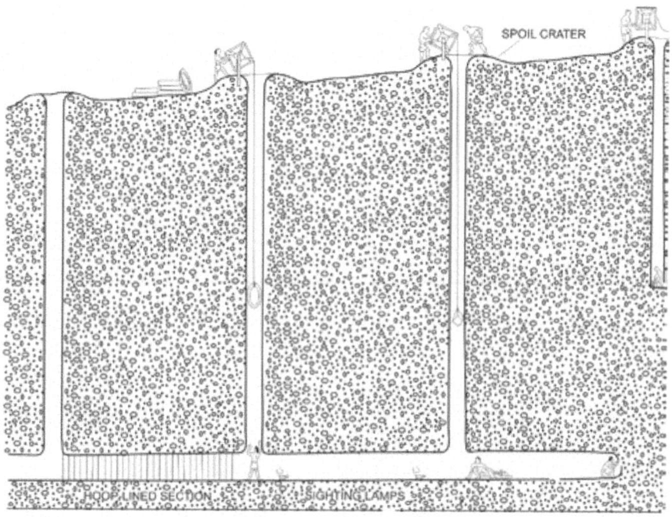

Figure 4.6 Constructing a qanat (Woulff, 1968 with modifications)

The expansion of Islam led to diffusion of qanats in Mediterranean countries (Spain, Italy, Cyprus and so on). In north and central Europe, water supply works constructed with qanat technology can be found in Germany (Trier, Saar-Mosel), Luxembourg, and the Czech Republic (Weingartner, 2007). In Figure 4.7, sites of qanats evidence in the Mediterranean are shown.

The technique of tunnelling had been used during the prehistoric period in ancient Greece. In the Kopais basin (Viotia, Greece), a tunnel of length 2.2 km with 16 vertical wells was discovered. This tunnel was constructed during 1450 BC by Minyes in order to drain the basin, which would flood during wet periods. The existence of qanats of the classical period has been recorded in Greece, for example, Samos and after the occupation by Ottomans in Serres, Chortiati (Thessaloniki), Thrace and so on. The largest tunnel of Greece is the tunnel of Samos (1 km), but the largest in total length (tunnel and transfer conduit) is that of Chortiati, Thessaloniki (20 km) with a water discharge of 80 m^3/h.

It is pointed out that there are two types of pipelines for transporting water: open and closed (Figure 4.8). The open conduits consist of a canal built with stones and covered with mortar to reduce the losses. They are gently sloping and the water flows by gravity. The closed conduits consist of linked earthen pipes and the water flows with pressure.

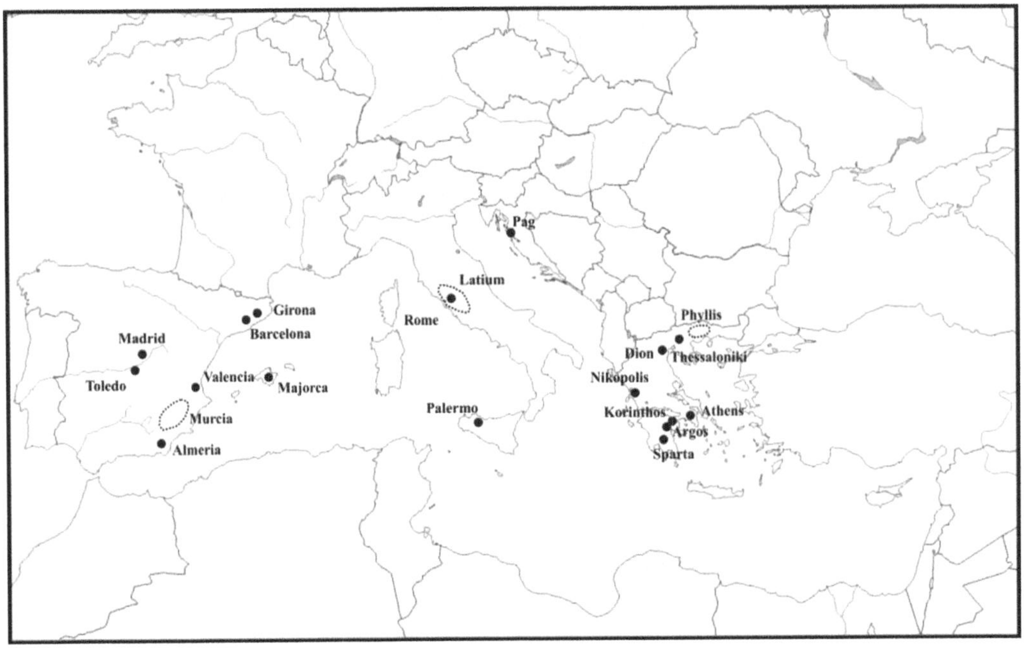

Figure 4.7 Sites of qanats evidence in Mediterranean countries (Weingartner, 2007 with modifications)

Figure 4.8 Open conduits in Cologne Germany (left). Closed conduits in Pergamos (right) (with permission of G. Karadedos, 2000)

In Figure 4.9 a representative qanat in the Fyllida area (Serres, north Greece) is shown. The initial (mother) well and the sequence of wells (shafts) were dug in alluvial sediments and the bottom of the tunnel at the contact of permeable and impermeable deposits (metamorphic rocks, marls). Consequently,

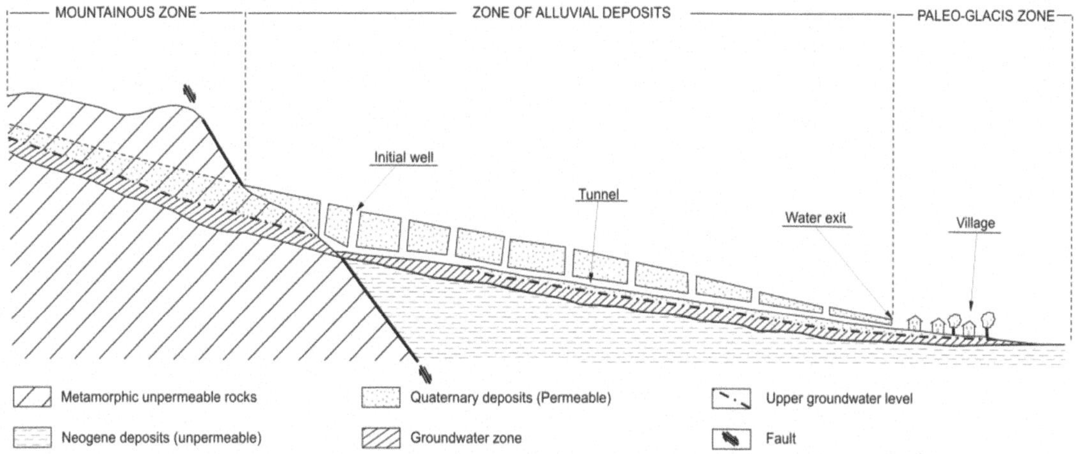

Figure 4.9 Sketch of a qanat in the Fyllida area, Serres, Greece (Vavliakis, 1989 with modifications)

Figure 4.10 Left: The initial well (shaft) of a qanat in Nea Zichni (with permission of K. Voudouris). Right: The tunnel of a qanat at Chortiati, Thessaloniki constructed in the 18th century (with permission of K. Voudouris)

the phreatic aquifer of the area is exploited. The initial well of the qanat at Nea Zichni (Serres) is shown in Figure 4.10. The tunnel of this qanat had a length of 1 km and is used today to cover the water demands of the village.

4.3.3 Tunnel of Samos

The oldest aqueduct in Greece is the tunnel of Samos, which is one of the greatest engineering achievements of ancient times. This tunnel was constructed by Eupalinos during the sixth century BC (*ca.* 2500 years ago), based on Persian technology. Herodotus mentions and describes the existence of the tunnel. It is a water tunnel 1036 m long that was excavated through mount Castro on the Greek island of Samos in the sixth century BC.

Delivering fresh water to growing populations has been an ongoing problem since ancient times. Therefore, Polycrates, the tyrant of Samos, engaged engineer Eupalinos of Megara to build a tunnel that would provide his city with a secure water supply. The goal was to transfer water into the town from a spring that existed at a village, Agiades, northwest from the city (Figure 4.11). The tunnel that was built for this purpose was dug through limestone by two separate teams advancing in a straight line from both sides of the mountain (Apostol, 2004).

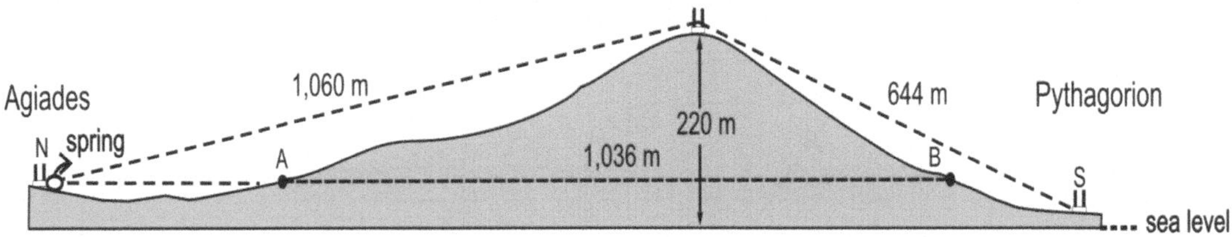

Figure 4.11 Longitudinal section of tunnel of Eupalinos (Adapted from Apostol, 2004, with modifications)

Furthermore, an inner, sloping, rectangular channel was excavated adjacent to the tunnel floor, along its eastern edge. Also, several shafts were built along the tunnel, for inspection and for helping with the excavation. In the system created, the water was brought from its source in Agiades to the northern mouth of the tunnel by an underground conduit that followed an 850 m sinuous course along the contours of the valley, passing under three creek beds en route. Once inside the tunnel, whose floor was level, the water flowed in the inner sloping channel. The water channel then left the tunnel a few metres north of the southern entrance and headed east in an underground conduit leading to the ancient city (Apostol, 2004). The tunnel, an outstanding engineering achievement representing the peak of ancient hydraulic technology, which was built 2500 years ago, has been preserved through time. Today, the tunnel is a popular tourist attraction and can be visited through its southern entrance (Figure 4.12). Koutsoyiannis *et al.* (2008) describe analytically the reasons for which the tunnel of Eupalinos is both remarkable and admirable.

Figure 4.12 The tunnel of Eupalinos in Samos (with permission of K. Voudouris)

4.4 DISCUSSION AND CONCLUSIONS

The socioeconomic development of a region is usually associated with the availability of water resources. Mediterranean countries are characterised by semi-arid climate conditions. After the last glacial period ended 12,000 years ago, climate changes were small and variable from region to region. Many areas today are facing the same problem of water shortage as they did in the past. An integrated strategy for water resources management, aimed at the mitigation of the impacts of water shortage, is therefore necessary at a local and regional scale. The ancient technologies and water management practices are a useful tool for current-day engineers.

At the earliest stage, people performed all their activities near rivers and streams and surface water was the main source of water supply. However, the surface water resources were influenced by the seasonal precipitation. For this reason people started to explore groundwater. Springs are the natural emergence of groundwater. Thus, many springs were used directly for water supply.

The ancient peoples devised wonderful hydraulic works, including tunnels, wells, aqueducts and so on, to supply the cities with water. These works are located in areas which today are faced with water supply problems. The Roman period is characterised by the construction of admirable aqueducts that carried water to the cities from a long distance. In the Byzantine period, the construction of aqueducts was abandoned and tanks and cisterns were constructed in order to collect rain water. The Ottoman period is characterised by the construction of qanats and the exploitation of springs. In the case of extended dry periods, the water supply was assisted by the exploitation of groundwater from shallow aquifers.

The greatest achievement in groundwater exploitation by ancient peoples in the Near and Middle East was in the construction of qanats. Many qanats in Greece were constructed during the Ottoman period; the largest in total length (tunnel and transfer conduit) is that of Chortiati at 20 km. The qanats consist of one underground tunnel and a sequence of shafts (wells) that convey water from shallow aquifers in highlands by gravity to lowlands and some of them are still in use today. In addition, the aqueducts consisting of channels, tunnels and water bridges, as well as small dams, cisterns and so on, are admirable hydraulic works of ancient people to transfer and store water. The oldest aqueduct in Greece is the tunnel of Eupalinos in Samos, which is one of the greatest engineering achievements of ancient times, representing the peak of ancient hydraulic technology.

Given the current climatic conditions, the period of dryness in Mediterranean countries ranges between 3 and 6 months. Water consumption has doubled during the last four decades. Agricultural demand for irrigation water is a major share of the total water demand in Mediterranean countries (60–85%). It should be mentioned that during the dry years the abstraction rate for agricultural purposes increases (up to 25%), because of the prolonged irrigation period, the land under irrigation and the amount of water applied per irrigated area (Chartzoulakis *et al.* 2001). Demand for domestic water also increases during drier periods. Droughts also contribute to the quality deterioration of surface water because they cause a decline in river and stream discharge (EEA, 2007). It is pointed out that coastal aquifer systems in Mediterranean countries are already threatened by seawater intrusion due to human activities and geological conditions.

The aforementioned works demonstrate that the ancient people in Mediterranean countries had an outstanding engineering knowledge of water supply. The study of these works will help solve the current water problem in many areas. The horizontal works (tunnels, galleries, and qanats) have the advantage of avoiding seawater intrusion problems in coastal areas. Furthermore, these water supply works have the advantages of saving energy (no pumping water) and reducing water losses from evaporation. Finally, these works are effective for water harvesting in arid or semi-arid areas of Mediterranean countries.

REFERENCES

Avgoloupis I. and Katsifarakis K. (2009). References on hydrology and hydraulics in the work of herodotus. *Hydrotechnika*, **18–19**, 21–34 (in Greek).

Apostol T. (2004). The tunnel of Samos. *Engineering and Science*, **1**, 31–40.

Bar-Matthews M. and Ayalon A. (1997). Late Quaternary paleoclimate in the eastern Mediterranean region from stable isotope analysis of speleothems at Soreq cave, Israel. *Quaternary Research*, **47**, 155–168.

Beaumont P. (1971). Qanat systems in Iran. *Bull. Int. Assoc. Sci. Hydrology*, **16**, 39–50.

Chartzoulakis K. S., Paranychianakis N. V. and Angelakis A. N. (2001). Water resources management in the island of Crete, Greece, with emphasis on the agricultural use. *Water Policy*, **3**, 193–205.

Crouch D. A. (1993). Water management in ancient Greek cities. Oxford University Press, Oxford, UK.

Daskalaki P. and Voudouris K. (2007). Groundwater quality in porous aquifers in Greece: A synoptic review. *Environmental Geology*, **54**(3), 505–513.

Davis S. N. and DeWiest R. (1966). Hydrogeology. John Wiley & Sons, p. 463.

De Feo G., Mays L. W. and Angelakis A. N. (2011). Water and wastewater management technologies in the ancient Greek and Roman civilizations. *Treatise on Water Science*, Chapter 4.01, 3–22.

De Menocal P. B. (2001). Cultural responses to climate change during the late Holocene. *Science*, **292**, 667–673.

European Environmental Agency (EEA) (2007). Climate changes and water adaptation issues. EEA Technical Report, No 2, Copenhagen.

Haut B. and Viviers D. (2007). Analysis of the water supply system of the city of Apamea, using computational fluid dynamics. Hydraulic system in the north-eastern area of the city, in the Byzantine period. *Journal of Archaeological Science*, **34**, 415–427.

Karadedos G. (2000). Hydraulic knowledge and technology: The case study of Dion. In: Water Relationships: Water as a Source of Life in Antiquity, K. Soueref (ed.), pp. 51–67 (in Greek).

Karas J. (1997). Climate Change and The Mediterranean Region. Report Prepared for Greenpeace International.

Keshtkar A. R. and Salajegheh A. (2005). Qanat: Traditional Method for Water Harvesting in Arid and Semi-arid Regions of Iran. *Proc. of 5th International Symposium on Management of Aquifer Recharge*. Berlin, Germany, June 2005, 887–894.

Konstantinou G. and Konstantinou K. A. (1999). Groundwater Exploitation in Cyprus Throughout its History. *Proc. of 5th Hydrogeological Conference*, Nicosia, Cyprus, 1–18 (in Greek).

Koutsoyiannis D. and Angelakis A. N. (2004). Agricultural hydraulic works in ancient Greece. In: The Encyclopedia of Water Science, B. A. Stewart and T. Howell (eds), Dekker, New York, pp. 415–417.

Koutsoyiannis D., Zarkadoulas N., Angelakis A. N. and Tchobanoglous G. (2008). Urban water management in Ancient Greece: Legacies and Lessons. *Journal of Water Resources Planning and Management*, **134**, 45–54.

Lightfood Dale R. (2000). The origin and diffusion of qanats in Arabia; new evidence from the northern and southern Peninsula. *The Geographical Journal*, **166**, Part 3, 215–226.

Lolos Y. (1997). The Hadrian aqueduct of Corinth. *Hesperia*, **66**(2), 271–314.

Lolos Y. (2010). The Hadrianic Aqueduct of Corinth and the Transportation of Water in Roman Times. Piraeus Bank, Group, Cultural Foundation, Athens, Greece, pp. 48.

Longfellow B. (2009). The legacy of Hadrian; Roman monumental civic fountains in Greece. In: Nature and Function of Water, Baths, Bathing and Hygiene from Antiquity through the Renaissance, C. Koss and A. Scott (eds), Publ. Brill NV, Leiden, The Netherlands, pp. 211–232.

Pana E. and Voudouris K. (2008). Water supplies in Korinthos town from the past until now. *Bulletin of the Geological Society of Greece Vol. XXXXI/I, 167–174. Proc. of the Conference "Land and Sea of Korinthos", Korinthos, Greece, May 2008.*

Papademos D. L. (1975). The hydraulic works in ancient Greece. TEE (ed.) Vol. C, Athens, Greece (in Greek).

Suppan P., Kunstmann H., Heckl A. and Rimmer A. (2008). Impact of climate change on water availability in the Near East. In: Climatic Changes and Water Resources in the Middle East and North Africa, Zereini Fathi and Höltz Heinz (eds), Springer, Dordrecht, The Netherland, pp. 47–58.

Vavliakis E. (1989). Qanat and other high-tech water systems in ancient Greece. In: "Water relationships: Water as a Source of Life in Antiquity", K. Soueref (ed.), pp. 47–50 (in Greek).

Voudouris K. (2009). Assessing groundwater pollution risk in Sarigkiol basin, NW Greece. In: River Pollution Research Progress, M. Gallo and M. Herrari (eds), Chapter 7, Nova Science Publishers Inc., New York, USA, pp. 265–281.

Weingartner H. (2007). Water supply by qanats: A contribution to water shortage in Mediterranean areas. *Proc. of the 10th International Conference on Environmental Science and Technology*, Kos island, Greece, 5–7 September 2007, Vol. A, 1555–1561.

Woulff H. E. (1968). The Qanats of Iran. *Scientific American.* http://users.bart.nl/~leenders/txt/qanats.html (Last accessed 10 October 2011).

Zhou Y., Zwahlen F. and Wang Y. (2011). The ancient Chinese notes on Hydrogeology. *Hydrogeology Journal*, **19**, 1103–1114.

Chapter 5

Water and water supply technologies in ancient Iran

Seyed Ali Mahmoudian and Seyed Navid Mahmoudian

5.1 INTRODUCTION

As a vital element, water has played a focal role from the very beginning of life and humans have always either tried to settle close to a watercourse or have attempted to bring water to their communities by some means or other. These moves required special efforts in each era as appropriate for each particular climatic condition.

Both the passage of time and the erosive quality of water have erased all traces of water supplies in long gone eras, but a look at history suffices to show us how our hard working ancestors endured difficulties to be able to divert water from its source, or to make it flow underground or, even by using primitive tools and procedures for digging tunnels, to convey it for a long distance to a specified point as we do today after thousands of years using the same information and similar tools and equipment.

During each period, in the area of the Iranian plateau with its diverse climates and conditions, humans have tried to access water by means of facilities available at their disposal. Those settled on river banks or along coastlines, solved their problems in their own special ways and met their demands using river or sea water, which was free of all pollutions. Furthermore, at the times of flooding and sea storms, they conducted excess water to settling ditches, which constituted the early initiatives for digging wells along river beds and the sea. Then they became aware of springs along the rivers and groundwaters at foothills and valleys and therefore moved inland to form small communities along these newly found resources. These societies grew in number and the need for water intensified to the extent that water from springs could no longer satisfy their needs. They were therefore forced to find the right solutions for survival.

At times, their disputes would lead to the migration of a group or tribe in search of new springs and thus the sequences and movements continued on. Sometimes the attacks looters seeking to dominate their environmental facilities would upset their tranquillity. After the killings and plunders, the defeated survivors took refuge in hills and deserts to save their lives. To be able to survive in the new environment, they had to exert an even greater effort in finding water. The harsh desert environments forced them to find solutions for supplying water and to focus all their attentions and ingenuity on conveying water from long distances. Therefore they brought surface waters by means of rivulets and canals and the groundwaters in tunnels and qanats to their settlements (Hashami, 2010).

The human societies gradually underwent intellectual evolution, while disputes and feuds were manifested differently and plunders appeared in different forms. Small communities also expanded to take the shape of civil societies and towns and cities. Naturally, during any urban expansion, water is the main concern and requires proper approaches. These problems that challenged our ancestors in meeting their basic demands, have continued on for thousands of years to the present times when actions to satisfy these needs have found new urgency and intensity. With the advent of the industrial era, and taking into account its relatively low age, we have witnessed overwhelming changes on earth. The speed of this momentum increases day by day and it can easily be claimed that in this context, evolutions occur every instant.

Cities have expanded and facilities have increased. In advanced countries, sanitation moves in step with social developments and population explosion manifests itself on a different level. Water is at the centre of all these changes and demands attention. The planet's potentials are limited and it is incapable of meeting all these basic needs. The passage through this period is extremely difficult and we must look for a sustainable solution.

Ancient Iranians settled in favourable regions and gradually established the foundations of today's civilization. However, they had to be ready to move in search of water. Despite all attributes described in the books of past prophets and in particular in the heavenly book of the Muslims, today, this divine element is seriously threatened.

The value of water, which is both life sustaining and the source of human purity and cleanliness, has never been ignored by humans, who have always considered it with reverence and respect: Anahita, the Goddess of Water[1] of the Iranians, also known as Ishtar before the Prophet Zoroaster, enjoyed a high status in pagan beliefs; while in the Koran, water is described as the essence and the life source of every creature.

Revering and valuing water and belief in its sanctity and purity is not limited to a specific region on earth. If the life sustaining powers of this pure element as described in all religions were to be studied, one would certainly reach a common conclusion.

It should be noted that the first part of the name of "Moses, the Great Israelite Prophet" in Hebrew is equivalent to the "*Ma*'" in Arabic or water in English. We only need to study the birth and life of this Prophet, to revise the Holy Koran and to read the book "The Son of the Nile" to realise this fact (Hashami, 2010).

5.2 PLATEAU OF IRAN, THE CRADLE OF PRESENT CIVILIZATION

The Plateau of Iran, one of the huge plateaus of the Asian continent, extends from the Persian Gulf on the Oman Sea in the south, to the Caspian Sea in the north, to the western part of the Pamir Mountains and the Arvan Rud basin in the west and to the Dorya (Jeyhoon) basin in the east. The area of this plateau is about 2,600,000 km^2. Since ancient times, this vast territory has possessed its independent language and dialect and a unique and rich culture. Due to its favourable climate, its diligent inhabitants, and central geographical position in the ancient world, and coupled with its strong links both to the East and the west, it has always played a determining role in the culture and civilization of the world.

According to Will Durant (1885–1981) in his book titled "Our Oriental Heritage: The story of Civilization" since the beginning of recorded history, the Middle East has contributed to at least half of

[1]The goddesses of Water in the antiquity: Anahita in Iran, Ishtar and Melita in Assyria, Ishtar in Babylon, Estradote in Phoenicia, Draktor in Caledonia, Poseidon in Greece and Anahid and Estakhik in Armenia. They enjoyed a greater respect than other deities among the people, and their statues and temples were constructed in places where water was abounded.

all evolutions which have taken place in human society and its culture and affairs. He noted that one may move one's finger over the map of Iran, from the river Shat in the eastern corner of the Persian Gulf to the city of al-Emaareh in contemporary Iran and then eastward to find the contemporary city of Shoosh at the site of an ancient land called "Eelaam" (Hashami, 2010). Prof. Ghirshman (1895–1979) said that the oldest urban settlement in the plains has been identified as Sialk, near the city of Kashan, just south of Tehran (Hashami, 2010). Prof. Iliffe says: *'there is no other eastern nation with so much influence on the composition of the present day civilization and human thought'* (Hashami, 2010). Albert Von Le Coq (1860–1930) in his book, "The Buried Treasures in Chinese Turkistan", also describes Iranians as the first civilized nation in Asia (Hashami, 2010). Prof. Ghirshman says (in Iran, from the Beginning to the Islamic Era): *"Iranians were the first people who established the global sovereignty together with promotion and the expansion of courage in justice and freedom"* (Hashami, 2010). A geographical map of modern day Iran is shown in Figure 5.1.

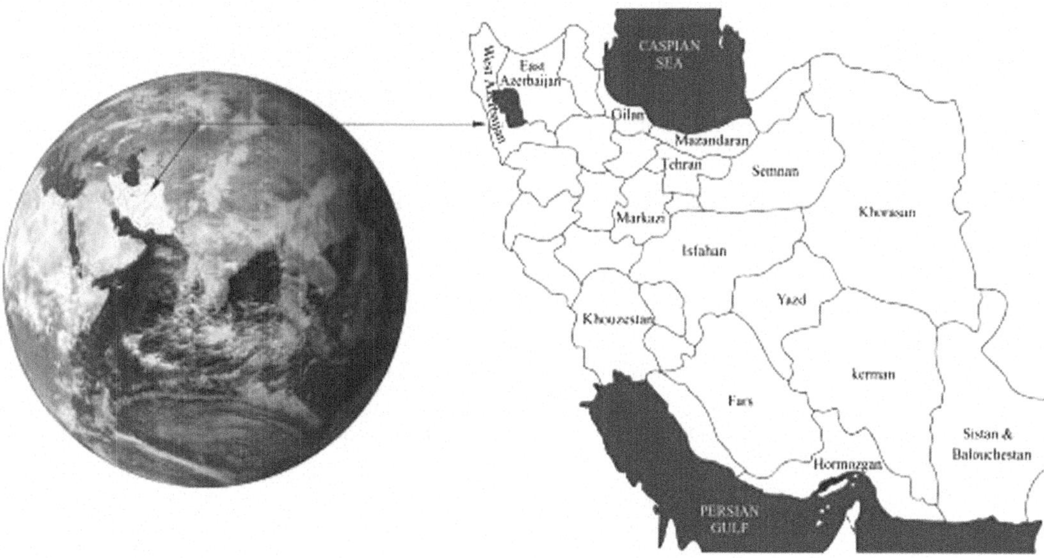

Figure 5.1 Map of Iran

5.3 HISTORICAL ASPECTS OF WATER AND WATER SUPPLY IN IRAN

5.3.1 Water in ancient Iran

Iranians became monotheists after recognising the regular movements of the stars, and the rotation of days and nights and the sun and the moon, and, thus the concept of theology and monotheism took root for the first time in Iran. According to the German Franz Anne Tim, 'Iranians called God "Izad" and attributed the human fears of the surroundings, particularly the natural disasters to Ahriman. They did not worship fire but held light, brightness and fire in high esteem.' Six to seven thousand years have passed since the Iranians first began believing in God, while this belief was established much later in India and Egypt. Therefore, from this perspective the Iranians have been the trend setters for other nations.

Thousands of years before the Roman civilization took shape, each Persian home had a well. Later on, when baths became common, in addition to offering a place for washing and swimming, they were equipped with eateries and bars as well as libraries.

In the ancient territory of Iran, water was the symbol of purity, because the Iranians had a great belief in maintaining the cleanliness of water. Before then, the Iranians used to discharge wastewater in wells. The French archaeologist Pierre Benoit (1906–1987) says, '*They had created sewers in coastal areas where they could not use wells*' (Hashami, 2010).

Iranians pioneered in urbanisation, as the first cities were built in Iran. According to Herodotus the father of historians, Cyrus the Great began building cities in Iran, where grade 1 roads measured 80 Zera or 40 m in width. The first exemplary city was built by Cyrus the Great near a village called Mashhad Morghab. The name given to this city was Pazargades, which was changed later on to Pazargad or Passargadae. Currently the ruins of the imperial palace measuring 2500 m in length and 200 to 800 m in width are all that remain. The Iranian engineers and architects created a city, which was quite similar to modern cities in design, including water supply and sewerage.

Esdras, the Jewish priest and scientist who lived in the fifth century BC, wrote that Susa had 10 neighbourhoods, and each neighbourhood had five north-southerly and five east-westerly streets. The city's structure had been engineered in a manner that allowed one to see the end of the opposite street from any road. There were three- and five-storey buildings in the city. He further writes, '*Susa has sewers and water is taken to upper floors of buildings by conduits*[2] (Hashami, 2010).

5.3.2 The establishment of water administration during the Achaemenid era

One of the greatest deeds of Darius the Great was the creation of "Water Organisation". The head of the organisation was called "Ao-Tar" or "Water Master" and he controlled the qanats, dams, rivers, etc. This great Achaemenid King initiated the construction of dams to prevent drought in India. Moreover, excavation of a canal to connect the Red Sea to the Mediterranean in Egypt figures among his inventions. In gratitude towards Darius's actions, the Egyptians called him a Pharaoh, since by digging qanats and other initiatives he had supplied the south of Egypt with irrigation water.

To be able to pass the river Sind, Darius dispatched Estilakis and a delegation to India to report on the regional facilities for construction of a dam. The latter presented a report, equal in quality to those prepared by modern day geography and hydrology experts. Mandorcles, another engineer of the Darius period, constructed a bridge over Begas Bosporus to allow the army to pass over. Bolts and nuts were used to fix the boards in its construction.

Darius ordered the reconstruction of the city of Sarod destroyed by the Greeks. Mendrokles presented to Darius a plan of the city, which was to be built over an area of 50 × 50 Ostad[3]. Piped water and sewers were considered in the plan.

Tridot or Tirdad, whose foresight rivalled that of Cyrus or Darius built a large city called Dara or Darium or Darius in the year 211 BC near the present day Abivard, to preserve the name of Darius the Great for prosperity. In this city water flowed in closed conduits and there were provisions for sewers. All houses

[2]Susa was an Achaemenid capital city during the rein of Cyrus. Before then and during the third millennium BC the Burnt City, which expanded over an area of 150 hectares and which is to be found at a distance of 56 km from Zabol, had a wastewater system. Quoting Iran Daily (January 2003), the city had streets 3 m in width and houses with gardens measuring 150 m^2. Units consisted of 6 to 10 chambers and there were even two-storey buildings. The large terracotta pipes found in the city are indications of the existence of a sewerage system in the city.

[3]Each Ostad is equivalent to 200 m.

were equipped with heaters and central heating, which brought steam from a hot water tank to the rooms via a piping system.

In the year 326 AD when the city of Susa was destroyed during an earthquake, Shapur ordered it to be rebuilt with all the urban facilities, including water flowing in every house, a sewer system and a laundry in each neighbourhood (Hashami, 2010). Advanced irrigation systems existed during Anushirvan's era. He facilitated the progress and development of these systems as well as the construction of dams, barriers and other relevant facilities, thereby taking effective steps to increase agricultural production.

Dam construction and qanat or tunnel excavation are among the inventions of Iranians. It is written of Shapur I in the Necropolis tabloid that Shapur constructed dams over rivers using funds from his treasury to save farmers from drought. Shapur has said, '*In Susa (modern day Khuzestan) I built so many dams to relieve farmers of a need for water.*'

The Sassanid kings took loans from Bankehs (Banks) and built dams on rivers, even class 2 ones. One must not consider the multi-purpose dams built in recent centuries as the outcome of modern engineers' inventions. Two-thousand years ago in Iran, dams were built that had dual and sometimes even triple purposes. A number of bridges built during Shapur I's reign had dual utility, meaning that the bridge's foundations were constructed in such a manner as to enable collection of water, while the main structure joined the two banks of the river. There were also triple-purpose water structures in Iran. The most outstanding example in subsequent eras is the Amir Barrier in Fars, since in addition to joining the two river banks together; it also used the two streams on either side of the rivers for water supply and irrigation. The third purpose of this water monument was to tap into water energy to turn the wheels of a millstone.

Grot Van Roken has reported the existence of a stream branching out of the Karoon River to the west of Shushtar, which flows in parallel to the Shahteiteh River. These types of special purpose qanats are called Safteh (Hashami, 2010). In its height of splendour and power, Rome lacked even latrines, whereas during the Sassanid era, wastewater systems were quite common. After the Sassanids, this trend was forgotten except in a number of urban areas such as Esfahan, a number of cities in the provinces of Guilan and Mazandaran and in Astarabad. Instead latrine wells were dug for this purpose. This is quite significant when one considers that not so long ago Versailles, the symbol of the French royal glory, did not have a single well. The French learnt the method of using toilet wells and bath tubs from Mohammad Reza Beg, the Iranian Ambassador to the court of Louis XIV (Hashami, 2010).

In any case, latrine wells existed in Iran well before. Such wells dating back to seven thousand and six thousand years ago were respectively found in Sialk Tepeh of Kashan and Hessar Tepeh of Damghan. Iranians, whose monotheism dates back to over six thousand year ago, believed that earth and water must be kept clean. They discharged water used in washing the body in wells. According to Kensias, the Achaemenids used to wash dead bodies, but not in a flowing water body as they held that it would be a profanity to pollute it (Hashami, 2010).

5.3.3 Water flow measurement in ancient Iran

Due to the expanse of the country and the variety of the water resources, different methods were developed in various regions to measure water flow. The smaller the volume of water, the more accurate was the measuring method. The "water distribution measurement device" consisted of a flat copper tub, a perforated cup and the stone (an old measuring scale). At periods when water was in abundance and there were no conflicts between the farmers, these devices, which were also known as water distributors, were used on rivers to measure and calibrate the flow.

An ancient Iranian scientist, Sheikhbahayee, designed the best way to measure water drawn from the Zayandeheroud River to be distributed for agriculture. The written process called the Sheikhbahayee's roll (Toumar-e- Sheikh Bahayee) was used for ending conflicts among farmers who lived close to the Zayanderoud River. Moreover, Amirkabir, the renowned Qajjar period's Prime Minister transferred water from the Karaj River (Karaj being a city located near Tehran) to Tehran (capital of Iran) and established a roll to distribute it among farmers. This roll is shown in Figure 5.2 (Garoosi, 2003).

Figure 5.2 Picture of a roll to distribute water from Karaj River drawn by Amirkabir

5.3.4 Water from the point of view of Islam

"And we created from water all things living"

The monotheist religions, the heavenly books and the adages, citations and traditions of people all underlined the importance, value and the status of water. Each religion has considered water as the source of life and originator of existence. The Koran, the religious book of Islam, pays special attention to this divine creation. The Heavenly book of Islam mentions water under different headings. In addition to Koran verses, numerous adages are left behind from the Holy Prophet and the prominent religious figures, stressing the importance of water and the need to respect it. The Holy book of Islam mentions water 63 times, pointing either to the purity of water or to its life giving powers for plants and animals. The Koran describes in detail how water first appeared on earth and the volume of waters available for use. It considers water as the vehicle for cleansing body and soul, and therefore, bans any extravagance and wastage. The religious leaders have also stressed the need to protect water from all pollutions.

Although water bodies have a faculty for self treatment and the components of a flowing water body have self cleaning effects, nonetheless, all recommendations stress the need to maintain the purity of water. Unfortunately, through scientific advances and industrialisation and the increased rate of discharges in rivers and seas, mankind has upset the order and cycle of nature and has brought the world to the brink of crisis (cf. the Koran, which says '*You shall bring upon your demise with your own hand*') (Hashami, 2010).

5.3.5 Water knowledge

Ancient nations had some knowledge of hydrology. In their studies Homer, Plato, Aristotle and centuries later the French Pierre Perole reached a conclusion that snow and rainfall were not enough to ensure the replenishment of rivers and their flows. Consequently, other resources were required to sustain the supply of water from rivers and springs.

Four thousand years ago the Assyrian kings built artificial lakes to transform arid lands around the Tigris and the Euphrates into lush gardens and conduits for irrigation. During the reign of the Assyrian king Senakhrib (705–681 AD) water was drawn from the well by a spool. Iranians were the inventors of qanats and after conquering Iran, the Arabs took Iranian diggers and specialists to North Africa to build similar systems. In his book on Iran, Professor Ghirshman announces that in pre-historical times, irrigation was a manual process. However, during the Achaemenid era there appeared an extensive system of underground networks known as qanats (Hashami, 2010).

The French Albert Chandour points to the fact that in the ancient fertile lands of Medes, Khuzestan and Pars irrigation and drainage systems were common features with an overwhelming number of qanats (Hashami, 2010).

There is a deep well of 4.20 m in diameter on Rahmat (Blessing) Hill[4]. It served as a reservoir, which was filled by floodwater in winter and spring and used in other seasons. Three other wells were also discovered in the year 1940 in the region. The remnants of clay and rocky gable roofs found in the Persepolis indicate that they were installed in the middle of the walls to conduct water from roofs to the underground conduits. Furthermore, there was a large pool measuring 55 m in length and 23 m in width with a depth of 2.15 m discovered in the south-west wing of the Persepolis complex, which was built out of rocks with lead and tar mortar.

Water purification in clay jugs, pots and copper tubs was common practice and these vessels were used to fetch water from spring, qanats, and so on. Some nomadic tribes in our country still maintain this tradition. However, these utensils have gradually left the scene, in favour of their plastic counterpart, to seek a position in museums. Boiling water and keeping it in silver vessels were two treatment processes practised during the Achaemenid period. This was also the method used to treat water in adequate volume for the king's use during his travels (Hashami, 2010).

5.3.6 An overview of water supply methods in ancient Iran

Water has always played a key role in the long history of Iran. Iranians are credited for creation of qanats and the invention of the Persian Wheel, two ancient water supply systems, which are well known in the world. According to Herodotus, the Greek historian, the technique for digging qanats was well documented and practised in the Achaemenid era (550–330 BC), some 2500 years ago. Ruins of reservoirs have been discovered along with water intakes, spillways and outlets and even the sewerage systems dating as far back as the Pre-Achaemenid and Assyrian (1500–600 BC) periods. The archaeological surveys suggest

[4]Rahmat was the hill on the levelled side of which Persepolis was built.

that Iranians enjoyed advanced culture and civilization some 7000 years ago. The civilization in the western part of the Iranian plateau flourished 5000 years ago with the invention of cuneiform writing. Discoveries prove that Iranians were peaceful and ingenious people in the third millennium BC who cultivated land and raised livestock. In general, there have been four methods of water supply in Iran. These were cisterns or water reservoirs, channels, canals, and weirs and dams.

Conventional wells were drilled vertically by means of manual tools and equipment and water was extracted for humans or farm animals. Ancient Iranians are attributed with processes, which resulted in artesian wells, and the means of bringing them under subjugation.

However, they still were unable to fully explore the above mentioned wells. Furthermore, excess precipitation was stored at certain locations known as cisterns or water reservoirs. Examples of these monuments, designed in the form of covered structures filled with water, are abundant in the vicinity of the salt deserts and the hot arid plains in the south of Iran. These were used to meet the water demands of villages and passing caravans.

The underground water channel, the so-called "qanat", was by far the most important method of water supply in many parts of Iran. In view of the scarcity of rivers in many parts of the country and given the scant number of perpetual rivers, approximately 3000 years ago, the ancient Iranians achieved an admirable invention known as "qanat" or "kareez". This highly important and incomparable initiative was later transferred from the Middle East to North Africa, Spain and Sicily for further exploitation. The records on the Iranian qanats in the Achaemenid era can be found in the writings of the Greek historians. Consequently, it can be asserted that this irrigation technique dates back to the Pre-Achaemenid period. There are also historical records of qanats during the Parthian and the Sassanid dynasties.

Water supply and irrigation by means of canals was achieved by erecting canals on the banks of rivers in order to draw certain quantities of water. In ancient Iran a multitude of canals branched off from large rivers such as the Tigris and the Hirmand to be diverted to the arid deserts. The ancient Iranians exerted their utmost effort and precision in canal construction. In conditions where water-channels were weak, strainer bricks were used in conjunction with mortar or hydraulic lime in an attempt to strengthen the floor of the canals.

The history of the ancient dams in Iran dates back 3000 years and shows that ancient Iranians were among the pioneers in dam construction in their efforts to develop water resources. The Darius Dam on the Kor River dates back 2500 years. The Bahman Dam in Shiraz which is over 2 millennia in age and the 1700 year-old Mizan Dam in Shushtar (500 m long with 40 spans) are still in place today (Figure 5.3).

Figure 5.3 Bahman weir dam in Fars province (2000 years old)

The Amir Dam constructed by the Buyids Dynasty, 35 km north of Shiraz is 1000 years old. This three-purpose dam (irrigation, bridge and mill) still exists and functions. During the Safavid Empire (1501–1736) water engineering progressed significantly and many storage and diversion dams and bridges were constructed in Esfahan and Mashhad, some of which still exist today (Hashami, 2010). The multi-purpose dam of Amir is shown in Figure 5.4.

Figure 5.4 Amir multipurpose dam in Fars province (1000 years old)

Two of the very famous and beautiful diversion dams and bridges called Khaju Bridge and diversion dam and Thirty-three Arch Bridge (See VO Se Pol) in the city of Esfahan, are shown in Figures 5.5 and 5.6.

Figure 5.5 Khaju Bridge and diversion dam in Esfahan province (354 years old)

Figure 5.6 Thirty-three Arch bridge in Isfahan province (398 years old)

5.4 WATER AND WATER SUPPLY TECHNIQUES IN ANCIENT IRAN

5.4.1 Qanat, a purely Iranian invention

A Qanat or kareez is a technical method used to provide a reliable supply of water for human consumption or for irrigation in hot, arid and semi-arid climates. The origin of this technology dates back to the ancient Persian Empire. The underlying idea of qanats was to access and transfer groundwater by sinking a series of wells and linking them underground. A scheme of qanats construction is shown in Figure 5.7.

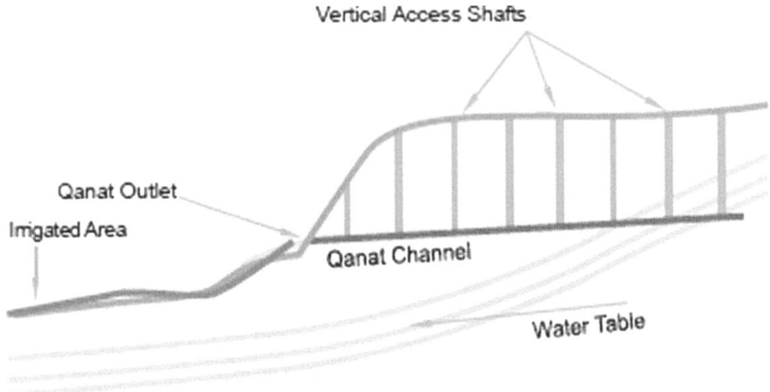

Figure 5.7 Qanat constructions

The construction of qanats by Iranians dates back to 3000 to 4000 years ago. In particular, the oldest and largest known qanat is located in the Iranian city of Gonabad. After some 2700 years it still supplies drinking and agricultural water. The depth of its main well reaches down more than 360 m while measuring 45 km in length (Behnia, 1988).

Persia is currently benefiting from 31000 active qanats producing some 9 billion m^3 of groundwater and accounting for approximately 15% of the aquifer discharge which is annually mined across the country. This huge amount of water is amazingly conveyed just by means of gravity. Since no electric or fossil energy is employed, this system is pollution free and bears no environmental contaminants.

Qanat technology is not only an Iranian system of water mining, but has also overshadowed our economic, social and cultural outlooks for the last three thousand years. However, due to the development of tube wells, some of the qanats are drying up. Even so there are some 50,000 qanats in Iran with close to 75% still in working condition. A view of a qanat in the Province of Yazd seen from ground level is shown in Figure 5.8 (Khorasanizadeh, 2008).

Figure 5.8 A view of a qanat in Yazd

All researchers believe qanats to be an ancient Iranian phenomenon. It is said that Kyaxares the Medes King who lived between the years 624 to 583 BC or approximately 2600 years ago, and who was the founder to the Ekbatan or the modern day Hamadan, found his fame after conquering Armenia and other regions, and this reputation reached many parts of the period. It prompted Ashurbanipal, the Babylonian king to ask for his daughter Nitokis' hand in marriage. This union led to a new method of irrigation in Babel.

R.J. Koldewey (1855–1925) the famous archaeologist discovered it during his survey of Babel. Despite the diversity of irrigation systems in this ancient city, the new technique was both interesting and quite unique. According to the report of Nebasus, the method was created by Ashurbanipal for his Iranian wife to allow her to live in a place which could remind her of her homeland. Studies have shown this method to be none other than the kareez or qanats, created by Iranians 2600 years ago (Hashami, 2010).

The qanat was developed in Iran and was gradually extended to other parts of the world. One can point to examples in the Near East, the Middle East, Africa, and South America on the Atlantic, Mexico and Europe in the Selb region of Germany, most of which originate from and are rooted in ancient Iran. The diffusion of the qanat from Iran to the Mediterranean and the Middle East is shown in Figure 5.9.

At the time of Egypt's occupation by Darius the Great, around 500 BC, Iranians dug qanats in sandstones; however qanats existed in Iran well before this period and during the Medes era. It appears that the French constructed a qanat in Algeria using modern methods, which has an adequate rate of discharge. Certainly the construction of such structures was made possible in Iran, given the climatic conditions and the availability of technical facilities (Hashami, 2010).

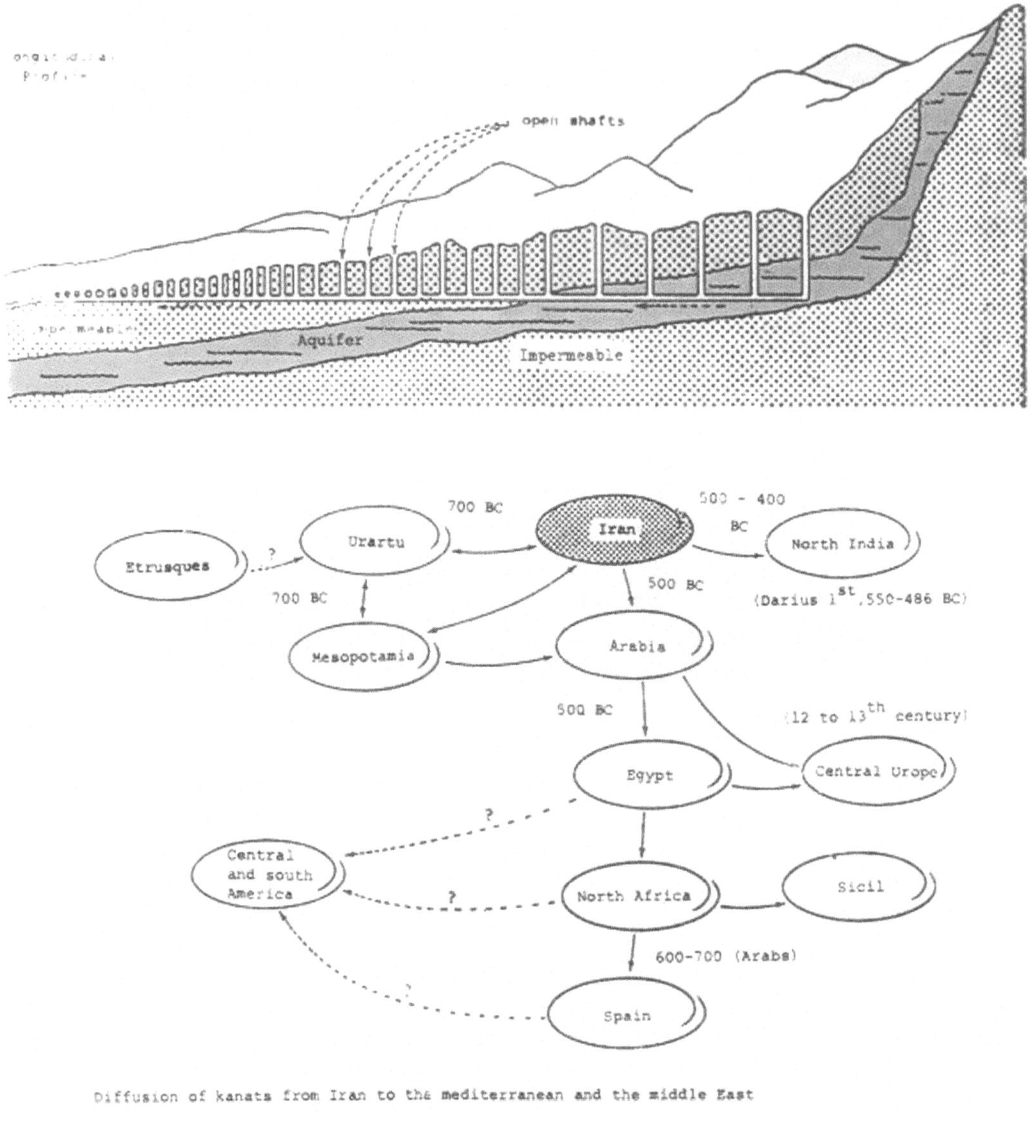

Figure 5.9 Qanat from Iran to the Mediterranean and the Middle East: Longitudinal profile (top) and diffusion schema (bottom)

5.4.2 The 3300-year-old water conveyance and physical treatment system in Chogha Zanbil, another symbol of ancient Iranian structure

The oldest physical water treatment system in the world, in Chogha Zanbil, was built *ca.* 3300 years ago next to the Ziggurat of Chogha Zanbil in the city of Susa (Figure 5.10). The Temple was built in the Middle Elamite period as a place for worship and benediction ceremonies. Assyrian records refer to this place as "Dur-Untash" or "Untash-Gal" city, a famous and glorious city, which was fully capable of competing with Susa as a political centre. The city of Untash-Gal with benediction buildings, paved brick roads, palaces and advanced water supply and irrigation systems, presents certain aspects of the Ilamid Culture, art and civilization. Chogha Zanbil spreads over an area of 4 km^2 and consists of three concentric brick ramparts with well known Chogha Zanbil Zigorat located at their centre. Water supply to temples, palaces and the city of Dur-Untash was among the components designed by the period's architects and technicians (Farhangi, 2004).

Figure 5.10 Chogha Zanbil Temple enlisted in the list of world culture Heritage by UNESCO

Water transmission: The Ilamid kings ordered the construction of a 50 km long canal to divert water from Karkheh River to Dur-Untash City via a treatment basin. This canal passed on the west side of Susa.

Hydraulic installations: Opposite Dur-Untash and enclosed in a wall, the ruins of hydraulic facilities comprising of a large reservoir outside the wall and a small basin on the inside can be observed. These facilities were created to transfer water through a set of small canals from the river, which, after a settlement process, was conducted to a small basin used for distribution of filtered water to the city's inhabitants. The reservoir is 10.70 m long, 7.25 m wide and 4.35 m deep and has an approximate capacity of 350 m^3 (Figure 5.11). Its floor is covered with fired bricks and an extremely hard calcareous material. The two side walls, composed of the aforesaid materials lie on the reservoir floor constructed out of soil materials. The inlet canal was probably built as an open canal all along its length. Facing the city, the fourth wall of the reservoir was also constructed using the same materials. A total of 9 cavities each distanced at intervals of 0.8, 0.15 and 0.8 m, are situated at the lowest part of this wall. The individual intakes were constructed using two layers of fired bricks and one layer of stones. Tar smeared from the floor to the stone layer ensured the sealing. Each of the nine intake gates, which are located beneath the city wall and supply water to the basin, consists of two oblique surfaces. Part of these surfaces contains 1.6 m long vertical sections.

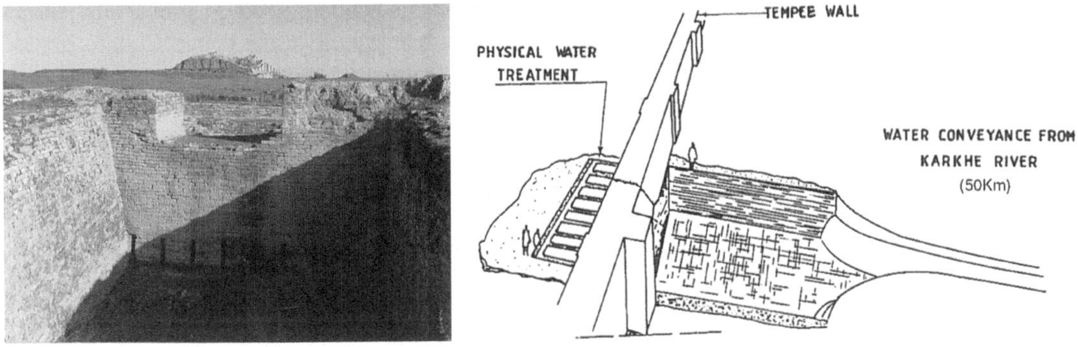

Figure 5.11 Chogha Zanbil Temple physical water treatment system: Today's view (left) and schematic layout (right)

The first diagonal surface, commencing from the large reservoir floor is 0.8 m in length, whereas the second surface measures 3.75 m in length. The difference between the water levels in the reservoir and the upper basin is 1.02 m. the basin is 7.6 m long, 9 m wide and 0.6 m deep, resulting in a holding capacity of 4.3 m^3. The abutment has been constructed using fragments of fired bricks joined with lime. They surround the periphery of the small basin, which has a width of 6 m on three sides, 4 m on the north-east and 1.5 m on the south side. This wall, which is buried 2.5 in the ground, was executed to form a platform around the basin. The operation of these hydraulic systems followed the principle of two connecting vessels. When the reservoir was filled up to the edge of the innermost wall, the filtered water would pass to the small basin through the nine canals (Farhangi, 2004).

5.4.3 Water supply, storage and construction of integrated surface water runoffs and sewerage networks (innovation in town planning dating back 2500 years)

Construction of the Persepolis Complex, one of the monuments listed by UNESCO as a world heritage site, 2500 years ago was an innovation in town planning and civil engineering, a masterpiece which was unprecedented.

5.4.3.1 The process of site selection, construction of the platform and preliminary works

Created on the rocky mount of Rahmat, Persepolis, as indicated by an inscription carved on stone, was built during the reign of Darius the Great (521–486 BC), of the Achaemenian dynasty, by his command. It demonstrated creativity in engineering and innovation in new art. Apart from the beauty and the magnificence of its structures, what doubled its splendour and protected it against the environmental influences such as rain and floods, was the construction of a platform and the preliminary works (Figure 5.12).

By reviewing the texts of the cuneiform inscriptions from the Achaemenian epoch, one realises the ingenuity and the deep thought exercised in every aspect of life. As the Persians had never experienced ruling on such a scale, the more astonishing issue is how they could advance new ideologies and take such great steps in innovation.

Figure 5.12 Persepolis. A view of 100 columns, Apadana Palace and the Garrison–Shiraz

Governing the vast regions under the reign of the Achaemenian dynasty and people of different backgrounds, cultures and customs was not a simple task, and it would not have been possible were it not for these innovations. It is now clear that the policies adopted by the Achaemenians were well organised and astonishing. Another example of the ideologies of this epoch is evidenced by the kind of buildings constructed in Pasargadae at the command of Cyrus the Great. At this place one can see the influences of artists and designers from Lydia, Mesopotamia, Anatolia, Babylon and, even further away, Phoenicia and Egypt. One cannot categorise Pasargadae as a mixture of simple and non-homogeneous pieces of art. It is obvious that the architecture of Pasargadae was following a special new pattern from the very beginning of the construction activities to the very last stone put in place.

The location of Persepolis was chosen because of its proximity to the flourishing town of Matzish. Matzish was in many ways valuable to Darius. It was a big, well known and a trading centre. In its interactions with Babylon and other large trading centres it could easily make Persepolis known beyond the domain of Achaemenian Empire.

In the construction of Persepolis, it was only Matzish that could provide food for the labourers and supply manpower and construction tools. Apart from being the origin of Persians, the location of the site at the foot of mount Rahmat, was unique from a strategic point of view. This place was cut off by a valley, which could easily divert floods from the mountain and prevent any damage being inflicted on Persepolis. Therefore, one can see that in the construction of Persepolis all technical aspects were carefully thought out and deployed. To level the land in the southern part of Rahmat mount, Darius the Great ordered the construction of a huge masonry wall around it, and filled the spaces in between with rubble stones of various sizes and thus created a high and level platform about 8 to 18 m higher that of the Marv-Dasht Plain. The same idea was also used in the construction of Apadana in Susa, with the exception of using earth-fill instead of rock-fill material.

In Persepolis, most of the efforts were concentrated on the excavation of the sloping ground. Darius himself, in an inscription carved on stone found in Susa, declares how the Apadana platform was constructed. The selected site was excavated to a depth where the rock surfaced out. Then sand and gravel were brought in to fill the excavated area. The stone aggregate was compacted.

Persepolis is located on a ridge in the southern part of mount Rahmat, sloping gently towards the west (Marv-Dasht plain). The slope and height of this ridge played an important role in the size and height of Persepolis. The foundation of the palaces were levelled in some places by cutting out parts of the mount and the rest by erecting large masonry walls and filling the spaces with rubble stones of different sizes.

The outer high walls of Persepolis were more or less constructed in alignment with this ridge. Therefore, the land where Persepolis is situated is a polygon with lengths of 450 m in the western, 300 m in the northern, 430 m in the eastern and 290 m in the southern directions. Its lengths along the northern-southern and eastern-western alignments are 428 m and 300 m respectively (Farhangi, 2004).

5.4.3.2 Water supply to the Persepolis complex

5.4.3.2.1 Jamshidi qanat

Compared with the other impressive activities, supplying water to Persepolis for the Persians, who were masters of qanat construction, was indeed a simple task. Seyyedan village, 19 km north-east of Persepolis, has numerous qanats. Jamshidi also known as the qanat is one such construction, which reveals its connection to Persepolis and was used for the supply of potable water. Qanat water reached Persepolis via a 19 km long canal, which passed through a plain field at the foot of mount Rahmat. Today only parts of this canal are still visible.

The canal had measured 70 cm in width and 40 to 80 cm in height. The canal disappears into the ground a few hundred metres from Persepolis. During the excavations, which were carried out in 1954 at the north-eastern part of Persepolis, another canal in alignment with the one mentioned above was discovered at a depth of 6 m. A. Sami, the Iranian historian, in his book "Capitals of Achaemenian kings" states: '*This deep canal is one of the many incorporated in the Persepolis platform. It is 6 m and 1 m wide and excavated in a rocky mount.*'

Due to time and labour constraints, he could not locate the extremities of this canal. It is believed that the more interesting points will be revealed after clearing this waterway. It is also thought that this deep canal may have been used to discharge water from the Persepolis Complex. However, it is certain that the major portion of potable water was supplied from the Jamshidi qanat, which crossed the plain fields and mount Rahmat. Furthermore, its alignment with the 6 m deep canal in the north of Persepolis strengthens the idea that these two were somehow connected with each other.

Given the lack of any opening in the Persepolis platform, the only remaining question is how this open canal was connected to the underground conduit. On the other hand, the elevation of the base of this 6 m deep conduit is much higher than the ground surface surrounding Persepolis. The connection may have been through a siphon.

In short, it may be stated that due to the dominating slope of the region from north to south and from east to west, potable water was supplied from the northern part of Persepolis. The high mount Rahmat is located in the eastern part of Persepolis and therefore the only remaining possible route to supply water is from the north, e.g. the direction of the canal and the 6 m deep conduit (Farhangi, 2004).

5.4.3.3 Surface water runoff collection network in Persepolis

5.4.3.3.1 Water runoffs around the town

Two different approaches were adopted to prevent any possibility of flood (from the west and the foot of mount Rahmat) damaging the mud-bricks of Persepolis:

5.4.3.3.2 Stone well

As mentioned earlier, the Persepolis stone well was dug out with a sufficient capacity to collect all the run-off floods, at the east of the treasury and at a distance of 80 m south of the Artaxerxes II mausoleum. The remains of the gutters and conduits, ending at the stone well at the edge of mount Rahmat, are still clearly visible. Next to the tomb of Artaxerxes II, there exists a narrow gutter, which reaches out the stone well. To make this system work properly, almost all of the gutters ending at this point were dug out perpendicular to mount Rahmat's slope.

5.4.3.3.3 The trench

At the western part of Persepolis there was a 10 m thick wall. The exploration carried out at the site indicates that there used to be some rectangular towers (19 m in height) made out of mud-bricks, connected to each other via the above mentioned wall thereby creating an impenetrable barrier against the invading armies. Behind the wall and next to it, a trench measuring 9.5 m in length, 6 m in width and 2.6 m in depth was excavated deep down in the mount. It created a means of protection for the palace against flooding. The general arrangement of the structure and the trench, which was built like a vertical wall, ensured the purpose. At the upstream of the trench a series of conduits were dug out on the foothill of the mount, which ended at the trench.

Furthermore, at the eastern edge of the trench there is a short masonry wall, which was built as a protective measure. Given its endurance it may not be wrong to presume that the trench was never filled up to its rim thereby allowing the construction of the perimeter palace wall with mud-bricks. The trench had been equipped with a bottom drain to discharge the incoming floods. This outlet could have been incorporated at its eastern or southern side.

A branch of the subterranean sewerage network passes next to the drain and it is not yet clear whether it was used to aid the trench in evacuating water or whether the water was just discharged from the surface weirs. The continuous drainage network of the towers and the walls was constructed on their perimeter in a manner that allowed water to be directed out of the compound.

5.4.3.3.4 Surface runoff within Persepolis

In the streets of Persepolis, regular pits were dug out to collect the surface water runoffs and direct them towards the surrounding ground surface and the subterranean water collecting networks (Figure 5.13). These networks were covered by masonry, brick or bitumen.

Figure 5.13 The general layout of the open air waterways at Persepolis

The rain water on the palace roof was discharged via the vertical conduits built either as a cantilever or embedded within the structures.

5.4.3.3.5 Cantilevered conduits

These types of conduits were one metre in length and made out of masonry or wood for discharging the water a safe distance away from the walls. The masonry conduits were shorter in length and the water could be discharged from the openings incorporated in the palace walls or other rocky locations.

5.4.3.3.6 Embedded conduits

Within the very thick walls of the Apadana Palace, there are rectangular passages 2 cm in length, which were used to collect and discharge rainwater into the sewerage system network.

Apadana Palace is rectangular in shape and measures 60 m in length. The roof of the palace is supported on 36 columns arrayed in six rows. The outer walls of the palace are 5.3 m thick mud-bricks, which lodged the embedded conduits. The conduits were made of the kiln fired bricks and bituminous mortar, to prevent water penetrating the interior compartments.

5.4.3.3.7 The sewerage network, a unique structure

Darius ordered the construction of the Persepolis sewerage network in parallel to the construction of its foundation. In most places the network was dug out of the sound rock in the foothill and then covered with huge stone pieces followed by the earth and rubbles to an appropriate elevation. At places where the bed rock level was very low, the ground was first elevated by means of the stones and rubbles to the proper level and then the network was constructed using boulders.

Figure 5.14 Underground sewers in Persepolis

The dimensions of the sewerage networks were not equal at all points as they depended mainly on the volume of water that had to be collected, even though the sewers' width was never smaller than 1 m. The depth of the conduit is variable and at times is as deep as 6 m. It is worth noting that no mortar was ever used in the construction of these networks. The bulk and weight of the stones on one hand and their smooth surfaces on the other led to the creation of a stable structure free of any joint or opening (Figures 5.14 and 5.15).

Figure 5.15 Underground sewerage networks

The access to this unique hydro-structure was via a stairway in the eastern part of the Xerxes palace. It consists of 20 steps with a length and width of 175 cm and 36 cm, respectively, and leads to the galleries in the basement of Persepolis (Figure 5.16). At the end this passage divides into two branches; one to the right (east) and the other to the left (west). Flushing this subterranean network was the soldiers' responsibility and its entrance was located at the barracks, which allowed a passage for labourers for cleaning. The western branch was used to drain the Hundred-Column palace, Apadana palace and other surrounding palaces. The eastern branch collected the surface water runoffs around the building next to the Rahmat mount. Therefore, along with laterals this network could drain an area of about 135,000 m^2 and transfer wastewater away from Persepolis. These networks had different characteristics, which among others included the following (Farhangi, 2004):

Figure 5.16 Access way to the sewerage networks

a) Characteristics of the Xerxes palace: Today, the foundation of the perimeter walls and the base columns of the palace are the sole structures standing up to the tests of time. The walls are made out of four huge cubical stones. The lower and the first one, which measured just a bit more than 2 m in width, distributed the wall load on the ground.

The second and the third stones were placed on the first one and the drainage conduit was lodged in between. The fourth stone was the cover and had the same dimensions as the first one and rested on the second and third stones. In the past, a mud-brick wall would stand out on this masonry foundation, which housed the brick and bituminous conduits limited in between the roof at the top and the cover stone at the bottom, with a built-in opening to let the water into the masonry foundation. This passage could also collect water from neighbouring conduits. The water was finally discharged into a branch of the subterranean sewerage network.

Due to the collapse of the mud-brick of Xerxes palace and demolition of its foundation, the characteristics of the conduits and the dimensions of their inlets are not clearly known today.

b) Characteristics of the sewerage network in the treasury: In the south-western part of Persepolis, next to the foot of mount Rahmat, where the museum is located today (the king's harem), there used to be a majestic building known as the Achaemenian treasury.

The water conduits of this building were outstanding from a number of aspects:

(i) they were brick-lined conduits with bituminous cover,
(ii) wooden covers were used over some conduits, and
(iii) they had open conduits, passing through patios.

The treasury consisted of two open courtyards. The first was in the initial part and the second in the next part of the building. The sewerage network was located in between the two courtyards, and collected the rainwater from the roofs, rooms and chambers of the treasury. In the first courtyard, only a part of the conduit was open and the rest was covered up. The water conduits were connected to the sewerage network through the walls and the openings in the stone cover.

c) Characteristics of the sewerage network in the Apadana palace: The largest and the most magnificent structure in Persepolis, by far, is the Hall of Receptions of Apadana. This building was constructed by the order of Darius I. From the study of the inscription on the steps and the tile works of the Apadana palace, it can be gathered that the construction of this structure took more than 30 years to complete.

This palace is located in the middle of the western side of Persepolis, overlooking the Marv-Dasht plain field. The area of the palace was 60.5×60.5 m^2 reaching a height of 20 m with roof resting on 36 columns. The rainwater from the roof and surrounding areas was discharged into the sewerage network via the conduits incorporated inside the walls. These conduits had the same characteristics as the ones mentioned in the section on Xerxes palace and the treasury. The surface of the brick works was covered by bitumen to prevent any leakage.

Four conduits within the walls were used to collect the rainwater of the Apadana palace roof and discharge it into the two branches of the sewerage network and out of the Persepolis compound. The first branch drained the northern half and the second branch did the same for the southern part of the Apadana palace. These two branches were joined at the outside of the palace, further away from the western stairway and then crossed below the Hundred-Column palace. The southern side of the Apadana palace had eleven columns overlooking the palace courtyard, which is next to the Darius palace and the Reception Hall. In this courtyard, there are two types of conduits. The eastern one is open for the purpose of collecting sewage. The western conduit, which was dug out in the bedrock of Persepolis, is a branch of the sewer in the southern half of the Apadana palace. This conduit is 0.67 m wide and 1.65 m deep. Below the southern wall of the Apadana palace a stone cover was discovered measuring 2.05 m in length, 1.20 m in width, and 0.5 m in thickness, pierced with two rectangular holes of 32×32 cm^2. It was positioned over another stone conduit 50 cm in width and 47 cm in depth. Since the depth of 47 cm was not enough for water to pass through, it was chipped in to increase the depth to more than 70 cm. The lower segments of the conduits at the sewage outlet are shown in Figures 5.17 and 5.18 (Farhangi, 2004).

d) Characteristics of the sewerage network in the Persepolis platform: Wherever there was a subterranean sewerage network, holes were incorporated to convey the surface runoffs out of the Persepolis platform by means of an appropriate slope. An example of this work can be seen in Hadish palace. At other places, gutters with proper slopes led the surface water into the subterranean network. A good example of this system is seen in the southern courtyard of the Apadana palace. In the eastern part of this compound there are several open conduits with irregular banks, which were dug in the bedrock. The widths of these gutters vary from 10 to 35 cm and their depths from a few to 18 cm. This gutter discharged into a bigger one, which ran parallel to the northern and eastern flanks of G palace (notation used by Ernst Herzfeld and Erich Schmidt representing the Oriental Institute of the University of Chicago to refer to the Throne Hall or the 100 Columns Palace) and at last, at the south-eastern side of

G palace discharged into one of the main branches of the subterranean sewerage network. In the northern courtyard of the Apadana palace, there is a curved gutter with a width and an average depth of 60 cm in the north-north-west to south-south-east alignment. The reason for construction of this gutter is not yet known. Sewers mainly refer to storm sewers and not to wastewater sewers.

Figure 5.17 Lower segments of the conduits at the sewage outlet

Figure 5.18 Lower segments of the conduits at the sewage outlet

5.4.3.3.8 Earthenware pipeline

Until a few decades ago in Iran earthenware pipes were still used for water conveyance. This system was also used in Persepolis for discharging sewage and, as it can be plainly observed, the method has not undergone many changes over the last 2500 years. An example of this is the sewerage system from the officers' quarters to the garrison street, where six pipes were deployed. Each pipe was 35 to 41 cm in length. To ensure water tightness, one end of the pipe was smaller (12 cm) than the other (17 cm). To protect the pipes against any damage, they were covered by brick works of 50 × 50 × 8 cm (Farhangi, 2004).

5.4.4 Subterranean water reservoirs

Subterranean water reservoirs (Ab Anbars) are historic Iranian hydraulic structures for drinking water supply that are still used in some regions of the country (Figure 5.19). These structures were built by philanthropists, to offer clean water to fellow humans. Subterranean reservoirs, with their considerable storage volumes, were built in the arid lands to supply water demands to passing caravans, as well as the people living close by. They were cylindrical in shape and covered with conical domes (Figure 5.19). Openings were provided at their apexes to act as vents. In some cases several openings functioned as the ventilation system of one subterranean reservoir. Access to draw water was in the form of a staircase going deep down to the lowest level of the reservoir. The width of the staircase was sufficient for people not to bump into each other during water collection. In some places several water taps were incorporated to speed up the collection. The storage capacities of these reservoirs vary from 300 to 3000 m^3. That is to say that, with a diameter of 20 m, one could have a depth of 10 m. In some places the reservoirs are much larger, and their volumes are as great as 100,000 m^3. In such cases, the roofs of these cylindrical structures had to be supported by several columns.

(a) (b)

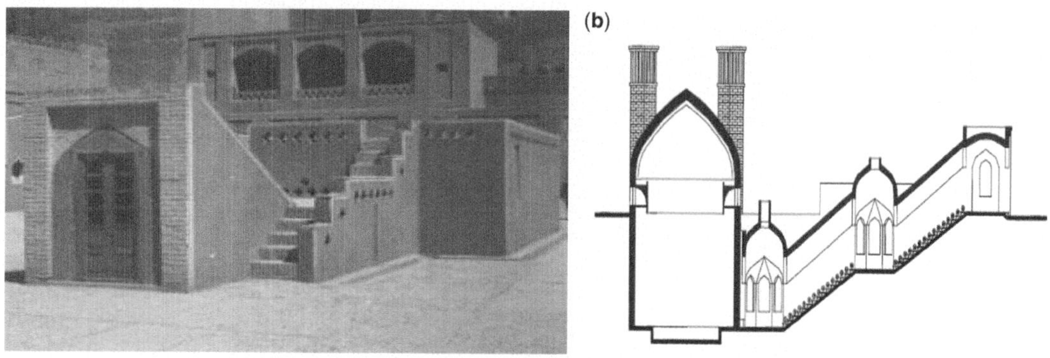

Figure 5.19 Water Reservoirs of Arg-E-Bam Ab Anbar: Entrance (left) and cross section (right)

The construction material used for Subterranean water reservoirs were very tough and made extensive use of a special mortar called Sarooj made of sand, clay, egg whites, lime, goat hair, and ash in specific proportions, depending on the location and climate of the city. This mixture was thought to be completely impermeable. The walls of the reservoirs were often 2 m thick, and special bricks had to be used in their construction. These bricks were called Ajor Ab Anbari (literally water reservoir bricks). Some reservoirs (Ab Anbars) were so large that they would be built underneath caravanserais such as the Haj Agha Ali water reservoir in Kerman. Sometimes they would also be built under mosques, like the Vazir water reservoir near Isfahan (Kooros *et al.* 1970).

5.4.4.1 Tekyeh Amir–Chakhmugh subterranean water reservoir

This reservoir is located in Yazd and was built around the 14th century AD. It has a doorway and 50 steps, which facilitated access for water collection. It has the following two distinctive features:

(a) The entire reservoir, including its ventilation system, is embedded in the ground, so the apex of the roof dome is at the ground level.
(b) Contrary to other reservoirs, where the stairways are located in the middle, this one has its stairway at one side so as not to weaken the overlaying structures (ventilation shafts).

This reservoir has a depth and diameter of 14 m and 9 m respectively. The height of the water intake structure is 9 m and the remaining space was used as the ventilation conduit. This reservoir had four rectangular ventilation shafts. Two of these have now collapsed and only 1.5 m of their lower portions still stand out of the bazaar's roof. Nevertheless, the reservoir was still in use up to 40 years ago. The entrance to the reservoir is shown in Figure 5.20 (Kooros *et al.* 1970).

Figure 5.20 The entrance of the Tekyeh Amir–Chakhmugh subterranean water reservoirs in Yazd province (ninth century AD)

5.4.4.2 Shesh Bud subterranean water reservoir

This reservoir is located at a site of the same name in the city of Yazd. It is a huge reservoir with a storage capacity of 2000 m^3 and equipped with two access stairways and taps.

This reservoir has six ventilation shafts, three of these were built originally and the rest were added later on. These ventilation shafts are octagonal in shape and designed according to the region's climate and the direction of the wind.

The special features of this reservoir are:

(a) The heights of the reservoir and the ventilation shafts;
(b) The attractive architecture and the construction materials used; and
(c) The brickwork decoration at the entrance.

The reservoir has 50 steps and the reservoir and the ventilation shafts measure 12.6 m and 10 m in height, respectively (Figure 5.21) (Kooros *et al.* 1970).

Figure 5.21 Shesh Badgir subterranean water reservoirs in Yazd province with a volume of 2000 m^3

5.4.4.3 Masjid-e-Jame-Kabir subterranean water reservoir

At the north-eastern corner of Masjid-e-Jame-Kabir in Yazd, a masterpiece of Islamic architecture from the fifth century BC, a stairway is located going deep into the ground. The reservoir is octagonal in shape, and has niches on seven sides located one metre above the ground level. The structure's roof was originally a clay dome. A new brick dome with no opening was built on top of the initial one about 50 years ago.

This reservoir is built on Zarch Qanat, which is still in use today. It is estimated that the construction of this monument dates back more than seven centuries. This reservoir was renovated during the last two centuries and lined with brick walls (Kooros *et al.* 1970).

5.4.4.4 Golshan subterranean reservoir

This reservoir is an example of water storage constructed in a populated area, located in the heart of the city of Yazd, which was still in use as recently as 30 years ago. It was equipped with two water taps. The reservoir has four rectangular ventilation shafts, reaching a height of 13 m, located at equal intervals on the perimeter of the cylindrical structure. This reservoir is of special significance as far as architecture, size, construction materials used, brickwork decorations at the entrance and the brick over the dome are concerned. The reservoir has a capacity of 2000 m^3 (Kooros *et al.* 1970). Water reservoirs were familiar sights in the province of Hormozgan near the Persian Gulf where the weather is very hot with frequent shortages of fresh water, and where water harvesting used to be the only means of water supply to urban and rural areas (Figure 5.22).

Figure 5.22 Water reservoirs in Hormozgan province of 300 (left) and 100 (right) years old

5.4.5 Ice – chambers

A Yakh-chāl (ice-chamber) is an ancient natural refrigerator which was mainly built and used in Iran to store ice, and sometimes food as well. A cross section of a Yakh-chāl ice-chamber is shown in Figure 5.23.

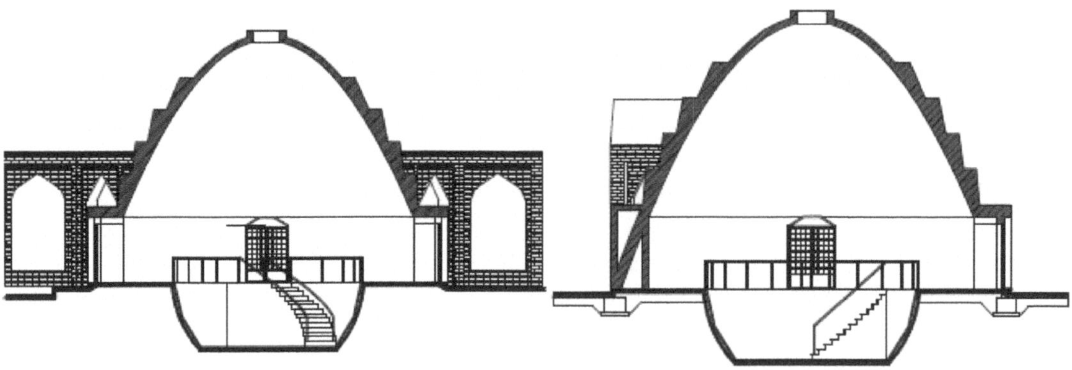

Figure 5.23 Cross section of a Yakh-Chāl

Ruins of huge conical reservoirs which were once used to store ice-blocks can still be seen in various parts of Iran. These reservoirs are called ice-chambers. In summer, people living in nearby villages used to obtain their ice from these structures. The subterranean space coupled with the thick heat-resistant construction material kept the outside heat from reaching the interior space throughout the year. From ancient times this ice was used for the making of Faloudeh (traditional Persian ice cream) (Khorasanizadeh, 2008).

In 400 BC Iran, Iranians had mastered the technique of storing ice in the middle of summer in the desert. Yakh-chāl was a large underground space (up to 5000 m^3) that had thick walls (at least 2 m at the base) made out of a special mortar called Sarooj, composed of sand, clay, egg whites, lime, goat hair, and ash in specific proportions, making it resistant to heat. This mixture was thought to be completely impermeable. The space often had access to a qanat, and contained a system of wind towers that could easily maintain temperatures inside the space at frigid levels during summer.

The method of ice production was also interesting. In winters, an appropriate amount of water was stored within the thick and high walls of the designated chambers, which had flat floors. The cold of the winter nights was sufficient to produce ice of a few centimetres thick. The high walls of the chambers, which were constructed in the east-west directions, protected the ice, which faced northward, against sunrays. This method needed a skilled approach in judging the rate of ice production at each stage. Once the thickness of the ice had reached to 30 to 40 cm, it was broken into pieces and then stored in the circular ice-chambers. A few wells were excavated at the bottoms of the ice-chambers to collect the ice melts during the hot seasons. The surfaces of the layer of ice were covered with straws to keep one layer separated from the adjacent ones. The ice-chambers were equipped with two corridors: one for storing ice blocks in the winter and the other, for taking them out in the summer (Farhangi, 2004).

Ice-chambers were mainly used in two short periods. Once during the freezing cold winter (to store ice blocks covered layer by layer with straws) and then during the hot summer season (to harvest the stored ice blocks and potable water). Arg-E-Bam ice-chamber in the province of Kerman in the south east part of the country is shown in Figure 5.24.

Figure 5.24 Arg-E-Bam ice-chambers in Kerman province

5.4.5.1 Meibod ice-chamber

The Meibod ice-chamber is the oldest one remaining. It is 50 km north of Yazd, next to an ancient caravan road and adjacent to a courier house, a subterranean water reservoir and a caravanserai. This ice-chamber is made out of sun dried bricks, most of which are fortunately still intact, and is one of the outstanding structures of its kind. The ice-chamber usually consists of four parts, namely the ice-production chamber, shadowing walls, ice-pit (ice-chamber) and the huge dome. The ice-production chamber in this case is a shallow pool, with a depth of almost 0.5 m, and an approximate area of 8000 m^2.

The three shadowing walls are 2 m thick, 8 m high and 20 m long. The eastern wall of the ice chamber has now collapsed. The chamber has a main entrance measuring 2.2 m in width and 2.0 m in height. It was excavated in a hard clayey ground at a diameter of 13 m and a depth of 6 m. Its huge dome was laid on the same hard ground. The dome has an area of about 300 m^2 and is about 15 m high. The thickness of the dome shell was skilfully reduced from 4.2 m at the base to 25 cm (e.g. brick width) at its apex (as a result of environmental elements, the north-western part of the dome has collapsed). A picture of the Meibod ice-chamber is shown in Figure 5.25 (Kooros *et al.* 1970).

Figure 5.25 Meibod ice-chambers

5.4.5.2 Mir-Fattah ice-chamber in Malayer

Construction of this ice-chamber dates back to the Qajar epoch. Its dome measures 12 m in height from the apex to ground level and the storage chamber is 4 m deep. There is a 3 m wide opening at the apex. The ice-chamber has a diameter of 12 m. The thickness of the dome decreases from 92 cm at the base to 60 cm at the apex (e.g. equivalent to three layers of brick works). The 4 m embedded part of the ice-chamber is made of masonry with the rest made out of bricks measuring 20 × 20 × 5 cm. From a technical point of view, this is a precise and valuable piece of work. This structure is located in farmland adjacent to the Malayer to Boroojerd connecting road (Kooros *et al.* 1970).

Two typical ice-chambers in the provinces of Kerman and Yazd are shown in Figure 5.26.

5.4.6 Dam construction

Iran is located in an arid-semi-arid region. Due to inequitable distribution of surface water, people had to endure many hardships to meet water demands and to counter fluctuation of yearly seasonal streams to be able to ensure better conditions for utilisation of water. Dam construction in Iran dates back to the ancient times. The multi- purpose dam of Amir built 2000 years ago is one such example (Figure 5.27). Iranians considered meticulously the three basic factors of site selection, the condition of foundations and the construction materials. In all cases, they fully observed all design and technical requirements and conditions for selection of the site and construction of the dam. The topography and the route for diverting the river flow during construction works were also duly considered in the design (Khorasanizadeh, 2008).

Figure 5.26 Ice-chambers: Moayeri in Kerman (left) and Abar Kuh in Yazd (right)

Figure 5.27 Dam, bridge and mill of Amir multipurpose dam (2000 years old)

All the ancient dams of Iran are masonry works of the following types:

Gravity dams: Studies indicate that all the criteria considered the design of ancient Iranian dams are exactly those that are also considered in the modern design of gravity dams. Saveh and Sheshtaraz dams which are over 700 and 900 years old respectively are examples of this type.

Arch dams: Kebar (one of the oldest arch dams in the world, with a height of 26 m, a crest length of 55 m, a thickness of 5 m and with an arch radius of 38 m, clearly demonstrating the skills of Iranians in dam construction) and Kerrit dams over 700 and 400 years old respectively are both arch dams. A view of Kerrit dam is shown in Figure 5.28.

Figure 5.28 Kerrit dam

Buttress dams: Akhlemad dam with a crest length of 230 m, height of 12 m and reservoir capacity of three million m^3 and the 400-year-old Fariman dam, are both buttress dams which are still in operation. A picture of Akhlemad dam is shown in Figure 5.29.

Figure 5.29 Akhlemad dam

Saveh arch dam and Durudzan fill dam are two modern structures that have been built very close to the old dam sites. The highest dam in Iran is Karun 3, while the largest dam, as far as reservoir capacity is concerned, is the Karkheh dam located in the province of Khuzestan (Khorasanizadeh, 2008).

5.4.7 Gavgard

The Gavgard (literary bull turning) is a device composed of multiple components mounted on the opening of a well. It used the force of a bull turning around a well to extract water for different purposes. They were usually installed in large wealthy gardens and mansions, which lacked qanat, streams or springs. It appears that the Gavgard was used extensively in the past.

Today it has lost its utility in Iran and most of these devices are now destroyed. There are three intact but abandoned specimens of these remaining in the province of Kerman. These are the Gavgard of the Late Yavar Nakhaee mansion and garden in the suburbs of the town of Ravar, the Gavgard of Harandi garden in the Kerman Museum and the Gavgard of Budagh Abad Garden (Zoroastrian) in Kerman. The Gavgards of Morsalin Hospital and the Mashallah Mokhtare' were standing until some decades ago in Kerman, but were subsequently destroyed for various reasons including road extension and construction. A smaller model of the Harandi Gavgard was constructed by the Kerman Water Museum and put on display in the headquarters of the Kerman Regional Water Company (Garoosi, 2003).

5.4.8 Water mills

One of the skills of ancient Iranians was to tap on water's hidden powers to rotate the stones of watermills. The roof of a watermill building was usually dome shaped. Light and air were supplied through a door. The watermills were powered by the river flow, springs or qanats. They were connected to the water sources by canals. A conduit just before the mill's shaft acted as a bypass when the mill was not in operation. The mill shaft was semi conical in shape and its diameter reduced from top to bottom. This shaft could be plugged by a wooden device accessible through a narrow gallery from inside the mill.

The wooden turbine consists of an axle with a top to bottom gradually increasing diameter. It has a lower iron tip housed in a pit in the lower millstone, acting as the turbine support. The upper iron tip of the axle is fixed in the upper millstone. This axle is surrounded by paddles and the whole system is known as the turbine wheel. The water jet impinges forcefully on the turbine wheel causing it to rotate. This in turn leads to the spin of the upper millstone. The lower millstone is stationary and the rotation of the upper stone on the lower grinds the grain. There is a hole in the middle of the upper stone, which discharges the grain into the space between the two stones. The two millstones are not truly horizontal, but are slightly inclined which helps the flour to be discharged into the bags. A cross section of a water mill is shown in Figure 5.30 and the paddles and upper millstone in a water mill are demonstrated in Figure 5.31 (Khorasanizadeh, 2008).

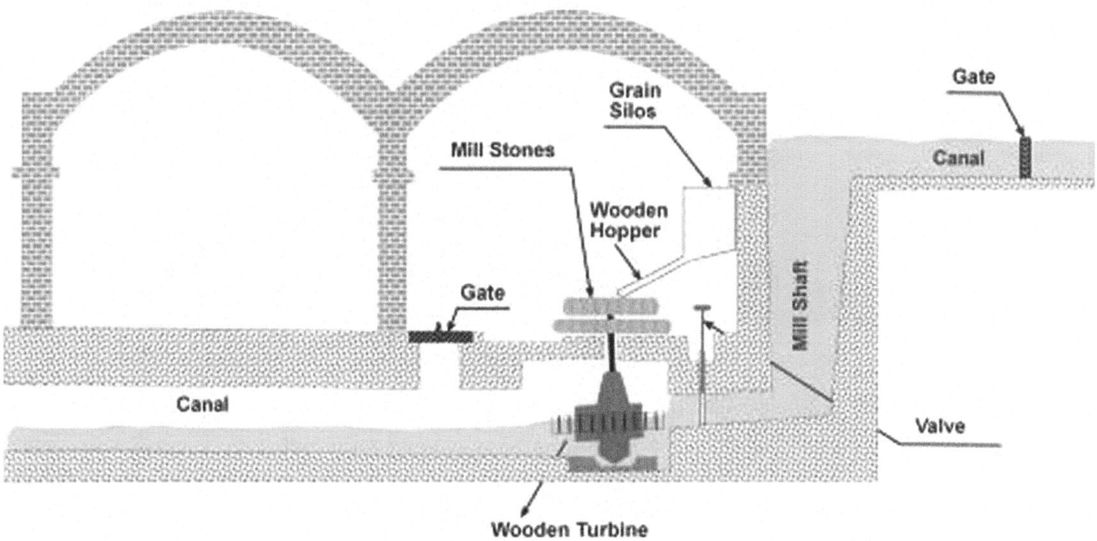

Figure 5.30 Cross section of a water mill

Figure 5.31 Water mills: paddles in Taft in the province of Yazd (left) and upper millstone in Kashan (right)

The grains were stored in a silo located at the top of the watermill and were discharged through a wooden hopper into the hole dugout in the upper stone. The grains in the silo discharged down through the rotation of the upper stone and the continuous strokes on the hopper. Water was let out through a conduit built below the watermill.

Some parts of the roofed mill compound are designated for the cleaning and screening of the grains and flour production. A space is also foreseen to lodge the animals used to transport the flour bags.

The Gar-Gar and Amir Weir mills were powered by reservoir water. The Fin-garden's watermill in Kashan was driven by a spring, whereas the Double Stone Water Mill of Mohammad Abbad (Meybod) (Figure 5.32), the ancient watermill of Ashkezar, 15 km away from Yazd, and the 200-year-old Taft watermill were all powered by the neighbouring qanats.

Figure 5.32 Water mills: Entrance of double stone of Mohammad Abad in Meybod (left) and the double stone of Mohammad Abbad in Meybod (right)

The Double Stone Water Mill of Mohammad Abbad in Meybod is an amazing hydraulic structure, which is absolutely unique. It has been created at a depth of 40 m on a qanat gallery for grinding grains. This 150-year-old water driven structure supplied all the flour requirements of the neighbouring oases.

No construction materials were used to establish this marvellous mill. It had a high vestibule to allow camels to move back and forth with ease and the lateral stables were large enough to accommodate the cows, donkeys or camels, which carried the produced flour. The mill is remotely situated in a deserted area at a distance of about 8 km from the village of Mohammad Abbad, 50 km away from Yazd. Gradually, over the course of time, due to flood waters, dust and slime had filled the mill's long narrow passage. It was recently unblocked by the regional water authorities who exerted great effort to excavate it. This mill is, at present, open to visitors (Khorasanizadeh, 2008).

5.4.8.1 The 1700-year-old Shushtar mills

Among a total of 57 outstanding phenomenal monuments remaining from past cultures and valued by UNESCO as mankind heritage, three Iranian architectural wonders, namely the Choqazanbil Temple (3250 years old), the Persepolis in Shiraz and the Naqshejahan Square in Esfahan are worthy of profound attention. The fourth under consideration by UNESCO for inclusion in this list is the Shushtar Mills Complex, whose age and precedence have not yet been determined with certitude (Figure 5.33).

Figure 5.33 The Shushtar watermills complex: General view (left) and two 1700-year-old mills (right)

It should be pointed out that in the Province of Khuzestan and in the prolific river basins such as Dez, Karkhe, Marun and Bahmansheer as well as in other sections of Karun River, the irrigation network of Minoo (Minab) stands out alone. The different structures and water facilities in this complex operated according to a unique plan and in full harmony and coordination with each other, to such an extent that all the discreet structures scattered in the arid desert areas can be considered collectively as part of the Shushtar irrigation network. The Shushtar waterfalls are among the components of this immense network.

An objective of this network, particularly as it relates to the beautiful Shushtar falls, was to enrich the water droplets and to exploit their inherent potential for spinning the millstones.

The key features of the south Karun River Stream are: a) The River is divided into two streams at the Mirzan Bridge Weir where the larger stream flows westward, b) The Mirzan Bridge (weir structure) is built with masonry materials, and c) Except for the structural disintegration of the priming coat at certain openings facing the Gar Gar River (Figure 5.34), which is now restored, the bridge weir has not sustained any damage and remains a solid and integrated structure (Farhangi, 2004).

5.4.8.1.1 Gar Gar Bridge, Weir and Mills

The Gar Gar Bridge is located 700 m south of the Mizan Bridge weir. A diversion canal diverts the excess flood water of the Gar Gar river from the north and east side of the bridge weir to the southern area of the array of mills and finally to the River (Figure 5.34). The Gar Gar Mills complex is situated to the south of the bridge and weir. The wells relating to the water reservoir, water supply network, water control gates and the space required for spinning the millstones were dug out of natural rocks on either side of the river bank.

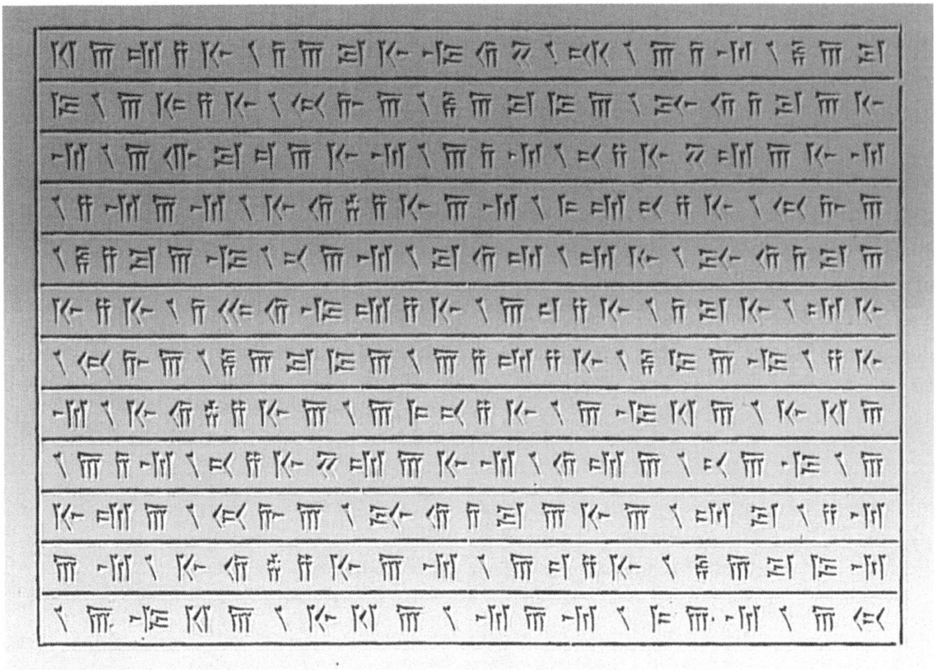

Figure 5.34 Darius Inscriptions saying '*Darius, the Achaemenid king declares: I am Persian. I conquered Egypt. I ordered this canal to be cut from the Nile that runs in Egypt to a sea in Persia. Then this canal was dug as I ordered, and these vessels set out from Egypt to Persia as I wished*' (Farhangi, 1998).

Until half a century ago, a total of 40 mills located behind the Gar Gar Weir produced the flour supplied to Shushtar and the surrounding towns. Then the site of the three mills was allocated to three power generation plants to meet the power requirements of Shushtar's citizens.

After the closure of the mills and the old power plants not so long ago, water circulation in this industrial complex remained uncontrollable. Currently water is let into most of the obliterated outlets of the obsolete mills, splashing out of the rocks and creating a remarkable scene (Farhangi, 200).

5.5 CONCLUSIONS

In this Chapter we mentioned the history of water in ancient Iran and the technologies used by Persians for water supply. Before concluding we intend to state the differences between water supply conditions in Iran and other ancient civilizations such as Rome (present day Italy), Greece and the Ottoman empire (Turkey).

Since Iran is located in a semi-arid region with a limited annual precipitation of 250 mm and faces occasional droughts, water supply was not only necessary to meet the drinking, sanitation and domestic demands of humans and animals, it also had to be of adequate quantity for agriculture (which accounted for 95% of the total water demands). Therefore in ancient times, Iranians living in regions close to rivers could easily satisfy their needs. However, at most places in the country, groundwater was the sole available resource. Therefore, the limited population had to resort to digging by manual means and to extract water using pails, wheel or animal force for their drinking purposes. However to meet the agricultural requirements, which were further intensified by the arid conditions of the country, Iranians needed to extract considerable volumes of water from the earth at a time when there were no modern-technology excavation machines to dig wells and essentially no pumps such as those we have available today. Therefore ancient Persians invented qanats, through which they were able to bring water from dozens of metres underground up to the surface without any energy, thereby solving the problem of agricultural water supply. To have a clear picture of the difficulty of the works, a rough calculation shows that digging a qanat of 30 km in length and with a mother well 100 m deep took about 40 years to complete. The discharge rate of such a qanat was about 80 to 100 L/s. Today, to dig a well or two to supply this volume of water takes no more than 20 days.

Another difference between Iran and the above mentioned countries concerns the climatic conditions from the aspect of temperature. In some parts of Iran the temperature in summer reaches up to 50° Celsius. Farmers had to work for hours under the sun in the desert and arid regions. They therefore needed to drink a large volume of water. To keep the water cool, in addition to earthenware, they built ice-chambers to produce ice during winters for consumption during summers. Such structures may not have been required in the temperate European countries.

The other difference in Iran is that most rivers are seasonal with no permanent water. Therefore their water, which flows during winter and spring, must be somehow stored for the summer. This resulted in construction of dams on these rivers by ancient Persians, and as we have already seen some of the dams are about 2000 years old. Due to availability of permanent rivers and numerous lakes in Europe, this was another need that was not felt in European countries in the past.

Given the population growth, the expansion of cities and the appearance of new technologies, conditions today are entirely different from the past. The development of well boring technologies and the production of different types of pumps have resulted the lowering of the groundwater tables have dried up a great many of these. Qanats were the symbol of sustainable development and operation of groundwater resources whereas, following expansion and multiplication of wells as well as the erratic operation of groundwaters, these resources, which should be preserved for future generations, are unfortunately facing serious threats.

The availability of fridges and freezers in homes and modern cooling systems has made the large public ice-chambers and wind towers obsolete, confining them to history books; and many such monuments have gradually been destroyed. A look at the past shows what difficulties and problems our ancestors had to deal with.

In conclusion the advantages of ancient technologies over modern solutions may be summarised as follows:

(a) The independence from the need for energy and fuel in water supply, agricultural production, cooling houses and milling cereals as well as for production of ice and building materials.
(b) No environmental pollution as, contrary to modern era, there was no need for fossil fuel to start up the pumps, and therefore the sustainable development and complete protection of the environment was ensured.
(c) Qanats did not deplete groundwater resources and maintained the balance of the basins. Contrary to the modern era where these resources are threatened by well technology, which has replaced qanats.
(d) The positive cultural impacts and the fostering of relations and exchange of information among the rural population, and particularly among women (as the qanat openings were the gathering point for village women, who would meet at this point each day to wash clothes and dishes and talk to each other, while doing their chores).

Despite all these advantages, as most works depended on human force, with machines having no role in the lives of people, life was essentially harder and more difficult than today, with people lacking modern welfare and conveniences. Now what answer are we to give if we were asked whether we would have preferred to live 1000 years ago or are we satisfied with living in the 21st century?

ACKNOWLEDGMENTS

The critical comments of Mr. M. A. Khan are very much appreciated.

REFERENCES

Farhangi B. (1998). Contemporary Dam Construction in Iran. Iranian Committee on Large Dams, Tehran, Iran.

Farhangi B. (2004). Water Melody in the Passage of Time (A Review of Hydro Structures of Iran from Ancient Era to Present Time). Iranian Committee on Large Dams, Tehran, Iran.

Kooros Gh., Reza A., Entezami A. A. and Imam Shooshtari M. A. (1970). Water and Irrigation Techniques in Ancient Iran. Ministry of Water and Power, Tehran, Iran.

Hashami M. (2010). Case History of Tabriz Water, East Azarbaiejan Province. Water and Wastewater Company, Tabriz, Iran.

Khorasanizadeh Sh. (2008). Novel Hydraulic Structures and Water Management in Iran, a Historic Perspective. Regional Centre on Urban Water Management, Tehran, Iran.

Behnia A. (1988). Qanat: Construction and Maintenance. Tehran University Publication Centre, Tehran, Iran.

Garoosi A. A. (2003). Water and Irrigation History in Kerman Province. Iranian National Committee on Irrigation and Drainage, Tehran, Iran.

Chapter 6

A historical perspective on the development of water supply in Egypt

F. A. El-Gohary

6.1 GEOGRAPHY, CLIMATE AND WATER RESOURCES

Egypt, a country of ancient genuine civilization and deeply rooted culture is situated at the north-east corner of the African continent in the heart of the Arab world. It lies between latitudes 22° and 32°N and longitudes 25° and 35°E. The country has a geographical area of 1,001,450 km^2 of which only about 5% is inhabited around the Nile valley and delta. The majority of the land is desert (Figure 6.1).

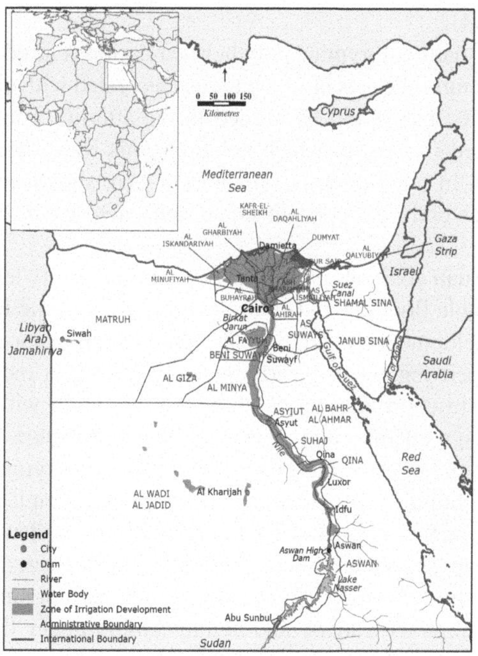

Figure 6.1 Map of Egypt

Hot dry summers and mild winters characterise Egypt's climate. Rainfall is very low, irregular and unpredictable. Annual rainfall ranges between a maximum of about 200 mm at Alexandria to 75 mm year at Port Said. It also declines inland to about 25 mm per year near Cairo. The rainfall occurs only in the winter season and in the form of scattered showers and therefore, it cannot be considered as a dependable source of water. Egypt's main and almost exclusive fresh water is Nile water, which accounts for 96% of its total water. The total amount of groundwater abstraction in the western desert and Sinai is estimated to be about 4.8 km^3/yr.

6.2 THE PREHISTORY OF EGYPT

Around 5000 BC, one of the most advanced and powerful civilizations arose on the banks of the Nile River called Egypt (www.123HelpMe.com). The name Egypt means "Two Lands", reflecting the two separate kingdoms of Upper and Lower prehistoric Egypt – the Delta region in the north and a long length of sandstone and limestone in the south. In 3000 BC, a single ruler, Menes, unified the entire land and set the stage for an impressive civilization that lasted 3000 years. He began with the construction of basins to contain the flood water, digging canals and irrigation ditches to reclaim the marshy land.

The history of ancient Egypt occurred in a series of stable *Kingdoms*, separated by periods of relative instability known as *Intermediate Periods*: the Old Kingdom of the Early Bronze Age, the Middle Kingdom of the Middle Bronze Age and the New Kingdom of the Late Bronze Age. Egypt reached the pinnacle of its power during the New Kingdom, in the Ramesside period, after which it entered a period of slow decline. Egypt was conquered by a succession of foreign powers in this late period. After Alexander the Great's death, one of his generals, Ptolemy Soter, established himself as the new ruler of Egypt. This Ptolemaic Dynasty ruled Egypt until 30 BC, when it fell to the Roman Empire and became a Roman province (Clayton, 1994).

Until the conversion to Christianity, Egyptian polytheistic religion centred on the afterlife. Pharaohs and rich Egyptians built elaborate tombs in caves or in pyramids, decorated with elaborate art on the interior and containing jewellery and objects that would be needed in the afterlife. The walls of the pyramids were decorated with elaborate stylised frescos, in which noble persons were shown as larger than slaves, and subjects were drawn in profile. In other periods, all subjects were shown in frontal view only. The king or noble person had his or her body wrapped in linen, and enclosed in an elaborate carved coffin as a mummy.

The success of ancient Egyptian civilization came partly from its ability to adapt to the conditions of the Nile River Valley. The predictable flooding and controlled irrigation of the fertile valley produced surplus crops, which fuelled social development and culture. The administration sponsored mineral exploitation of the valley and surrounding desert regions, the early development of an independent writing system, the organisation of collective construction and agricultural projects, trade with surrounding regions, and a military intended to defeat foreign enemies and assert Egyptian dominance.

The many achievements of the ancient Egyptians include the quarrying, surveying and construction techniques that facilitated the building of monumental pyramids, temples, and obelisks; a system of mathematics; a practical and effective system of medicine; irrigation systems and agricultural production techniques; the first known ships; Egyptian faience and glass technology; new forms of literature; and the earliest known peace treaty (Clayton, 1994). Motivating and organising these activities was a bureaucracy of elite scribes, religious leaders, and administrators under the control of a Pharaoh who ensured the cooperation and unity of the Egyptian people in the context of an elaborate system of religious beliefs (James, 2005; Manuelian, 1998).

Egypt left a lasting legacy. Its art and architecture were widely copied, and its antiquities carried off to far corners of the world. Its monumental ruins have inspired the imaginations of travellers and writers for centuries. A new-found respect for antiquities and excavations in the early modern period led to the scientific investigation of Egyptian civilization and a greater appreciation of its cultural legacy (James, 2005). The unique history of ancient Egypt and visible monuments to that history helped Egyptians to preserve a distinct national consciousness, and to remain a separate entity during the years of Arab, Mameluke and Ottoman conquest. Ancient Egypt boasted considerable achievements in art, medicine, astronomy and literature, and was the hub of civilization in much of the Near East and North Africa.

6.3 WATER RESOURCES IN ANCIENT EGYPT

6.3.1 General

With the onset of the last great ice age about 30,000 years ago huge glaciers formed on the high African mountains of Ethiopia, Kenya and Uganda. When the great global meltdown began about 12,000 years ago these huge glaciers sent massive volumes of water to the north. The gigantic discharges flowed out of Lake Victoria and down the Blue and White Nile valley basins. Catastrophic floods filled the lower Egyptian valley, washing away all villages and burying their shattered remnants in sediment, thus breaking the continuity of human life in Egypt. Archaeologists today call this the "Wild Nile". Then, as the glacial melting slowed, the valley became suitable for human settlement once again (1).

Ancient Egypt was an agricultural estate and mainly relied on the flooding of the Nile for fertile soil. Inundation or flooding is the yearly gradual overflow of the Nile water. Each year, in June and extending to the end of November, land would be covered with water which would slowly drain and flow back leaving behind highly fertile soil. This annual cycle of flooding and depositing of silt created a new layer of topsoil each year. This topsoil was rich in organic nutrients and basic elements for plants such as phosphorus and nitrogen. Beside this, when the water receded in October, it left behind pools of water in depressed areas which was stored for some time until the soil could absorb more water therefore acting as a reservoir. On the other hand, the mud left by the flooding would be the best medium for planting their crops.

6.3.2 The Nile

The fertility of the Nile River is well known and hardly requires mentioning here. This river helped sustain Egyptian civilization for three millennia. The ancient Greek historian Herodotus wrote that Egypt is the gift of the Nile, meaning that it flourished on the top soil that was formed from silt brought by the recurrent floods of the Nile. He states:

> *.... the water begins to rise at the summer solstice, continues to do so for a hundred days, and then falls again at the end of that period, so that it remains low throughout the winter until the summer solstice comes round again in the following year.*

The floodwaters of the Nile come as a result of the rainy season in Ethiopia, which erodes the silt of the Ethiopian highlands, and carries it towards Egypt along the Blue Nile and other tributaries. The River Nile in Egypt receives 90% of its water during a 100-day flood period every year. No appreciable amount of water comes to Egypt via the White Nile, which starts from Central Africa. Indeed, Egypt is the gift of the Nile, and it was founded and developed around that river, for there is no rain in Egypt. However, each year flooding caused disasters to Egyptians presented in the risk of their villages being damaged and their crops destroyed. So, Egyptians suffered many losses due to this yearly flooding. As an effect of that, the first form of government appeared when the Egyptians organised their efforts under one leadership to avoid these disasters and the yearly flooding following the concept of authority (decision making). As time passed, the ruler (pharaoh) became more important with more power and influence on the Egyptians. The government dealt with many problems that Egyptians were looking forward to controlling like the irrigation systems, storage of food surplus, the harvest cycle and many other tasks. But these acts were not carried out for free and the government would not stand without having a resource or income so they had to follow the tax method. Over time this form of government started to become more and more complex (2). As the state grew and more complex religious and political systems started to emerge, the need for a system to record events was growing too. The papyrus paper was found to satisfy that need. Ancient Egyptians used a phonetic-pictograph writing called hieroglyphics by the Ancient Greeks. This system evolved from portrayal of pictures of objects to using stylised representation of objects to represent sound combinations and compose words, to a phonetic alphabet much like our own. In general, all the scientific, political and agricultural advancements were a direct result of the existence of that river.

6.3.3 Water resources management in Ancient Egypt

Egypt is (and was) one of the most arid areas in the world. The Ancient Egyptians managed their limited water resources efficiently, and became the best dry-weather agrarians in the world. Ancient Egypt was renowned worldwide for its dry-weather irrigation and farming techniques. In 3000 BC, Menes, began with the construction of basins to contain the flood water, digging canals and irrigation ditches to reclaim the marshy land. From these earliest of times, so important was the cutting of a dam that the event was heralded by a royal ceremony. King Menes is credited with diverting the course of the Nile to build the city of Memphis on the site where the great river had run. By 2500 BC, an extensive system of dikes, canals and sluices had been developed. It remained in use until the Roman occupation, circa 30 BC–641 AD. Diodorus spoke of the efficient Egyptian farming system, he wrote:

> *. . . being from their infancy brought up to agricultural pursuits, they far excelled the husband-men of other countries, and had become acquainted with the capabilities of the land, the mode of irrigation, the exact season for sowing and reaping, as well as all the most useful secrets connected with the harvest, which they had derived from their ancestors, and had improved by their own experience.*

Several entities were formed along the Nile Valley to manage the gushing floodwaters by observing, recording, and regulating the water flow to the whole Nile Valley. As a consequence, a highly organised communal irrigation system was developed and used since time immemorial. The limited available water resources in Ancient Egypt were managed most efficiently by utilising organised methods of water conservation and diversion. According to Strabo, the Egyptian communal irrigation system was so admirably managed:

> *. . . that art contrived sometimes to supply what nature denied, and, by means of canals and embankments, there was little difference in the quantity of land irrigated, whether the flood was deficient or abundant.*

6.3.3.1 Nilometers

The annual flooding of the Nile had a big impact on Egyptian agriculture, so taxes were exacted based on flood levels. In the beginning, a portable tool, called a Nilometer, was placed vertically into the Nile to measure the flood levels. It was probably a long reed stick on which different levels were marked. To assess taxes fairly, the Ptolemaic rulers built temples along the Nile and installed Nilometers in them. The Nilometer uncovered on Philae Island is a staircase with reliefs of Nilometers with arms carved on its internal walls, along with the timing and duration of the flood. During the Roman period in Egypt, the Roman rulers showed interest in the monuments built along the Nile with Nilometers, but did not construct any new buildings themselves.

Until the time of the Emperor Constantine, a portable Nilometer was kept in the temple of the god Serapis. The ancient Egyptians believed that they were in debt to Serapis for bringing them the annual Nile flood. After every measurement that showed a rise in the Nile waters, they would return the portable Nilometer to the temple of Serapis. This became a ritual. The portable Nilometer was called "the arm, or branch, of the Nile".

When Constantine ordered the Nilometer to be placed in the Church of Alexandria, chaos ensued in Egypt. The people thought Serapis would be angry and not allow the Nile to rise that year. However, the Nile did rise. The Emperor Julian the Apostate later ordered the Nilometer be returned to the temple of Serapis. It remained there until the time of The Emperor Theodosius the First, who ordered the entire temple to be destroyed.

Nilometers were built in various places along the river. They had three different formats: a slab or pillar, a well or a series of steps (Figure 6.2). But all three were calibrated using the same unit of measurement, the cubit (slightly shorter than 5 cm). These Nilometers' readings were taken by priests and then studied by Egyptian architects and astronomers.

Figure 6.2 Different types of Nilometers

The most important Nilometers, built and studied, were located at Elephantine Island, Philae Island, and at Edfu between Luxor and Aswan. Knowing that the Nile flooded every year, a yearly calendar was made based on the flooding. This calendar consisted of twelve months and divided each one into thirty days and it was the most accurate calendar of its time.

The elevations at the Nilometers throughout Egypt were all tied to a single common datum. Regulating the flow amounts and duration was controlled by knowledgeable officials, using sluice gate(s) to control the flow of water to a determined height and duration. Diodorus affirms:

'. . . at flood-time it might not form stagnant pools over the land to its detriment, but that the flood-water might be let upon the countryside, in a gentle flow as it might be needed, through gates which they[Egyptians] had built'

The water of the inundation was managed differently in various districts. This depended on many factors, such as the relative heights/elevations of the adjoining lands, and what the crops they happened to be cultivating at the time, etc.

6.3.3.2 Water projects

The Ancient Egyptian waterworks and land reclamation projects were huge, even by our present-day standards of projects that use heavy equipment. Below are a few examples.

A major waterway diversion project was carried out over 4000 years ago. The project began at present-day Asyut, where they dug a long canal called Bahr Yousuf to bring water from the Nile to the present-day Fayoum, located about 65 miles (100 km) southwest of Cairo for irrigation (Figure 6.3). The Fayoum Oasis lies below sea level, and contains Lake Qarun.

Figure 6.3 Fayoum Project

The lake was originally used as a catchment basin for the Nile overflow, and once filled the entire region. This water carried with it, and deposited, the fertile Nile silt on the bottom of the lakebed. This ancient major project caused the diversion of millions of gallons of Nile water that was wasted at the deserts around the Fayoum region. The flow of water into the lake was reduced. As a result, about 80% of the original lake area was reclaimed and the rich soil was cultivated. A series of waterwheels were used to raise water to the banks along this branch of the Nile. Additionally, more water was available along the Nile Valley north of Asyut, increasing arable lands.

There is archaeological evidence of major public projects in Kush, which were built in order to establish a permanent presence in the area during the Middle Kingdom. King Senwasret III (1878–1844 BC) established (modern-day) Semna. The area above the Third Cataract was fertile and supported a large population. During the Middle Kingdom, an artificial dam blocked the channel. A portion of this dam is still visible, today, at Semna East. The dam construction raised the level of the Nile for hundreds of miles to the south, enabling trading expeditions to navigate far into the interior of Africa. There are about 25 inscriptions on the rocks below the channel fortresses of Semna East and Semna West. They represent Nile flood levels recorded during the Middle Kingdom, and all of them show a level about 25 ft (8m) higher than the maximum water levels of today.

6.4 AGRICULTURE

6.4.1 Farming

A combination of favourable geographical features contributed to the success of ancient Egyptian culture, the most important of which was the rich fertile soil resulting from annual inundations of the Nile River. The ancient Egyptians were thus able to produce an abundance of food, allowing the population to devote more time and resources to cultural, technological, and artistic pursuits.

Land management was crucial in ancient Egypt because taxes were assessed based on the amount of land a person owned (Manuelian,1998). Farming in Egypt was dependent on the cycle of the Nile River. The Egyptians recognised three seasons: *Akhet* (flooding), *Peret* (planting), and *Shemu* (harvesting) (Figure 6.4). The flooding season lasted from June to September, depositing on the river's banks a layer of mineral-rich silt ideal for growing crops. After the floodwaters had receded, the growing season lasted from October to February. Farmers ploughed and planted seeds in the fields, which were irrigated with ditches and canals. Egypt received little rainfall, so farmers relied on the Nile to water their crops (Nicholson, 2000). From March to May, farmers used sickles to harvest their crops, which were then threshed with a flail to separate the straw from the grain. Winnowing removed the chaff from the grain, and the grain was then ground into flour, brewed to make beer, or stored for later use.

Figure 6.4 The three farming cycles

The ancient Egyptians cultivated emmer and barley, and several other cereal grains, all of which were used to make the two main food staples of bread and beer. Flax plants, uprooted before they started flowering, were grown for the fibres of their stems. These fibres were split along their length and spun into thread, which was used to weave sheets of linen and to make clothing. Papyrus growing on the banks of the Nile River was used to make paper. Vegetables and fruits were grown in garden plots, close to habitations and on higher ground, and had to be watered by hand. Vegetables included leeks, garlic, melons, squashes, pulses, lettuce, and other crops, in addition to grapes that were made into wine (Nicholson, 2000).

6.4.2 Irrigation systems

The Egyptians depended on the annual flooding of the Nile to cover their fields with black silt and to irrigate their crops. They measured the flood to determine their taxes. To make sure their crops had enough water, the ancient Egyptians came up with ways to deal with the seasonal flooding of the Nile, which lasted about six months. During this time the farmlands were under water, so they developed an elaborate irrigation system to distribute the waters of the Nile to places far from the banks of the Nile for irrigation to convert their arid land into the bread basket of the Near East.

They dug irrigation canals and built catch-basins, created by building small dykes, to bring water to lands not reached by the flooding. Water was also carried in pails that hung from a yoke carried across the back.

Raising water from the level of the Nile to the surface of the farmlands was a very important activity in Egypt. An invention called the tanbour (Figure 6.5) made this task easier. The well-known scholar, Archimedes, invented the tanbour during his stay in Alexandria and named it the "Archimedes screw". It consists of a piece of wood in the form of a screw surrounded by an inched disk. The bottom part of the tanbour is placed in the water and rotated, causing the water to rise to the higher level; the tanbour was passed down by many generations of Egyptians to the present. Egyptian peasants still use it in times of low water levels.

Figure 6.5 The tambour

The shadouf (Figure 6.6) was invented in the New Kingdom. It was made up of a long pole either balanced on an upright pillar or post or suspended from a timber framework. The pole could pivot from side to side and up and down. At one end was a bucket, while a counterweight was at the other. A worker would pull the pole down to fill the bucket with water, and then the counterweight would raise the water to where it could be emptied into an irrigation channel. This helped irrigate higher land and increased the quantity of the grain harvest.

Figure 6.6 The shadouf

Another tool used since ancient times is the saqiyya (Figure 6.7). It consists of many small pots along a wheel, which is attached to a branch. A cow is tied to the branch which turns the wheel. The Greco-Romans continued to use it and it was improved by the Muslims. After the Islamic conquest of Egypt, kings and sultans undertook many projects to build canals, channels, and dams to bring water from the Nile for agriculture and drinking water. When Sultan Salah al-Din built the walls around Cairo, the roofs were used as channels for water. Pottery pipes led from these channels into private residences, distributing water to kitchens, bathrooms, and basins. Sultan Al-Nasir Mohammad Ibn Qala'un built an aqueduct, a structure for carrying a large quantity of flowing water, and four water wheels.

Figure 6.7 The saqiyya

A highly regarded occupation in Islamic Egypt was that of the Saqqa. The Saqqa was a person responsible for bringing water from cisterns to mosques, schools, and public drinking fountains. Like the Greco-Roman rulers before them, Islamic rulers used the level of the Nile's flooding to determine the amount of taxes to collect from farmers. They also used Nilometers and had an employee at Roda Island who recorded the level of the Nilometer.

6.5 HISTORICAL DEVELOPMENT OF WATER SUPPLY SYSTEMS

6.5.1 General

Evidence of activities concerned with human health and water supply has been found in civilizations throughout human history (James, 1998). The human search for pure water supplies must have begun in prehistoric times. Thousands of years passed before our more recent ancestors learned to build cities and enjoy the convenience of water piped to the home and drains for water-carried wastes (see Table 6.1).

Table 6.1 Historical development of water supply systems.

4000 BC	Water supply tunnels in Middle East
2000 BC	Water purification in Egypt and Iraq – they learn the benefits of filtration
312 BC	Roman aqueducts built (Aqua Appia, 18 km) – they learn that lead in water is toxic
300 BC	Storage cisterns used in cities (e.g. Istanbul)
1100 AD	Polluted water supplies in Europe = plagues
1183	Paris aqueduct built
1235	London makes same mistake as Romans – using lead pipes
1619	London provides house connections
1804	Sand filters used in Scotland, 4 millennia after Mesopotamia
1835	Charles Storrow writes Treatise on Water-works
1850s	Polluted water again = major cholera outbreaks in London
1860	Hamilton water works
1890	Chlorine disinfection
1993	Cryptosporidium infects 400000, Milwaukee

Earliest archaeological records of central water supply and waste water disposal date back about 5000 years. In the ruins of Nippur there is an arched drain with each stone being a wedge tapering downward into place. Wells, canals, aqueducts, reservoirs, cisterns and distribution pipes were built as a community effort to bring water to a central supply. An extensive system of drainage conveyed the wastes from the palaces and residential districts of the city. These waterworks were built by the Egyptians, Babylonians, Mesopotamians, Persians and Phoenicians centuries before the birth of Christ.

Sites excavated in the Indus Valley and in Punjab show that bathrooms and drains were common in Indian cities 4000 years ago. Streets were drained by covered sewers 2 ft deep and made of moulded bricks cemented with a mortar of mud. Within the houses, drain pipes were made of pottery embedded in gypsum (Rosen). Even two millennia BC, the Greeks and Egyptians had adequate supplies of drinking water for their cities, drained streets, had bathrooms in their houses, and, in Crete, water flushing arrangements for toilets. The Incas also had impressive sewerage systems and baths (James, 1998).

6.5.2 Tunnels

The first aqueducts in the form of tunnels or "qanats" originated in ancient Persia (Iran) probably as early as the fourth millennium BC. Their purpose was to bring water from the foothills of the northern mountains to the southern plains region for irrigation and domestic use. This photo (Figure 6.8) was taken in 1990 in the deserted village of Tanuf in Oman. The qanat is essentially a tunnel constructed for the conveyance of water from an underground aquifer to a point at ground surface some distance away. The water source is the tunnel which reaches down and into the water table. The other shafts provide ventilation and give access for cleaning and repair of the conduit tunnel below (Figure 6.9). Qanats carry an average of 10^5 gallons of water.

Figure 6.8 The Qanat

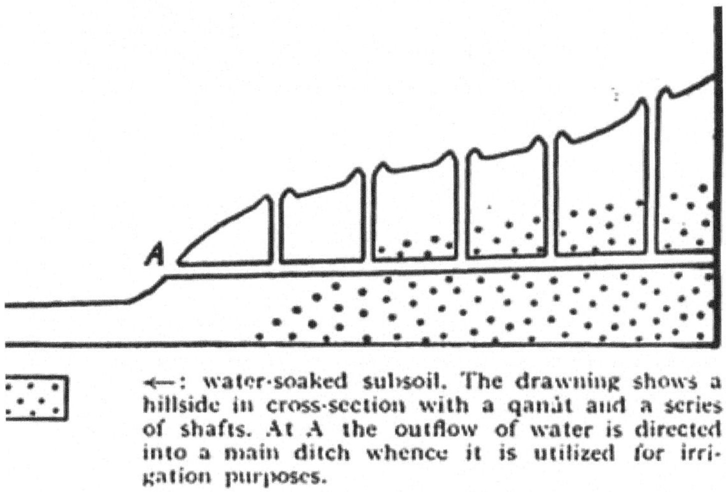

Figure 6.9 Qanat diagram

6.5.3 Cistern

One of the most amazing discoveries is actually the system of cisterns found in Alexandria. Napoleon's Scholars had counted only 400 after his invasion of Egypt, but a few decades later Mahmud el-Falaki had listed 700. By 1990, there was only one ancient cistern that could be visited, called el-Nabih (Figure 6.10). The others were lost to us, only memories of a distant past. However, 10 years later, thanks to the efforts of some very dedicated archaeologists, many of these forgotten wonders were once again found.

Figure 6.10 Inside El-Nabih Cistern

These subterranean canals were frequently described as underground cathedrals. Sometime between 1710 and 1712, Francois Paumier, a member of the third order of St. Francis wrote: 'nothing more superb than the pieces of marble with which they are surrounded.' In a report of the events of 1882, the English periodical, the Graphic, in an article concerning the shelling of the city centre claimed with admiration, 'there is nothing more beautiful and complete than the vaults; nothing better constructed than their apertures'. The British Navy gives an illustrated account of adventurous visitors, armed with hurricane lamps, exploring these huge subterranean spaces, 'Shrouded in darkness, they balance precariously on the arches which link the rows of columns rising out of the depths'.

When Khedive Ismail instructed el-Falaki to draw up a plan of the ancient city, the resulting book, published in 1872, explains that Alexandria was 'superimposed on another city of cisterns, the streets of which are subterranean canals (Dunn, 2011).

Distribution pipes were built as a community effort to bring water to a central supply. In the ruins of Nippur there is an arched drain with each stone being a wedge tapering downward into place.

6.6 WATER TREATMENT

The human search for pure water supplies must have begun in prehistoric times. Even two millennia BC, the Greeks and Egyptians had adequate supplies of drinking water for their cities, drained streets, had bathrooms in their houses, and, in Crete, Greece, water flushing arrangements for toilets (5). Earliest water treatment records come from Sanskrit writings and Egyptian inscriptions. The earliest record of the beginning of water treatment is linked to Hippocrates (known as the "Father of Medicine") in the fourth century BC. He advised boiling water to improve taste and then filtering through a cloth bag. Pliny the Elder (23–79 AD), in his *Historian Naturalis*, discusses the characteristics of potable water. In Egypt and China they put alum in tubs to clarify the water.

The earliest known apparatus for clarifying liquids was pictured on Egyptian walls in the 15th and 13th centuries BC (Figure 6.11).

Figure 6.11 Apparatus for clarifying liquids

The first picture, in a tomb of the reign of Amenhotep II (1447–1420 BC), represents the siphoning of either water or settled wine. A second picture, in the tomb of Ramses II (1300–1223 BC), shows the use of wick siphons in an Egyptian kitchen. The ancient Egyptian operators allowed impurities to settle out of the liquid, siphoned off the clarified fluid using siphons and finally stored it for later use (Jesperson, 1996).

But, early humans thought that the taste of the water determined its purity, and they did not consider that even the best tasting water could contain disease-causing organisms. We know now that just because water tastes good, it is not necessarily safe to drink. However, the efforts of these water treatment pioneers were not in vain. It was through their trials and errors that we now know how to make water safe to drink.

In more recent times, treatment of water to remove pathogenic organisms began around 1892 after Dr. Robert Koch traced the cholera epidemic in Hamburg, Germany, to its unfiltered water supply. Since

1855, in London, England, a parliamentary statute required its water supply to be filtered through slow sand filters. Slow sand filters were introduced into the United States around 1870 and a modern rapid-sand filtration plant was built in 1902 at Little Falls, NJ. In 1909, liquid chlorine was developed for disinfection of water supplies.

6.7 SANITATION

Proper sanitation is an important factor in any city in order to address the problems of health and sanitation. These issues were also important in the ancient world. The ancient Egyptians practised sanitation, but in the widest sense of the word as modern technologies were not available to them. However, they have what appears to be a workable, viable sanitation system. The degree of sanitation available to certain individuals varied according to their social status.

Bathrooms were built right in their homes. There is evidence that in the New Kingdom the gentry had small bathrooms in their homes. In the larger homes next to the master bedroom there was a bathroom that consisted of a shallow stone tub that the person stood in and had water poured over him. The waste water ran into a large bowl in the floor below or through an earthenware channel in the wall where it was emptied into still another bowl outside. Then that bowl was bailed out by hand. There is no evidence that the common people had bathrooms in their homes. Remains of early earth closets with limestone seats have been also discovered. The disposal evidently took place in the sandy soil.

By 2500 BC the Egyptians were pretty adept at drainage construction, accentuated by the significance that water played in their priestly rituals of purification and those affecting the burial of the kings. According to their religion, to die was simply to pass from one state of life to another. If the living required food, clothing and other requirements of daily life, so did the dead. Thus, it's not surprising that archaeologists have discovered bathrooms in some tombs.

Excavators of the mortuary temple of King Suhura at Abusir discovered niches in the walls and remnants of stone basins. These were furnished with metal fittings for use as lavatories. The outlet of the basin was closed with a lead stopper attached to a chain and a bronze ring. The basin emptied through a copper pipe to a trough below. The pipe was made of 1/16 inch beaten copper to a diameter of less than 2 inches. A lap joint seam hammered it tight.

Also found within a pyramid temple built by King Tutankhamen's father-in-law at Abusir, was a brass drain pipe running from the upper temple along the connecting masonry causeway to the outer temple on the river.

Excavators have discovered a tomb which supposedly contains the body of Osiris. It contains the dividing line between Life and Death, that is, a deep moat containing water that surrounds all sides of the figure of the god on his throne. After 5000 years, water still fills the canal through underground pipes from the River Nile.

6.8 THE HISTORY OF PLUMBING

The ancient Egyptians were early developers of pipes and the techniques of making copper alloys. In the beginning, of course, their pipes and fittings were very crude. Like the Mesopotamians, they used clay pipes made from a combination of straw and clay. First it was dried in the sun, and then baked in ovens. As they improved their clay sewer pipes, the Egyptians were able to drain the low-lying portions of the Nile Valley, and gradually the entire region evolved into a fertile garden (theplumber.com). The Greeks made their earthenware pipes in curved sections as well as straight, and of tapering shape so that each fitted into the next. This method was adopted by the Romans. In the sixth century BC the city of Athens was served by two aqueducts terminating in a reservoir from which water was distributed to the city, initially by a stone masonry channel (Figure 6.12), and later by pipes of earthenware and lead materials.

Figure 6.12 Stone channels

The Egyptians were also skilled in working metals. They melted metal in a crucible over a super-hot fire, provided by men fanning the fire with blowpipes made of reeds tipped with clay. The molten metal was poured out and allowed to cool, then beaten out with smooth stones into sheets of the required thickness. It was then cut to shape. One explanatory picture in a tomb chapel describes the process as "causing metal to swim" (theplumber.com). Very ancient metal pipes are known to have been used in Egypt. They were of hammered copper 1.4 mm thick. These pipes were cemented into grooves cut in solid flagstones.

Other examples of their craftsmanship are found in bowls of beaten copper on which they cast double spouts. Originally copper basins were used only by the pharaohs (theplumber.com)

6.9 WATER RESOURCES MANAGEMENT IN MODERN EGYPT

6.9.1 Aswan Dams

Before the dams were built, the Nile River flooded each year during late summer, as water flowed down the valley from its East African drainage basin. These floods brought high water and natural nutrients and minerals that annually enriched the fertile soil along the floodplain and delta; this made the Nile valley ideal for farming since ancient times. Because floods vary, in high-water years, the whole crop might be wiped out, while in low-water years widespread drought and famine occasionally occurred. As Egypt's population grew and conditions changed, both a desire and ability developed to control the floods, and thus both protect and support farmland and the economically important cotton crop. With the reservoir storage provided by these dams, the floods could be lessened, and the water could be stored for later release.

6.9.1.1 Construction history

The earliest recorded attempt to build a dam near Aswan was in the 11th century, when the Iraqi polymath and engineer Ibn al-Haytham (known as *Alhazen* in the West) was sent to Egypt by the Fatimid Caliph, Al-Hakim bi-Amr Allah, to regulate the flooding of the Nile, a task requiring an early attempt at an Aswan Dam. After his field work convinced him of the impracticality of this scheme, and fearing the Caliph's anger, he feigned madness. He was kept under house arrest from 1011 until al-Hakim's death in 1021, during which time he wrote his influential *Book of Optics*.

6.9.1.2 Old Aswan Dam

The history of modern water resources management in Egypt begins with the construction of the Old Aswan Dam in 1902 (Figure 6.13) and barrages on the Nile in the 19th and early 20th century. The Old Aswan Dam partially stored the water of the Nile to allow the growing of multiple crops per year in the Nile Delta, while the barrages raised the water level of the Nile so that water could be diverted into large irrigation channels running parallel to the river.

Figure 6.13 Old Aswan Dam

Following their 1882 invasion and occupation of Egypt, the British began construction of the first dam across the Nile in 1898. Construction lasted until 1902, and it was opened on December 10, 1902, by HRH the Duke of Connaught and Strathearn. The project was designed by Sir William Willcocks and involved several eminent engineers of the time, including Sir Benjamin Baker and Sir John Aird, whose firm, John Aird & Co., was the main contractor. The Old Aswan Dam was designed as a gravity-buttress dam; the buttress sections accommodate numerous gates, which were opened yearly to pass the flood and its nutrient-rich sediments, but without retaining any yearly storage. The dam was constructed of rubble masonry and faced with red ashlar granite. When constructed, the Old Aswan Dam was the largest masonry dam in the world. The design also included a navigation lock of similar construction on the western bank, which allowed shipping to pass upstream as far as the second cataract, before a portage over land was required. At the time of its construction, nothing of such scale had ever been attempted. Despite initial limitations imposed on its height, due to concern for the Philae Temple, the initial construction was soon found to be inadequate for development needs, and the height of the dam was raised in two phases, 5 metres between 1907–1912 and 9 metres between 1929–1933, and generation of electricity was added. With its final raising, the dam is 1950 m in length, with a crest level 36 m above the original riverbed; the dam provides the main route for traffic between the city and the airport. With the construction of the High Dam upstream, the Old Dam's ability to pass the flood's sediments was lost, as was the service ability provided by the locks. The previous Old Dam reservoir level was also lowered and now provides control of tail water for the High Dam (7).

6.9.1.3 Aswan High Dam

Construction of the High Dam (Figures 6.14 and 6.15) became a key objective of the Egyptian Government following the Egyptian Revolution of 1952, as the ability to control the flood waters, and harness the hydroelectric power that it could produce, were seen as pivotal to Egypt's industrialisation. The High Dam was constructed between 1960 and 1970. It aimed to increase economic production by further regulating the annual river flooding and providing storage of water for agriculture, and later, to generate hydroelectricity. The dam has a significant impact on the economy and culture of Egypt.

Figure 6.14 Aswan High Dam

The Aswan High Dam is 3830 m long, 980 m wide at the base, 40 m wide at the crest and 111 m tall. It contains 43 million m^3 of material. A maximum of 11,000 cubic metres per second of water can pass through the dam.

Figure 6.15 Aswan High Dam (NASA satellite photo)

There are further emergency spillways for an extra 5000 m^3/sec and the Toshka Canal links the reservoir to the Toshka Depression. The reservoir, named Lake Nasser, is 550 km long and 35 km at its widest with a surface area of 5250 km^3. It holds 111 km^3 of water (Figure 6.16). The water regime of the river changed fundamentally in 1970 when the Aswan High Dam was completed, eliminating the annual Nile flood.

Figure 6.16 Lake Nasser behind the Aswan dam, Egypt, (5250 km)

The dam brought major benefits such as increased water availability for Egyptian agriculture during drought years leading to higher income and employment, hydropower production, flood control, improved navigation, and the creation of fisheries in Lake Nasser. But, it also had negative environmental and social impacts including resettlement, loss of fertile silt that now accumulates in the reservoir behind the dam, water logging and an increase in soil salinity, and increased coastal erosion.

6.9.2 Implementation of new projects

Water resources management in modern Egypt is a complex process that involves multiple stakeholders who use water for irrigation, municipal and industrial water supply, hydropower generation and navigation. In addition, the waters of the Nile support aquatic ecosystems that are threatened by abstraction and pollution. Egypt also has substantial fossil groundwater resources in the Western Desert.

A key problem of modern water resources management in Egypt is the imbalance between increasing water demand and limited supply. In order to ensure future water availability, the Government of Egypt launched three mega-projects to increase irrigation on "new lands". They are located in the Toshka area, (the "New Valley"), on the fringe of the Western Nile Delta, and in the Northern Sinai. These projects all require substantial amounts of water that can only be mobilised through better irrigation efficiency and reuse of drainage water and treated wastewater (7).

6.10 CONCLUSIONS

The Nile River has played an extremely important role in the civilization, life and history of the Egyptian nation. One of the most well known river Nile facts is the river's ability to produce extremely fertile soil, which made it easy for cities and civilizations to spring up alongside the banks of the Nile. The fertile soil is contributed to by the annual flooding, when the Nile River overflows onto the banks. Much of the Egyptian nation consists of dry desert land. Throughout most of the year, very little rain falls on Egyptian deserts. This has remained true for thousands of years. The abundant Nile River provided much needed irrigation, even in ancient times. This waterway also provided a source of drinking water, and source of irrigation for farming as well as papyrus reeds that could be used for a variety of purposes such as paper and building materials.

In modern Egypt, water poses a different set of challenges. As a result of population increase, industrialisation and rapid economic development, freshwater resources are on the decline, with demand far exceeding the available supply. Significantly greater efforts are therefore, required to formulate and implement national flexible and sustainable water development strategies and to monitor the state of the water environment within and across countries.

REFERENCES

Ancient Egypt – Land of the River. 123HelpMe.com. http://www.123HelpMe.com/view.asp?id=23403 (accessed 13 January 2012).

Clayton and Peter A. (1994). Chronicle of the Pharaohs. London, England: Thames and Hudson. ISBN 0-500-05074-0, http:en.wikipedia.orgwikiAncient_Egypt (accessed 7 January 2012)

Dunn J. (2011). The Cisterns of Alexandria. www.touregypt.net/featurestories/Alexandriacisterns.htm

James T. G. H. (2005). The British Museum Concise Introduction to Ancient Egypt. Ann Arbor, Michigan: University of Michigan Press. ISBN 0-472-03137-6. http:en.wikipedia.orgwikiAncient_Egypt

James W. (1998). A Historical Perspective on the Development of Urban Water… www.soe.uoguelph.ca/webfiles/wjames/homepage/…/wj437hi.htm

Jesperson K. (1996). Search for Clean Water Continues http://www.nesc.wvu.edu/old_website/ndwc/ndwc_dwh_1.html

Manuelian P. (1998). Egypt: The World of the Pharaohs. Bonner Straße, Cologne Germany: Könemann Verlagsgesellschaft mbH. ISBN 3-89508-913-3. http:en.wikipedia.orgwikiAncient_Egypt

Nicholson P. T. *et al.* (2000). *Ancient Egyptian Materials and Technology*. Cambridge, UK: Cambridge University Press. ISBN 0-521-45257-0, http://en.wikipedia.org/wiki/Ancient_Egypt www.theplumber.com. The History of Plumbing – Egypt

Chapter 7

The impact of climate changes on the evolution of water supply works in the region of Jerusalem

A. S. Issar and M. Zohar

7.1 INTRODUCTION

Various investigations were carried out on the ancient water works of Jerusalem yielding a long list of articles and books that were published towards the end of the 19th century which still continues to grow to this day[1]. These investigators were mostly archaeologists, hardly ever geologists or hydro-geologists specialising in the hydro-geology of limestone and dolomite terrains which characterise the region of Jerusalem.

The outstanding deficiency of most prior works is the neglect of the impact of climate changes on the history of the region in general and on that of Jerusalem and on its water works in particular. A look at the precipitation map of Israel (Figure 7.1) reveals that, although at present it is situated at the annual average precipitation line of 500 mm, its eastern border is immediately on the Judean Desert where the annual precipitation declines abruptly to the desert precipitation border line of 200 mm about 15 km to the east. To the south this desert line is some 60 km away. Thus the movement of this desert line north, due to periods of global warming, as happened throughout history (Issar & Zohar, 2007) has caused aridisation and the desertion of cities. During such extreme periods of droughts, springs, particularly those fed by a local perched water table as in the region of Jerusalem, may dry up and lead to the abandonment of the site. Such events have been observed throughout the past in the region.

On a short time scale, fluctuations in the precipitation may be rather extreme as in the following example. The wettest year of record since 1870 was 1992 after the explosion of the Pinatubo volcano in the Philippines. The rainfall during that year was six times more than 1960 which was the driest year on record. In this connection it should be mentioned that the impact of the volcano's explosion causing reduction of solar illumination and the influence of solar-spots on the short term changes in precipitation in the Middle East has not yet been thoroughly investigated. The wettest 10-year period on record, 1889–98, had 1.8 times more average annual rainfall than the driest 10-year period, 1925–34.

The impact of the precipitation regime on the hydro-geology of the region of Jerusalem is of primary importance. The present moderate annual quantity of about 550 mm is not sufficient to bring about the

---

[1]A partial list of authors and their dates of publication include Wilson & Warren (1871), Warren & Conder (1884), Vincent (1912), Amiran (1951), Ussishkin (1976), Issar (1976), Shiloh (1984), Gill (1991), Rosenberg (1998), Reich & Shukron (1998, 2006), Frumkin *et al.* (2003).

development of an advanced system of "karst", characterised by sinkholes and underground rivers as in the Dinarids Alps. Thus the general regime of flow is in narrow solution channels in the saturated part of the aquifer and the groundwater flow through a certain section of the saturated part of the aquifer is according to D'arcy's Law, namely decided by the permeability coefficient, and the gradient of the water table. Yet, underground "karstic" channels still exist in the region of Jerusalem, which speaks for more humid palaeo-climates, during which these subsurface flow systems developed. The higher the annual precipitation the more intense are the "karst" forming processes of the limestone and dolomite rocks. In brief the term "karst" used in this context refers to the dissolution processes of the carbonate rocks (limestone and dolomite). These processes form cavities and surface and subsurface shafts and passages. The dissolution is due to the fact that the rainwater becomes acidic when in contact with carbon dioxide in the atmosphere and the soil. This light acidic water infiltrates into fractures in the underlying rock and dissolves it thereby creating a network of passages and cavities. Over time, as water continues to flow through the dissolution channels, it continues to dissolve and to erode the rock and thus enlarge the passages. Caves are formed where water runs as streams and dissolves and erodes the rocks. Eventually the water issues as "karst" springs once the subsurface stream reaches a topographical low formed by surface erosion.

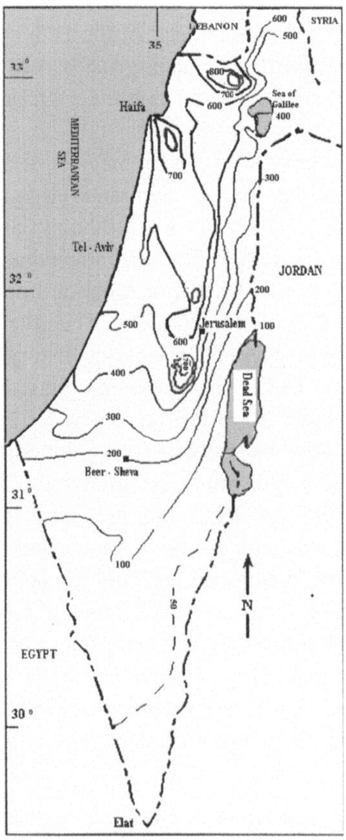

Figure 7.1 Precipitation map

Another deficiency in the archaeological works on the ancient water works of the region of Jerusalem is the absence of any field observations related to the ancient methods still used in a few countries in the Middle East. Thus, one finds no reference to the water galleries in alluvial fans and in bed rock (*qanats*) still practised in Iran and other parts of the Middle East for the development of natural and artificial springs (Beaumont *et al.* 1989). Such galleries, typified by a series of shafts dug at a certain distance each from the other, to supply air and remove the excavated material, are found in the region of Jerusalem and will be discussed later. When subsurface galleries are dealt with, no reference was made by the investigators who discussed ancient galleries, to the problem of the supply of air to the digger. The personal experience in deep mines and tunnels made it clear that a tunnel which is not continuously supplied by air turns into a suffocation chamber. This convinced Issar (1976) and Issar & Zohar (2007) that the famous Hezekiah's Tunnel, connecting the Gihon spring to the Siloam Pool, was carved from both sides along a natural karstic water passage resulting in the serpentine configuration of the tunnel. This suggestion was supported by the geologist Dan Gill (1991). The fact that this passage of water also supplied air was important as will be discussed in more detail in Section 7.5.

7.2 PRESENT CLIMATE

The climate of the Middle East is decided by the Mediterranean climate belt as well as by that of the Saharo-Arabian desert climate belt. The Mediterranean climate is part of the climatic westerlies system which brings cold air masses from the Atlantic during winter. While travelling over the relatively warm Mediterranean these air masses become saturated with moisture. The entrance into the higher area of land causes adiabatic cooling due the rising of the air masses and their expansion (lower atmospheric pressure) resulting in precipitation. The Saharo–Arabian desert belt is dry during all seasons. During summer this belt moves northward and the Near East becomes part of this belt. The rate of the north and south movement of these two belts affects the mean annual quantity of precipitation as well as its variability from year to year.

Within this regime the specific climate of Jerusalem is decided by it being partly situated facing the Mediterranean Sea and thus in the path of the cold winter rain storms enhanced by its topographical position, at an average altitude of 700 m above sea level, intercepting these storms. The average annual precipitation is about 550 mm. The rainiest months are between November and February. Light snow may occur during the winter months while heavy snow fall happens only during very cold winters. January is the coldest month with an average temperature of 8°C while July and August are the hottest months with an average temperature of 23°C. Evenings and nights during summer are cool and cold to very cold in winter. During the transitional periods of spring (April–May) and autumn (September–October), dry eastern-south-eastern winds are common, blowing from the desert, (*khamsin* in Arabic, *sharav* in Hebrew, *sirocco* or *föhn* in Europe) which may bring hot air up to 40°C as well as dust storms. Dew may fall during warm summer nights.

7.3 CLIMATES OF THE PAST

Although the present chapter deals with the impact of climate changes on the evolution of the water supply works in the region of Jerusalem during its more recent history, the climates of a few hundred to ten thousand years ago still have to be considered as they are crucial to understanding the special features of the development of its karstic groundwater system, as mentioned above.

Various investigations have shown that the region passed through a series of severe climate changes connected to the global glacial and inter-glacial phenomena at the onset of the Quaternary Age starting about 2.5 million years ago. This period was also one of the uplift of the mountainous backbone on both side of the Great Rift Valley. The uplifted zone comprised the Judean Mountains. In general, the climates during glacial periods of the Quaternary Age were cold and humid, whereas inter-glacial periods and the present post glacial period were warm and dry (Issar & Zohar, 2008; Issar, 2008). Evidence for the high humidity rate of the last glacial period can be deduced from the fact that the Dead Sea extended over most of the Jordan Valley during this period. This body of water was called Lake Lisan, i.e. "tongue" in Arabic, for the marls deposited at its bottom which had formed a peninsula and projected into the Dead Sea on its eastern shore (Picard, 1943). The humid period began around 70,000 years before present (BP) and, after a few fluctuations, the lake had reached its maximum level of about 164 m below MSL by *ca.* 25,000 BP. It stayed at this level for roughly 2000 years and then began to drop. By *ca.* 10,000 BP it had reached the level of *ca.* 325 m below MSL (Bartov, 1999). During the Holocene, namely the last 10,000 years, cold and humid climates caused the level of the Dead Sea to rise while warm and dry periods caused its drop (Bookman *et al.* 2004; Issar & Zohar, 2007).

Biblical texts as well as historical documents and the archaeological evidence leave no doubt that throughout its past, the southern portion of the land of Israel has always been more arid than its central part, not to speak of its northern part. This difference clearly indicates that the westerlies regime was the main climate system responsible for transmitting precipitation during these late pre-historic and historic periods. Consequently, the southern part of Israel not facing the sea, due to its being south of the path of the cyclonic-lows, has always received less precipitation than the area facing the sea. Additional evidence can be derived from the stalagmites of the Soreq Cave south-west of Jerusalem and thus on the slope facing west. The altitude of the cave is about 400 m above MSL. It is about 40 km inland from the Mediterranean Sea and gets an average annual precipitation of about 500 mm. The cave contains stalagmites, which are still being formed due to water dripping from its roof. The dating and analysis of the isotopes of the rings of the stalagmites correlated with that of present average precipitation dripping into the cave enabled the reconstruction of the palaeo-precipitation regime of this area (Bar-Matthews *et al.* 1996).This indicates the order of magnitude of the precipitation during the last 8000 years in the Jerusalem region (Issar & Zohar, 2007).

7.4 GEO-HYDROLOGY

The region of Jerusalem includes the city and its municipal territory which spreads over the area in a radius of *ca.* 10 km from the centre of the city. It forms part of the central ridge of the mountainous "backbone" of the country formed by a series of anticlines, divided by synclinal topographical saddles in one of which Jerusalem is situated. To the south extends the anticline of Hebron while to the north is that of Judea. This ridge is also the watershed, surface and sub-surface. The flow eastward from the watershed is towards the Dead Sea whereas the western flow is towards the Mediterranean Sea. While the surface flow is seasonal in the mostly dry riverbeds (*wadi* in Arabic, *nahal* in Hebrew) flowing in both directions, the regional subsurface flow is perennial in the saturated part of the limestone-dolomite aquifers and perpendicular to the direction of the strike of the anticlines. The geological structure gives a SE direction in the eastern watershed and a NW direction in the western watershed. The flow in the perched aquifer is according to the local geological configurations, lithology, dips and faults.

As mentioned above, the aquiferous rocks under most of the area of the region of Jerusalem are limestones and dolomites of the Judea Group of the Cenomanian Turonian, formed during the Cretaceous age. The permeability of the rocks is a function of dissolution processes by rain and subsurface flow, either free or confined. In certain areas local conditions cause higher permeability of the rocks. These special conditions depend on the lithological character of the rock (coarse crystalline not tightly consolidated) or fracturing due to folding or faulting. In some cases special high permeability can be traced to karstic dissolution processes forming channels and caves. Such are found frequently at the outlets of springs. Impermeable layers of marls and clays are situated between the permeable layers which cause the formation of perched water tables, local aquifers and springs. The western and eastern flanks of the anticlines are built of the impermeable chalks of the Mount Scopus Group (Menuha and Mishash Formations, Figures 7.2 and 7.3).

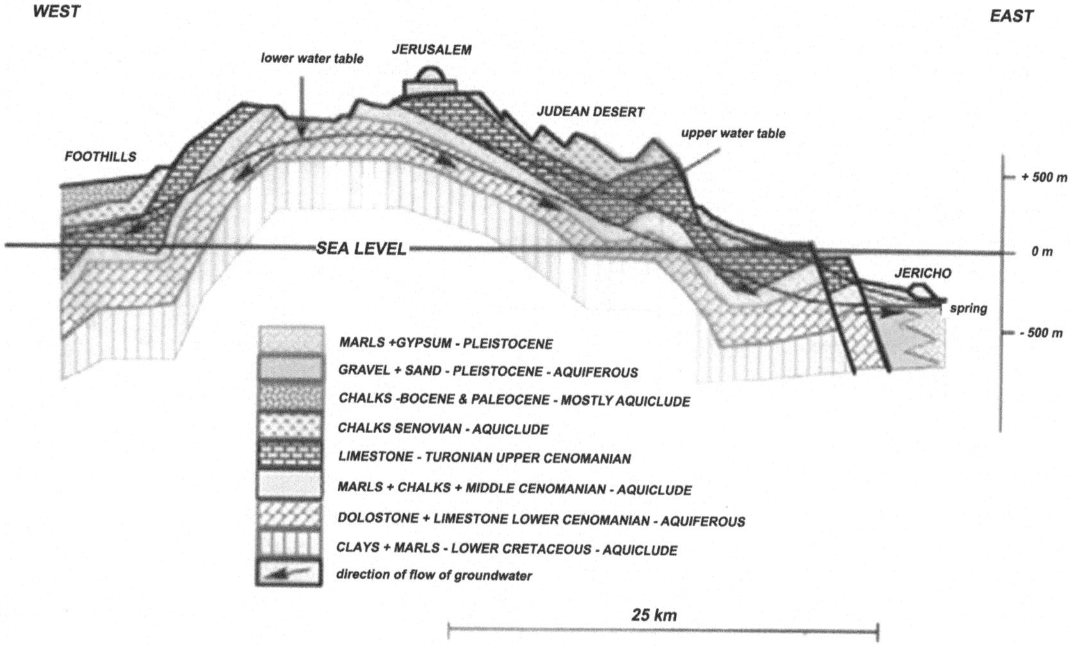

Figure 7.2 Hydro-geological cross-section of Central Israel and Jerusalem (According to Issar, 1990)

From top to bottom the uppermost aquiferous unit of about 120 m thickness is built of limestone of the Bina Formation resting on dolomite of the Veradim Formation. This aquifer rests on a marl layer of about 20 m thickness, the Kefar Sha'ul Formation, on which a local perched water table may be formed. Beneath it is found the next aquiferous layer, the Amminadav Formation. It consists of a fine crystalline dolomite rock which is a permeable sequence of about 150 m thickness. Underlying this sequence is an impermeable marly, chalky layer with a thickness of about 20–40 m, the Moza Formation. This impermeable layer is rather extensive and forms a regional perched water table which feeds a series of springs. The Bet-Mei'r Formation, about 100 m thick, is built of alternating layers of dolomites and marls and is semi-permeable, forming in part of the region an aquitard allowing the passage of water, and in other parts an aquiclude, an enclosed and sealed body of water. In the areas where it is an aquiclude, it contributes to the perching character of the Moza Formation, while in some areas the dolomite is more permeable, and the marl layers between the dolomitic layers form local perched water tables. The underlying sequence is built mainly of dolomite of the Lower Cenomanian period. They consist of the Kessalon, Soreq, Givat-Ye'arim and Kefira Formations which together are about 200 m thick. This mainly permeable layer is situated on the Katana Marls of the Lower Cretaceous age which forms the basis of the whole regional aquifer of the Judea Group.

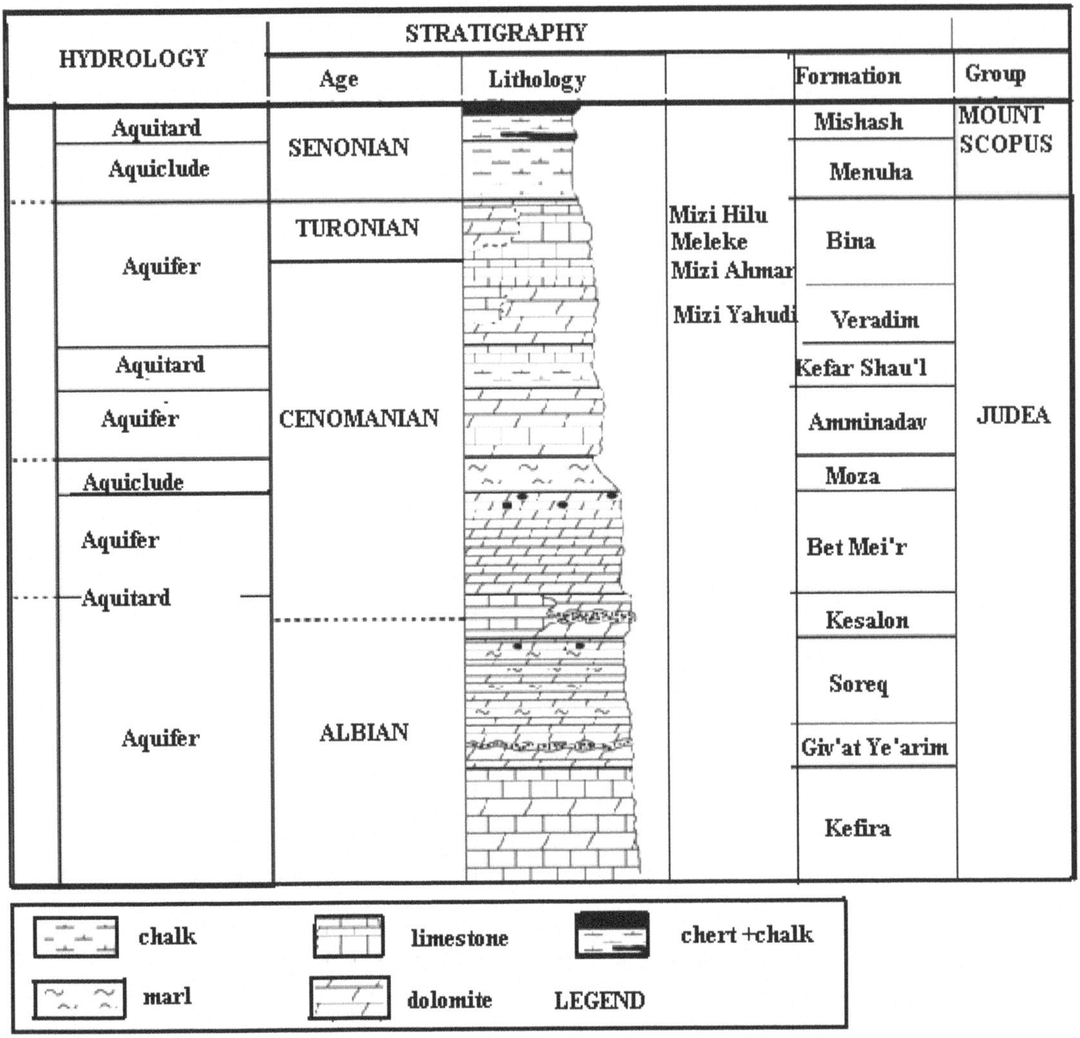

Figure 7.3 Hydro-geological columnar section (according to Gavrieli *et al.* 2002)

7.5 THE IMPACT OF CLIMATE CHANGES ON THE ANCIENT WATER WORKS

Three main factors have dictated the nature of the water works in Jerusalem from its earliest history until today.

(a) The short term climatic factor, namely the dry summer months and occasional years of drought.
(b) The long term climatic factor of the past periods of wet and dry climates.

(c) The need for security which dictated the building of the often fortified settlements on the peaks of hills and thus creating the need for a secure access to the nearest water resource in the valleys.

The short term climatic factor, of rains during winter and dry summer months, dictated, in the first place, the need for collecting the rainwater. This called for the excavating of cisterns either by private people in their yards collecting the water from the roofs of their houses, or by the authorities creating reservoirs or pools as public projects. This device of water supply, known since ancient times was not, however, always practical in Jerusalem due to the fact that the oldest part of city (variously known as the Eastern Hill, Ophel, Jebusite City, City of David, Siloam and Silwan) was built on the rather permeable Bina limestone. Thus the construction of cisterns had to wait until the invention of impermeable plaster sealing the bottom and the walls of these structures. The production of plaster involved the smelting of limestone to produce lime, slaking it with water and producing some kind of paste. William F. Albright suggested in 1940 in his monumental work *From the Stone Age to Christianity* that it was the appearance of iron tools for digging water cisterns in bedrock and the invention of impermeable plaster which enabled the early Israelites to settle in the hills. Zertal (1990) suggested that during the early occupation of the Iron Age/Israelite period, of the 13th to the 11th century BC, water was stored primarily in large earthen jars (*pithoi*), a working solution for a limited population. The excavations carried out in Jerusalem have shown that already the pre-Israelite inhabitants during the Middle Bronze Age were able to hew and cut horizontal and vertical passages through the bedrock in order to reach the springs below their city. However, as they seem to have lacked the knowledge to produce plaster to stop the leakage, there is, so far, no example of pre-Israelite cisterns or tanks, although such cisterns may have existed but were re-excavated or have not yet been found.

The other method to guarantee water supply during the summer months was to locate the fortified city above a natural perennial spring and construct sheltered access passages to reach the spring in the valley, particularly in times of siege. This solution became practicable after the introduction of bronze technology around 2000 BC as well as of iron tools by 1000 BC. It would involve searching for a passage to approach the spring itself or to draw or divert water from it to the site where it was needed. Such passages existed as natural solution channels, the work needed was to enlarge them and build openings, staircases and tunnels to allow access to the spring.

7.5.1 Impact of climate changes during prehistory

The climate during the Pleistocene, the time of the arrival of the Hominins stone tool makers, was a function of the changes between glacial and inter-glacial periods. Glacial periods, as evidenced by the Mediterranean Sea regressions, were cold and humid, while inter-glacial and post glacial periods, evidenced by transgressions, were warm and dry. During glacial periods in the region of Jerusalem terra-rosa type soils developed, while during inter-glacial periods erosion took place. The climate during the last glacial period (*ca.* 70,000 to 15,000 BP) was not different, namely, generally cold and humid. During the last glacial period the palaeo Dead Sea extended over most of the Jordan Valley (Bookman *et al.* 2004; 2006). At the end of this glacial period at *ca.* 15,000 BP, warmer conditions prevailed over the Middle East.

Tools of Palaeolithic hunters and gatherers have been discovered and dated most probably to glacial periods in the north and thus humid periods in the Middle East, as these tools were found in heavy red loams characteristic of these periods. A site of the Middle Palaeolithic age (*ca.* 200,000–50,000 BP) was excavated at Ramat Rachel, on the southern outskirts of Jerusalem. Two other sites of the Palaeolithic period were found in its southern part (Emeq Rephaim Street) and in the vicinity of Mount Scopus.

So far no remains of the Neolithic periods, *ca.* 9000 to 5000 BC, have been found around Jerusalem. Neolithic settlements seem to have been concentrated in the valleys and in the coastal plain of the

Mediterranean Sea. Small amounts from the Chalcolithic period, *ca.* fifth and early fourth millennia BC, were found throughout most of the excavations of the area of ancient Jerusalem on the eastern hill, particularly around the Gihon Spring (Shiloh, 1978–1984).

Remains of the Early Bronze Age, dated to the third millennium BC, were reported by the archaeologist Yigal Shiloh as well as Reich and Shukron (2002a) on the eastern hill. The cemetery explored by de Vaux on the central section of the upper slope of the Ophel was dated to the second phase of the Early Bronze Age I period (de Vaux, 1971). The size of the cemetery indicates a modest village. The evidence for the subsequent Early Bronze Age II and III period, in general the first urban phase characterised by monumental architecture in the Levant, is extremely thin on the ground, again pointing to a village or small but possibly fortified town. The water supply of a settlement this size during the short term dry season in an overall cold and humid climate of the Early Bronze could easily be solved by storing water in jars or basins, sherds of which were found at the lowest levels of the excavations and in crevices of the rock. Whatever the case, the massive construction work of the following periods eliminated any archaeological evidence of these early periods. During this period a karstic gallery, enlarged to a tunnel may have been excavated from the small walled city on the Eastern Hill to reach a small perched spring. This spring dried up during the Intermediate Bronze Period and the gallery was abandoned. This project will be discussed in detail below.

The Early Bronze Age ended around the middle of the third millennium BC followed by the Intermediate Bronze Age (IBA), which was a dry period, the desiccation reaching its maximum around *ca.* 2200 BC. The cities were abandoned or destroyed and the economy of the entire region shifted accordingly to a more pastoral subsistence economy. There are no remains of this period on the eastern hill due to the massive reworking of the area during the following periods, as mentioned above. However, rock-cut tombs, common during this period and regarded as a sign for a dominantly transhumant way of life, are found scattered throughout the hill country indicating that some sort of settlement did exist.

7.5.2. Impact of climate changes during history

7.5.2.1 Middle to Late Bronze Age period

During the Middle Bronze Age from 2000 to 1550 BC the climate became relatively humid again, which enabled a cultural and material renaissance. During this period the introduction of bronze and other technological innovations took place. The excavations on the Eastern Hill revealed that the most impressive visible remains were of a fortress-city of the Middle Bronze Age dated to around 1900 to 1800 BC. Most probably Jerusalem and its territory had then become an independent city-state ruled by a king similar to her neighbours such as Shechem (Tell Balata) to the north, Hebron (Tel Rumeida) to the south and Gezer to the west. During that period different variations of the city name "Jerusalem" were found in the Tell el-'Amarna letters in Egypt (*ca.* 13th century BC) and the Sennacherib inscriptions (*ca.* 7th century BC). The meaning of the name is not clear, but it possibly originates from the Sumerian word for town or city *uru*, or the root *yrw* to found, establish, combined with the name of a West Semitic deity *Salimu*, the god of sunrise.

Now we come to the question of why the low topographical extension of the Eastern Hill (Figure 7.4a) was chosen for a settlement throughout history rather than the higher level of the Temple Mount (Moriah), or any of the other neighbouring hills such as Western Hill across the Tyropean Valley (Mount Zion) or the Mount of Olives to the east (Figures 7.4a and b).

The conventional answer to this question is that ancient Jerusalem was nearer to the perennial Gihon spring. As mentioned before, we suggest that there existed a small perched spring high up flowing from a karstic channel at the bottom of the porous Meleke crystalline limestone on top of the Mizi Ahmar denser dolomite (Issar, 1976). We suggest that the most probable reason for deciding to settle at this site for the first time, as well as for the location of the Middle Bronze walled city, was the presence of this spring which, flowed during the Early Bronze Age, dried up during the Intermediate Bronze Age and may have started to flow again during the more humid Middle Bronze Age period. It dried up again during the dry Late Bronze Age period. Moreover, about 1000 years later the existence and development of the karstic horizontal channel leading to this spring helped the discovery and development of the system known today as Warren's Shaft which enabled the people of the city to reach the spring of Gihon from inside the city.

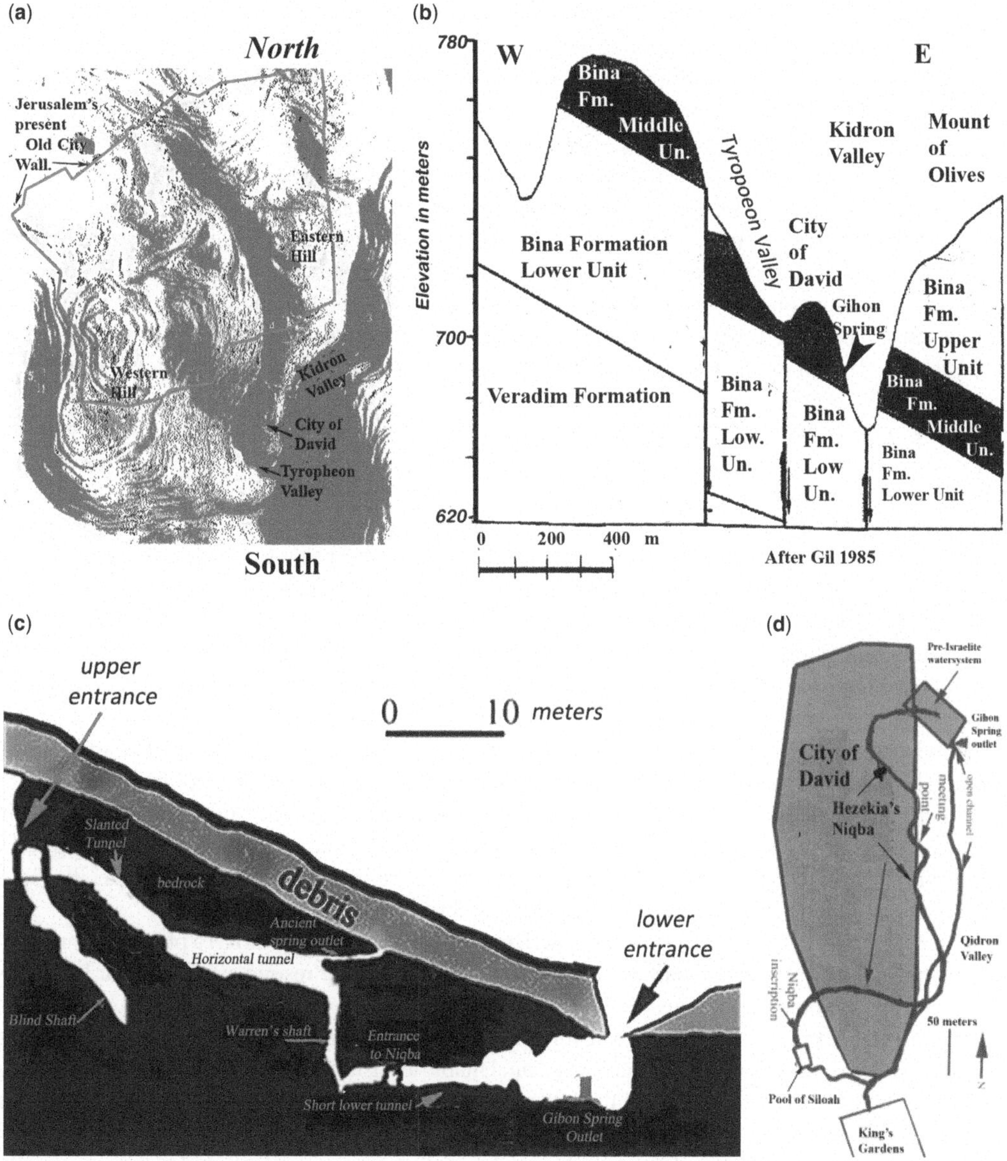

Figure 7.4 (a) Relief map of Ancient Jerusalem. (b) Geological cross-section through City of David. (c) Water supply system of Jebusite City and City of David. (d) The City of David, the Gihon outlet and the "Niqba"

The existence of such a spring was suggested by Issar (1976) and the excavations of Reich and Shukron confirmed, in our opinion, Issar's suggestion. In their report of 2002(a) they ask why the roof of the tunnel of approach to Warren's Shaft is 6.5 m high. When standing at the upper opening of that shaft looking south, an opening about 3 m higher than the floor of the tunnel can be seen. Reich and Shukron cleaned this opening and discovered that the tunnel turns east and leads to the top section of a rock-cut pool. By accident they found remains of the expedition of Vincent and Parker of 80 years before. This tunnel follows the contact line of the Meleke crystalline permeable limestone and the dense dolomitic Mizi Ahmar rock. We suggest that the natural spring, then still flowing, was fed by the karst passage, which was followed by the tunnel diggers of the Middle Bronze period, who may have widened a more ancient tunnel. This perched spring received its water from the permeable Meleke layer. Shukron and Reich maintain that when this tunnel, the height of which was about 2.5–3 m, was cut during the Middle Bronze Age, the workmen were unaware of Warren's Shaft, a natural sink hole just below their feet (Figure 7.4c).

As the climate during the second millennium BC was less humid and more varied than that of the preceding period, the supply of the upper perched spring may have diminished or even dried up in summers of aridity. Consequently, collecting pools on the slope below the outlet of the small spring as well as surface water flow from the near environment had to be constructed. In addition, the entrances to the upper spring and the pools which collected its water, as well as that of the lower lying spring Gihon, were protected by massive towers discovered by Reich and Shukron in the late 1990's (2002a). They measured 13.7 × 16.8 m with a wall thickness close to 4 m and were appropriately called by the discoverers "Pool Towers".

The spring of Gihon (from *g-y-h* to burst-out) is a typical pulsating karstic spring issuing from the dolomitic rocks of the lower part of the Bina Formation and the upper part of the Veradim Formation. The karstic system has developed, most probably, on the semi-permeable layer of the Kefar Shaul chalky-marl layer. The pulsation is due to a siphon structure in one of the karstic channels: when the level of the water table reaches above the "knee" of the siphon, the water emerges with a throb, adding to the basic flow of the spring coming from the other sub-channels. The duration of each pulse used to be about 30–40 minutes, occurring each 4–6 hours during winter and about 8–10 hours in the dry summer. The storage capacity of the spring varies due to its limited local recharge basin. Accordingly, after years with high precipitation such as, for instance, winter 1983 the daily output reached about 5000 m^3/d, whereas in autumn 1979, a relatively dry year, it reached only 700 m^3/d. We assume that during humid periods the spring reached the surface and overflowed into the Kidron Valley. The farmers living down the valley utilised the water by digging channels and diverting it to irrigate their terraced fields. When the spring disappeared in dry years, attempts were made to enlarge the outlet. Eventually, a curved stepped slope was cut to facilitate fetching water by jars or skins. When the yield of the spring fell to a minimum, or even dried up during extremely dry years, attempts were made to renovate its yield by driving tunnels into the fissures wherefrom the water used to come out.

From *ca.* 1500 to 1300 BC namely over the Late Bronze period (1500–1200 BC) a phase of warming and aridity took place. The result was a marked negative impact on the water resources of the Near East causing political and socio-economical changes in its wake.

Towards the end of the period, a phase of severe desiccation most probably caused the pastoralists east of the Jordan and in the southern deserts to invade Canaan. Unable to sustain their herds they were forced to settle in the more humid parts of the country. (In the letters found at Tell el Amarna in Egypt the kings of Canaan complain about the threat of the "piru/Habiru" pastoral nomads who were gaining a hold in the region.) Whether these are the forerunners of the "Children of Israel" is a question still under debate. Parallel to this invasion from the east and south, and again probably linked by similar environmental changes, a series of sea-borne invasions of the so-called Sea Peoples, among them the Philistines, arrived from the west.

In Jerusalem the arid climate caused the small upper perched spring discussed above to dry up. The level of the Gihon subsided and put the former Middle/Late Bronze Age water works out of use. Whether this fact brought about the desertion of the city by its former inhabitants and its occupation by other tribes, such as the Jebusites from Anatolia, can only be surmised. And so, Reich and Shukron's (2002a) claim that the famous Warren's Shaft was a post-Middle Bronze Age. Project appears to be substantiated but with the question of its date remaining open.

As the main effort of the new people occupying the city was to ensure their supply from the Gihon spring, especially in times of siege, it brought them to develop the karst channels leading to the outlet of the now dry small perched intermittent spring mentioned above and utilise a karst shaft to arrive at the level of the main spring. This feature is the famous Warren's Shaft, a 40 m long vertical sink-hole. At the bottom of the shaft is a small solution hole where containers lowered by a rope could be filled (Figure 7.4c). Whether this shaft is the Zinnor which helped Joab to conquer Jebus (II Samuel V:8) is outside the subject matter of the present work. While enlarging the cave from which the Gihon emerged, a solution channel was found connecting the Gihon to another karstic spring known as Siloam. The connection between the two springs could be observed either by the similar pulsation or simply by throwing straw into water at the Gihon and seeing it appearance at Siloam. This gave the idea for a unique project which will be discussed later (Figure 7.4d).

Two other sophisticated water works were found in the Biblical city of Gibeon (present village al-Jib), which is situated about 10 km north of Jerusalem, and its water works also show the impact of the dry period of the Late Bronze period. (The root of Gibeon is "*gev*", namely water-hole and in Arabic *Jib*). These water works comprise two independent systems (Pritchard, 1961): (a) a large pool with its spiral stairway (Figure 7.5) and (b) an inclined gallery leading to the spring outside the walls and horizontal galleries feeding the spring (Figure 7.6).

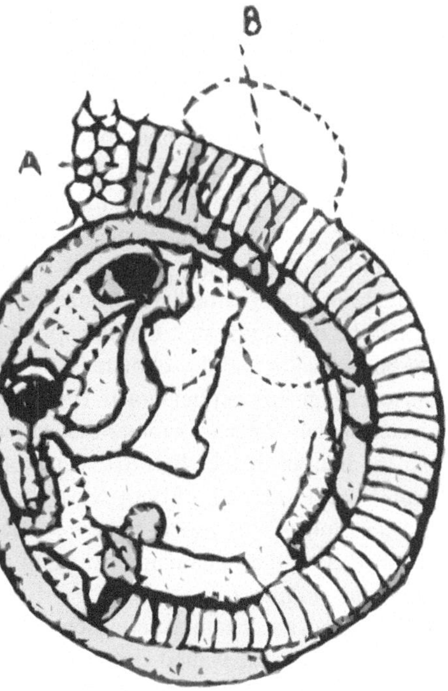

Figure 7.5 Gibeon, large pool with spiral stairway (After J. Pritchard: The Water System of Gibeon, 1961)

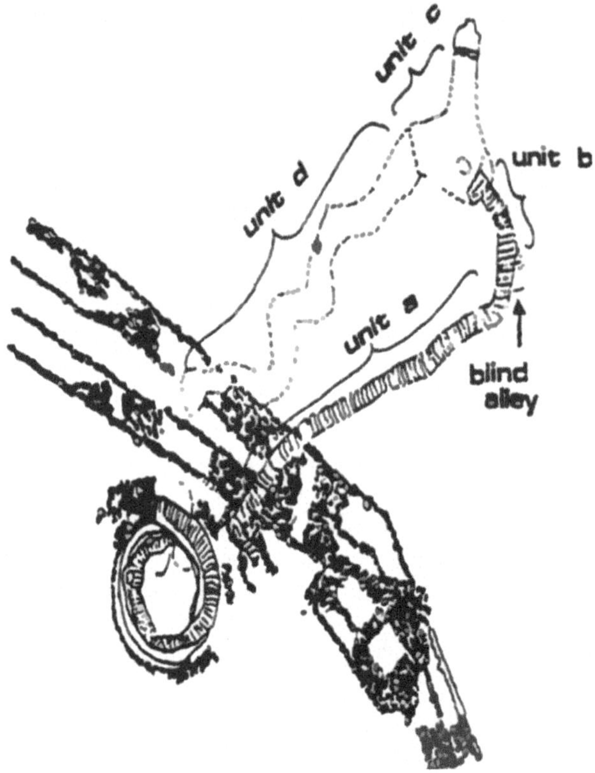

Figure 7.6 The gallery water system of Gibeon (After J. Pritchard: The Water System of Gibeon, 1961)

The first system consists of a pool and stairway to the bottom of the pool and another steep passage from the bottom of the pool to a small cavern with a dripping spring. Issar (1976) suggested that the general evolution of this system was like that of Jerusalem, namely that in the first stage a passage to a small perched spring was developed. This passage was enlarged and widened to a shaft with a staircase so that water could be carried with less effort. It is now suggested that the flow of the small perched spring dried out as a result of the dry climate of the Intermediate Bronze period. (In 1976 Issar was not aware on the severity of climate change and he thus suggested that the spring dwindled 'due to the covering of the drainage basin'. The new data on the warm and dry climate of the Intermediate and Late Bronze Age (Issar and Zohar op.cit.) led him to change his explanation). The shaft was then excavated during the Middle Bronze Age to look for water. When water was found the staircase of the shaft was developed to enable the carrying of water from the bottom of the shaft. (It seems probable that this system, after the second system leading to the spring was developed, did not serve for daily supply but as a ritual place connected with water, though a discussion of this aspect is beyond our scope here). When the climate dried up again during the Late Bronze Age, the inhabitants extended the shaft downwards into a lower karstic cave containing a dripping spring (Figure 7.5).

The second system is like that of the Jebusite system of Jerusalem, which was built to reach the perennial spring emerging at the foot of the hill on which the walled city was situated. Also in the case of Gibeon the changes due to the climate changes can be seen. The whole system leading to the spring is composed of four units (Figure 7.6):

(a) A gallery, moderately inclined, from the inside of the wall reaching a small blind alley. It is suggested that this gallery led to a small perched spring, which dried up during the Intermediate Bronze Age.
(b) A steep gallery deviating from the lower end of "unit a" and opening into "unit c". This gallery may have been constructed during the Middle Bronze Period.
(c) A spring cave-room.
(d) A feeding gallery of the existing spring, opening into the cave-room cistern. And

The whole system is a combination of a typical spring-cave-tunnel galleries and may have been excavated during the Middle and Late Bronze Periods and extended during the Iron Age.

7.5.2.2 The Iron Age period (ca. 1200–600 BC)

The Iron Age which began around *ca.* 1200 BC was again cold and humid. So far, there is no archaeological or Biblical evidence about water works, or anything else being built in the time of King David who was occupied most of the time in establishing and later enlarging his kingdom. His wars were against his opponents from within and enemies abroad who included the Arameans in the north and the Philistines in the west. It seems that during his time the population grew and Jerusalem, enclosed within the old walls, became overcrowded. For this reason or the reason given by the Biblical scribes, namely the sin of David's census of his kingdom, there was pestilence. In order to appease the Almighty and at the same time enlarge the city, David bought a threshing-floor on the higher ridge above the City of David from the former Jebusite King Araunah (*erevna* means *king* in Hurrian) where he built an altar, made offerings and secured the site for the building of the future First Temple. The building of this temple was accomplished by David's son Solomon, who inherited the kingship. The struggle for succession between David's sons took place in connection with the springs of Jerusalem. As David became old, one of his sons tried to ensure his right to the throne by giving a great feast near the spring of 'En-Rogel, near today's well of Bir 'Ayub. In order to counter this move, another son, Solomon, was declared the successor, this time at the probably more prestigious spring of Gihon.

The humid climate of the Iron Age, brought prosperity to the region and according to the Biblical text, King Solomon could engage in extravagant building projects, for example a temple. He participated in trade across the desert of Arabia bringing spices and incense from the kingdom of Sheba (Saba), shipping them through the port of Tyre.

After the death of Solomon the heavy yoke of taxes on the people of Israel led to rebellion and the kingdom was divided into two hostile states, the northern Kingdom of Israel with its capital in Samaria and the smaller Kingdom of Judea centred on Jerusalem. Invasions of neighbouring kingdoms were the natural consequence and after years of fighting each other and everybody else, both were eventually conquered by the Assyrians. It is noteworthy however, that Jerusalem was not conquered by the Assyrians although they laid siege to it. This might be attributed to the foresight of King Hezekiah who built new massive fortifications and dug a tunnel from the Spring of Gihon, the city's only water source then, to a pool within the city called the Siloam Pool.

The story of this water project is found twice in the Bible, once in the book of Kings II (20:20):

> *'And the rest of the acts of Hezekia and all his might, and how he made a pool, and a conduit, and brought water into the city, are they not written in the book of chronicles of the kings of Judah' (Kings II 20:20)*

This story is also told in the book of Chronicles II (32:2,3,4)

> *'And when Hezekiah saw that Sennacherib was come, and that he was purposed to fight against Jerusalem he took counsel with his princes and mighty men to stop the waters of the fountains which were without the city,*

and they did help him. So there was gathered much people together, who stopped all the fountains, and the brook that ran through the midst of the land, saying, why should the kings of Assyria come and find much water'.

The third written evidence on this project was found as an engraved inscription on the wall of the tunnel, and written in Hebrew letters characteristic of the period of the late First Temple. This inscription describes how the two crews of diggers working from each end of the tunnel met axe to axe after hearing each other for some time. (Figure 7.7)

Figure 7.7 The Niqba inscription

'*This is the story of the boring through: whilst (the tunnellers lifted) the pick each towards his fellow and whilst three cubits to (be) bored (there was heard) the voice of a man calling his fellow, for there was a way in the rock on the right hand and on (the left): And on the day of the boring through, the tunnellers struck each in the direction of his fellows, pick against pick, And the water started to flow from the source to the pool twelve hundred cubits. A hundred cubits was the height of the rock above the head of the tunnellers.*'

The length of the tunnel is 533 m, the difference in altitude between the two interconnected springs is 2.1 m and the height 1.1 to 3.4 m. The distance between the two interconnected springs is only 320 m. This means that the tunnel makes a rather long serpentine (Figure 7.4d) which puzzled the archaeologist. It was suggested that the wide detour was made in order to avoid passing the water under the tombs of the kings of the house of David due to the Jewish tradition that the remains of the dead defile everything, even flowing water. The present authors suggest that this serpentine course was the result of the diggers following the karstic channel connecting the Gihon and Siloam springs. Opposing this view, Reich and Shukron (2002b) investigated the tunnel and claim that there was no karstic channel or water flow and the twisting of the tunnel was planned in order to reduce the angle of the slope (which as stated is just 1/250 m). This claim ignores the basic fact that the investment of work with chisels and mallets in a tunnel driven a long a straight line is much lower than a curve and improves the meeting of the workers advancing from opposite directions. (One of the ancient methods, which the qanat diggers still use, for keeping a straight direction underground is to place a few oil candles in the straight direction of the tunnel, and the miner when looking back had to see one light.) Until now, Iranian qanat diggers are strict to dig an aeration shaft at a certain distance from one to the other, even if it means a considerable depth from surface. Sails are erected over the opening of the shaft for catching and directing the wind into the shaft and tunnel (Issar and Zohar, 2007: 247–249) as the cutting of a tunnel cannot be carried out without a constant supply of air. A karstic channel would answer the question of the meeting mentioned in the text despite the absence of a straight line as well as the unnecessary height of the tunnel in the southern part of the tunnel. The late archaeologist Yigal Shiloh, who also investigated these water works aided by the geologist Dan Gill, agreed with our explanation (Shiloh, 1996). The age of the tunnel was confirmed by radiometric dating (Frumkin *et al.* 2003)

In addition to this tunnel, remains of at least two other aqueducts beginning at the Gihon spring are known, one deeply cut in rock and the other a built channel. Both ran in a very winding way below the eastern city walls along the Kidron Valley towards the Siloam Pool. Their openings show that their primary function was to irrigate the fields in the bottom of the valley and the so-called "King's Gardens" at the junction of the Kidron with the Tyropheon Valley.

The water supply of Jerusalem during the time of the First Temple consisted also of pools, which collected rain and flood waters. The collection reservoirs are mentioned in the Bible, such as the "Upper Pool":

> *'Then said the Lord unto Isaiah: Go forth now to meet Ahaz, thou, and Shearjashub thy son, at the end of the conduit of the upper pool in the highway of the fuller's field'* (Isaiah 7:3);

or the "Lower Pool":

> *'Ye have seen also the breaches of the city of David, that they are many: and ye gathered together the waters of the lower pool'* (Isaiah 22:9);

and the "Old Pool":

> *'Ye made also a ditch between the two walls for the water of the old pool'* (Isaiah 22:1).

The location of these reservoirs is not certain but they could have been connected with the Gihon project. In any case, these citations show that in those times the Jerusalemites made use of all water resources available in the vicinity of their city to ensure safe supply in times of peace as well as war. That could not prevent internal strife and a weakened Jerusalem fell in 586 BC to the Babylonians under King Nebuchadnezzar.

7.5.2.3 The Persian period (ca. 500–300 BC)

The climate during the Persian period was again dry. Neither archaeology nor history throws light on when the ingenious method of tunnelling to reach the ground water and thus to cope with the desert conditions of the Iranian highlands was developed or who developed it. The early Persians were the archetypal raiders on horseback when they emerged in the early first millennium BC from central Asia. They soon continued their conquests, defeated the Babylonians by the late *ca.* sixth century BC and conquered the whole Near and Middle East, including Egypt. The Judeans in exile were allowed by King Cyrus the Great to return to Jerusalem and rebuild the city and its temple. Springs in the region of Jerusalem developed by the Iranian method of a gallery and shafts, may perhaps be the finger print of Persian-Iranian water engineers.

7.5.2.4 The Hellenistic–Hasmonean humid period (163–67 BC)

During the Hellenistic period the climate became cooler and wetter. Once the Persian Empire fell into the hands of Alexander the Great, the Hellenistic culture was adopted by nearly all inhabitants of the new empire, including many Judeans and in due time Jerusalem became a Greek city, with gymnasia, theatres and temples. The heirs of Alexander divided his empire and indulged in a series of wars between them.

Palestine became a battle field between the heirs of Ptolemy and Seleucus. The latter enforced Hellenisation and introduced foreign rituals which gave rise to the Maccabean Rebellion. Jerusalem became the capital of the newly-founded Hasmonean Kingdom and a centre for pilgrims from all over the Diaspora, especially during the three main holidays of Passover, Shavuot (Pentecost) and Sukkoth (Feast of the Tabernacles). During the week of Sukkot the ceremony of water libation took place when water from the Siloam pool was carried to the temple in a public and joyous procession.

As the requirement for water grew, an aqueduct was built carrying water from the spring of 'Ein Etam, about 10 km to the south, into the city. The aqueduct follows the topography and winds its way accordingly (Figure 7.8) with a length of 39 km. The inlet is only 30 m higher than the outlet, and the inclination is about 1:1000. At two sites, one near Bethlehem and the other near Jerusalem, where the curves dictated by too lengthy spurs was excessive, tunnels were excavated. In order to regulate the flow in the aqueduct during the various seasons, one reservoir was constructed at the inlet and two at the outlet near the site of the Temple. The aqueduct brought 3–5 m^3/d of water to Jerusalem. Another water supply project recently discovered and attributed to the Hasmonean period collected the flood water from the valleys of Tyropheon in the southern part of the city and Beit Zeita (Bethesda) in the northern part of the city. Rock-cut channels and tunnels led the water into subsurface reservoirs below the fortified area of the Temple Mount.

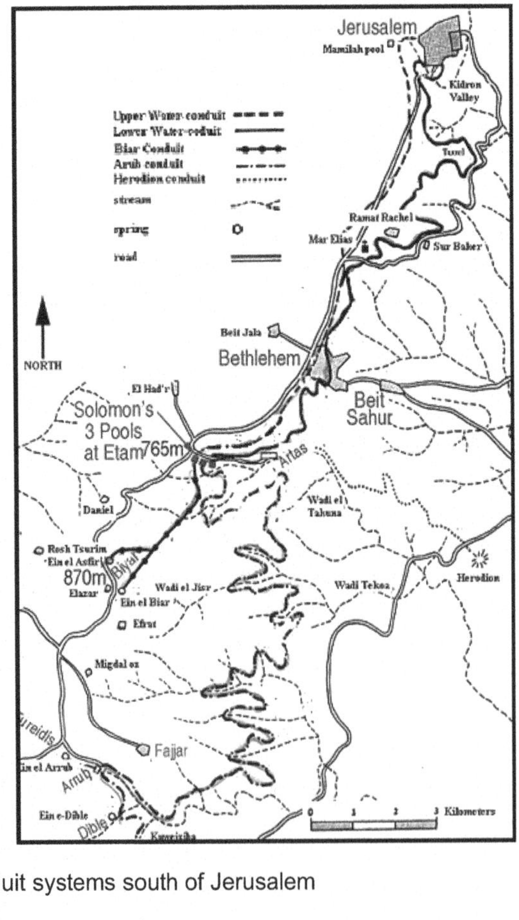

Figure 7.8 The water conduit systems south of Jerusalem

The Hasmonean royal family ended as usual by feuding each other and some members of the family called in the Romans who took over the country and made it a protectorate-kingdom under the rule of Herod, a king of Edomite origin. Agriculture flourished as the climate turned colder and more humid, particularly in the Jordan valley, where Herod owned extensive estates. The economic prosperity enabled Herod to engage in large-scale construction projects at home and abroad. In order to gain favour with the Jews, he re-built the modest temple built by the returning exiles from Babylon a few centuries earlier and replaced it with a magnificent building.

Today two aqueducts are known which transported water from a series of springs about 10 km south of Jerusalem to the city. The upper aqueduct was built by the Romans as mentioned by Josephus Flavius as well as from inscriptions found on the aqueduct. However, it is still unclear which of the Hasmonean kings was responsible for building the lower aqueduct. It can be assumed that while it was begun earlier, it was completed by Herod who had also built an aqueduct from the area of Solomon's pools, part of system of the aqueduct to Herodion, his fortress on a partially artificial hill in the Judean Desert.

7.5.2.5 The Roman–Byzantine period (67 BC–600 AD)

The climate during this period continued to remain humid and the problem was no longer the low yield of the springs in the region of Jerusalem but the transport of their water to the city. This the Romans accomplished by adding the upper aqueduct already mentioned from the springs south of Jerusalem. They also added pools and reservoirs to regulate the flow. This they did stage by stage as they were faced by revolts of the Jews, who did not accept the rule of the Romans who, following Herod's death at *ca.* 4 BC, transformed the whole area into the Roman province *Judea* ruled by a Roman prefect and then procurator. According to Josephus Flavius, the prefect Pontius Pilatus of New Testament fame, spent money from the temple's treasure to build an aqueduct to Jerusalem, probably the "Upper Aqueduct" mentioned above. When the Jews protested this use of sacred money, scores were killed in the riots.

The tension reached its climax with the First Revolt in 66 AD. The Romans dispatched reinforcements, put the city under siege, broke into the city, wrecked it and burnt down the temple (Yakir *et al.* 1994). Jerusalem was left in ruins for some time until the emperor Hadrian promised to rebuild it. When he changed his mind, the Second Revolt broke out in 132 AD when Jerusalem was liberated for three years. After crushing this revolt, Hadrian re-founded Jerusalem under the name "Aelia Capitolina". The Romans refurnished the upper aqueduct to Jerusalem by adding siphons, tunnels, bridges and another reservoir. The work was accomplished by the 10th Legion, a Roman version of an engineer corps stationed in Jerusalem.

While the first aqueduct takes its water from the spring of 'Ein 'Arub, the upper aqueduct takes it water from the parallel riverbed springs of El Biar. The length of this aqueduct is only 4.7 km due to the use of tunnels which were quarried by using shafts (in total 23 shafts, the deepest 33 m). Both perched spring systems in the limestone-dolomite aquifer of Jerusalem region meet at the so-called "King Solomon" pools from where they divide again. The pools, which are built on an impervious marly layer, are in a descending altitude, and thus overflow from the highest to the lowest. During the first stage of the building of the aqueducts during the Hellenistic and Roman periods there existed two pools, while the third pool was probably added much later during the Mameluk period, as the earlier travellers mentioned only two pools. The storage volume of the three pools is about 29,000 m^3. Part of the siphon system of the Roman upper aqueduct which can be seen near Bethlehem (some stone pipe sections on view in the Israel Museum) carried names of officers in the Roman 10th Legion. Recently, archaeologists of the Antiquities Authority of Israel discovered a well-built aqueduct believed to be the entrance of the upper aqueduct to King Herod's palace some 40 m south of today's Jaffa Gate. According to the excavators, the aqueduct was built in the first century BC and was in use until the second century AD. Again, it contained stamped roof tiles of the Roman 10th Legion. It is 1.5 m high and 60 cm wide and built with large, flat stones. Every 15 m a shaft connected the aqueduct to a road above it. The last 40 m stretch of the aqueduct ends just before it reaches the walls of the Old City where it is blocked, apparently by a collapsed shaft. The existence of an aqueduct has been known for about a century thanks to a map by the German architect and archaeologist Conrad Schick who unearthed a short stretch of it. It was never excavated as this area is one of the city's busiest intersections.

Another end of the Upper and the Lower Aqueduct met on the western escarpment of the Tyropheon Valley [up the slope of the western hill (Mount Zion today)] opposite the Temple Mount. It is assumed that the water crossed the valley on Wilson's Bridge and was stored in the huge subsurface reservoirs underneath the Herodian temple platform.

Towards the end of the Roman direct rule, *ca.* 250–300 AD another series of dry years coincided with civil wars ripping the empire. After its division into an eastern and a western empire with Constantine the Great declaring himself emperor (324 AD), *Provincia Judea* came under the rule of the Byzantines. During the early years of their rule the climate had turned mostly humid and they managed to maintain the aqueducts and reservoirs in good order providing enough water for the many pilgrims reaching Jerusalem, now a sacred city to the Christians. In 1920, an inscription was found near 'Ein 'Arub in which Plavius Anias Silenterius, a Byzantine commander announced to the landlords, tenants and farmers who owned lands along the aqueduct: 'Let it be known that the divine pious emperor ordered that no one is allowed to plant or sow in the distance of 15 (about 4.5m) feet from the aqueduct. If anyone will disobey the order he will be sentenced to death and his property will be confiscated'.

7.5.2.6 The Arab period

When in 637/8 AD Jerusalem was surrendered to the Arab army, the climate was still humid. Around 800 AD the climate again turned warm and dry with the peak of aridity around 900 AD, an event which may correspond with the beginning of the Medieval Warm Period of Europe (950–1250 AD).

Jerusalem was a holy city also to the Moslems and had become the destination of pilgrimages. The growing number of pilgrims increased the demand for water. However, the dry period had affected the springs, their flow severely diminished. In addition, the aqueducts were damaged by the farmers living in the vicinity who diverted the water into their fields. In spite of the many cisterns in Jerusalem, the situation went from bad to worse and it is reported that during years of drought the poor died of thirst.

7.5.2.7 The Crusaders and Mameluk periods (end of the 11th to the 15th century AD)

The Crusaders and Mameluk periods were again cold and humid and the aqueducts to Jerusalem and its water supply were renovated. During the period of their rule the Mameluks engaged in various construction and reconstruction works, including water works. These were carried out especially by Muhammed Ibn Keilan and Qait Bey during the 15th century. Also the lower third pool of the "Solomon's pools" near Bethlehem was constructed, as mentioned by the Arab geographer Al-Muqaddasi who visited this area during the 10th century describing two pools, whereas the traveller Felix Fabrey who visited the place in the 15th century, reports three.

7.5.2.8 The Ottoman period (from the beginning of the 16th to the 20th century)

During most of the time of the Ottoman Empire, the climate was again warm and dry. The Soreq Cave stalagmites record and the long-term annual rainfall record for Jerusalem show that the generally dry period occasionally switched to colder and more humid phases, for instance around 1650, 1770 and 1850 which corresponds to events during the Little Ice Age in Europe. The Ottomans invested great effort to renovate the Lower Aqueducts and transport water to the public drinking fountains (*sebil*). There is no report about the Upper Aqueduct which seemingly stopped functioning. The reason for this may have been the drop of the water table of the springs whereas the Lower Aqueduct could get its water from lower springs. It was, however, vandalised by the neighbouring villagers who broke it, once and again to get some water in order to survive. The government appointed guards and introduced clay pipes, but the aqueduct failed notwithstanding all the efforts. Sultan Osman the 3rd built a fort near Solomon's pools in 1617 in order to protect these crucial installations, but with little success.

The archaeologist Warren refers to a project by the Turkish government to renovate the Lower Aqueduct some time prior to his visit 1870, yet the renovation failed again, and this aqueduct supplied a small quantity of water which was restricted to be used only by Moslems. A document included in a set of documents prepared by a British company, as basic data for their proposal, explains that the 1870 project failed because the "fellahin" along the aqueduct broke it to supply water to their herds. The British plan included the reconstruction of the whole aqueduct from 'Ein 'Arub to Jerusalem, including Solomon's pools. This program was not accepted because these pools were registered property of the Moslem Waqf Foundation and could not be renovated by non-Moslems for the fear of defilement.

The last renovation was carried out by the Ottoman government and ceremonially inaugurated in 1902. The project was partly based on introducing iron pipes instead of clay pipes. The introduction of a new reservoir on the outskirts of Jerusalem reduced the pressure in the pipeline. Jerusalem received only 180 m^3/d which went mainly to the area around the Omar Mosque, namely The Temple Mount and the Christian Quarter. This quantity of water was not sufficient and the population of Jerusalem depended mainly on water accumulated in private cisterns, where the water from their roofs was collected, and public subsurface reservoirs. After relatively dry winters water was brought by train and neighbouring villagers carried water skins to town for sale.

7.5.2.9 The British mandate

Jerusalem, whose population at that time had reached 50,000 people, was conquered by the British Army in 1917 during World War I and came under British military rule until 1922, followed by the League of Nation Mandate (1922–1948). Soon after the conquest, the Royal Engineers got involved in solving the problems of supplying water for the inhabitants and the army forces stationed in the region. The first source to be tapped was again the springs of 'Ein el-'Arub. This time the water was pumped from the pool and pressured via a steel pipe to a tank at a high altitude from where an amount of about 15,000 m^3/d flowed by gravity in pipes to Jerusalem. After 1920 and another drop in the precipitation, the British Government repaired the Solomon Pools in 1924 and built a dam for collecting flood waters all of which were diverted to the pools and filtered, parallel to the restoration of the Lower Aqueduct. Then a portion of the water was pumped from the pools in the above mentioned steel pipe, the other flowed in the Lower Aqueduct, a system which worked until 1967 supplying water to East Jerusalem. Also in 1924 the British Government built a pipeline from the springs of 'Ein Far'a and 'Ein Qelt in Wadi Qelt, a canyon about 7 km NE of Jerusalem. In 1936 a pipeline was constructed from the springs of Rosh ha'Ain, about 62 km NW of Jerusalem. This project included four pumping stations and a large storage tank at a high point in western Jerusalem from where the water was supplied to the houses by pressure. This project supplied about 4 million m^3/yr of water.

7.5.2.10 State of Israel

As war between the Jewish and the Arab population had started already on November 30, 1947, the local authorities nationalised and sealed all private cisterns in the houses' yards in order to ensure the water supply. Likewise, the water tank of the pipe line which ensured 115,000 m^3 for the Jewish population of 100,000 people was nationalised. When the British Mandate in Palestine ended in May 1948 and the State of Israel was declared, immediately followed by the invasion of the combined armies of the Arab world, Jerusalem came under siege by the Arab Legion and the Iraqi Army who cut the water line to Jerusalem from Rosh ha'Ayin. Once the siege started water was rationed to 10 litres per capita in the Jewish western part of the city. As the eastern part was occupied by the Arab Legion, it continued to get its water from the Lower Aqueduct. The precipitation during 1947 and 1948 was low and the water supply during summer 1948 was precarious. During this summer a new road to Jerusalem called "the Burma Road" was cut through the mountains taken by the Israeli forces. By means of this rough road, supplies of water, food, ammunition and petrol were transported by cars in order to relieve Jerusalem's besieged Jewish population. Soon after the road was constructed, the national water company Mekorot started to install a pipeline and pumping stations fed by wells in the coastal plain. The new pipeline started from Kibutz Hulda to the section of the former British pipeline which was in the hands of the Jewish forces. The water reached Jerusalem just in time before the local cisterns were empty. Since then, additional deep wells were drilled around Jerusalem, locating water in the limestone and dolomites of the Cenomanian Turonian age.

At present the population of Jerusalem is about 750,000 people. The various lines from the National Water Carrier and local wells amount to the quantity of 50 million m^3/yr.

7.6 CONCLUSIONS

The archaeological investigations of the ancient water works of Jerusalem began in the middle of the 19th century and are still continuing. The most recent excavations proved that remains of the oldest water works date to the Middle Bronze Age (*ca.*1800–1500 BC). These water works were altered and extended in later periods. Jerusalem is situated not far from the border of the desert which moved north and south as well as east and west due to climate changes causing variations in the yield of the local water resources. Yet, no reference to this factor is found in the numerous archaeological reports, although it is obvious and to be anticipated that these changes have had a decisive impact on the environment and history of the Near-East in general, and Jerusalem in particular. In our case, small perched springs diminished and even dried up during dry periods forcing the inhabitants to deepen the works and develop the more perennial springs in order to safeguard their water supply. As the population increased, so the demand for water grew during the urban phase of the Hellenistic, Roman, and Byzantine periods. The main effort shifted from the concept of securing internal local water supplies to integrating the immediate surroundings and importing water from the outside. The need for transportation from greater distances led to the construction of elaborate aqueducts and water reservoirs similar to other cities in the classical world. During the dry phases of the early Arab period and the periods following it most of these aqueducts ceased functioning due to the subsidence of the groundwater table feeding the springs and to neglect. The industrial revolution arrived in the early *ca.* 20th century when increasing efforts were made to improve the general water shortage which plagued Jerusalem. By integrating some of the pre-existing structures with the improvements of modern technology, an efficient water supply for the city was finally achieved with the establishment of the State of Israel.

Considering the importance of Jerusalem for the three main religions and its richness in archaeological sites and remains, the knowledge of the impact of climate changes on the availability of water to its inhabitants is of special importance. In the first place for a better understanding of the past and in the second place for forecasting the future impact due to the present warming of the global climate.

The main conclusion from the study of the past was that warm climates spelled dryness and the first to be influenced were the small springs with a limited recharge area and a small storage capacity. The discharge of bigger springs was also reduced once the warm dry period extended over a longer period. Thus the bigger the storage capacity, the better are the chances that the storage will mitigate the negative impact of dry periods.

As described in the present article, the optimal storage of water in the region of Jerusalem is found in the limestone-dolomite aquifers of the Judean Group. Indeed these aquifers are utilised by a series of deep wells and this source of water is backed by a pipeline bringing water from the national water supply system of Israel.

In case of a lengthy and extreme period of dryness there will be no alternative but to pump and desalinise the brackish water stored in the sand layers of the Lower Cretaceous below and separated from the limestone-dolomite aquifers. Once desalination is carried out by the reverse osmosis method, the remnant brine can be piped to the Dead Sea. The preliminary estimate of the potential quantity is about 70 million m^3 for a period of about 100 years (Greitzer and Issar, 2001).

Yet the future long term and safe supply of water for the inhabitants of Jerusalem will come only after the final solution of the Israeli-Palestinian conflict. 'The sad question now is how much more blood will be spilled over the stones of Jerusalem before enough people come to the conclusion that this city—whose archaic root, *Ir Shalem*, means the "city of peace"—should become a cornerstone of peace on earth, rather than in heaven' (Issar, 1998).

REFERENCES

Albright W. F. (1957). From the Stone Age to Christianity: Monotheism and the Historical Process, 2nd edn. Doubleday Anchor Books. Publisher, p. 432.

Bar-Matthews M., Ayalon A., Matthews A., Sass E. and Halicz L. (1996). Carbon and oxygen isotope study of the active water-carbonate system in a karstic Mediterranean cave: implications for palaeoclimate research in semiarid regions. *Geochimica et Cosmochimica Acta*, **60**, 337–347.

Bartov Y. (1999). The Geology of the Lisan Formation in the Massada Plain and the Lisan Peninsula, M.Sc. Thesis, The Hebrew University of Jerusalem, (in Hebrew with English abstract).

Beaumont P., Bonine M. and McLachlan K. (eds) (1989). Qanat, Kariz, and Khattara: Traditional Water Systems in the Middle East and North Africa. London, School of Oriental and African Studies: London, UK.

Bookman (Ken-Tor) R., Enzel Y., Agnon A. and Stein M. (2004). Late Holocene lake levels of the Dead Sea. *Geological Society of America Bulletin*, **116**, 555–571.

de Vaux R. (1971). Palestine in the Early Bronze Age. In I. E. S. Edwards, C. J. Gadd and N. G. L. Hammond (eds) The Cambridge Ancient History of the Middle East, **1**(2), Cambridge University Press: Cambridge, UK.

Frumkin A., Shimron A. and Rosenbaum J. (2003). Radiometric dating of the siloam tunnel, Jerusalem. *Nature,* **425**, 169–171.

Gavrieli I., Burg A. and Guttman J. (2002). Transition from confined to phreatic conditions as the factor controlling salinization and change in redox state, upper subaquifer of the Judea Group, Israel. *Hydrogeology Journal,* **10**, 483–494.

Gill D. (1991). Subterranean waterworks of Biblical Jerusalem: adaptation of a karst system. *Science* **254**, 1467–1471.

Gill D. (1996). The geology of the city of david and its ancient subterranean waterworks, In: D. T. Ariel and A. De-Groot (eds), Qedem 35: Excavations at the City of David 1978–1985 (Directed by Yigal Shiloh), The Hebrew University of Jerusalem. Tel Aviv, Israel.

Greitzer Y. and Issar A. S. (2001). Investigating the pre-feasibility of the influence of the brackish water storage in the Lower Cretaceous aquifer in the Judean mountains on the Cenomenian Turonian aquifer and the possibilities of desalinisation of the brackish water in the region. An unpublished report to the Water Commissioner, Govt. of Israel (in Hebrew).

Issar A. (1976). The Evolution of the ancient water supply system in the region of Jerusalem. *Israel Exploration Journal*, **26**, 130–136.

Issar A. S. (1998). A geologist looks at the stones of Jerusalem. *American Scientist*, **86**(5), September–October 1998.

Issar A. S. (2008). The impact of global warming on the water resources of the Middle East: past, present, and future. In: F. Zereini and H. Hötzl (eds) *Climatic Changes and Water Resources in the Middle East and North Africa*, Heidelberg, Springer.

Issar A. S. and Zohar M. (2007). Climate Change – Environment and History of the Near East. Springer Berlin, Heidelberg and New York, USA.

Issar A. S. and Zohar M. (2008). Climate change impacts on the environment and civilization in the near east. In: H. G. Brauch, U. O. Spring, J. Grin *et al.* (eds) Facing Global Environmental Change: Environmental, Human, Energy, Food, Health and Water Security Concepts, Volume IV. Springer–Verlag, Berlin Heidelberg, Germany.

Issar A. S. (1990). Water Shall Flow from the Rock. Hydrogeology and Climate in the Lands of the Bible, Springer–Verlag, Berlin Heidelberg, Germany.

Karlén W. (1991). Glaciers fluctuations in Scandinavia. In: Starkel, K. J. Gregory and J. B. Thornes (eds) Temperate Palaeohydrology: Fluvial Processes in the Temperate Zone During the Last 15,000 Year. University of Jerusalem, Jerusalem, Israel.

Lemcke G. and Sturm M. (1997). $d^{18}O$ and trace element as proxy for the reconstruction of climate changes at Lake Van (Turkey): Preliminary results. In: N. Dalfes, G. Kukla and H. Weiss (eds) Third Millennium BC Climate Change and Old World Collapse, NATO ASI Series, **149**, 654–678. Springer Verlag, Berlin Heidelberg, Germany.

Picard L. (1943). Structure and Evolution of Palestine with Comparative Notes on Neighbouring Countries. Geology Department Bulletin no 4, University of Jerusalem, Jerusalem, Israel.

Pritchard J. (1961). The Water System of Gibeon, University of Pennsylvania Press. Philadelphia, USA.

Reich R. and Shukron E. (2002a). Jerusalem, City of David, Hadashot Arkheologiyot, Excavations and Surveys in Israel, **114**, 77–78.

Reich R. and Shukron E. (2002b). Reconsidering the karstic theory as an explanation to the cutting of Hezekiah's tunnel in Jerusalem. *Bulletin of the American Schools of Oriental Research*, **325**, 75–80.

Shiloh Y. (1984). *Qedem 19: Excavations at the City of David*, 1, 1978–1982. The Hebrew University of Jerusalem, Israel.

Yakir D., Issar A., Gat J., Adar E., Trimborn P. and Lipp J. (1994). ^{13}C and ^{18}O of wood from the Roman siege rampart in Masada, Israel (AD 70–73): evidence for a less arid climate for the region. *Geochimica et Cosmochimica Acta*, **58**, 3535–3539.

Zertal A. (1990). 'In the Land of the Perizzites and the Giants' The Israelite settlement in the hill country of Menasseh. In N. Naaman and I. Finkelstein (eds) From Nomadism to Monarchy – Archeological and Historical Aspects of Early Israel, Yad Ben Zevi Press. Jerusalem, Israel (in Hebrew).

Chapter 8
History of water supply in pre-modern China

P. Du and A. Koenig

8.1 INTRODUCTION

Starting from the Eastern Zhou dynasty (770–221 BC), the water resources developments of ancient China surpassed all other ancient civilizations in number, scale and technological advancement (Chi, 1936; Needham & Wang, 1965; Needham *et al.* 1971). Besides the more well-known extensive irrigation works and man-made transport canals linking up the major rivers, the provision of water supply to its cities formed the third important element in the development of China's ancient water civilization. This chapter describes the siting and layout of cities with regard to favourable water supply conditions, comprehensive water system planning, water supply facilities, associated hydraulic works and evolution of water supply technology as exemplified mainly by the ancient imperial capitals. These cities were built by successive dynasties in accordance with cosmographic principles and had up to one million inhabitants. Water supply systems which developed under different topographical and climatic conditions in the lower Yangtze plain (canal towns) and in frontier lands are also presented. After thousands of years of uninterrupted linear evolution, largely in isolation from other civilizations, China was forced to open to Western powers in 1842, at the conclusion of the Opium War with Great Britain. The year 1842 is therefore taken as the date when the era of pre-modern China ended. Soon thereafter modern water supply systems were introduced when the traditional methods were slowly replaced by western technology.

For a better understanding of Chinese history, the chronology of Chinese dynasties and historic periods as well as the names of the capitals are shown in Appendix A.1. The location of the capitals and other cities discussed in Sections 8.2 to 8.5 are shown on a map in Appendix A.2. For further geographical information, the reader is referred to the large number of available books, descriptions and atlases.

References written in the Chinese language are cited in pinyin, the official system used to transcribe Chinese characters using Roman letters, followed by the English translation in brackets.

8.2 ANCIENT CAPITALS

8.2.1 Siting and layout of Chinese capitals

The history of China can be traced back through many dynasties. Beijing, Nanjing, Luoyang and Xi'an have been capitals for a long time during different dynasties, hence they were called the "Four Great Ancient

Capitals of China". Kaifeng and Hangzhou are also famous cities, which had been capitals once upon a time in history. The chronology of Chinese ancient capitals is shown in Appendix A.1.

When selecting the site of a capital, many factors should be considered. They may include the economic base, military defence, transportation and natural conditions, etc. Since the ancient Chinese cities originated on the basis of agricultural settlements, most of the early capitals like Yangcheng, Chang'an and Luoyang, were sited in the middle and lower reaches of the Yellow River, an area with a developed agricultural economy. Following the economic rise of south-eastern China, capitals were then moved south-eastward, and canal transportation became another important factor in site selection. Some dynasties even set their capitals directly in the south-eastern region of China like Nanjing and Hangzhou. Cosmographic principles on capital site selection have been mentioned since the early Warring States period. The book *Guan Zi* mentioned that "the capital should be built either at the foot of a great mountain or near a big river; stay away from dry lands at high elevation in order to get enough water, and stay away from water at low elevation to prevent flood and save drainage and embankment".

The book of *Zhou Li* written in the Western Zhou Dynasty described the ideal layout of ancient Chinese capitals (Figure 8.1). However, in actual construction, the layout and pattern of ancient cities differed according to their varying geographical and natural conditions. In the northern plain area, most capitals were built in square and symmetrical patterns, with the roads intersecting each other at right angles. However, in southern China, which abounds in hills, rivers and lakes, the cities were always built in accordance with the shape and direction of the waters and mountains. Therefore, cities in southern China had variable shapes. Most of the time the river and canal courses formed the core of their layout, with the roads vertically and horizontally crossing the rivers.

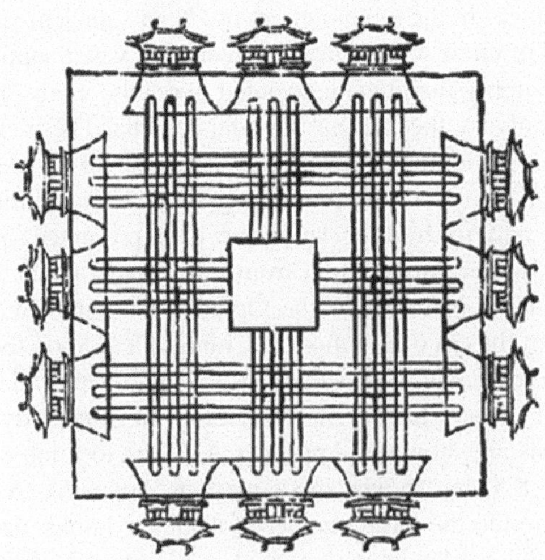

Figure 8.1 Planar diagram of the ideal capital in the Western Zhou Dynasty. Taken from the *San Li Tu* (Dong, 1988)

The inner patterns and layouts of ancient Chinese capitals reflected the idea of zoning. The imperial palace was always located in the centre of the city protected by walls and moats. Around it were royal gardens, temples and administration buildings. The residential areas were separated from the above structures and were under strict control, commonly organised in *Li-Fang* (wall-enclosed blocks) units. For example, in Chang'an City of the Sui and Tang Dynasties, the city was divided into 110 Fang, with the opening and closing times of the gates of those Fang announced by the beating of a drum (IHNS and CAS, 1985). There were also several markets called *Shi* in the capital. From the Song Dynasty onward, because of the flourishing of commercial and handicraft industries, the division between residential and business areas gradually disappeared and the *Li-Fang* units were replaced (IHNS and CAS, 1985). The ancient capitals also had multiple rings of city walls and moats. Outside the walls and the *Guo* (outer city wall), water canals and dykes were built selectively according to the topography of the city.

8.2.2 Yangcheng (Eastern Zhou Dynasty)

The city of Yangcheng in the Eastern Zhou Dynasty was located in Dengfeng, Henan Province. It was an important western military base of the States Zheng (Spring-Autumn Period) and Han (Warring-States Period). The entire underground water supply pipeline system of Yangcheng was discovered in archaeological excavations (Figure 8.2), providing important physical evidence on early water supply of cities in ancient China. More detailed information is given by Du (1998).

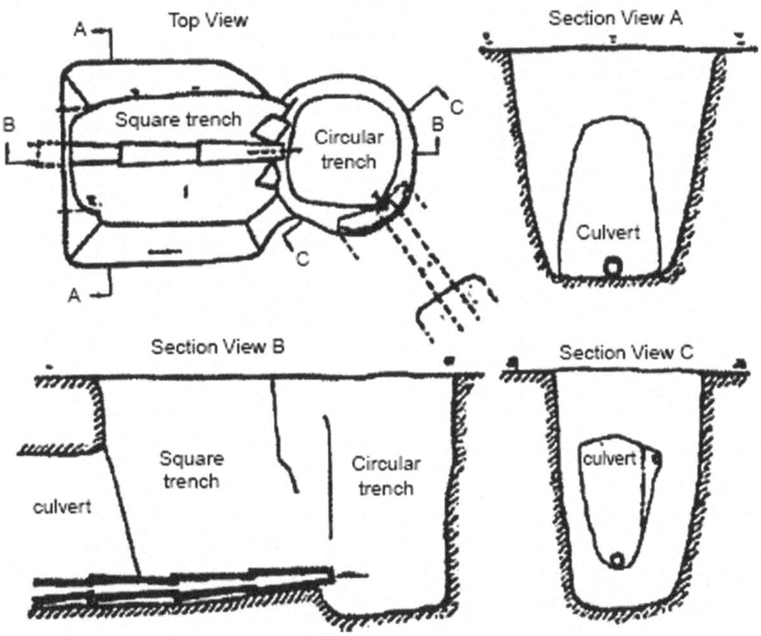

Figure 8.2 The "Valve trench" of Yangcheng (Li and An, 1982)

8.2.3 Chang'an/ Xi'an (Han, Sui, Tang Dynasties)

Located in the middle of the Central Plains of the Yellow River Basin, Chang'an (now called Xi'an), which means long-term stability, is the initial starting point of the famous Silk Road and was, for over 1200 years, the capital for 13 Chinese dynasties. The Western Han and Tang Dynasties, which were two of the most powerful dynasties in Chinese history, both built their capitals here. In these two dynasties, Chang'an city was famous for its huge scale and city planning.

8.2.3.1 Chang'an of Han Dynasty

In 206 BC, Liu Bang established the Han Dynasty and made Chang'an its capital (202 BC). That was the first time the name of Chang'an was mentioned in history. The location of the ancient Chang'an was to the northwest of the modern city of Xi'an. The city was in the Guanzhong Plain with the Wei River passing to its north. To the south of Chang'an, there were the prosperous areas of Ba and Shu (modern Sichuan Province), to the north, there was profitable commerce with foreign tribes. The defence of the surroundings had always been strong (Wu, 1995), hence many dynasties used Chang'an as their capital.

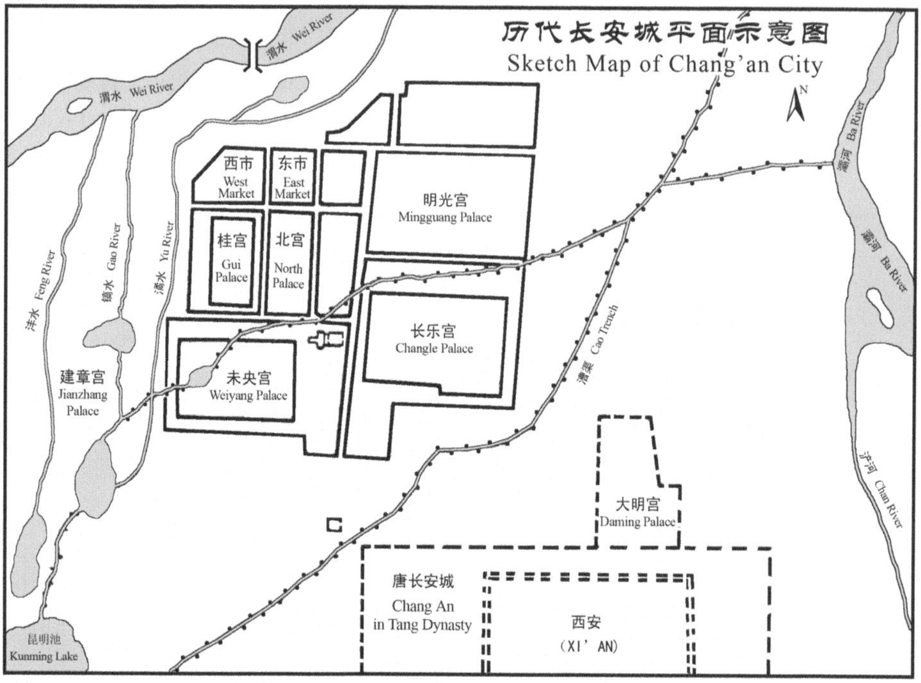

Figure 8.3 Map of channels around Chang'an in the Han Dynasty (Huang, 1958)

Chang'an of the Han Dynasty was established on the base of the palaces of the Qin Dynasty. The old palace-wide water-supply system was put into use. In the early years of the Han Dynasty, this water supply in Chang'an was adequate. However, when the Wu Emperor (7th emperor of Han Dynasty) rapidly expanded the size of the capital, the water supply system became insufficient and the Kunming Reservoir was dug to serve as a water source in Chang'an (Liu, 1988). A map of the water-transfer channels of Chang'an in the Han Dynasty is shown in Figure 8.3. The water in the Kunming Reservoir originated from the Jiao River. *The Book of Water (Annotation)* recorded that there was a "Shi Ta Yan" (stone tablet weir) built at the point where the Jiao River entered the Kunming Reservoir. The purpose of the weir was to block or transfer water into the reservoir. During the flood season, extra water from the Jiao River could flow over the weir into the Feng River, thus protecting the Kunming Reservoir from the flood. The Kunming Reservoir supplied water downstream through channels on both the east and north side. The one in the east, called the Old Kunming Channel, was built specifically for supplying water to canals. The one in the north was called the Kunming Reservoir Water and was built specifically for supplying water within the city. There was another adjusting reservoir connected to the Kunming Reservoir water before it entered the city. The water was divided into two distributaries after this: one to the north into the palace heading for the Wei River and the other one to the northeast into the city (Xiong and Guo, 1989).

The Kunming Reservoir in Chang'an City was designed as an integrated hydraulic project which combined the function of water storage, diversion, drainage and flood control. It also included weirs, water gates, canals, embankments and other hydraulic works. In addition, a special aqueduct called the "flying canal" was built to bestride the low-lying areas (Zheng, 1985). Wu (1995) estimated the capacity of the Kunming Reservoir at 35.497 Mm^3, which is equivalent to a medium-size modern reservoir. Kunming Reservoir and its channels ensured water supply in Chang'an for the remainder of the Han Dynasty. The population of Chang'an reached about 300,000 during its most prosperous period. Further details on the water supply structure and technology of Chang'an City in the Han Dynasty (202–208 BC) can be found in Du (1998).

8.2.3.2 Chang'an of Sui-Tang Dynasties

The channels and water supply system in Chang'an were severely damaged by civil wars at the end of the Han Dynasty. Water in the city became unsuitable for drinking due to salinisation. This was one of the key reasons why the Sui and Tang Dynasties sought new locations for their capitals. The selection of the new location was closely related to water sources, with the Ba and Chan Rivers to the East, and Jue and Jiao Rivers to the South (Du, 1998). Different from the Han Dynasty, the urban water supply of Chang'an City in the Sui-Tang Dynasties relied mainly on canals and wells (Figure 8.4). The canals could be divided into three groups according to their different functions.

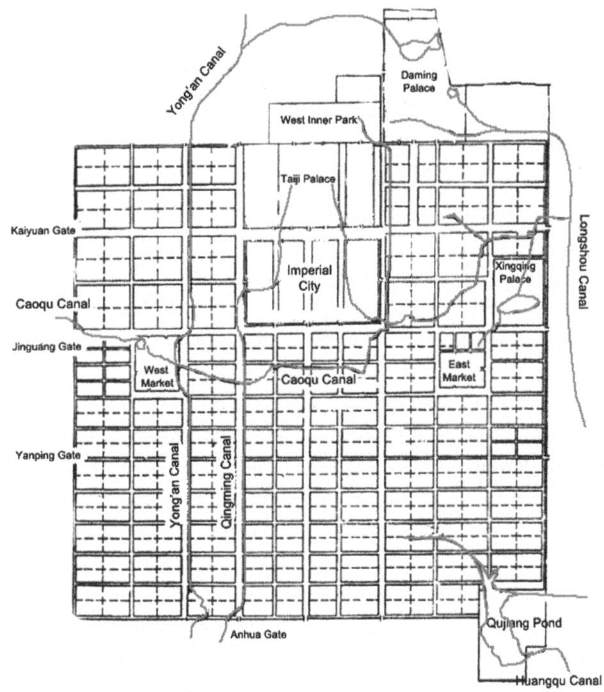

Figure 8.4 Direction and distribution of canals in Chang'an during the Sui-Tang Dynasties. *Source*: http://www.hudong.com/wiki/%E9%95%BF%E5%AE%89

First, the three inner city canals, Longshou Canal, Yong'an Canal and Qingming Canal, were mainly responsible for providing water for washing, drainage and imperial use. Diverted from the Chan River and flowing into the city from the East side, Longshou Canal was built in 591 AD. It forked outside the city with one branch flowing through the east city wall, then running into the imperial city, and proceeding northward into the East Sea (man-made pond) in the Taiji Palace. The other branch flowed north outside the city and entered the Forbidden Garden in Daming Palace where it broadened into Longshou Lake (Source: http://wiki.china.org.cn/). The two other canals mentioned above were led into the city from the south. Yong'an Canal originated from the Jiao River. It meandered through the West Market and flowed into the Forbidden Gardens and then headed for the Wei River to the north of the capital. The canal from the Jue River, known as Qingming Canal, lay to the east of Yong'an Canal, then entered the imperial palace from the southwest corner (Zheng, 1985). It was recently discovered that the section on the inner side of the Xinghua *Fang's* west wall had a width of 9.6 m (Shaanxi History Museum, 1972).

Second, a transportation canal, Cao Canal was built in the second year of the Tianbao reign (742 AD) by Han Chaozong, who was the mayor of Chang'an City at that time. It led water from the Jue River through the western city wall, flowed by the West Market and ended in a large pool for timber storage. In 766 AD, it was extended eastward and then turned to the west part of the imperial palace with a width of about 2.6 m, and a depth of 3 m (Wang, *Tang hui yao*). Therefore, the Cao Canal inter-connected the three inner city canals and added another water source to the city and palace.

Finally, the Huang Canal was built to supply water for scenic spots. It delivered water to the Qujiang Pond in the south-eastern corner of the city. There were also several diversion canals, which supplied water for the southeast part of the city from the Qujiang Pond.

Water was usually diverted by weirs or dams, led into canals and conveyed to Chang'an. The canals were dug in conformity to the natural terrain, which was high in the south and low in the north, letting the water flow by gravity controlled by the slope. Culverts and aqueducts were also built to overcome topographical obstacles. A relic of the Qingming Canal, discovered both in Xinghua *Fang* and Taiping *Fang*, revealed the technology of canal construction in the Sui and Tang Dynasties. According to the archaeological discovery, the shape of the main canal was an inverted trapezoid or half moon with a large top and a small bottom. The top of the canal was about 10 m wide, the bottom about 3 m deep and 2 m wide. The cross-sectional area of this segment of canal was about 30 m^2, which indicated quite a large water flow. All the canals excavated were well maintained: the walls were smooth and neat, with some places being reinforced, the bottom was covered with fine sand. Such maintenance could minimise erosion of the canal while at the same time reducing permeation and keeping the water clean (Teng, 2003).

Most of the branch canals were built with bricks in a trapezoidal shape. Some narrow canals were covered or built underground. Baffle walls were built in the branch canals such that, when diverting water from the main canal, they could prevent damage caused by excessive water flow. At the confluence point, the canals often became narrow to speed up the flow and to avoid blockage (Teng, 2003).

The layout of Chang'an City in the Sui-Tang Dynasties was like a great chessboard divided by longitudinal and transverse streets. There were 11 streets running north to south and 14 streets east to west. According to the historical records and archaeological discoveries, there were ditches on both sides of the streets. Those along Zhuque Street were 3.3 m wide and 2.1 m deep. The ditches called "Yang Gou" were flanked by neat rows of trees (Ma, 1963). The five main canals, along with many branched canals and underground drainage ditches constituted a complicated water network throughout the city. In addition to ensuring water supply, the network also played an important role in transport, drainage, fire control and landscaping.

8.2.3.3 Wells in Chang'an

The imperial palace and gardens took priority in using the water from the canal system. Then the water might be used for washing, transport and landscaping. The residents in Chang'an got their drinking water mainly from wells in the *Fang* (wall-enclosed blocks). Abundant records of wells could be found in the ancient records. The rich and freemen had special private wells, while most city dwellers and workshops got water from public ones.

Most of the wells discovered in Chang'an were made from bricks and earth, with a depth of 2–6 m. The well in Ximing Temple, which was the deepest one excavated, was less than 8.4 m deep, which indicated that during the Sui-Tang Dynasties, the groundwater resource of Chang'an city was abundant and easily accessible (Zhao, 1994).

8.2.4 Luoyang (Han, Wei, Tang Dynasties)

Located in the central plain of China and serving as a pivot connecting the east and west of China, Luoyang has the Yellow River to the north and Luo River to the south. The origin of Luoyang can be traced back more than 4000 years, beginning from the Xia Dynasty (21st–17th century BC). It served as the capital through 13 dynasties, for a time of 1529 years. The water conservancy projects of Luoyang in the Sui-Tang Dynasties were relatively complete and advanced (Figure 8.5).

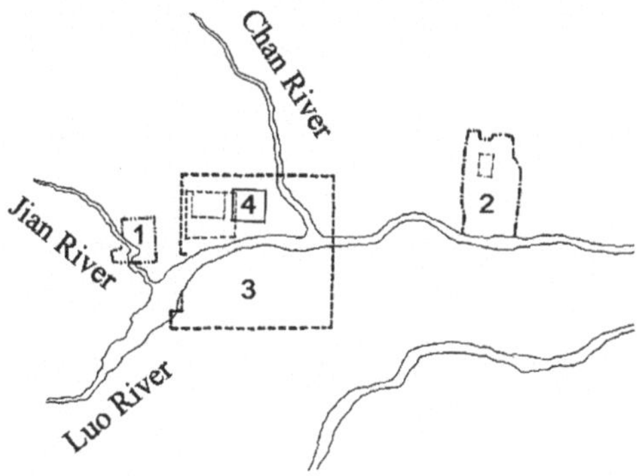

Figure 8.5 Location of Luoyang through history. 1. Eastern Zhou Luoyang, 2. Han and Northern Wei Luoyang, 3. Sui-Tang Luoyang, 4. Luoyang today [Drawn by Huang Yunsheng] (Steinhardt, 1990)

Canal transportation was the main reason for the prosperity of Luoyang in the Sui-Tang Dynasties. Because the economic development in southern China had already surpassed that in the north, the political centre and economic centre of China became geographically separated. In order to maintain the normal operation of the administration and support the troops, a large amount of grain and goods had to be transported from the south through the Grand Canal, which was built during the Sui Dynasty. Since Luoyang was at the centre of the Grand Canal network, it was selected as the Eastern Capital of the Sui and Tang Dynasties, and replaced Chang'an as the political centre during the reigns of Emperor Yang in the Sui Dynasty, and Emperor Gao Zong and Empress Wu Zetian in the Tang Dynasty (Zheng, 1985). The population of Luoyang at its most prosperous time was more than a million.

Luoyang in the Sui-Tang Dynasties was located to the southwest of the same city in the Han-Wei Dynasties. The Luo River was its main water source and the city was also connected by canals to the Gu River, Yi River and Chan River. The Luo River flowed across Luoyang and divided the city into a northern and southern part.

At the time it was said that the 'Luo River crosses Luoyang city, just like the Milky Way crosses the sky'. Due to the terrain features, the water supply condition in the city greatly differed between the northern and southern part (Figure 8.6). The south bank of the Luo River was relatively high, and there were five canals transferring water into the city: Tongji and Tongjin Canals drew water from the upstream Luo River, whilst the Yun Canal and two branch canals diverted water from the Yi River. After entering the city of Luoyang, all the canals mentioned above flowed back into the Luo River. Since the north bank of the Luo River was low-lying, water was often diverted from the river instead of draining into it. The Cao Canal was also a major canal in the area. It formed the upper course of the Tongji Canal, which was an important section of the Sui-Tang Grand Canal. The Cao Canal diverted water from the Luo River by building an overflow weir in its centre. A spillway and auxiliary structures like weirs and water gates were set up to control the water flow and canal transport. In addition, the Gu River entered Luoyang from the northwest and flowed directly into the imperial palace as the main water source of the Royal Gardens. In short, the Luo River was the key river or Luoyang's water supply. In order to maintain and protect it, many embankments and waterfront structures were constructed during the Sui and Tang Dynasties (Zheng, 1985).

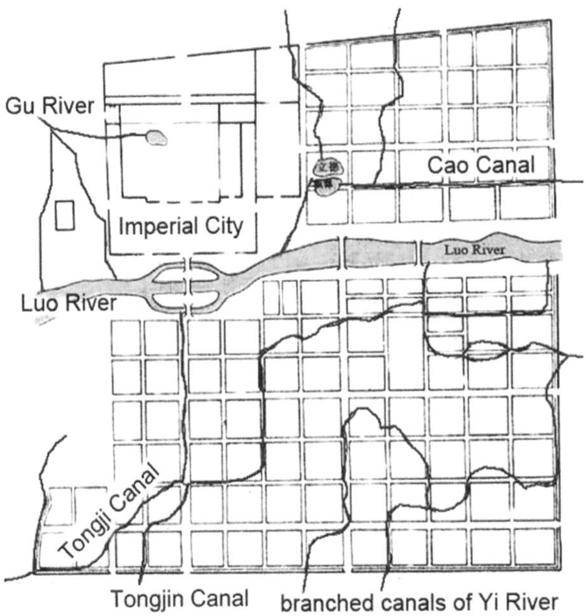

Figure 8.6 Distribution of canals in Luoyang City (Sui-Tang Dynasties) (Chen, 1983)

Branch canals were also built in Luoyang, such as the Huangdao and Chongjin canals near the Tianjin Bridge, the Ganquan, Ganshui and Longlin canals to the southeast of Jianguo Gate, and so on. Private ponds and pools also prevailed in Luoyang at that time. There were 13 ponds in the palace and 21 in the outer city. The pond of Prince Wei and the one in Bai Juyi's (poet in Tang Dynasty) residence were among the most famous ones (Xu, Qing Dynasty). Moreover, the use of hydraulic machinery and water lifting devices was very common in the city. *Shui nian* (watermills) were built in some main canals to process rice and grain. The Si Nong Watermill on Cao Canal and the Fu Xian Temple Watermill on Yun Canal were very large. The waterwheel was also invented at that time. Chen Tingzhang described the power of waterwheels in his literary work *Shui Lun Fu*.

The Luo River crossed the city of Luoyang and effectively promoted the city's water conservancy, whilst at the same time bringing flood risk to the city. A total of 27 floods occurred in Luoyang during the period of the Sui and Tang Dynasties (Luoyang di li zhi, 1992). Besides inadequate dykes, the poor arrangement of the canal system was another factor that caused flooding. Although there were many canals and watercourses in Luoyang, the Luo River and Cao Canal were the only two rivers flowing out of the city. With all the seven canals pouring water into the Luo River, overloading and flooding were inevitable. In addition, the overflow weir in the Luo River, which diverted water into the Cao Canal also aggravated the situation in the river by increasing the water level in the urban section.

Luoyang was devastated during the An-Shi Rebellion (755–763 AD). Historical records report that nine out of ten palaces and houses were burned during the wars. Although the city recovered after the war, it could not reach its pre-war prosperity. The economic and political status of Luoyang gradually declined and gave its place to the new water transport centre, Kaifeng.

8.2.5 Nanjing (Wu, Southern Tang, Ming Dynasties)

Located in the western part of the lower reaches of the Yangtze River Delta, Nanjing is a famous ancient city. It was the capital of the Eastern Wu State during the Three Kingdom Period (220–280 AD), then the capital of the Eastern Jin Dynasty (317–420 AD). The States of Song, Qi, Liang and Chen of the Southern Dynasties (420–589 AD), the Later Tang of the Five Dynasties and the Ming Dynasty all established their capitals there.

The Qinhuai River is the biggest river in Nanjing. It is a branch of the Yangtze River, about 110 km long, and forms the major watercourse around and within the city. During the Later Tang Dynasty, the river was divided into two branches: the inner and outer Qinhuai River. The outer one served as part of the city moat while the inner one, which was called "Ten-Mile Qinhuai River" in history, ran through the city from east to west. The areas along the inner river used to be the most thriving places of the city.

In 1368 AD, Nanjing became the capital of the early Ming Dynasty, and was renamed as Yingtian Fu (prefecture). The city was built upon the base of the capitals of past dynasties and expanded greatly reaching a perimeter of 67 *li* (=33.5 km). In the early Ming Dynasty, the complicated mountain and water characteristics around Nanjing directly influenced the construction of the city wall. As Gu Yanwu described, 'with Xuanwu Lake in the north, Qinhuai River in the south and the flowing Yangtze River on its right, Nanjing had a natural moat around'. The early Ming Dynasty thus created the unique feature of a winding, free style city wall by building it in compliance with the shape and direction of the natural waters around Nanjing (Quan, 2007). It is the longest surviving city wall in the world.

The inner Qinhuai River was the main source of water supply to Nanjing. It regulated the water inflow and outflow through the city and supplied water for both the urban demand and canal transport. Also, an effective water supply system, including the outer Qinhuai River, Xuanwu Lake, Qingxi River, Yundu Canal, Yangwu City Moat, Pearl River and the Jinxiang River was formed to guarantee the urban water needs (Wu, 2005) (Figure 8.7).

Many water gates and sluices were built to control the water level of the inner city rivers. For example, the East Water Gate (Tongji Gate), and the West Water Gate (Sanshan Gate) were set up at the place where the inner Qinhuai River enters and exits the city. The East Water Gate (Figure 8.8), which is 16 m high and still preserved well today, was at the entrance of the Qinhuai River. It comprised distribution gates, a bridge way and caverns to hide soldiers. There were two gates and 11 arched watercourses under the bridge way. Twenty-two soldiers-hiding-caverns were chiselled into the wall for defence purposes and to alleviate the load of the wall. The gates were opened for water transport during the day and locked at night; soldiers guarded them all the time.

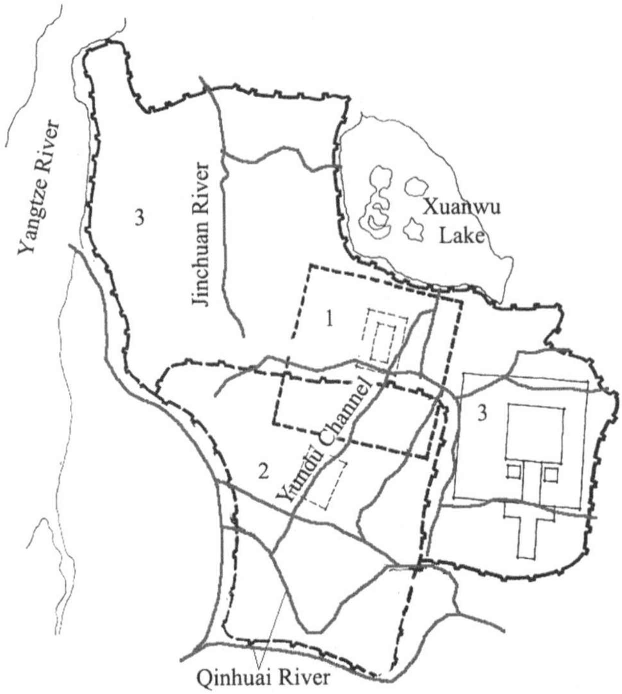

Figure 8.7 Nanjing through history showing the distribution of rivers and channels (in blue) in the Ming Dynasty. (1 = Jianye-Jiankang, 2 = Southern Tang Nanjing, 3 = Ming Nanjing) *Source*: revised from (Steinhardt, 1990)

Figure 8.8 East water gate of Nanjing today (Tongji Watergate) (*Source*: http://china.eastday.com)

Of all the surviving water sluices in Nanjing, the Wumiao Sluice to the south of the Xuanwu Lake is the most famous one. In front of the sluice, there was a 40 m curved inlet trough to slow down the flux into the sluice. After crossing the sluice, water flowed through the city wall into the Pearl River through a 143.9 m underground metal pipeline. The pipeline consisted of 107 copper tubes (D = 95 cm, L = 104 cm, W = 1.5 cm) and 43 iron tubes (D = 98 cm, L = 81 cm, W = 2.0 cm). Two layers of arched bricks were paved above the metal pipeline to alleviate the load of the city wall and lake embankment above the pipe. The sluice gate on the inner city side was made of two overlapping copper valves. The lower valve with five round holes was secured while the upper valve had five round tenons and could be lifted. When the sluice was closed, the tenons were in the holes, which would stop the water flow; when the upper valve was lifted, water would flow through the holes in the lower valve (Guo, 1997) (Figure 8.9).

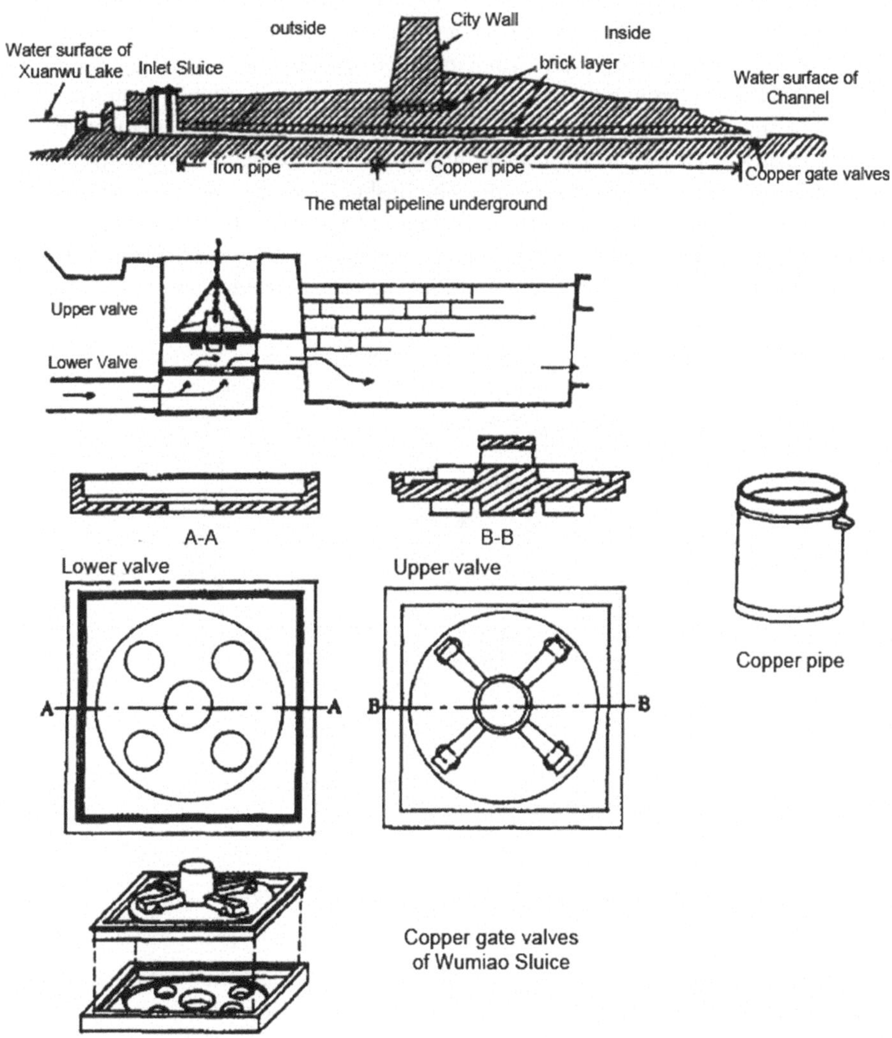

Figure 8.9 Pipeline of Wumiao Sluice of Xuanwu Lake and its copper gate valves, Ming Nanjing (Guo Husheng, 1997)

8.2.6 Dongjing or Bianjing/Kaifeng (Northern Song Dynasty)

Kaifeng was the capital of the Northern Song Dynasty (960–1127 AD). It was large in scale and surrounded by three rings of walls with a population of more than one million between the years 1102–1106 AD. Located in the alluvial plain in the middle and lower reaches of the Yellow River, Kaifeng was in the centre of a canal network, with more than 32 rivers flowing in its 100 km^2 drainage area. It was exactly because of the well-developed canal network that Kaifeng satisfied its water supply and transportation needs (IHNS & CAS, 2000).

The canal network of Kaifeng was comprised of mainly four artificial canals: Bian River, Cai River, Wuzhang River and Jinshui River (Figure 8.10). The Bian River, which was also called Tongji Canal and built by the Emperor Yang in the Sui Dynasty, was an important section of the North-South Canal connecting the water passage of the Yellow River and the Huai River. It entered Kaifeng from the West Gate and exited through the East. The prosperous scene of the Bian River is manifested in the famous scroll painting *Qing Ming Shang He Tu (Riverside Scene at Qingming Festival)* which depicts the scene before and after the Bian River flows through the city (Figure 8.11). Being the most important transport canal, the Bian River had an extraordinary impact on the Song government. Large amounts of grain were shipped from the Yangtze-Huai River Basin to the capital through the Bian Canal. The Cai River and Wuzhang River were also important canals for water transport. The Jinshui River was built for domestic, royal garden and landscape water supply. It diverted water from the Jing and Suo rivers from Xingyang city, and crossed over the Bian River by aqueduct before entering Kaifeng and then heading for the imperial palace (Zheng, 1985). The Jinshui River was then extended southward into residential districts for citizen use.

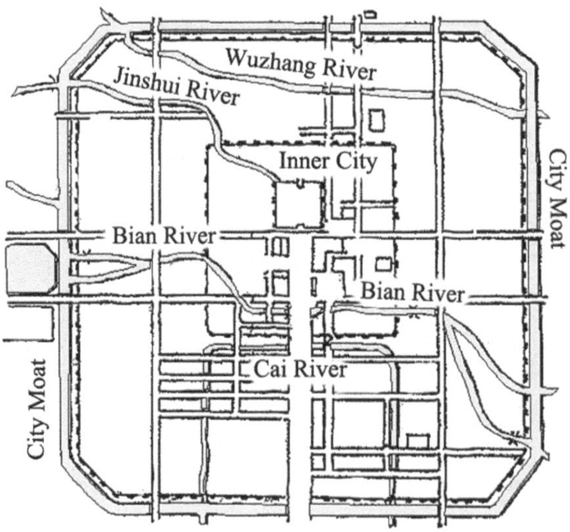

Figure 8.10 Canals of Kaifeng in the Northern Song Dynasty (Zhang, 2003)

High river density and storage capacity were major features of Kaifeng in the Northern Song Dynasty. The urban water system of Kaifeng City consisted of the four above mentioned artificial canals, three city moats, ditches along the streets as well as several lakes and pools. According to Wu's estimation (1995), the total length of the four canals was 30 km, and the total water storage capacity was about 866,300 m^3. In addition, Kaifeng possessed wider city moats, which were named as "Hu Long River". Wu calculated that the total length of the three rings of moats was 47.4 km, with a width of 80 m and depth of 4.8 m. Their water storage capacity was about 17,656,000 m^3. So the total water storage capacity of Kaifeng was 18,522,300 m^3, and the river course density was 1.55 km/km^2 (Wu, 1995). The planning, design and construction of the entire city water supply and drainage system of Kaifeng reflected a high level of technology (Du & Qian, 1999).

Figure 8.11 Bian River: Water gate on Bian River, part of Riverside Scene at Qingming Festival (Qing Ming Shang He Tu), a painting about urban life of Kaifeng City in the Northern Song Dynasty by Zhang Ze Duan

After the fall of the Northern Song Dynasty, Kaifeng was severely damaged during the Yuan and Jin dynasties. Flood disasters were frequent after 1194 AD, when the Yellow River changed its course and flowed towards the southeast. From the Jin to the Qing Dynasty, the Yellow River flooded Kaifeng six times and the nearby areas 40 times. All the disasters caused great harm to the water supply system. The four main canals silted up because of the floods and the water transportation decreased. The once prosperous Kaifeng gradually fell into oblivion (Chen, 1983).

8.2.7 Hangzhou (Southern Song Dynasty)

Situated on the lower reaches of the Qiantang River, Hangzhou was the southern end of the Grand Canal and had become an important commercial centre since the Sui Dynasty. Later it was chosen as the new capital of the Southern Song Dynasty and remained its capital from the early 12th century until the Mongol invasion of 1276 AD. The city was called Lin'an then.

8.2.7.1 The West Lake and Six Wells in the Tang Dynasty

The water for domestic use, irrigation and the canals of Hangzhou mainly came from the West Lake. Due to the presence of saltwater in the area formed by the retreat of the sea, the underground water of Hangzhou was not suitable for drinking. Therefore Li Mi, the governor of Hangzhou, built, during the years 780–784 AD, the Six Wells to divert water from the West Lake for domestic use. Instead of drawing water from underground, the Six Wells used the approach of "Yin Dou", which introduced water through bamboo pipes buried underground, to divert water from the West Lake (Xiong & Guo, 1989). The Six Wells, from north to south, were Xiaofang Well, Baigui Well, Fang Well, Jinniu Well, Xiangguo Well and West Well. They were situated in the western part of Hangzhou, near the West Lake and along the Qinghu River (Figure 8.12). The West Lake became thus the ideal water source to ensure the drinking water supply of the urban residents.

In 824 AD, Bai Juyi (then Prefect of Hangzhou, but more famous as a poet) heightened and elevated the dykes of the West Lake to increase its water storage. After that, while ensuring domestic use, the water from the West Lake could also support the irrigation of the farmlands in the Qiantang and Yanshan area. Bai described other facilities on West Lake in his records "*Qian tang hu shi ji*": there were 'stone culverts on the north side and a bamboo pipeline system on the south (of the West Lake)'. The stone culverts could store and discharge the lake's water. The bamboo pipe system, which was similar to a modern pipeline, was buried underground and diverted the water to the Six Wells. Bai Juyi also mentioned the use of a flood spillway (Que'an) near the West Lake. The function of these three facilities could be combined together to control the water volume of West Lake. From the Tang Dynasty onward, this water supply method consisting of the West Lake and the Six Wells was kept up until the Qing Dynasty (Zhang, 2007).

Figure 8.12 The Six Wells in Hangzhou (*Source*: Zhongguo Liuda Gudu, Chen Qiao Yi, 1983)

8.2.7.2 Urban river network in the Five Dynasties and Song Dynasty

During the period of Five Dynasties, the Wuyue Kingdom made great efforts to build up the urban canal system. The four main rivers of the inner city, Qinghu River, Shi River, Yanqiao River and Maoshan River all relied upon the water from the West Lake. In addition, water sluices like Longshan Gate and Zhejiang Gate were built to block the intrusion of saltwater and sediment from the Qiantang River into the city canals.

In the Northern Song Dynasty, some weirs and water gates were destroyed, and water from the Qiantang River entered the canal system. 'The West Lake became narrow and shallow and the Six Wells were gradually damaged' (Su Shi, Petition to Dredge the West Lake). Su Shi, a famous writer and poet of the Song era, implemented therefore a large-scale dredging project during his tenure as prefect of Hangzhou.

Su's dredging project was divided into three phases. First, in order to solve the siltation problems of the canals, he put up several dams to guide water into the Maoshan River for sediment settling. Second, "stone tanks" (small ponds) were built along the diversion canals, to restore water for citizens' use and for fire prevention (Zhang H., 2007). Third, Su assembled over 200,000 workers to dredge the West Lake and construct a pedestrian causeway across the lake. He abolished the farmlands in the lake, replaced the bamboo pipes with tiled ones and laid stone grooves with a thicker bottom (Su, Records of the Six Wells in Qiantang).

Hangzhou had become unprecedentedly prosperous when the Southern Song Dynasty moved its capital there. According to *Meng Liang Lu*, a detailed description of the society and economy, as well as the urban life of Hangzhou, the population of Hangzhou at that time was over one million. The urban water supply still depended on the West Lake. In addition, canals built in the Northern Song Dynasty were expanded. According to historical records, Hangzhou City in the Southern Song Dynasty had a total of 22 rivers, which included the four major inner city rivers Maoshan, Yanqiao, Shi and Qinghu rivers, 10 in the eastern city, 6 in the northern city and 2 in the western city. These rivers connected the Qiantang River in the south with the Jiangnan Canal and Tai Lake basin in the north. The West Lake together with those canals comprised the entire layout of the water supply system of Hangzhou in the Southern Song Dynasty (Figure 8.13).

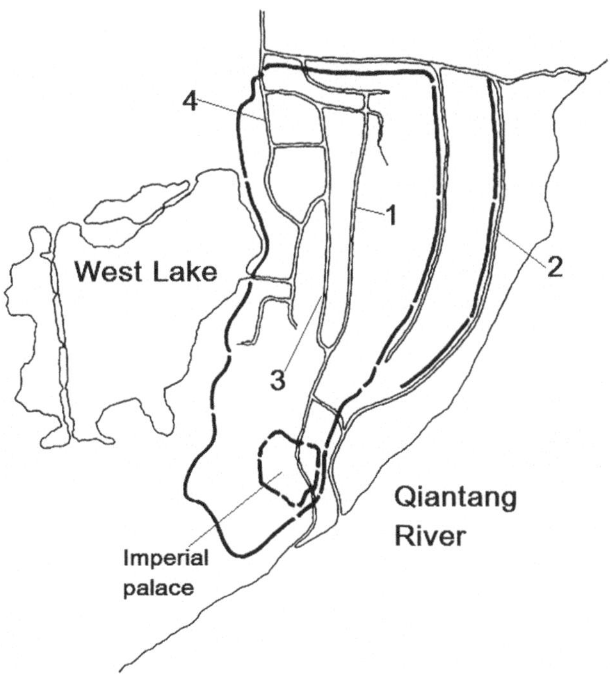

Figure 8.13 West Lake and channels in Hangzhou (Southern Song Dynasty) 1. Yanqiao River, 2. Caishi River, 3. Shi River, 4. Qinghu River (Steinhardt, 1990)

8.2.8 Zhongdu/Dadu/Beijing (Liao, Jin, Yuan, Ming, Qing Dynasties)

The City of Beijing has a long history. In 936 AD, the Taizong Emperor of the Liao Dynasty (Kitan Empire) named Beijing as the second capital of the empire and then renamed it Nanjing (Southern Capital) or Yanjing. In 1153 AD, the Jin Dynasty (Jurchen Empire) moved its capital to Yanjing and renamed the city Zhongdu (Central Capital). Since then, Beijing served in succession as the capital of the Jin, Yuan, Ming and Qing Dynasties and became the centre of power of Imperial China for more than 700 years (Chen, 1983) (Figure 8.14).

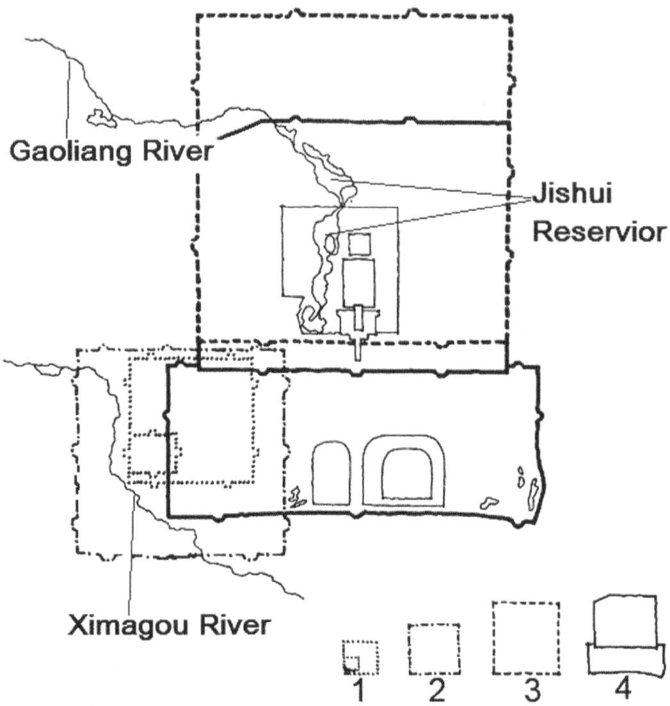

Figure 8.14 Beijing through history [Drawn by Huang Yunsheng]. 1. Liao Nanjing, 2. Jin Zhongdu 3, Yuan Dadu, 4. Ming-Qing Beijing (Steinhardt, 1990)

Situated in the northern part of the North China Plain, Beijing is surrounded by mountains on the north, northwest and west sides. Major rivers flowing around the city include the Yongding and Chaobai rivers. Beijing is at the northern end of the Grand Canal, which was built across the North China Plain starting in Hangzhou. Because of the uneven distribution of runoff, high silt content and steep slopes, it was difficult for Beijing to maintain stable water sources for a long time. That is the reason why many hydraulic structures were constructed to exploit the water resources in and near Beijing, together with exploitation and utilisation of groundwater (Zheng, 1985).

8.2.8.1 Zhongdu of Jin Dynasty

Zhongdu was situated in the south-western part of today's Beijing, near the Lianhua (Lotus) Reservoir watershed. Sources of water in the city were wells and transfer of surface water from the Lianhua

Reservoir. The Lianhua Reservoir was originally an underground spring and was called Xi Hu (West Lake) (Duan, 1989). According to *The Book of Water (Annotation)*, the water was conveyed from the east side of the reservoir via the Xima Canal and entered the city from the south gate. After the Jin Dynasty moved its capital here, the Xima Channel became a part of the city, crossing into the imperial palace and becoming an important water source of it (Hou, 1979). The City of Zhongdu expanded rapidly and the population reached one million. As the demand for water also increased rapidly along with the growing population, the Lianhua Reservoir quickly became insufficient. Then, the Xishan spring was developed to solve the water shortage problem. Cai (1987) inferred that during the Jin Dynasty there might have been canals located upstream of the Lianhua Reservoir that transferred water from the Xishan spring. In addition, in the early Jin Dynasty the Zha River was dug to divert water for canal transport from the Gaoliang River in the north-eastern outskirts of Beijing to Tongzhou in the east. Along the canal several gates were built from which the name Zha (Gate) River was derived. However, because of the limited flow of the Gaoliang River, the issue of canal transport could not be satisfactorily resolved.

8.2.8.2 Dadu of Yuan Dynasty

Dadu, the capital of the Yuan Dynasty, was located to the north-east of Zhongdu. The city proper then moved from the Lianhua Reservoir watershed to the Gaoliang River watershed (Figure 8.14). One reason for the relocation was that Zhongdu had been severely burned down during the preceding war, and the palaces were in ruins. Another, more important reason for the selection of the new site was the relative abundance of water resources, because the Gaoliang River watershed could provide more water than the Lianhua Reservoir for purposes of water supply and water transport (Chen, 1983). Under the careful design and supervision of Liu Bingzhong and his student Guo Shoujing, the construction of Dadu took 18 years and was completed in 1285. A few years later, additional water from Baifu and Xishan Springs was transferred into the Gaoliang River. The water supply and canal transport needs were thus satisfied at the time (Du, 1998).

Two water supply systems were constructed in Dadu City. One, specifically serving the imperial palace, consisted of the Jinshui River and Taiye Pond. The Jinshui River collected spring water from Yuquan Mountain and entered the city through the Heyi Gate (now Xizhi Gate) to flow into the Taiye Pond. To maintain and control the water quality, the river was not allowed to mix with other water sources. Independent water passes were built in the places where the Jinshui River had to cross other rivers. According to legal documents from the Yuan Dynasty, bathing, washing, dumping and watering livestock were prohibited (Du, 1998).

Another water supply system was the canal transport system consisting of Gaoliang River, Haizi and Tonghui River. A transport canal had already been constructed in Zhongdu during the Jin Dynasty, but since the water source could not be guaranteed, canal transport was not effective. In the Yuan Dynasty, Baifu Spring water was transferred to fill the canal by means of a hydraulic project proposed by Guo Shoujing. The History of the Yuan Dynasty recorded that Guo Shoujing built a weir and channels to divert water from the Baifu Spring to Wengshan Lake (now Kunming Lake) and then to the Jishui Lake. The outflow from the city entered the old course of the Zhahe River of the Jin Dynasty. Several water gates were set up in this section. The entire project was completed in 1292 AD. Cargo ships and boats could now enter the city directly through the canal, which was named Tonghui River (Figure 8.15) and is still in use today. The Baifu Spring water transfer project was one of the most innovative solutions to the water resource problems in the history of Beijing (Hou, 1979).

Figure 8.15 Tonghui River port near Beijing (*Source:* http://www.bjmem.com/bjm/)

8.2.8.3 Beijing of Ming-Qing Dynasties

In the early Ming Dynasty, the old canal system was damaged during the reconstruction of Beijing. The palace-only canal, the Jinshui River was also abandoned. There were efforts made to restore the Tonghui River canal in the Ming Dynasty, but none of them were effective due to lack of water source (Zheng, 1985). Water supply in the city could only rely on spring water from Yuquan Mountain. Spring water was collected at Wengshan Lake (now Kunming Lake) and transferred into the Jishui Reservoir within the city. Then the water was divided into two distributaries: one into the Forbidden City, and another into the Tonghui River which significantly raised the water level and the capacity of Kunming Lake, making it the first artificial reservoir in Beiji (canal). Therefore, the water supplied to the Ming imperial palace was no different from the water in the canal. This was very different from the scenario in the Jin and Yuan Dynasties. But the situation has remained unchanged since then (Hou, 1979).

During the Qing Dynasty, water canals in Beijing were mainly based on the old canals left by the Ming Dynasty. However, lack of water was still the main problem for canal transport. In the Qianlong Reign (1736–1795), the government finally decided to manage the water canals in the west of Beijing for landscape and transportation purposes. The first step was to utilise Wengshan Mountain as a site for a landscape garden. The second was to deepen Wengshan Lake and construct a levee (Figure 8.16). Wengshan Mountain then was renamed Wanshou (longevity) Mountain. The east levee significantly raised the water level and the capacity of Kunming Lake, making it the first artificial reservoir in Beijing. Also, three sluices were built under the east levee and at the north and south ends of the lake to control the water level. The condition of water supply was then much improved (Zheng, 1985). In addition, stone troughs were built to divert water from the West Mountains and Xiang Mountain into the Kunming Lake as a supplement (Hou, 1998).

Figure 8.16 Kunming Lake and Wanshou Mountain in the Summer Palace. Photo was taken by S. Yamamoto, in 1900–1906 (*Source:* http://images.china.cn/)

8.2.8.4 Wells in the city of Beijing

In Beijing, dwellers had, for generations, been using wells to draw groundwater for their daily life. The alleys in Beijing are called "hutong" which means "well" in Mongolian. In Dadu of the Yuan Dynasty, the wells were dug in the alleys, and the alleys were named after the wells. Therefore "hutong" has the meaning of "alley". When the City was reconstructed in the Ming and Qing Dynasties, many "hutongs" were left without a well. According to contemporary records from the Yuan, Ming and Qing Dynasties, the quality of water in the wells had been low due to salinisation and this led to two results: water was supplied at three levels of quality (for washing, cooking, and drinking tea); and, selling water became a profession in Beijing (Duan, 1989). As of 1885, there were still 1245 wells recorded (Beijing zhi, 2003) but their water had mostly a salty and bitter taste.

8.3 CANAL TOWNS ON LOWER YANGTZE FLOOD PLAIN

The middle and lower reaches of the Yangtze River Plain are situated between the Yangtze Delta and Taihu Lake. The region is called "Jiangnan" which means south of the Yangtze River. This region was a high-yield agricultural area in China and it was said, that 'once the grain in Jiangnan were ripe, then it would be enough to feed the whole country'. Especially after the excavation of the Grand Canal, transport and trade were greatly promoted and many cities in this area became important and prosperous.

Different from the cities in the north, the south of the lower reaches of the Yangtze Delta showed a general image of "water country". Throughout the region waterways were laid out in a crisscross pattern. Yet, each city in this region had its own characteristics. For example, Suzhou had a canal system in a chessboard fashion; Wuxi had a rhombic moat and a fish skeleton-like water network; Shaoxing's seven-canals-system looked just like seven strings of a Chinese musical instrument; Jiading, Shanghai and Songjiang had round moats, but the exact layouts of canals differed from each other, forming unique features of their own (Wu, 1991) (Figure 8.17). Although there were many rivers and lakes in and around the "canal towns", floods barely occurred here because the location of the cities was selected on a relatively higher relief. The canals were very important to these cities as they served as water transport arteries and sources of domestic water supply.

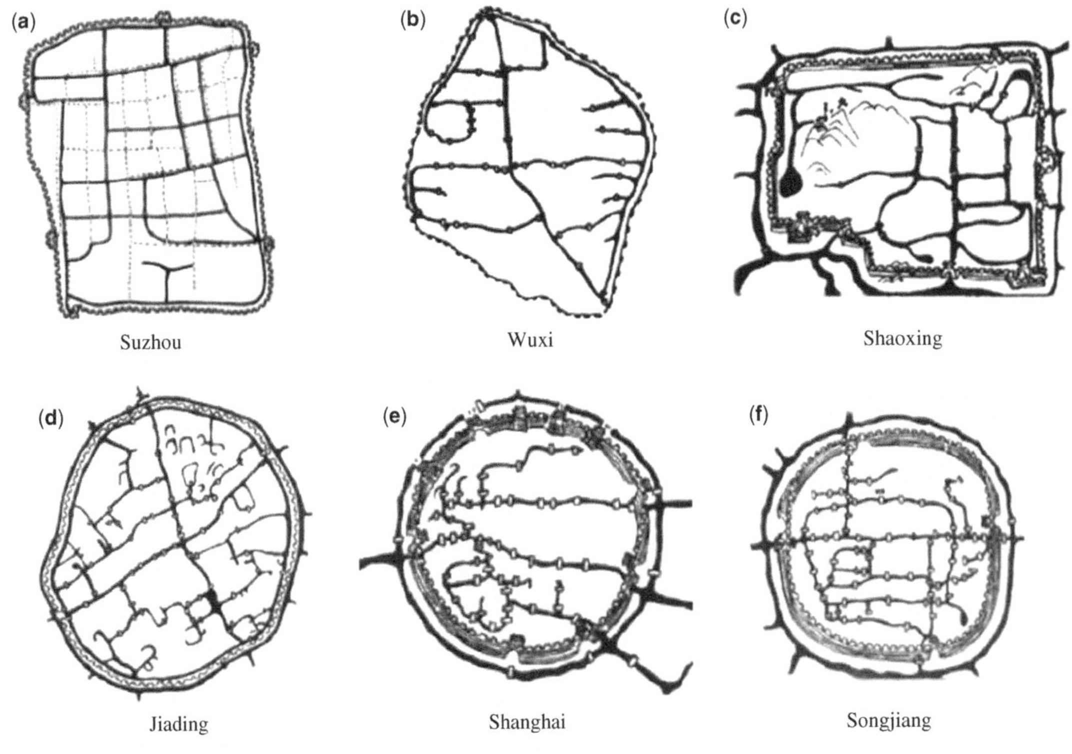

Figure 8.17 Canal systems of various patterns (Wu, 1991)

8.3.1 City of Suzhou

Situated in the lower reaches of the Yangtze River and on the shores of Taihu Lake, Suzhou was founded in 514 BC and was the capital of the State Wu in the Spring and Autumn Period. The city had eight water gates and two sets of transport systems: canals and roads. Named Pingjiang during the Song Dynasty, Suzhou was the trade and canal transport centre in the region after the construction of the Grand Canal in the Sui Dynasty (Zheng, 1985).

The famous stele of the *Pingjiang Map* (1229), 2.76 m high and 1.45 m wide, is the earliest city map of Suzhou dating from the Song Dynasty. It depicts in detail the distribution of roads, canals, waterways and bridges of Suzhou City during the Southern Song Dynasty (Figure 8.18), showing about 100 watercourses with a total length of about 82 km as well as 359 bridges. Different from imperial capitals, which often abandoned their former sites, the location of Suzhou did not change much through different dynasties. This was due to the framework function of the river network. Although the buildings on the land were several times destroyed in wars, the river and canal network infrastructures were preserved and could be put back into use after some repair.

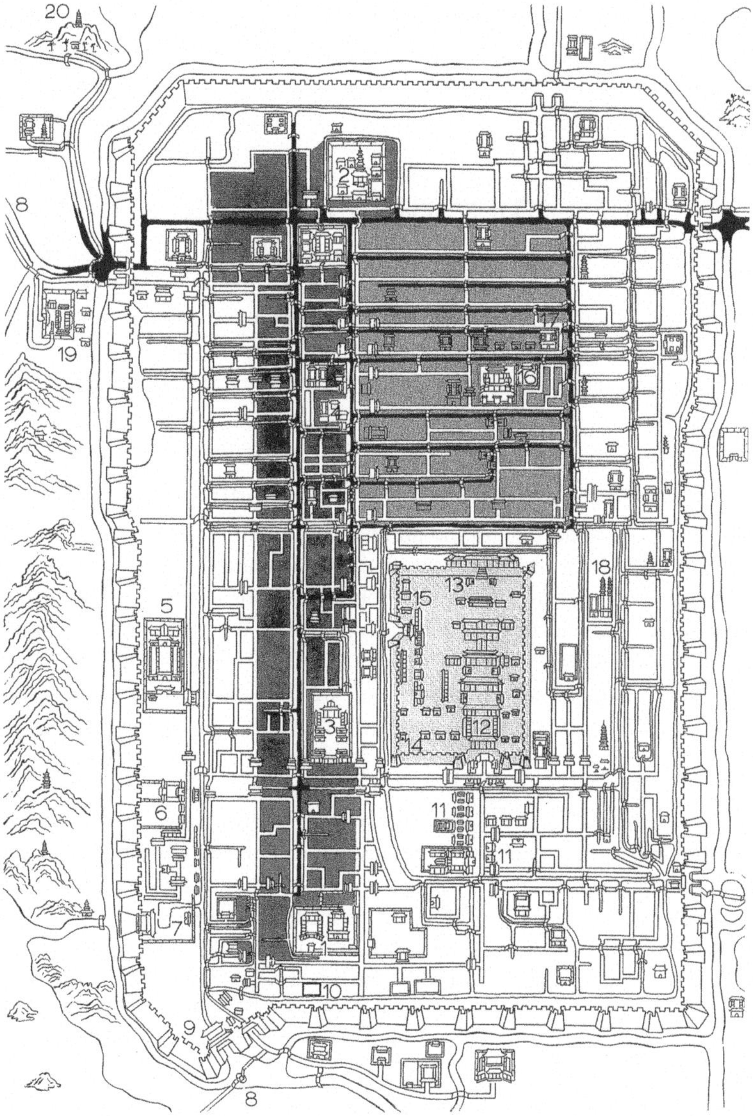

Figure 8.18 Map of Pingjiang City (Suzhou City) of Song Dynasty (*Source:* Golany, 2001, p. 216)

Suzhou was a typical canal network city, with a characteristic water supply pattern. In the city, canals were similar to streets and were built in a chessboard fashion. According to the *Wu Di Ji (Record of Wu*

Area), there were seven main canals in the city, with three east to west, and four north to south, known as "three longitudinal and four transverse". Small branch canals extended to every lane and corner in the city like blood vessels. Dwellings in Suzhou were built in the same form with the front door facing the street and the back door against the canal, just like a poem described: 'Houses are all pillowed on water's edge'. Most canals were excavated artificially with neat embankments and crossing bridges, which provided convenience for transportation. The canal system could also meet the demand of urban water supply, drainage and fire prevention, etc. Well water was another important water source. According to data collected in 1959 by the Suzhou government, the city had about 2000 wells, some of them tracing back to the Song Dynasty. Most of the wells of the Southern Song Dynasty were made of soil and brick (IHNS & CAS, 1985).

8.4 CHINESE FRONTIER CITIES WITH ETHNIC MINORITIES

Large areas of the present day territory of China such as the autonomous regions of Xinjiang, Tibet, Inner Mongolia, Ningxia and Guangxi, were originally settled by non-Chinese speaking peoples such as Uighurs, Tibetans, Mongolians and others. Chinese-speaking Han people constitute about 95% of the current population of China and live mainly in the fertile plains (approx. 60% of the territory), whilst the minority peoples often live in the frontier areas, which are mostly mountainous and arid (approx. 40% of the territory).

8.4.1 Lijiang (branched canal system in mountainous region)

Lijiang City is located at 2400 m altitude in the north-west of Yunnan Province, China. Starting from the 13th century AD, the Old Town of Lijiang has built up a unique water supply system, with spring collection works, a reservoir, spillway, water distribution network, weirs, gates and sluices, and other stone structures which met the needs of the inhabitants until the end of the 20th century. With the advent of modern developments since the 1970s and the declaration of the Old Town of Lijiang as a UNESCO World Heritage Site in 1997, the ancient water supply system changed its function and became an important part of the successful tourism industry. Lijiang's old town differs from other Chinese historic and cultural cities in its irregular street pattern, architectural style, and lack of city walls. Unlike other large cities in China, which are inhabited by Han people, the old town of Lijiang is an ancient frontier city inhabited mainly by the Naxi ethnic minority as well as Han, Bai, Yi, and Zang (Tibetan) nationalities. Lijiang's water system originates from the Black Dragon Pool (Figure 8.19), which is fed by dozens of springs on the mountainside. Water flows from here through the Jade River to the Jade Dragon Bridge, where it branches into three tributaries, known as the East, Middle and West Rivers (Figure 8.20). These subdivide further into a network of canals and culverts along natural contours to supply every house in the town with water. The Middle River is the old natural watercourse in the middle of the city, whilst the West River and the East River are artificial rivers dug during the Yuan Dynasty and the Qing Dynasty respectively. Besides the rivers, many small springs emerge in the city which are developed in a series of small pools called three-eye wells (Figure 8.21). Figure 8.22 shows the complex water network of the rivers and branching canals (Koenig and Fung, 2009). The width of the canals varies from 8–10 m in the Jade River to about 2–3 m in the three main branches, and then progressively decreases in the secondary branches to as little as 0.1 m. Similarly, the depth varies from less than 1 m in the Jade River and the three main branches to about 0.1 m in the secondary branches. The swift flow (0.5–1 m/s) and shallowness of the waters prevents any navigation.

Figure 8.19 Black Dragon Pool

Figure 8.20 Trifurcation of Jade River into the West River, Middle River and East River

Figure 8.21 Typical three-eye water well

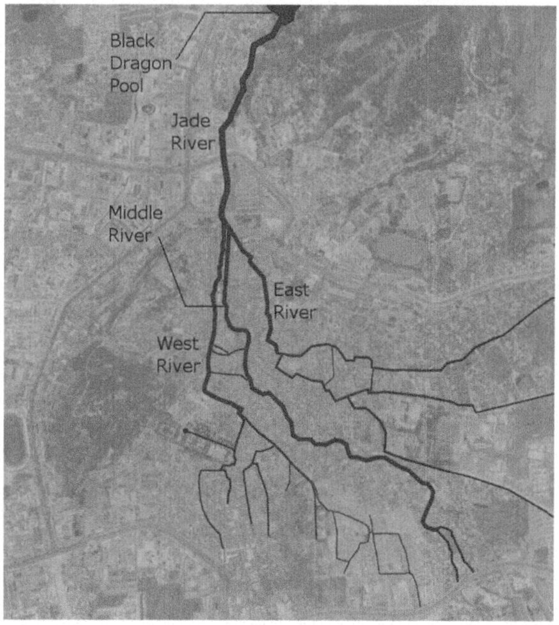

Figure 8.22 River and canal network of the Old Town of Lijiang

City residents drew drinking water and cooking water from the canals early in the morning, washed food in the canals late in the morning, and washed clothes and other household items in the canals during the afternoon (Murphy & Chen, 2007). Similarly, the water of the three-eye wells (Figure 8.21) was used for public drinking and cooking water (upper pool), food washing (middle pool), and clothes washing (lower pool) (Murphy & Chen, 2007).

The origin of the ancient water supply system in one of the remotest corners of China remains unknown. Since no similar systems are found anywhere else in China, indigenous development, transfer of technology from Central Asia after the Mongol conquest in 1257, or perhaps a combination of both is suggested. Despite its unknown origin, the water supply system existed for hundreds of years with little change. The long-term sustainability of Lijiang's unique water system seems to have depended on the sloping topography, which provided for swift flow, and on the relatively stable flow of water from springs, which minimised flow variations. These two necessary factors did not apply to the cities of China proper, which are mostly located on plains.

8.4.2 Turpan (Karez system in desert region)

Turpan, also known as Turfan or Tulufan, is an ancient oasis town in the Xinjiang Uygur Autonomous Region, formerly the north-west frontier region of China. It is located on the northern side of the Turpan Depression, the only part of China below sea level (maximum at −154 m), at an elevation of about 30 m above sea level. It has an extreme desert climate with an average annual precipitation of about 15 mm. Its inhabitants are mostly Muslim Uygur people. Turpan was a strategic stop along the Silk Road's northern route to provide traders with food and water. Due to the paucity of perennial streams, the Turpan people developed a unique underground tunnel system to harvest water from aquifers in alluvial fans of the Tianshan Mountains, making use of the natural hydro-geological conditions. Since all water flows by gravity, sloping land is an essential condition for this system. The tunnels are called karez in the local Turkic language, or kanerjing in Chinese, and can be compared to the Persian qanats. Some Chinese scholars claim that the karez technology is one of the three great engineering achievements of ancient China, comparable to the Great Wall or Grand Canal.

Research has mostly focused on construction aspects of karezes and their utilisation for agricultural irrigation as well as their origin and history (Hansen, Kobori, 2010 [with extensive bibliography]; Lein & Shen, 2006). However, another important purpose is the provision of drinking water for the oasis' inhabitants and animals as well as for domestic uses. A schematic of the karez system for village water supply is shown in Figure 8.23. Figures 8.24 and 8.25 show the inside and outlet of a karez. A small storage pond at the outlet of the tunnel usually served for equalisation of water flow (Figure 8.26) prior to distribution in open, branched canals through the town (Figure 8.27), and further to agricultural fields.

Karezes can also be found in other oasis towns in Xinjiang, but their highest concentration has always centred around the Turpan Region (consisting of Turpan City and the counties of Piqan and Toksun) and Hami. A survey in 2002/03 recorded 512 karezes in Turpan City, of which 253 produced water and 259 were dry, with a total length of 1535.558 km of underground channels. Corresponding numbers for the Turpan Region were 1091 karezes (of which 404 were with water) with a total length of 3724.11 km, whilst the whole of Xinjiang counted 1,784 karezes with a total length of 5,272 km of underground channels, of which 614 still produced water (Karez in Xinjiang, 2004, p.1134). A typical karez usually measures from several hundred metres to more than 10 kilometres in length.

Since the early records of Turpan dating back to the Han Dynasty, the first appearance and number of Turpan's karezes has remained largely in the dark, but most likely expanded over time with population growth, changing social economy as well as agricultural production. The total number of karezes in Turpan till the Daoguang Reign of the Qing Dynasty (1820–1850) was reported as nearly 300 (Karez in Xinjiang, 2004, p. 1136). Consolidation and expansion of Qing rule during the second half of the 19th century, especially due to the efforts of Lin Zexu and general Zuo Zongtang, led to the construction of many water conservancy projects including the repair and excavation of numerous karezes.

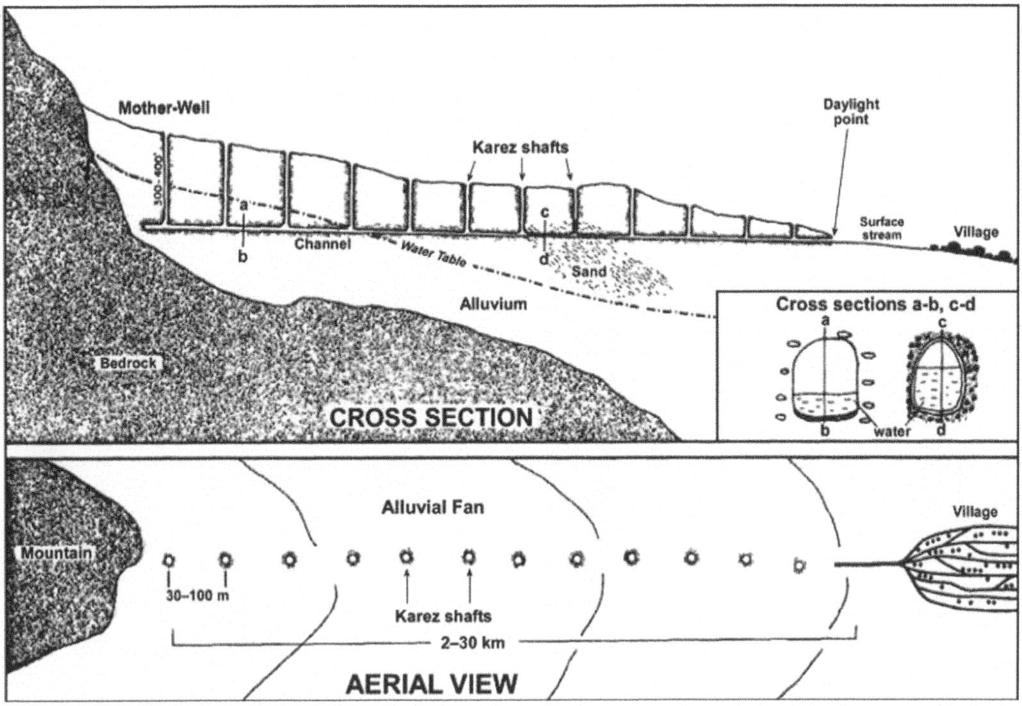

Figure 8.23 Typical features of karez for provision of water to a village (English, 1966, p. 31)

Figure 8.24 The inside of a karez (Photo taken in 2004 by HHP Fang)

Figure 8.25 The outlet of a karez (Photo taken in 2004 by HHP Fang)

Figure 8.26 Storage pond near outlet of karez (Karez in Xinjiang, 2005, p. 38)

Figure 8.27 Open channel flowing through Turpan city providing water (Karez in Xinjiang, 2005, p. 20)

A survey at the beginning of the Republic of China (1912–1949) mentioned about 600 karezes in Turpan City, about 360 in Piqan and about 100 in Toksun. River water accounted for 30% and karez water for 70% of the total water flow (Karez in Xinjiang, 2004, p. 1137). By the end of 1949, the number of karezes in the Turpan Region was 1084, with an annual water output of 508.1 million m^3 and an irrigation area of 19,300 ha. That amounts to 3510 m^3/year of available water per capita for the population of 144,754 recorded in 1949 (Tulufan di qu zhi, 2004, p. 119), which is far in excess of domestic needs and clearly shows that more than 90% of the water was used in agriculture. The maximum number of karezes was reached in 1957 at 1237, with an annual output of 562.6 million m^3 irrigating 21,400 ha (Karez in Xinjiang, 2004, p. 1122). With fast agricultural development and concomitant overuse of electrically pumped groundwater, the importance of karezes declined and many ran dry because of lowered groundwater levels. Since the end of the 1950s modern water conservancy projects such as storage reservoirs, diversion works, and pumping wells increased the amount of available water and began to slowly replace the ancient, labour-intensive karez systems. However, sustainable water use and management has become an increasingly critical issue for the further development of the Turpan Region.

The origin of Turpan's karez technology has been discussed quite controversially, with three theories proposed: (i) local development by indigenous people, (ii) imported from Persia where qanats were historically first encountered, and (iii) imported from elsewhere in China, which is the currently prevailing theory among Chinese scholars who see many similarities between the karez underground tunnels and the water tunnel in the Longshou canal project of 128–117 BC (Figure 8.28). On the other hand, Trombert (2008) is of the view that karez technology was rarely applied in Xinjiang in ancient times and became widespread only during the 19th century Qing administration. Whilst karez tunnels were used to collect underground water and had only one outlet without inlet, the Longshou canal, in contrast, was conveying river water through the inlet and outlet of the tunnel, without collecting additional water from the underground. So the mystery of the origin of the karez in Turpan remains, until more pertinent documents are found and analysed (see also Kobori, 2010).

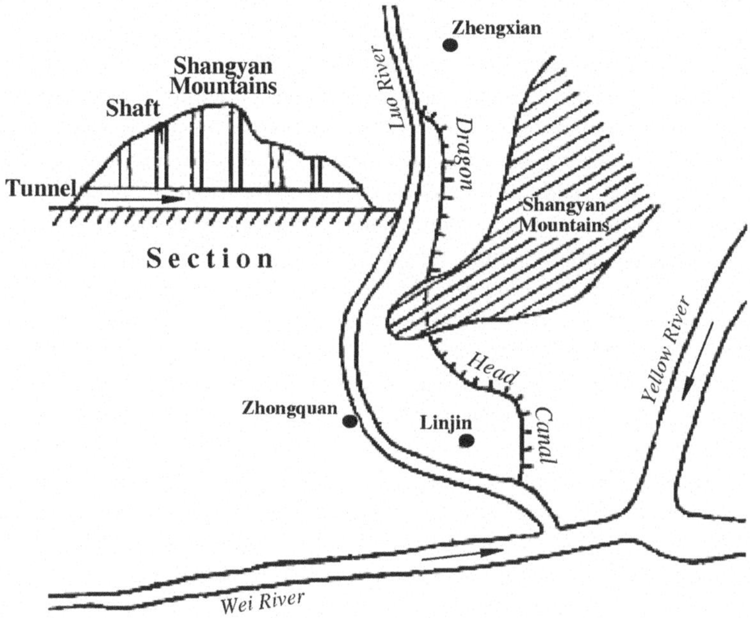

Figure 8.28 Water tunnel for Longshou (Dragonhead) canal (128–117 BC) (Zheng Liandi, 2004, p. 147; Trombert, 2008)

8.5 MULTI-PURPOSE WATER WORKS

Some urban water supply works formed part of larger, integrated water management schemes, which were designed for multiple purposes such as water supply, irrigation, flood control, navigation, and prevention of tidal inflow. In the following, three outstanding examples are described.

8.5.1 Dujiang weir – water supply for Chengdu City

The Dujiangyan Irrigation System, briefly Dujiangyan or Dujiang Weir, is the most famous ancient water-control project of China and was named in December 2000 as a UNESCO World Heritage Site for the following reasons: (i) 'The Dujiang Irrigation System, begun in the 3rd century BC, is a major landmark in the development of water management and technology, and is still discharging its functions perfectly, and, (ii) the immense advances in science and technology achieved in ancient China are graphically illustrated by the Dujiang Irrigation System'. (WHC, 2000).

Dujiangyan is located in Dujiangyan City (formerly called Guanxian County) in the Province of Sichuan, where the Min River exits the mountains and enters the alluvial plain, approximately 59 km from the provincial capital Chengdu. Dujiangyan was built in 256 BC during the reign of King Zhaowang of Qin, in order to consolidate and develop the newly conquered territory of Shu (Sage, 1992). Its main functions were (i) diversion of the Min River to prevent flooding, (ii) provision of irrigation water for food production in the Chengdu plain, (iii) provision of drinking water to the villages and towns of the Chengdu plain, in particular the capital city of Chengdu, and (iv) navigation for transport of soldiers and military supplies further downstream to the Yangtze River and the Yangtze Plain. The success of the Dujiangyan system may have greatly contributed to the conquest and first unification of China by emperor Shihuang in 222 BC.

The headworks of Dujiangyan were built by Shu County magistrate Li Bing. They comprised three major components, namely Yuzui (Fish-mouth) diversion structure, Feisha overflow weir, and Baopingkou intake. The Yuzui structure divided the surging Min River flow into the Inner and Outer river, while the Feisha weir controlled the flow of the Inner river into the Baopingkou intake by discharging excess water and sediments back into the Outer river, which followed the natural course of the Min River. This ingenious, dam-less design was able to control the flow in the Inner river through the Baopingkou intake in such a way that during the rainy season it received 40% of the Min River flow, but during the dry season 60%. The Baopingkou intake was cut by thousands of workers through a hard-rock cliff on the bank of the Min River. Since most of the construction materials were locally procured, like bamboo cages, timber and stones, which are short-lived and perishable, annual maintenance needs including sediment removal required a large work force. A schematic diagram and overview of the Dujiang Weir headworks is shown in Figures 8.29 and 8.30.

After flowing through the Baopingkou (Figure 8.31), the Inner river is divided into three tributaries: Puyang River, Botiao River and Zouma River, which further subdivide into secondary and tertiary branch canals forming an extensive irrigation network for agricultural crop production, mainly of rice. Two downstream rivers, Pi River (or Fu River) and Jian River (or Jin River), which flow through the city of Chengdu, were the main water resource of Chengdu. In the course of time, the water supply system of Chengdu was frequently modified. Water was diverted to the city moats, new canals were dug, artificial storage ponds were built for water supply in the dry season, which could also be used for aquaculture, and sediment settling basins were installed (Wu, 2008). Besides for water supply, the rivers also served for navigation and changed the formerly isolated place of Chengdu into the transportation and economic centre of south-western China.

The Dujiang Weir not only brought hundreds of thousands of hectares of fertile farmland in the Chengdu Plain under irrigation and increased the prosperity of the land, but also protected Chengdu City and the surrounding counties from droughts and floods for over 2300 years. To honour the builder Li Bing and his son and collaborator Er Lang, the magnificent Erwang temple was erected on the riverbank next to Dujiangyan.

The hydraulic engineering achievements of Dujiangyan were introduced in the West mainly through the efforts of Needham and his research group at Cambridge (Needham & Wang, 1965; Needham *et al.* 1971; Ronan, 1995), though many earlier travellers had already reported on it. In recent years, the literature on historic, technical, scientific, operational and management aspects of the Dujiangyan Irrigation System and its management has greatly increased (Cao *et al.* 2010; Jin, 1988; Kono, 1997; Li & Xu, 2006; McColl; Willmott, 1989; Tan, 2009).

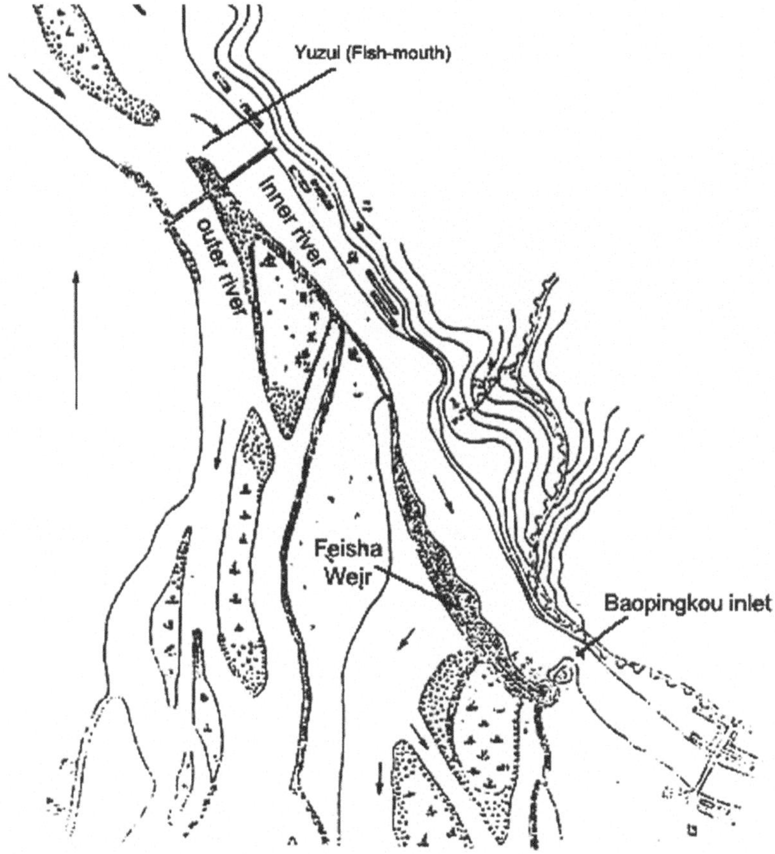

Figure 8.29 Schematic diagram of the Dujiang Weir headworks (Zhou, 2008)

Figure 8.30 (a) Aerial photo of Dujiang Weir on Min River (b) The Yu Zui (Fish Mouth) structure of Dujiang Weir (Photo taken in June 2010)

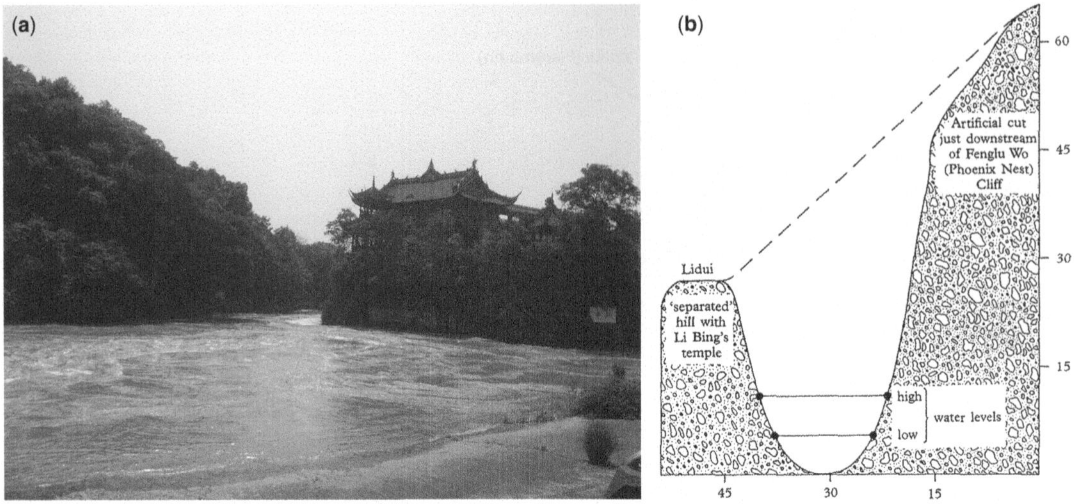

Figure 8.31 (a) The Baopingkou intake of Inner River (Photo taken in July 2010), (b) Cross section of the Baopingkou intake (Ronan, 1995, p. 210)

8.5.2 Tashan Weir – Water supply for Ningbo City

The Tashan Weir (also read as Tuoshan Weir) is located south-west of Yinjiang Town to the South of Ningbo City, Zhejiang Province. It was built by Wang Yuanwei, administrator of Yin County, in the year 833 AD during the Tang Dynasty. Its purpose was (i) to prevent the inflow of tidal sea water, (ii) to conserve, store and distribute fresh water during the dry season for irrigation and water supply to Ningbo City, and (iii) to prevent flooding. It is considered one of the four major water conservancy projects of pre-modern China. In 1988, it was declared a historical site under State protection.

The weir diverted the water of the upstream Zhang River towards the Nantang River and after passing several small towns emptied into the Sun Lake (no longer existent now) and the Moon Lake outside Ningbo City, a typical southern water town. From there it flowed 'into the city through the sluices near the west and south gates. Small canals carried the water through the city to two small lakes, and water from the lakes flowed out through sluices near the two east gates' (Shiba, 1977). The stable water supply greatly promoted the development of Ningbo (Shiba, 1997, Xu, 1997). The Nantang River, with further diversions and branch distribution networks, irrigated thousands of hectares of farmland in seven towns in the Yinxi Plain. The design of the weir was such that in time of floods, 70% of the water was discharged through the Zhang River (outer river), with the remaining 30% diverted to the Nantang River (inner river); whilst in times of drought, 70% of the water entered the Nantang River, and the remaining 30% the Zhang River. This was achieved through the construction of three additional water gates between the inner and the outer rivers and in the downstream of the Nantang River.

A schematic diagram of the headworks is shown in Figure 8.32. 'The weir, 134.4 m long and 4.8 m wide, was built up with 80 and a half pieces of stone slabs measuring 2 to 3 metres long and 0.2 to 0.35 metres wide. Thirty six stone steps were provided on both the right and left. The weir body was of wood and stone structure'. (http://nbgov.nbcom.net/en/page09-08.html) (Figure 8.33). Other sources give a length of 113.73 m (Shang *et al.* 2006).

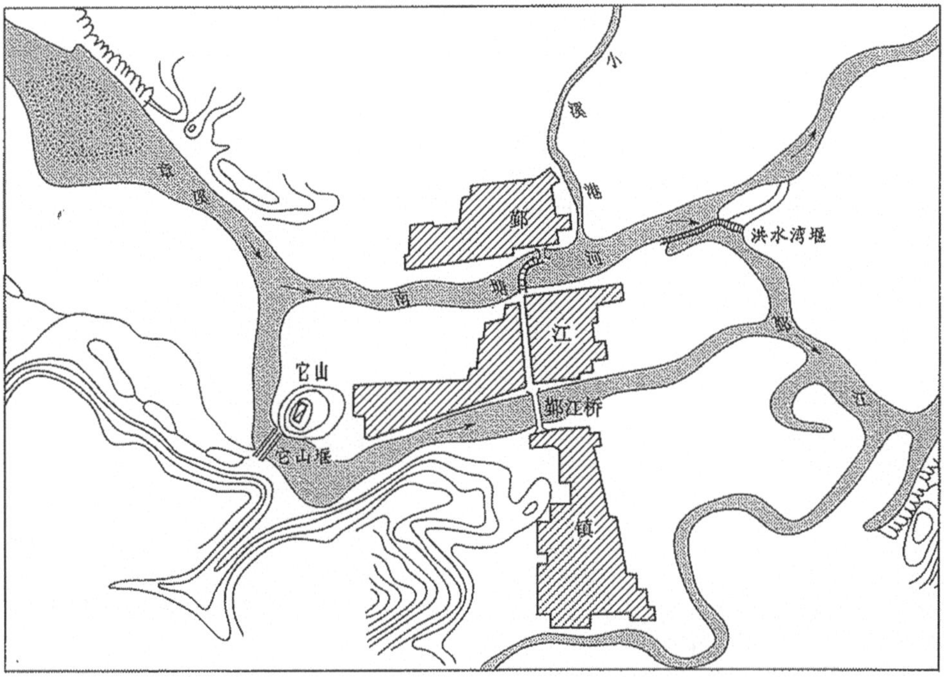

Figure 8.32 Schematic diagram of the Tashan Weir headworks (Zheng Liandi, 2004, p. 168)

Figure 8.33 Overview of Tashan Weir (Photo taken on 22 May 2011)

During the 1991–1993 renovation of the weir, scientific and engineering studies were carried out which revealed, among others, the following interesting characteristics (Shang *et al.* 2006): (i) the weir acts as a bridge, even when the water is overflowing, because of the stepping stones placed on top which are connected to the top of the weir with iron nails (see Figure 8.34 for details); (ii) the weir form is convex towards the upstream and thus can disperse water pressure; (iii) the weir bottoms dip towards the upstream which prevents slippage of the weir; (iv) a seepage control clay wall inside the weir; (v) laying of stripped or blocked stones inside and outside the weir for weir reinforcement; and, (vi) use of wood piles to reinforce the weir foundation.

Figure 8.34 View of the weir from the outside. The apron of boulder strips in the front toe and block stones are set up in downstream side. The top left plate shows floods discharging over the weir (Shang *et al.* 2006)

8.5.3 Mulanbei weir – water supply for Putian City

The Mulanbei weir is located near the mouth of the Mulan River in Putian Prefecture, Province of Fujian. Similar to the Tashan Weir, its purpose was (i) to prevent the inflow of tidal sea water, (ii) to conserve, store and distribute fresh water during the dry season for irrigation and water supply to Putian City and the towns in the Putian Plain, and (iii) to prevent flooding. As one of the best preserved ancient large-scale water conservancy projects, in 1988 it was declared a historical site under State protection.

The water of the Mulan River is led from the east and south ends of the weir into a southern and a northern canal, from where it is distributed through a network of branch canals into the "southern plain" as well as the "northern plain" including Putian prefectural city, respectively (Clark, 1991) (see Figure 8.35). The southern plain's irrigation system alone has seven major channels and 109 secondary channels, with an overall length of 113 km, whilst the canals of the northern plain irrigation system extend for 173 km (Dean and Zheng, 2010a, b). The weir itself is 219.31 m long, with 32 sluice gates (see Figure 8.36). The canals are able to irrigate about 14,000 hectares of land. The return flows into the Mulan River and to the nearby sea are controlled by sluice gates to prevent intrusion of tidal saltwater which, in combination with a sea wall constructed earlier during the Tang Dynasty, protects the low-lying lands from tidal invasion. Hence the Mulan Weir water conservancy project is also counted as one of the largest coastal reclamation projects in Fujian Province (Clark, 2007). (http://www.fzu.edu.cn/fujian/eput.html) Similar to the Tashan weir, the Mulan weir also acts as a bridge.

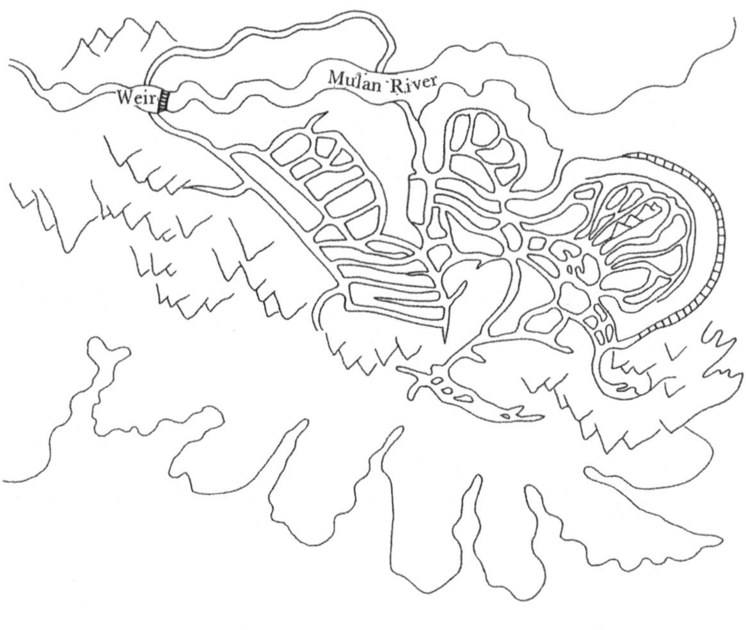

Figure 8.35 Schematic diagram of the Mulan Weir water distribution and irrigation network showing seawall for protection of the lowland from tidal invasion (Clark, 1991)

As pointed out by Lin (1997), the Mulan Weir provided the drinking water for the inhabitants of several hundred villages in the Putian Plain, covering an area of 424 km^2 (Dean & Zheng, 2010a, b). Without the canals, only saline groundwater could have been found which could not meet the needs of the inhabitants. Since the 1980s, economic development and industrial growth in the Putian plain gradually contaminated the canal water and rendered it unsuitable as drinking water. Now piped water from uphill reservoirs supplies the drinking water.

Figure 8.36 Views of Mulanbei Weir (http://www.amoymagic.mts.cn/Putian/putian34.jpg)

The management of complex water distribution and irrigation systems, such as the Mulan irrigation system, has been subject to extensive studies, including the role and analysis of various village temple associations and their multi-village ritual alliances (Dean & Zheng, 2010a, b).

8.6 MAIN METHODS OF WATER SUPPLY IN PREMODERN CITIES

8.6.1 Wells and water lifting devices

8.6.1.1 Wells

In the ancient Chinese literature, the word "*Shi jing (market and well)*", which means market place, is often mentioned. Zhang Shoujie, a scholar in the Tang Dynasty, explained the word in his *Shi ji zheng yi (A commentary of Shi ji)*. 'At first, people got together at wells to draw water while selling their goods. That was the origin of the market'. Wells played an important role in the formation and development of markets and cities in ancient China and were an important auxiliary resource for urban water supply. Considering the poor sanitary condition of ancient China and the relatively stable well water quality, it remained the preferred choice for drinking for the citizens, even with multiple water resources available (Hong, 2006).

The earliest written records about wells are documented in the inscriptions on oracle bones in the Shang dynasty (16th–11th century BC). There are also legends in ancient history records about the men who first built wells like Bo Yi (about 22nd century BC) or the Yellow Emperor (26th–24th century BC). Though the exact origin of wells cannot be conjectured from these tales, it shows, at least, that wells had been invented before the time of Bo Yi, and their improvement and popularisation had gone through a long period in history.

The first well found in field archaeological research in China was in the Hemudu Neolithic Ruins (in Zhejiang Province), which was built approximately 5600 years ago (Jiao, 2007) (Figure 8.37). It was built using more than 200 piles of wood and logs and was covered by a simple pavilion for protection. Some pots with rings were also excavated from its bottom mud. They were obviously used to take water from the well (Yang, 1987). The history of wells in the Central Plains of China began about 4000 years ago, a little later than that in the South of China. However, in terms of technology, the North was more advanced; the ancient wells in the Central Plains had a larger diameter and deeper bottom.

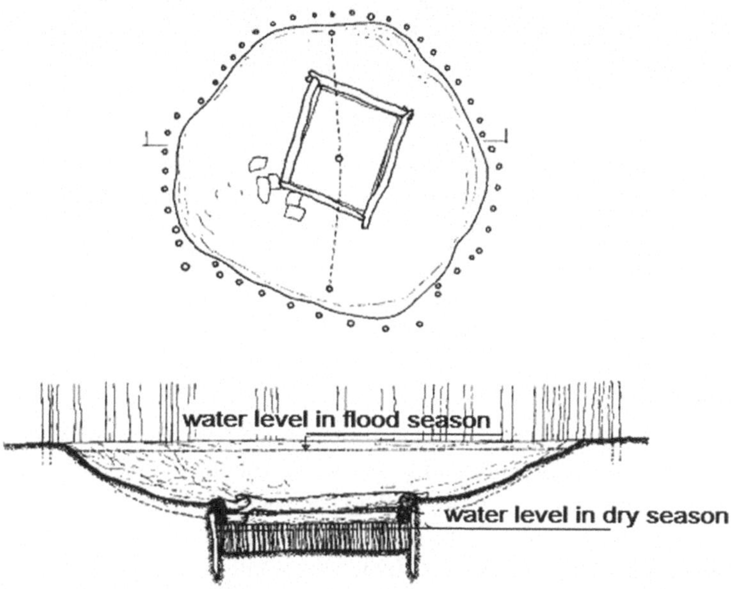

Figure 8.37 Reconstruction drawing of timber structure well at the ruins of Hemudu (Yang, 1987)

In the Xia, Shang and Zhou Dynasties, the use of square or round wells was very common. The wells were mainly made of earth. By the time of the Warring States, the use of wells became more popular. Wells in this period were mostly made with pottery clay rings, and the drilling technology was greatly improved as well (Zhang, 1996). Since the Qin and Han Dynasties, wells and the developments of cities became more closely related. Besides earth and pottery clay, bricks and rocks were also used as materials to build wells (Du, 1998).

Many ceramic wells used as sacrificial objects in burials have been excavated from tombs of the Eastern Han Dynasty (Figure 8.38). They are precious models of ancient wells. Some well models have a very complete structure including sidewalls, curb, head frame, well pavilion as well as a pulley. In addition, the large amount of excavated well models reveals the importance of wells in people's daily life at that time (Wu, 2002; Jiao, 2009). During and after the Sui and Tang Dynasties, although complicated canal networks were dug to divert water to big cities, well water still remained an important drinking water source in urban life for a long time. In the Chang'an city of the Sui and Tang Dynasties, many well relics have been excavated near the canals and rivers.

The rise and development of an ancient city is necessarily accompanied by a large number of well drilling projects. Many wells constructed in ancient times have been retained in some cities. Only after the wide spread of modern water supply systems have the water wells in cities gradually disappeared.

Figure 8.38 (a) Ceramic model of square well unearthed in Guangdong Province. Later Eastern Han (77–220 AD), height 22.2 cm. (b) Ceramic model of round well unearthed in Guangdong Province. Eastern Han, height 25 cm, bottom diameter 18.4 cm (both photos taken by J. Jiao in 2008)

8.6.1.2 Lifting devices

Xu Guangqi (Paul Hsu) evaluated different water lifting methods in his book *Nong Zheng Quan Shu (Complete Book of the Regulation of Agriculture)*. He mentioned that the most primitive method to draw water from wells was to use a cord and corresponding containers such as barrels or buckets. In order to reduce the labour intensity, the ancient Chinese had invented other advanced lifting tools such as "Jie Gao" (shadoof or counterweight lever), "Lu Lu" (windlass or pulley wheel) and elevated cylinder wheel (Figure 8.39). The invention of the vertical waterwheel further improved the efficiency in drawing water from wells. This kind of waterwheel has existed at least since the Tang Dynasty.

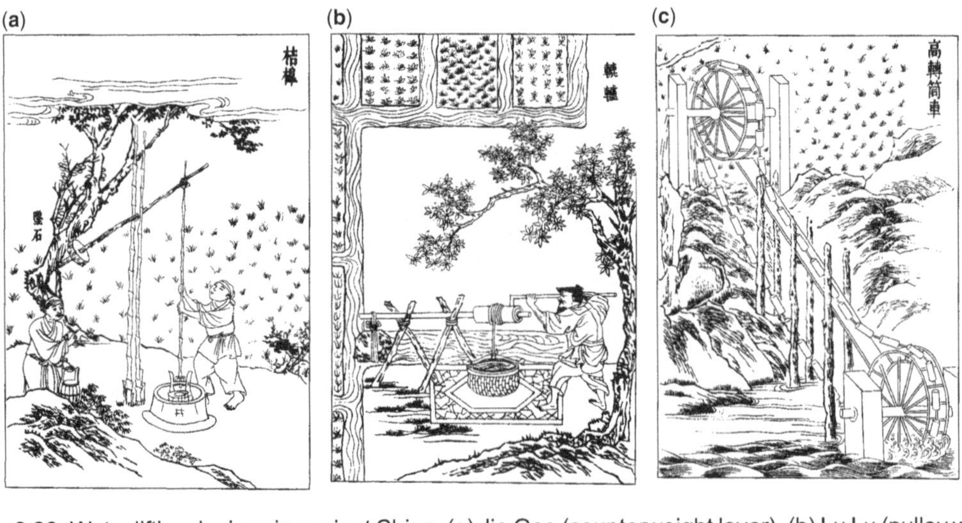

Figure 8.39 Water lifting devices in ancient China. (a) Jie Gao (counterweight lever), (b) Lu Lu (pulley wheel), (c) elevated cylinder wheel. [*Source:* Sung (1966)]

8.6.2 Transfer and conveyance of water through canals

The location of ancient cities usually followed the rule 'Stay away from dry lands at high elevation; stay away from water at low elevation'. Considering the topographical and flooding conditions, the cities should not be built too close to water. Therefore canals must be built to transfer water into the cities. Chang'an, Luoyang, Chengdu and Dadu had all built canals to divert water into the city.

Canals were important construction works for water delivery. Canals, which are denoted as "*Qu*" in Chinese characters, first appeared in the *Historical Records of Canal Works (Shi ji he qu shu)* written in 91 BC. The earliest form of canal was the furrow system in the Western Zhou Dynasty. In the Warring States period, many large-scale irrigation canals were constructed such as the 12 Canals of the Zhanghe River and the Zhengguo Canal. Most of the ancient canals were made of soil with a trapezoidal cross section; some were made of stone. However, most of them had the defect of no seepage control, which resulted in large amounts of water lost along the transmission route.

The headwork of a diversion canal could be built either with or without a dam. Generally, dams or weirs were constructed across a river to raise the water level and then to divert the flow in full, or in part, into a water supply canal. Diversion structures without dams were mostly built during the Qin and Han Dynasties. They were built of local materials and often placed in the middle of the river. Such method was mainly applied to large rivers. The Dujiang Weir on the Min River is one of the famous examples.

Canal route design is another important factor for successful delivery of water. *Kao Gong Ji (The Records of Examination of Craftsman)*, an important work of science and technology before the Qin Dynasty, had already put forward that 'ditches and canals should always adapt to the water potential, water in good canals will always flow smoothly and never be blocked', indicating that the slope always had to be controlled. Even before the Qin and Han dynasties, canal route design had already reached a high level. Examples are the Zhengguo Canal (246 BC) and the Dujiang Weir irrigation system.

Apart from general canals, the ancient Chinese also designed many special crossing structures to overcome terrain obstacles. Aqueducts, culverts and inverted siphons were most commonly used. For the Longshou canal a tunnel was excavated applying the shaft method for construction (see Figure 8.28, Section 8.4.2).

8.6.3 Conveyance of water through pipes

Piping systems were also used for urban water supply in ancient China. Many historical records mention the use of bamboo pipes, involving also the application of siphons and inverted siphons. In ancient China, people used the stem of bamboo called "Jian" to convey water (Figure 8.40). Bamboo pipes were first used in Sichuan Province to convey brine, then the method was gradually applied to deliver fresh water. However, frequent replacement of the pipes was required due to the perishable nature of bamboo.

The largest bamboo piping systems seem to have been built by the great poet-official Su Shi. Under his inspiration, water mains of large bamboo trunks were installed at Hangzhou in 1089 AD and at Guangzhou in 1096 AD. In the latter system there were five parallel mains. Holes were provided at intervals for freeing blockages, and ventilation taps for the removal of trapped air (Needham, 1965). In addition, tile and clay pipes were also used in urban water supply. Copper and iron pipes were used in Nanjing City in the early Ming Dynasty to transport water from Xuanwu Lake into the inner city canal. Metal pipes made of copper or bronze were used only from the 15th century onwards (Needham & Wang, 1999).

(a) (b)

Figure 8.40 (a) Conveyance of water through bamboo pipeline in a village of the Li ethnic minority group, Hainan province. (http://www.hinews.cn/), (b) Bamboo aqueduct in Hong Kong Island painted by French painter Auguste Borget in 1838, before Hong Kong became a colony (Ho Pui Yin, 2001)

8.6.4 Storage of water in reservoirs

Building dams and levees to block and store water and form large capacity reservoirs was an important method for solving problems regarding city water supply. The Kunming Reservoir of Chang'an in the Han Dynasty, the West Lake of Hangzhou in the Tang and Song Dynasties, and Kunming Lake of Beijing in the Qing Dynasty are all well-known examples. Other than these, some cities artificially dug large ponds to store water from springs or rivers and used them as supplemental water sources, such as Daming Lake in Jinan, Shandong Province. Reservoirs not only served as water sources, but could also manage and prevent floods effectively (Du, 1998).

Impounding reservoirs were developed very early in China. Their origin is traced back to the prehistoric time of Yu the Great (Dayu in Chinese), the ancient hero known for controlling floods. After the Western Zhou Dynasty, a historical record mentions "Zhu" as a kind of irrigation reservoir. By the time of the Spring and Autumn period, large scale reservoir projects already existed such as the Shaobei Reservoir in Anhui Province (Xiong & Guo, 1989). Impounding reservoirs are integrated projects including a retaining structure, spillways and regulating pondage, and so on. Besides that, small ponds or reservoirs, which are called "Bei chi" or "Shui gui", were also built on the halfway point of the delivery route, to regulate and control the flow of water from the main reservoir to the city.

8.6.5 Water gates

Canals and rivers supplying cities with water required water gates for passage of the water under the city walls. For example, the Ming dynasty city walls of Beijing (25 km long) included 10 water gates for water flowing into and out of the walled city (http://en.wikipedia.org/wiki/Beijing_city_fortifications). The 35 km long Ming dynasty walls of Nanjing, the longest wall in China, had seven water gates for water entering and 10 water gates for water leaving the city (Yang & Wang, 2008). For security

8.6.6 Delivering water by water carts and water carriers

In the absence of piped water distribution systems, and because only a few households possessed their own well, water was generally distributed by professional water carriers. They collected the water from streams, canals or storage ponds within the city and either carried the water in buckets suspended on a bamboo pole over the shoulder, or transported it in buckets or tanks loaded onto carts or wheelbarrows to the customers (Figure 8.44).

Figure 8.44 (a) A water coolie of the late 19th century (Photo taken in the 1880s) (Ho Pui Yin, 2001); (b) Water carrier with a wheelbarrow in Beijing (after Spicq, 2009)

Due to their large numbers, water carriers usually formed powerful guilds of considerable local influence (Rogaski, 2004; Rowe, 1989). They often opposed the introduction of piped water distribution systems in the late 19th and early 20th century for fear of losing their jobs (Hart, 1890; MacPherson, 1987). Wheelbarrows have been used in China for more than 2000 years, with the Chinese design of central wheel and axle greatly facilitating the transport of heavy loads (Needham, 1986, Vol. 4, Part 2, 258). For larger quantities of water, animal-drawn water wagons were utilised, for example to supply the imperial palace in Beijing with large quantities of spring water from the nearby Jade Fountain (Yuquan) Mountain.

8.6.7 Water quality control techniques and rules

The quality of drinking water was directly related to the health of its consumers. For this reason, the ancient Chinese paid great attention to it. Generally, water was drunk without any treatment because its sources were usually protected from pollution. An example of protecting the water quality was the construction of independent aqueducts bridging over the Jinshui River Canal in Dadu as previously mentioned, thus preventing the mixing of clean water supply with surface water. This method was widely used in ancient China. In some cities where the water source was muddy, "Deng Cao" (similar to modern water sedimentation troughs and purifying pools) were built to treat the water before consumption. For example, in the Northern Song Dynasty, the Jinshui River of Chengdu was equipped with purifying channels for the sediments to deposit before the water entered the city. Also, the residents would put some alum into the water and stir it strongly then let it deposit. This method, which is similar to the mechanism of the modern coagulation process, could quickly purify the water. In 2004, a complete multi-tank water purification system of the late Ming period was excavated in a family compound of Xikou Village in Zhejiang Province (Wenzhou, 2005). It is claimed to be the first water purification system in China, consisting of five tanks in series and with the water being transported by overflowing (Figure 8.45).

Figure 8.45 Plan and cross-sectional view of multi-tank water purification system of Xikou Village, Zhejiang Province (Wenzhou, 2005)

Local governments would also enact regulations and laws to prevent water bodies from becoming polluted. In the Song Dynasty, the Hangzhou city government promulgated a regulation to separate water resources into different types serving different purposes in order to keep the drinking water clean. In the Yuan Dynasty, Dadu issued a strict rule to forbid people washing their hands in the Jinshui River. Huizhou City set up regulations during the Qing Dynasty to protect E Lake, 'That residents living nearby the lake should not dry up the lake, nor break the bank or dispose rubbish into the lake. Punishment would be made in case of violation' (Wu, 1991).

8.7 INTRODUCTION OF WESTERN WATERWORKS STARTING IN LATE 19TH CENTURY

8.7.1 The opening of China

Until the end of the 19th century, China's contact and trade with foreigners was very limited. In the wake of the colonial expansion of Western powers, a small Portuguese settlement was established in 1557 in Macau, conducting highly regulated trade with nearby Canton (now Guangzhou). From 1760 until 1842 other foreign ships were also allowed to enter the port of Guangzhou, but trading was legally restricted to the trading factories during the Trading Season. After centuries of isolation and limited contact with foreign countries, China was finally opened by force to foreign trade at the conclusion of the First Opium War with Great Britain. The Unequal Treaty of Nanjing in 1842 forced China to cede the island of Hong Kong and to establish five treaty ports at Shanghai, Canton (now Guangzhou), Ningbo, Fuzhou and Amoy (now Xiamen). Following the Second Opium War of 1856–1860 and the Unequal Treaty of Tianjin in 1858 as well as the Convention of Beijing in 1860, another series of treaty ports and extraterritorial concessions or settlements under the control of foreign powers (initially United Kingdom, France and the USA) were established. After 1860, foreigners were legally free to travel anywhere in China, but direct commercial activities, residence and property rights were restricted to the designated treaty ports and other specified cities open to foreign trade. By the end of the 19th century, more than 80 treaty ports and other cities existed along the Chinese coast and navigable waterways where specified areas were outside the control of local Chinese governments, enjoying legal extraterritoriality, but still under Chinese sovereignty. Under direct foreign administration, outside Chinese sovereignty, were the colonies of Macau (Portugal), Hong Kong (Great Britain) and Taiwan (Japan) as well as the leased territories of Kiautschou (now Qingdao, German), the New Territories of Hong Kong (Great Britain), Guandongzhou (now Dalian; successively under Japan, Russia, and Japan), Weihai (now Yantai; under Great Britain) and Guangzhouwan (now Zhanjiang; under France). From 1917 on, the treaty rights were gradually relinquished by the foreign powers, with the last ones formally relinquished in 1943. Only Hong Kong and Macau remained as colonies until 1997 and 1999, respectively.

8.7.2 Modern waterworks until 1911

During their existence between 1842 and 1943, the treaty ports and foreign concessions/settlements formed not only the starting points for the penetration of Western commerce and technology into China but also provided an important platform for exchange of new ideas. Compared to the surrounding Chinese cities, the foreign settlements and concessions introduced efficient public administration and built reliable utilities, parks and well kept streets, including modern Western water supply systems. Modern systems are defined as including one or more of the following components: mechanical pumping (steam-driven), water treatment, clean water storage reservoir, and high-pressure piping system for distribution. The main reasons behind the building of modern waterworks were improvement of public health through provision of clean water, fire protection, demand by foreign residents accustomed to tap-water supply, profit motive of private investors, and probably show-casing of superior Western technology by colonial governments. In contrast to other countries at that time, the unusually fragmented political and economic conditions of China did not allow planned, coordinated development of waterworks. This resulted in various projects in different cities, by various initiators, and for different reasons, but no detailed overview is available.

In the following, the establishment of modern waterworks in China up to the end of the Qing Dynasty in 1911 is briefly summarised. The main sources of information are Hong *et al.* (2006) for cities on the mainland, and Liu & Liu (1992) for cities in Taiwan. On North-eastern (Manchurian) cities only limited information was available (A Pictorial Record, 2009). In addition, many city gazetteers and other less known Chinese as well as foreign sources were consulted, which are not cited here. Detailed descriptions of the Shanghai, Hong Kong, Hankou and Shantou waterworks are found in the Minutes of the Proceedings of the Institution of Civil Engineers (Hart, 1890; Orange, 1890; Moore, 1910; Ough, 1913). Specialised studies deal with the history of waterworks in Hong Kong (Ho, 2001), Shanghai (MacPherson, 1987), Tianjin (Spicq, 2003, 2006) and Beijing (Sternfeld, 1997, 2006). Limited information on Guangzhou and Qingdao was obtained from Lee (1936) and Schnee (1920) respectively.

8.7.2.1 Historical and geographical development

The first Western waterworks in the territory of China was completed in 1863 in the British colony of Hong Kong. The Pokfulam Reservoir scheme comprised a 9080 m^3 impoundment reservoir, a 15,090 m long

8.7.2.2 Overview

The first modern waterworks, with the exception of the fortification Lüshunkou (Port Arthur), were all designed and built exclusively by foreigners, with foreign materials, in foreign colonies, leased territories or concessions. The first truly Chinese-owned waterworks did not open until 1902 in Shanghai and Tianjin, followed by Guangzhou, Hankou and Shantou.

With the exception of Hong Kong and Dalian all waterworks relied on surface water (rivers) and/or groundwater as the raw water source. Steam-driven pumps were required for lifting and conveying the water or to generate sufficient pressure in the mains of the distribution system. In the absence of rivers and with only limited groundwater resources, but because of its hilly topographical characteristics, Hong Kong relied on impounding reservoirs to collect rainwater. Initially gravity alone was used for conveying and distributing the water, but, in the later water supply schemes, steam-driven pumps were employed for raising water from low-level to high-level reservoirs or for pumping limited quantities of groundwater. Similarly to Hong Kong, Dalian relied on an impounding reservoir, whilst Lüshunkou (Port Arthur) initially had water supplied by gravity.

A preliminary overview suggests that by the end of the Qing Dynasty in 1911 an estimated 42 modern waterworks with separate distribution systems were completed or under construction in 32 different cities. They are grouped according to political status, with the year of opening, design capacity and ownership given, if available. These were:

(1) Treaty-Ports (with foreign concessions): 10 waterworks in 5 cities

Waterworks	Year	Capacity	Ownership
Shanghai:			
Shanghai Waterworks Company (International settlement)	1883	13,093 m^3/d	private (GB)
French Waterworks (French concession)	1902	9090 m^3/d	public (F)
Inland Water Supply Co. (Chinese districts)	1902	n.a.	private (Ch)
Zhabei Waterworks (Chinese districts)	1911	9090 m^3/d	private (Ch)
Tianjin (Tientsin):			
Tianjin Waterworks Company (TWC)(British concession)	1899	1363 m^3/d	private (GB)
Tianjin Native City Waterworks Company (TNCTW) (Chinese city and some foreign concessions)	1903	27,258(?) m^3/d	private (Ch, I)
Guangzhou (Canton):			
Guangdong River Water Supply Company (Chinese city)	1909	1350 m^3/h	public-private (Ch)
Shamian (British-French concession)	1913	n.a.	private (GB, F?)
Hankou (Hankow):			
Jiji Water and Electric Company (Chinese city)	1909	27,727 m^3/d	private (Ch)
Shantou (Swatow)			
Swatow Waterworks (Chinese city)	1914	8133 m^3/d	private (Ch)

(2) South Manchuria Railway Zone (in today's Northeast provinces): 11 waterworks in 11 cities

Liaoyang	1902	n.a.	public (J)
Shuangmiaozi	<1905	n.a.	public (J)
Sipingjie	<1905	n.a.	public (J)
Fushun (Mining city)	1908	4000 m^3/d	public (J)
Dashiqiao	1908	n.a.	public (J)
Wafangdian	1909	n.a.	public (J)
Changchun	1911	2000 m^3/d	public (J)
Kaiyuan	1911?	n.a.	public (J)
Gungzhuling	1911?	n.a.	public (J)
Shenyang (Mukden, Fengtian)	1911?	n.a.	public (J)
Dandong (Andong)	1913	n.a.	public (J)

(3) Leased territories: 4 waterworks in 3 cities

Qingdao (Kiautschou Bay and Tsingtau, Leased Territory of 515 km^2, German 1897–1914, Japanese 1914–1922): 2 waterworks in 1 city

Hai Po Waterworks	1901	400 m^3/d	public (G)
Litsun Waterworks I	1908	2000 m^3/d	public (G)
Litsun Waterworks II	1913	2000 m^3/d	public (G)

Guandongzhou (Kwantung, Leased Territory of 3,463 km^2, Japanese 1894–1898, Russian 1898–1905, Japanese 1905–1945): 2 waterworks in 2 cities

Dalian (formerly Dalny, Dairen)	1901	n.a.	public (R, J)
Port Arthur = Lüshunkou	1879	n.a.	public (R, J)

(4) Colonies: 16 waterworks in 12 cities

Hong Kong (British colony 1842–1997): 7 separate waterworks for 3 independent service areas

Pokfulam Reservoir (Hong Kong Island West = HKW)	1863	4500 m^3/d	public (GB)
Tai Tam Reservoir (Hong Kong Island East = HKE)	1889	9000 m^3/d	public (GB)
Wong Ngai Chung Reservoir (HKW)	1899	n.a.	public (GB)
Tai Tam Byewash Reservoir (HKE)	1904	n.a.	public (GB)
Tai Tam Intermediate Reservoir (HKE)	1908	n.a.	public (GB)
Yau Ma Tei Pumping Station (Kowloon)	1895	1800 m^3/d	public (GB)
Kowloon Reservoir (Kowloon)	1910	3375 m^3/d	public (GB)

Taiwan (Japanese colony 1895–1945):

Danshuei	1899	3000 m^3/d	public (J)
Keelung	1902	19,200 m^3/d	public (J)
Changhua	1908	5000 m^3/d	public (J)
Taipei	1909	52,788 m^3/d	public (J)

Jinshan	1910	477 m^3/d	public (J)
Beitou	1911	2500 m^3/d	public (J)
Dajia	1912	555 m^3/d	public (J)
Shilin	1912	695 m^3/d	public (J)
Douliu	1912	1180 m^3/d	public (J)
Kaohsiung	1913	13,360 m^3/d	public (J)
Sanhsing	1914	450 m^3/d	public (J)
Chiayi	1914	8350 m^3/d	public (J)
Pinglin	1914	n.a.	public (J)
(5) Other cities (without foreign concessions): 2 waterworks in 2 cities			
Lüshunkou (fortress and city)	1879	n.a.	public (Ch, R,J)
Beijing (Jingshi Tap Water Company)	1910	3300 m^3/d	private (Ch)

Note: GB = Great Britain, F = France, Ch = China, R = Russia, G = Germany, J = Japan, I = International, and n.a. not available.

8.7.3 Modern waterworks after 1911

Following the establishment of the Republic of China in 1912, unstable political and financial conditions slowed down considerably the development of further waterworks until 1927, when China was again unified with Nanjing as the capital. During the next few years of stable government many waterworks were built, but the creation of the Japanese puppet state in Manchukuo in 1931 as well as the second Sino-Japanese war (1937–1945) halted any progress yet again. By 1949, only 72 Chinese cities owned modern waterworks, with 6589 km of distribution pipes and a total production capacity of 2.406 million m^3/d for a population of 541 million (Hong *et al.* 2006). After the establishment of the People's Republic of China in 1949, urban water supply was rapidly developed and expanded. By 2003, the corresponding values were 3479 municipal waterworks, with 186,000 km of distribution pipes and a total production capacity of 167.44 million m^3/d for a population of 1292 million (Hong *et al.* 2006, p. 13).

8.8 CONCLUSION

Cities in ancient China were supplied with water to satisfy the needs of daily life, goods production and transport, landscaping, plant irrigation, fire fighting, and so on. Urban sites, especially for imperial capitals, were carefully selected to ensure dependable supplies, generally relying on natural rivers, lakes, artificial canals, channels, storage ponds, but also on groundwater and wells. Many dynasties implemented large-scale water supply projects for their capitals, often containing more than a million inhabitants, to ensure long-term development and economic prosperity. Some projects even comprised integrated multiple purpose schemes requiring advanced hydraulic design and management techniques. In regions with unusual ethnic, topographic or climatic conditions, such as in mountainous, arid or coastal areas, ingenious water supply technologies were developed like karezes (qanats), branched distribution systems, or entire canal towns. A wide range of hydraulic structures, water conveyance methods, lifting devices, and water quality control methods were applied; however, only a few piping systems for water supply or distribution were found. In many cities, drainage facilities were integrated with the water supply system. Many ancient water supply schemes are still in use, testimony to their remarkably advanced design and examples of true sustainability over hundreds of years. In sum, ancient China attained significant achievements with regard to methods and facilities for water supply and controlling water quality, which are still worthwhile to consider and use as reference in modern city planning.

The search for and development of new water sources has always been a major challenge for ancient cities. As long as the problem of water source was solved, city development would be rapid and smooth. On the other hand, if the problem could not be solved, then development would be limited. In the modern era, China vastly expanded water supply facilities to reach almost all of its large urban and rural population. But, paradoxically, this success led to serious water shortages and concomitant water pollution, which may severely impair modern city development. As of today, out of about 600 large cities in China, the number of cities that have water shortage problems already passed 400, whilst many water sources have grossly deteriorated. The questions of how to ensure water for development and how to supply water in a sustainable fashion have become of national concern. In a situation like this, looking back to the history of water supply in ancient times and summarising the experience of developing water sources in ancient cities is highly relevant.

APPENDIX A.1: TABLE OF DYNASTIES AND PERIODS

Dynasties/Periods	Years	Capital city
Prehistoric Times	1.7 million years ~ the 21st century B.C.	
Xia (Hsia)	21st ~ 16th century BC	
Shang	16th ~ 11th century BC	Anyang (Yin)
Zhou (Chou)	Western Zhou (11th century BC ~ 771 BC) Eastern Zhou - Spring and Autumn Period (770 BC ~ 476 BC) - Warring-States Period (476 BC ~ 221 BC)	Xi'an (Zongzhou, Hao), W Luoyang (Chengzhou), E Luoyang (Chengzhou)
Qin (Chin) (first unification)	221 BC ~ 206 BC	Xi'an (Xianyang)
Han	Western Han (206 BC ~24 AD) Eastern Han (25 ~ 220)	Xi'an (Chang'an) Luoyang
Three Kingdoms Period (first partition)	Wei (220 ~ 265) Shu Han (221 ~ 263) Wu (229 ~ 280)	Luoyang Chengdu Nanjing (Jianye)
Jin (Tsin) (second unification)	Western Jin (265 ~ 316) Eastern Jin (317 ~ 420)	Luoyang Nanjing
Northern and Southern Dynasties (second partition)	Northern Dynasties - Northern Wei (386 ~ 534) - Eastern Wei (534 ~ 550) - Western Wei (535 ~ 556) - Northern Qi (550 ~ 577) - Northern Zhou (557 ~ 581) Southern Dynasties - Song (420 ~ 479) - Qi (479 ~ 502) - Liang (502 ~ 557) - Chen (557 ~ 589)	 Datong (Pingcheng), Luoyang Ye Xi'an (Chang'an) Ye Xi'an (Chang'an) Nanjing (Jiankang) Nanjing (Jiankang) Nanjing (Jiankang) Nanjing (Jiankang)
Sui (third unification)	581 ~ 618	Xi'an (Dongdu)

(Continued)

Dynasties/Periods	Years	Capital city
Tang	618 ~ 907	Xi'an (Chang'an) Luoyang, E
Five Dynasties and Ten States (third partition)	Five Dynasties – Later Liang (907 ~ 923) – Later Tang (923 ~ 936) – Later Jin (936 ~ 946) – Later Han (947 ~ 951) – Later Zhou (951 ~ 960) Ten States (902 ~ 979)	 Kaifeng (Dongdu) Luoyang (Dongdu) Kaifeng (Dongjing) Kaifeng (Dongjing) Kaifeng (Dongjing)
Song (Sung) (fourth unification)	Northern Song (960 ~ 1127) Southern Song (1127 ~ 1279)	Kaifeng (Dongjing) Hangzhou (Lin'an)
Liao (Kitan Empire)	916 ~ 1125	Harbin (Shangjing) Beijing (Yanjing) S
Jin (Kin) (Jurchen Empire)	1115 ~ 1234	Harbin (Shangjing) 1115–1153 Beijing (Zhongdu) 1153–1214 Nanjing 1214–1234
Yuan (Mongol Empire)	1271 ~ 1368	Shangdu 1264–1276 Beijing (Dadu) 1276–1368
Ming	1368 ~ 1644	Nanjing 1368–1421 Beijing 1421–1644
Qing (Ching) (Manchu Empire)	1644 ~ 1911	Shenyang (Shengjing) 1636–1644 Beijing 1644–1912
Republic of China	1912 ~ 1949	Beijing, Nanjing, Chongqing
People's Republic of China	1949 ~ Present	Beijing

Note: W = Western capital, E = Eastern capital, S = secondary capital, ancient names of capital cities are in brackets.

APPENDIX A.2: MAP OF CHINA SHOWING LOCATION OF CITIES MENTIONED IN SECTIONS 8.2 TO 8.5

REFERENCES

A Pictorial Record of the Qing dynasty: Manchurian Railway (2009) compiled by the Lüshun Museum. Cengage Learning Asia. Singapore, China Renmin University Press, China.

Bai Ju Yi (824 AD, Tang Dynasty). Qiantang hu shi ji (Records of the Lake and Stone in Qiantang), Bai shi chang qing ji (Collected Works of Bai Juyi). *Edition of si bu cong kan*, **59**, 326.

Beijing city fortifications. http://en.wikipedia.org/wiki/Beijing_city_fortifications (accessed 02 October 2011).

Beijing Liao Jin City Wall Museum, undated brochure.

Beijing zhi, Shi zheng juan, Gong shui zhi, Gong re zhi and Ran qi zhi (2003). City of Beijing Gazetteer, Section on City Administration. Water Supply, Heat Supply, Gas Supply. Beijing chu ban she, Beijing.

British Water Tower Appreciation Society (2008). The 1881 metal tower of Shanghai. British Water Tower Appreciation Society: The Journal of the only British society dedicated to these structures, Friday 18 July 2008.

Cai Fan (1987). Beijing gu yun he yu cheng shi gong shui yan jiu (Research of Ancient Canals and City Water-Supply of Beijing). Beijing chu ban she (Beijing Press). Beijing, China, pp. 173–189.

Cao S., Liu X. and Er H. (2010). Dujiangyan irrigation system – a world cultural heritage corresponding to concepts of modern science. *Journal of Hydro-environment Research*, **4**(1), 3–13.

Chen Qiaoyi (1983). Zhongguo liu da gu du (Six Major Ancient Capitals of China). Zhongguo qing nian chu ban she (China Youth Publishing House), Beijing, China, pp. 125–160, 171–204, 257–294.

Chi Chao-ting (1936). Key Economic Areas in Chinese History as Revealed in Development of Public Works for Water-Control. Georg Allen & Unwin, London, UK/New York, USA.

Clark Hugh R. (1991). Community, trade, and networks: Southern Fujian Province from the third to the thirteenth century. Cambridge University Press, Cambridge, UK.

Clark Hugh R. (2007) Portrait of a community: Society, culture, and the structures of kinship in the Mulan River Valley (Fujian) from the Late Tang to through the Song. The Chinese University Press, Hong Kong.

Dean Kenneth and Zheng Zhenman (2010a). Ritual alliances on the Putian plain. Vol. 1. Historical introduction to the return of the gods. Brill; Leiden, Boston, Germany.

Dean Kenneth and Zheng Zhenman (2010b). Ritual alliances on the Putian plain. Vol. 2. A survey of village temples and ritual activities. Brill; Leiden, Boston, Germany.

Dong Jianhong (1988). Zhongguo gu dai cheng shi jian she (City Construction in Ancient China). Zhongguo jian zhu gong ye chu ban she (China Architecture and Building Press), Beijing, China, pp. 158–188.

Du Pengfei (1998). Zhongguo gu dai cheng shi gei shui (Water Supply of the cities in Ancient China). *China Historical Materials of Science and Technology*, **1**(19), 3–10.

Du P. and Qian Yi (1999). Zhongguo gudaide chengshi paishui. *China Historical Materials of Science and Technology*, **18**(2), 136–146.

Duan Tianshun (1989). Yan shui gu jin tan (History of Waters of Yan). Yan shan chu ban she (Yanshan Press), Beijing, China, pp. 1–5.

English Paul Ward (1966). City and Village in Iran: Settlement and economy in the Kirman Basin. The University of Wisconsin Press, Madison, USA, 31.

Golany Gideon S (2001). Urban design ethics in ancient China. E. Mellen Press, Lewingston, New York, USA.

Guangzhou Water Supply Company Co., Ltd. Website: http://www.gzwatersupply.com (accessed 02 October 2011).

Guo Husheng (1997). Ming Nanjing (Nanjing city of Ming dynasty). *Jian zhu shi (Architect)*, **8**(77), 36.

Hansen Roger D. Karez (Qanats) of Turpan, China. http://www.waterhistory.org/histories/turpan/turpan.pdf (accessed 02 October 2011).

Hart John William (1890). Shanghai water-works. *Minutes of the Proceedings,* ICE, **100**, January 1890, pp. 217–245.

Ho Pui Yin (2001). Water for a barren rock: 150 years of water supply in Hong Kong. Commercial Press, Hong Kong.

Hong Jue Min (2006). *Zhongguo cheng zhen gong shui ji shu fa zhan shou ce* (Handbook of water supply technological development of China's cities and towns). Zhong guo jian zhu gong ye hu ban she (China Architecture and Building Press), Beijing, China.

Hou Renzhi (1979). *Li shi di li xue de li lun yu shi jian* (Theory and Practice of Historical Geography). Shang hai ren min chu ban she (Shanghai People Press), Shanghai, China, pp. 272–307.

Hou Renzhi (1998). *Hou Renzhi wen ji* (Collected work of Hou Renzhi). Beijing da xue chu ban she (Beijing University Press), Beijing, China, pp. 78–93.

Huang Shengzhang (1958). *Xi'an cheng shi fa zhan zhong de gei shui wen ti yi ji jin hou shui yuan de li yong yu kai fa* (Water-Supply Problems in Xi'an City Development and Present Water Source Use and Development). *Di li xue bao (Journal of Geography)*, **24**(4), 406–426.

Institute for the History of Natural Science (IHNS) and Chinese Academy of Science (CAS) ed. (1985). Zhong guo gu dai jian zhu ji shu shi (History and Development of Ancient Chinese Architecture). Science Press, Beijing, China, pp. 400–401.

Jiao Jiu J. (2007). A 5,600-year-old wooden well in Zhejiang Province, China. *Hydrogeology Journal*, **15**, 1021–1029.

Jiao Jiu J. (2009). Ceramic models of wells in the Han Dynasty (206 BC to AD 220). *China. Ground Water*, **46**, 782–787.

Jin Yongtan (1988) Main experiences on design and management of the Dujiangyan irrigation system. *Irrigation and Drainage Systems*, **2**, 173–184.

Karez in Xinjiang [Xinjiang kan er jing], in Uygur, Chinese, and English, Vol. 1 and 2 (2005). Xinjiang Karez Research Society [Xinjiang Weiwuer Zizhiqu kan er jing yan jiu hui]. Ed. Xinjiang People's Publishing House, Education and Cultural Press Ltd. [Xinjiang ren min chu ban she], Urumqi, China, 1146.

Knapp Ronald G. (2000). China's walled cities. Oxford University Press, Oxford; Hong Kong, 32.

Kobori I. 2010. Notes from the Turpan Basin: Pioneering Research on the Karez. In: Water and Sustainability in Arid Regions: Bridging the Gap Between Physical and Social Sciences, G. Schneier-Madanes and M.-F. Courel (eds.). Springer Science+Business Media, Netherlands, pp. 139–149.

Koenig A. and Fung S. C. C. (2010). Ancient water supply system of the Old Town of Lijiang, Yunnan Province, China. *Water Science & Technology: Water Supply*, **10**(3), 383–393.

Kono Yasuyuki (1997). Post-1949 development of the Dujiangyan irrigation system, South China: Bridging a gap between the government and farmers. *Water Resources Development*, **13**(1), 111–122.

Lee Bing-Shuey Edward (1936). Modern Canton. Mercury Press, Shanghai, pp. 62–70.

Lein Haakon and Shen Yuling (2006). The disappearance of the karez of Turfan. Report from the project 'Harvest from wasteland. Land, people and water management reforms in the drylands of Xinjiang'. *Acta Geographica – Trondheim.* Series A, No. 15. Geografisk institutt, NTNU, Trondheim, 18.

Li Jinhua and An Jinhuai (1982) Excavation of Water Storing and Transferring Facilities in Dengfeng Yangcheng from Warring-States Period. *Zhong yuan wen wu (Cultural Relics of Central China)*, **1982**(2), 1–8.

Li Keke and Xu Zhifang (2006) Overview of Dujiangyan irrigation scheme of ancient China with current theory. *Irrigation and Drainage*, **55**, 291–298.

Liang Sicheng (ed.) (1983). Ying zao fa shi zhu shi [by Li Jie (1065–1110)] (Annotaded Yingzao fashi [State building standards]) Zhongguo jian zhu gong ye chu ban she. Xin hua shu dian Beijing fa xing suo fa xing, Beijing, China (in Chinese).

Lin Guoju (ed.) (1997). Mulanpo shui li zhi (Mulanbei weir waterworks). Fang zhi chu ban she, Beijing, China.

Liu Qingzhu (ed.) (1988). Chang'an chun qiu (Spring and Autumn in Chang'an). Ren min Chu ban she, Beijing, China.

Liu Ts'ui-jung and Liu Shi-Yung (1992). Water supply and water drainage: a study on the history of settlement environment on Taiwan. *Academia Economic Papers*, **20**(2), 459–504.

Luoyang di li zhi (Geographical history of Luoyang) (1992). Luoyang shi di fang zhi bian zuan wei yuan hui (Committee of local history office of Luoyang) ed. Hongqi Press, Luoyang, China.

Ma Dezhi (1963). *Tang dai Chang'an cheng kao gu ji lue* (Record of the archaeology of Chang'an City of Tang Dynasty). *Kao gu (Archaeology)*, **1963**(11).

MacPherson K. L. (1987) A wilderness of marshes: the origins of public health in Shanghai 1843–1893. Oxford University Press, Hong Kong.

McColl Robert W. (1987) A new look at China's Dujiangyan irrigation scheme and its possible Central Asian origins. *Scottish Geographical Magazine*, **103**(3), 150–155.

Moore Richard S. George (1910). Hankow waterworks. Minutes of the Proceedings, ICE, **180**, 298–306.

Murphy William M. and Rui Chen (2007). Water Chemistry in the Ancient City of Lijiang, Yunnan Province, China. *The International Journal of Environmental, Cultural, Economic and Social Sustainability*, **3**(2), 71–78.

Needham J. and Wang L. (1965). Science and civilization in China, Volume 4. Physics and Physical Technology, Part II: Mechanical Engineering. Cambridge: Cambridge University Press, Cambridge, UK.

Needham J., Wang L. and Lu G.-D. (1971). Science and civilization in China, Volume 4 Physics and physical technology. Part III: Civil engineering and nautics. Cambridge University Press, Cambridge, UK.

Needham Joseph and Wang Ling (1965). Science and Civilization in China. Mechanical Engineering Cambridge University Press, Cambridge, UK, **4**(2), 129, 133, 507.

Needham J. and Wang L. (1999). Water-pipes, fountains and clocks. In: The Pre–industrial Cities and Technology Reader, C. Chant (ed.), 187–190. Routledge in association with The Open University, London, New York.

Orange James (1890). Tytam water-works, Hong-Kong. *Minutes of the Proceedings, ICE*, January 1890, **100**, 246–276.

Ough Arthur Henry (1913). Swatow waterworks. Minutes of the Proceedings, January 1913, **191**, 300–311.

Quan Wei (2007). *Ming chu Nanjing shan shui xing shi yu cheng shi jian she hu dong guan xi yan jiu* (Interactive Relations Studies on Locations of Mountains and waters with Nanjing City Construction in Early Ming Dynasty). Master thesis, Shaanxi Normal University, Xi'an, China (in Chinese).

Rogaski Ruth (2004) Hygienic modernity: meanings of health and disease in treaty-port China. University of California Press, Berkeley, USA.

Ronan Colin A. (1995). The shorter science and civilization of China: an abridgement of Joseph Needham's original text. vol. 5. Cambridge University Press, Cambridge, UK.

Rowe William T. (1989). Hankow: conflict and community in a Chinese city 1796–1895. Stanford University Press, Stanford, USA.

Sage Steven F. (1992). Ancient Sichuan and the unification of China. State University of New York Press, Albany NY.

Schinz Alfred (1996). The magic square: cities in ancient China. Axel Menges, Stuttgart, Germany.

Schnee Heinrich (ed.) (1920). Deutsches Kolonial-Lexikon. Quelle & Meyer, Leipzig, Germany, 3 volumes, Keyword Kiaotschou.

Shaanxi History Museum (1972). Tang Chang'an cheng xing hua fang yi zhi zuan tan jian bao (Report of the Xing hua fang ruins excavation of Tang Dynasty). *Wen wu (Cultural Relics)*, **1972**(1), 43.

Shang Y. J., Yang Z. F., Li L. H., Wang S. J. and Chen S. G. (2006). Scientific and engineering highlights of an 1171-year-old weir in Eastern China. *Agricultural Water Management*, **81**(3), 371–380.

Shiba Yoshinobu (1977). Ningpo and its Hinterland. In: The City in Late Imperial China (1977), G. W. Skinner (ed.), 391–439. Stanford University Press, Stanford. 1977.

Shiba Yoshinobu (1998). The Case of the Southern Hangzhou Bay Area from the Mid-Tang Through the Qing. In: Sediments of Time: Environment and Society in Chinese History (1998), Elvin, Mark and LiuTs'ui-jung (eds). Cambridge University Press, Cambridge, Eng.; New York, pp. 135–164.

Spicq Delphine (2003). L'hydraulique urbaine et la politique de l'eau dans la plaine du Nord de la Chine: le cas de Tianjin. 1900–1949 (Urban hydraulics and the politics of water in the North China Plain: the case of Tianjin, 1900–1949). Ph.D. Thesis, Université PARIS 7 Denis Diderot.

Spicq Delphine (2006). The Tianjin water supply and foreign-Chinese relations. Paper presented at City, Space and People International Conference, 2006 Tianjin.

Spicq Delphine (2009). Chinese urban water supply and the role of foreign companies. Le Réseau Asie, Editorial 31.08.2009.

Sung Ying-Hsing (1966). T'ien-kung k'ai-wu: Chinese technology in the seventeenth century; translated by E-tu Zen Sun and Shiou-chuan Sun. Pennsylvania State University Press, University Park. [also cited as Song, Yingxing, Tian gong kai wu (Exploitation of the Works of Nature)].

Steinhardt Nancy S. (1990). Chinese Imperial City Planning. University of Hawaii Press, Honolulu.

Sternfeld Eva (1997). *Beijing: Stadtentwicklung und Wasserwirtschaft; Soziooekonomische und oekologische Aspekte der Wasserkrise und Handlungsperspektiven* (Ecological and socio-economic determinants of the Beijing water crisis and strategies for water management). TU Berlin, Berlin, Germany.

Sternfeld Eva (2006). *Neunzig Jahre Fliesswasserversorgung in Beijing: Zur Geschichte der Beijinger Wasserversorgungsbetriebe* (Ninety years of tap water supply in Beijing: history of the Beijing waterworks company). In: Ueber Himmel und Erde. Festschrift fuer Erling von Mende, Th.R. Kolb and M. und Siebert (eds). Harrasowitz, Wiesbaden, Germany, pp. 399–418.

Su Shi [official and scholar during Northern Song Dynasty] (1976). Qi kai nan jing xi hu juan (Petition to Dredge the West Lake). In: Su Shi ji (Su Shi work collection). Tao Wenpeng (ed.), Vol. 57. Fenghuang chu ban she (Phoenix Press), Nanjing, China (in Chinese).

Su Shi [official, scholar during Northern Song Dynasty] (1976). Qiantang liu jing ji (Records of the Six Wells in Qiantang). In: Su Shi ji (Su Shi work collection). Wenpeng Tao (ed.), Vol. 35. Fenghuang chu ban she (Phoenix Press), Nanjing, China (in Chinese).

Tan Xuming (2009). Dujiangyan shi (History of Dujiang Weir). Zhongguo shui li shui dian chu ban she (China Waterpower Press), Beijing, China.

Teng Zhihong (2003). Geologic textual research of relics of Qingming Canal during the Sui and the Tang Dynasties in campus of Northwest University). *Xi bei da xue xue bao (Journal of Northwest University)*, **33**(3), 307–310.

Trombert Eric (2008). The karez concept in ancient Chinese sources: myth or reality? T'oung Pao, **94**, 115–150.

Tulufan Diqu zhi (Gazetteer of Turpan Prefecture) (2004). Tulufan Diqu di fang zhi bian zuan wei yuan hui bian; zhu bian Bo Xiao. Xinjiang ren min chu ban she, Wulumuqi, China (in Chinese).

Wang Pu (ed.) (Song Dynasty, 961AD). Important history record of Tang Dynasty. Vol. 87, Cao yun (canal transportation) (in Chinese).

Wang Yan (2009). Bei song Bianliang shui li gong cheng kao (Research of hydraulic projects on Bian Canal of Northern Song Dynasty). *Anhui nong ye ke xue (Journal of Anhui Agri. Sci.)*, **37**(2), 907–910.

Wenzhou Municipal Institute of Ancient Monument Protection and Archaeology and Yongjia County Cultural Center (2008). Excavation of the Ming Period Water-Clearing Pool at Xikou Village in Yongjia County, Zhejiang. Kaogu (Archaeology) No. 8, 2008, 49–54.

Willmott W. E. (1989). Dujiangyan: Irrigation and society in Sichuan, China. *The Australian Journal of Chinese Affairs*, **22**, 145–153.

WHC [World Heritage Committee] Nomination Documentation (2000). Mount Qincheng and the Dujiang Irrigation System. http://whc.unesco.org/uploads/nominations/1001.pdf.

Wu Qingzhou (2005). Study on Citywall and Moat Military Defence System of Nanjing City in Ming Dynasty. *Jian zhu shi* (Architect), **2**, 86–91.

Wu Qingzhou (1995). *Zhongguo gu dai cheng shi fang hong yan jiu* (Research on City Flood-Defence in Ancient China). Zhong guo jian zhu gong ye chu ban she (China Architecture and Building Press), Beijing, China.

Wu Qingzhou (2005). Study on Citywall and Moat Military Defence System of Nanjing City in Ming Dynasty. *Jian zhu shi (Architect)*, **2**, 86–91.

Wu Qingzhou (2008). Chengdu: A Successful Example of Urban Flood Control in Ancient Chinese Cities. *Nan fang jian zhu (South Architecture)*, **6**, 9–13.

Wu Yucheng (2002). The Culture of Wells of China. Tianjin ren min chu ban she (Tianjin People's Press), Tianjin (in Chinese).

Xiong Dacheng and Guo Tao (1989). *Zhongguo hui li kexue ji shu shi gai lun* (Introduction to History of Hydrological Sciences and Technology in China). Chengdu University of Technology Press, Chengdu, China, pp. 50–62, 84–104, 219–227, 387–421.

Xu Mengguang (1997). The influence of Tashan weir on the development of Mingzhou [now Ningbo] during the Song Dynasty. In: Proceedings of Symposium on Water Conservancy History of Tashanyan and East Zhejiang. Zhongguo kexue jishu chubanshe, Beijing, pp. 45–48 (in Chinese).

Xu Song (Qing dynasty) (1985). *Tang liang jing cheng fang kao* (Record of the two capitals' city guard of Tang Dynasty), Vol. 5 (Dongjing). Zhong hua shu ju (Chinese Publishing House), Beijing, China.

Yang Hongxun (1987). *Hemudu yi zhi mu gou shui jing jian ding* (An Identification of Timber Structure Well at the Ruins of Hemudu). Jian zhu kao gu xue lun wen ji (Essay of Archaeology of Architecture in China). Wen wu chu ban she (Cultural Relics Press). Beijing, China, pp. 55–59.

Yang Guoqing and Wang Zhigao (2008). Nanjing Chengqiangzhi (Record of the City Wall of Nanjing). Fenghuang chubanshe, Nanjing, China.

Zhang Huiru (2007). Study on the Interactive Relations between the Hydroenvironment in Cities and the Urban Development of Hangzhou in the Southern Song Dynasty. Shannxi Normal University, 2007, 4.

Zhang Song (2007). Conservation and adaptive reuse of industrial heritage in Shanghai. Frontiers of Architecture and Civil Engineering in China. **1**(4), 481–490.

Zhang Yuhuan (2003). History of Chinese cities. Bai hua wen yi chu ban she (Baihua Literature and Art Publishing House). Tianjin, 164 (in Chinese).

Zhang Ziming (1996). *Qin han yi qian shui jing de kao gu fa xian he zao jing ji shu* (Archeological findings and technology of Wells before Qin and Han Dynasties). *Wen bo (Relics and Museology)*, **1**, 10–25.

Zhao Qiang (1994). *Lue shu sui tang Chang'an cheng fa xian de jing* (Report of wells of Sui and Tang- Chang'an City). *Kao gu yu wen wu (Archaeology and Cultural Relics)*, **1994** (6).

Zheng Liandi (1985). Gu dai cheng shi shui li (Water Resources in Ancient Cities). Shui li dian li chu ban she (China Waterpower Press), Beijing, China, pp. 24–36, 48–64, 69–72, 73–86, 91–105.

Zheng Liandi (ed.) (2004). *Zhongguo shui li bai ke quan shu* (China water conservancy encyclopaedia), 2nd edn., Volume 2, Shuili shi fen ce (History of water conservancy). Beijing: Zhongguo shuili shuidian chubanshe.

Zhou Kuiyi (ed.) (2008). *Zhongguo ke xue ji shu shi shui li juan* (A History of Science and Technology in China, Volume of Water Conservancy). Ke xue chu ban she (Science Press), Beijing, 2008.

Chapter 9

Evolution of water supply technologies through the centuries in Crete, Greece

A. N. Angelakis, E. G. Dialynas and V. Despotakis

9.1 INTRODUCTION

The overall theme of this Chapter is not the exhaustive presentation of what is known today about hydraulic works, related technologies and their applications in water supply in Crete since the Minoan era. Instead, it will present a number of characteristic paradigms in selected sites, ranging chronologically from the early Minoan period, through to Hellenistic and Roman and to the Venetian and Ottoman periods. Most of those water supply technologies are characterised by their efficiency, integration, and mechanical compliance. In some examples, the hygienic and the functional features were so advanced that they can only be paralleled by the modern urban water systems in use in the second half of the 19th century in the developed world. In this way, the historical development of water supply technologies is effectively traced up to the modern era. In addition, the location, the prevailing climate conditions and the hydrology of the Crete Island are briefly discussed.

In order to glean a more integrated perspective, the ancient water systems discussed in this Chapter, will be assessed in terms of their relevance to modern systems, thereby highlighting their contribution to a broader development (Mays, 2010; Mays *et al.*, 2007). The impact of ancient works will be examined in terms of the evolution of technology, technological advancement, homeland security, and management principles. With a few exceptions, the basis for present day progress in water transfer is clearly not a recent development, but an extension and refinement of past achievements (Angelakis & Spyridakis, 2010).

Our investigation will begin with the rise of the palatial systems of Minoan Crete. The term Minoan, applied to this chronological period, was coined by Sir Arthur Evans, the director of the excavators of the "palace" at Knossos, and relates to the legendary King Minos (Evans, 1921–1935; Driessen & MacDonald, 1997). In this period Crete became the cradle of one of the most important civilizations of mankind and the first major civilization in Europe. Minoan technological developments in water management principles and practices are perhaps not as well known as other notable achievements of ancient Greek civilization particularly in the fields of poetry, philosophy, science, politics and the visual arts. Nonetheless, the hydraulic and architectural function of the water supply systems in palaces and cities are regarded as one of the defining characteristics of Minoan civilization (Angelakis & Spyridakis, 1996).

Archaeological, and indeed complementary, evidence indicate that, in Bronze Age Crete, advanced water management and sanitary techniques were practised in several settlements. One of the major achievements

9.2.3.5 Potential evapotranspiration

Potential evapotranspiration (ET) as estimated using the Penman-Monteith method, a system which provides the most accurate estimates, varies from 1240 mm/yr to 1570 mm/yr. Within the annual circle, the monthly ET rate changes from about 25 mm in winter to 225 mm in summer. The mean annual actual ET has been estimated to represent 75% to 85% of the mean annual precipitation in low elevation areas (less than 300 m asl) while dropping to 50% to 70% in high elevation areas (Paranychianakis *et al.* 2011).

In conclusion the estimated hydrological balances for Crete in billions m^3/yr for three hydrological conditions (namely a normal year, a wet year and a dry year) are: a normal year returned a period equal to or exceeding 50%, a wet year returned a period equal or exceeding 10%, while a dry year indicated a period equal or exceeding 90%. The average yearly precipitation on Crete is about 930 mm, which correlates to approximately 7730 Mm^3. Less than 10% of the precipitation percolates through the ground, whereby evapotranspiration and surface runoff to the sea accounts for 63% and 27%, respectively (Paranychianakis *et al.* 2011).

9.3 THE SIGNIFICANCE OF WATER SUPPLY FOR ANCIENT CRETAN CIVILIZATIONS

9.3.1 General

Advancements in water supply technologies on Crete, since the Minoan era were such that they can be compared to urban water systems developed in the second half of the 19th century in Europe and North America (Angelakis & Spyridakis, 1996). While it should be noted that hydraulic technologies in ancient Crete are not limited to urban water systems the progress in urban water supply was certainly noteworthy, as attested by several aqueducts and piped supplies, cisterns, wells, and other water facilities discovered, including the famous Minoan piped supply system of Knossos and Tylissos, the cisterns of Zakros, Archanes, Myrtos-Pyrgos and Tylissos, and the wells of Paleokastro, Zakros, and Itanos (e.g., Koutsoyiannis *et al.* 2008). This knowledge of advanced Minoan technologies was exported to the Greek mainland in later periods of the Greek civilization, for example, in Mycenaean, Classical, Hellenistic and Roman periods (Angelakis & Spyridakis, 2010).

This technological progress was accompanied by a solid understanding of the water related phenomena. Thus, *ca.* 600 BC, Greek philosophers developed the first scientific views on natural hydrological and meteorological phenomena (Koutsoyiannis *et al.* 2007). During the Hellenistic and Roman periods, further significant developments were made by Cretans in hydraulic works, such as the construction and operation of aqueducts, cisterns, and wells in much large-scale than those of Minoans used. Several of such projects are presented in this Chapter such as the aqueducts of Chersonisos, Elyros, Gortys, Hierapytna, Lappa, Lyttos, and Kissamos, the cisterns in Aptera, Dreros, Eleutherna Kissamos, Lato, Minoa, and Polyrrhenia, the harbour in Phalasarna, the water distribution systems in Kissamos, and the thermae in Aptera, Eleutherna, and Kissamos. The Byzantine and Venetian periods on Crete were witness to further improvements in hydrotechnology. Several sophisticated defence structures, incorporating complex water supply systems were constructed in those periods (Koutsoyiannis *et al.* 2008).

9.3.2 Minoan civilization (*ca.* 3200–1100 BC)

9.3.1.1 General

The island of Crete, Greece, was first inhabited shortly after *ca.* 6000 BC but it was only during the Bronze Age, that the Minoan civilization developed and reached its pinnacle as the first Greek cultural Phenonnese, of the Aegean world (Alexiou, 1964). The Minoan and Mycenaean settlements (in Crete and Peloponnesus, respectively) developed and applied various technologies for collecting, storing, transporting and using surface water and groundwater resources (Angelakis & Spyridakis, 2010; Koutsoyiannis *et al.* 2008). At that time the island was probably divided into four political units, the north being governed from Knossos, the south from Phaistos, the central eastern part from Malia and the eastern part from Zakros. Smaller settlements have been found in other places.

The evolution of urban water management in ancient Greece began in Crete during the Early Bronze Age (*ca.* 3200–2100 BC). A great variety of remarkable developments characterise the different stages in Minoan civilization, involving various scientific fields of water resources such as wells and groundwater hydrology, aqueducts, cisterns, water distribution networks and domestic water supply, construction and use of fountains, and even recreational uses of water. One of the salient characteristics of the Minoan civilization was the architectural and hydraulic function of its water supply systems in the palaces and other settlements. It can therefore be deduced that a specific group of technicians or "engineers" living in Bronze Age Crete were aware of some basic principles of what we call today "the water and environmental sciences" (Koutsoyiannis *et al.* 2006).

Over the past century archaeological investigations have brought to light impressive water engineering technology dating from the Minoan era on the island of Crete (Webster & Hughes, 2010; Angelakis & Spyridakis, 2010; Koutsoyiannis *et al.* 2008; Antoniou & Angelakis, 2011; and others). From the early Minoan period (*ca.* 3200–2300 BC) issues related to water supply were considered of great importance and were accordingly developed. Archaeological and other evidence indicate that during the Bronze Age advanced water management and sanitary techniques were practised in several settlements in Crete, such as sewers, drains, bathrooms, and toilets (Angelakis & Spyridakis, 1996; Angelakis & Koutsoyiannis, 2003; Antoniou & Angelakis, 2011). Best known are the drainage and sewerage networks such those as located at the Knossos and Phaistos palaces and Aghia Triada town.

Minoans had strategic advantages when defending a settlement on the top of a hill. Therefore villages, towns or palaces often had to be built on the isolated mountains since the Minoan era. Water supply for human beings and animals at these settlements was also compulsory, especially in respect of the possibility of being besieged for long periods of time. Thus, Minoans gave careful thought to water supply, by developing three major technologies: wells to reach groundwater, cisterns to store rainfall and aqueducts to transfer spring or surface sources.

9.3.1.2 Aqueducts

In ancient Crete, the technology of transporting water long distances through pipes and/or channels was very well developed, due to the mountainous terrain (Mays, 2007; Mays *et al.* 2007). Although the Minoan inhabitants of Knossos depended partially on wells, they mainly relied on water provided by the Kairatos River to the east of the low hill on which the palace was built. It has also been suggested that originally the water supply system for the palace tapped the spring of *Mavrokolybos* (called so by Evans) (Angelakis *et al.* 2007; Evans, 1921–1935). Subsequently, in response to an increase in population, other springs at further longer distances were also incorporated into this system. Thus, a conduit made of terracotta pipes probably traversed a bridge on a small stream to the south of the palace which carried water from a perennial spring on the Gypsadhes hill (Graham, 1987; Mays, 2007).

It has similarly been argued that water was conveyed to Knossos from as far away as Mt. Juctas, at a distance of *ca.* 10 km. Today this particular area, located a short distance from the modern village of Archanes, is richly watered by the springs of Karidaki, Paradissi, and Foundana. Etymologically, the root of the name of the village, "Ar(c)h", often presented in transposition as "A(c)hr", relates to water elements, as in the cases of the Acheloos river or the Acheroussia lake (Sakellarakis &

Sapouna-Sakellaraki, 1997). It should be noted, however, that, like other sophisticated water-related technologies in Minoan times, these supplies would have been supplemented by the more traditional means of transportation by animal or human labour. Hence, the supply of water was largely dependent on animal and human haulage, as clearly depicted on later black or red-figured vases (of the Archaic and Classical periods), motifs which often dedicated hydria or other water containers. Men or women, with or without the help of animals, would carry containers of water made of clay or other materials such as leather or wood, though these activities are not easily documented in the archaeological record (Angelakis & Spyridakis, 2010).

A second example of a long-distance piped supply was discovered at Tylissos where sections of a stone conduit, designed to hold pipes, were traced near the entrance to a complex of houses while other secondary systems led the water to a cistern dated to *ca.* 1425–1390 BC (Mays *et al.* 2007). Moreover, the water supply in the palace at Malia was dependent on a long-distance conduit, which is also attested at the settlements at Gournia and Mochlos.

However, long-distance systems discovered in Minoan Crete reveal a plethora of information involving matters pertaining to water and the sophisticated urban life of this period. The remnants of long-distance water supply systems have been detected at the Minoan sites at Knossos, Tylissos and Malia. These conduits carried water from mountain springs to Minoan palaces and towns, utilising a combination of open channels and closed pipes (Angelakis *et al.* 2007). The open channels were typically rectangular stone channels or U-shaped terracotta channels. Like Greek pipes of later periods, Minoan pipes were manufactured in sections with the ends shaped to provide neatly fitting joints (Webster & Hughes, 2010). Thus, Minoan supply systems are a combination of two principals of water conduction: (a) the open/natural gravity flow system and (b) the closed/pressured pipe system (Angelakis & Spyridakis, 2010).

9.3.1.2.1 Open/natural gravity flow system

Aqueducts with open channel sections typically comprised rectangular stone channels or U-shaped terracotta channels. Sections of open type conduits are attested near water springs, such as those at *Mavrokolymbos* in the Knossos area (Evans, 1921–1935). *Mavrokolymbos* was a pure limestone spring associated with the Vlichia River at a distance of about 400 m southwest of the palace of Knossos (Figure 9.4). Its original elevation lay at about 150 m above sea level, whereas Knossos lies at an elevation of 85 m. Below the spring Evans found sections of a system that could be categorised as an open flow system. The conduit was comprised of adjacent vertical carved slabs and a horizontal one with a channel in the middle used as a raceway. The system was one section of a system which conducted water from the spring to the palace. When Evans was excavating in the early twentieth century, the spring was found to be lower than the elevation of the palace, a fact that led him to the assumption that the original spring of *Mavrokolymbos* was located further up and that the water table had decreased considerably since the Minoan period. Recent investigations suggest that Evans was correct since the original spring was identified at a higher elevation (Angelakis *et al.* 2007). The initial location of the spring could thus eliminate problems of gravity flow towards the palace area.

A similar system has been discovered in the ancient town of Tylissos, located west of Knossos. Here, at the spring of Agios Mammas, in the vicinity of Tylissos, the water of the spring was filtered by a multi-layer terracotta device before entering the conduit. The filter was conical in shape with a height of 0.80 m, a base diameter of 0.62 m, and an upper diameter of 0.22 m, allowing for a total volume of 0.136 m^3 (Angelakis *et al.* 2007). Terracotta devices used as water filters were connected with domestic water supply systems and/or reservoirs for providing suitable quality water and, from the evidence at the spring of Aghios Mammas, long-distance conduits and public distribution systems could also have been equipped with similar refinement devices.

palace at Knossos supplied the latter with water from the low hill at Gypsades (Evans, 1921–1935; Hutchinson, 1950; Graham, 1987). The water first descended and then ascended through a bridge with an estimated inclination rate of 5% suggesting an application of the principle of communicating vessels. If this is the case, then the inference must be made that the Minoan master craftsmen were somewhat aware of this principle. In the Knossos palace, water supply was provided through a network of terracotta piping located beneath the floors in depths that varied from a few centimetres up to 3.0 m (Evans, 1921–1935). These pipes with their expertly shaped, tightly interlocked sections date from the earliest days of the construction of the building during the middle period of the first palace and are quite up to modern standards. The best examples from Knossos can be found at the levels below the South Porch and in the Draught Board area.

9.3.1.2.3 Minoan terracotta pipes design and energy efficiency

The pipes at Knossos were made 76 to 82 cm long with unequal ends and a thickness of 1–2 cm (Figure 9.4, bottom). The smaller ends had interior dimensions 7.5–8.3 cm and the thicker ends were 15–17 cm (Angelakis *et al.* 2007). The particular shape of these pipes meant that could be tightly interlocked with plaster rings; each section was rather strongly tapped in order to increase the rate of water flow, thus helping to flush any sediment through the pipe. Some of them had handles to facilitate transportation and strengthen the joints. The sections of the clay pipes resemble those used in Greece in Classical times, though Evans (Platon, 1974) considered the Minoan version as a superior design, since it tapered toward one end in order to increase the velocity of water flow (Angelakis *et al.* 2007). Furthermore, recent studies suggest that earthquakes could have damaged such systems resulting in increased permeability of the limestone at higher elevations (Perles, 2001). The geological structure can be highly vulnerable to a series of earthquakes that could cause a decline of groundwater access for a long period of time and affect the transportation and supply of water to the palaces and other settlements from high distances, and/or even destroy the supply lines (Angelakis *et al.* 2007).

A very important consideration regarding the Minoan terracotta pipes with respect to science is, that in the course of the pipeline, a pressurised part existed for first time in human history. The fact that the principle of communicating tubes was applied in practice can also be proved by a fresco found in the palace (Fahlbusch, 2008). It is clearly shows a fountain, which postulates the application of a pressure pipeline.

Another important hydraulic consideration for the conical pipe design is energy loss. Recently, Webster & Hughes (2010) measured losses greater than 90% which were shown to be almost entirely due to the geometry of the joints. The loss per pipe section is high (0.05 to 0.30 m of water for flow rates ranging from 8 to 14 L/s) relative to the loss through a cylindrical pipe. Losses increased with flow rate in a predictable manner (Webster & Hughes, 2010). Intuitively the design would induce large head losses, particularly through the joints where sudden changes in diameter occur. The maintenance of pressure in water supply networks is usually desirable for: (a) preventing accumulation and deposits of sediment by increasing water velocity and (b) minimising losses that occur due to friction and burst of the pipes in the cases of steep terrains. Such reasoning might apply to some sites, but it does not explain the appearance of the conical pipes under the floor of the Knossos palace. At Knossos, closed pipes were used within the palace but not in the aqueduct itself. Only open channels were used to bring the water from the *Mavrokolybos* spring to Knossos (Angelakis *et al.* 2007). Several reasons can be suggested for the shape of the pipes that are not based on hydraulics but on manufacturing and assembly. In conclusion possible reasons for the conical shape of the Minoan pipes are the following:

(a) The conical shape was easier to manufacture than a cylindrical one
(b) The conical shape served the joint design.

(c) To control pressures in steep terrains.
(d) Curved alignments were implemented easier by the conic shape.
(e) To avoid deposit of sediments.

When water pipes were later reintroduced in Crete as well as in the Greek mainland their shape was never similar to that used in Minoan Crete. Perhaps that is an indication of insufficiency. The Minoan pipe shape facilitated the join between pipes whereby the particular conical shape overcame the problem of hydraulic mortars, which are not attested archaeologically in Minoan Crete. Greek and later Roman pipes were cylindrical and maintained almost constant diameter. However, precipitation of $CaCO_3$ from spring water, in the walled pipes, was a serious issue for later Greek and Roman aqueducts. The major aqueducts implemented in Minoan Era are shown in Table 9.1.

Table 9.1 Characteristics of major Aqueducts in Minoan Crete.

Aqueduct Name	Location	Construction	Reconstruction	Length (km)
Gournia	Faneromeni, Asari	Minoan	na	7
Karphi	Karphi, Lassithi	Minoan		na
Knossos (*Mavrokolympos*)	Knossos	Minoan	Roman	0.7
Malia	Profitis Ilias, Malia	Minoan	Hellenistic, Roman	0.85 or 1.15
Mochlos	Mochlos, Lassithi	Minoan	na	na
Tylissos	Tylissos	Minoan	na	1.4

na: not available

9.3.1.3 *Water collection (harvesting) systems*

From the early civilizations, people in arid and semi-arid regions have relied on collecting (harvesting) surface water from rainfalls and storing the water in cisterns. Not only were cisterns used to store rainfall runoff, they were also used to store water from aqueducts. Cisterns during the ancient times have ranged from construction of irregular shaped holes (tanks) dug out of sand and loose rock and then lined with plaster (stucco) to water proof them, to the construction of rather sophisticated structures (Gorokhovich *et al.* 2011).

Minoans were very well practised in the construction of water collection systems. They developed remarkable technologies for collecting and transporting water to settlements. In Crete, due to very dry summers, rainfall collection was accomplished from both roofs of the buildings and larger court areas. Hydraulic structures associated with the rainfall collection were found in Knossos, Phaistos, Tylissos, Aghia Triadha, Chamaizi, Myrtos Pyrgos and Zakros. These hydraulic structures include large stone conduits with branches that were used to supply collected water to cisterns such as those found in Knossos (Figure 9.5a). Terracotta pipes were also used to convey rainwater to cisterns. In Myrtos-Pyrgos the terracotta pipe of rectangular shape (Figure 9.5b) supplied the nearby cistern system with storm water collected from the rooftops (Cadogan, 1978). Also, alongside a stairway in Knossos is a small stepped channel consisting of a series of parabolic-shaped step chutes that was used to convey rainwater from terraces down to a sedimentation (desalting) basin. The same components of rainfall harvesting system, for example, cistern, channel and sedimentation tank, also existed in other settlements (Angelakis & Spyridakis, 1996; Gorokhovich *et al.* 2011).

ashlar masonry and were probably roofed. Both have steps that facilitated their water supply. Another feature shared by them is the enclosure of the spring as the water came from the lower levels in a manner recalling the traditional Majahir cisterns found in Syria.

A cistern belonging to the post palatial era is that of Tylissos dating to *ca.* 1330–1200 BC (Hazzidakis, 1934). Its size and shape remind us of the earlier examples from Zakro and Archanes with their similar steps and plastered walls. A small tank cut out of stone was used as a sedimentation device for the pre-treatment of water, mainly the removal of sediments and/or of suspended solids. Similar basins for the treatment of the drainage, though somewhat smaller, were used in the East Bastion of the palace at Knossos. The overflow of this small tank was channelled to the main cistern through a scaled spout which estimated the rate of flow.

The rainwater was collected in cisterns from rooftops and open courts. Special care must have been given to secure clean surfaces in order to maintain the purity and quality of collected water by: (a) cleaning the surfaces used for collecting the runoff water and (b) by the use of other filtering devices or coarse sandy filters. The water collected in the cistern was primarily used in crafts (e.g. pottery, metallurgy), in domestic activities and in gardening irrigation (Angelakis & Spyridakis, 1996). It would have been used for drinking only in case of drought or siege.

Finally, it should be noticed that eight or perhaps nine wide and/or deep round structures similar to those described above were found in the palaces of Phaistos and Knossos. These structures are now called "Kouloures", denoting their round shape. Despite their uncertain function, their possible relation to water storage should not be excluded.

9.3.1.5 Groundwater and wells

Archaeological investigations confirm the fact that wells were used in Crete since Neolithic times (Perles, 2001). Also, Knossos, the best known Minoan settlement, had wells since the Early Minoan times (Muller 1996). During the period of the first palaces (*ca.* 1900–1700 BC), several wells were used for drawing drinking water. At least six such wells have been reported (Evans, 1921–1935). Their depth did not exceed 20 m and their diameter was no more than 5 m (Buffet & Evrard, 1950). Several others were used for drawing drinking water, but their date is uncertain. The most interesting one with a depth of about 12.5 m and a diameter of about 1.5 m is found in the north-west part of the little palace area, in the basement of House A, which belongs to the middle period. Its upper circuit was mostly of rubble masonry dated probably to the Roman period. Below, the well was dressed with a series of terracotta collars (Angelakis & Spyridakis, 1996). Each collar had three parts, imitating ashlar masonry, and an upper rim. Triangular holes to enter the well for cleaning purposes were occasionally made. A similar but of later date and slightly smaller terracotta drum was also found at Phylakopy, Melos (Evans, 1921–1935).

The most important and best known has depth of about 12.5 m and a diameter 1.0 m. In the palace of Zakros, a well-spring is located near the south-east corner of the central Court; steps, now damaged by the visitors' feet, led down into the chamber (Angelakis & Spyridakis, 2010). The wood of the windlass was found in the water, along with an offering cup containing, among other offerings, perfectly preserved olives and raisins. Naturally, the olives maintained their relative freshness only for a few minutes after they were taken out of the water (Angelakis & Spyridakis, 1996).

The water supply in the Minoan town of Palaikastro was also dependent on groundwater (Figure 9.6b). Here several wells have been discovered to date with depths ranging from 10 to 15 m (Angelakis *et al.* 2011). The sites of Palaikastro and Zakros are situated in the eastern part of the island which in antiquity appears to have been exceptionally rich in groundwater; today, however, its water is saline. Other wells were found north of the Minoan "Villa of the Lilies" at Amnissos, constructed probably during the early

Minoan terracotta pipes were used for both the distribution and collection of water for storage (Figure 9.4, bottom). However, in most cases, they did not serve a domestic water supply network but a public one. They were rather feeding a public fountain or an open area from a spring, as shown in the case of the town of Zakro. In a building there, two handmade tubes about 30 cm long with a narrow opening on a zoomorphic side were found (Figure 9.7b). The other side is open with traces of plaster whose presence has been explained as indicative of its incorporation on a built wall (Angelakis & Spyridakis, 1996). The artefacts are dated before the end of the late period (*ca.* 1450 BC). It has been argued that the two zoomorphic artefacts were waterspouts, part of a fountain in the area, which was fed with water by an aqueduct. The fountain was constructed in a public area which, if combined with the zoomorphic waterspouts and their possible meaning, may be suggestive of an occasional ritual function of the fountain. The animals roughly depicted on the waterspouts, most likely lions, had the apotropaic function of protecting and securing the quality and the constant flow of water (Platon, 2001). Also, in several other places (e.g. the Pediada region) most of the Minoan sites identified have been built by natural springs but it is impossible to tell now if these springs were used in their simplest form or they were adorned with fountain houses (Panagiotakis, 2006).

Most craftsmen in Minoan Crete knew how to built open U-shaped terracotta or stone conduits. Bearing this in mind, the closed terracotta pipe must have been designed to perform other functions. These would be: (a) underground transportation of water or (b) transportation of good quality water. The former (underground=unseen) may have been intended for use by a restricted group of people or for a specific activity, which in turn may be indicative of a ritual use of water (Angelakis & Spyridakis, 1996). Moreover, the evidence suggests that the pipes probably fed fountains in open areas of the palaces and they were not part of a domestic distribution system, as most scholars think (Platon, 2001). However, the piped system is attested to in areas of palatial architecture and not in simpler constructions. Thus, it may be argued that these systems along with other sophisticated techniques were also manifestations of prestige indicating that their users were people of a higher status.

9.3.1.7 Recreational uses

As already mentioned, Minoan engineers had developed techniques for the recreational use of water. In the Minoan palaces water jets, fountains, aquaria and other related facilities have been found.

9.3.1.7.1 Minoan fountains

Perhaps the most remarkable of all the fragments from the "House of Frescoes" are those restored in an attempt to present their original form. Although the actual summit of the fresco was not found, the object depicted in the upper part of the field is clearly some type of fountain or "jet d'eau" with the spout of the water made to rise from a forked base (Evans, 1921–1935). Moreover, the fragment below, with the same forked base and falling drops seems to be the base of another column of water drawn in similar conventional manner, but with a small section of the contour of the ground projecting an undulating bank beneath it. The background here is white, the central column of water and the falling drops on either side a deep blue, while the drops falling in front of the main jet are painted white, thus becoming quite distinct and visible. Remnants of it are exhibited in the Archaeological Museum of Iraklion (Figure 9.8, left).

Since no geysers or sulphurous ebullitions exist in Crete, like those in Palici of Sicily, it is highly unlikely that these designs were copied from any natural fountains found on the island. Here we are clearly dealing with artificial "jets d'eau", of a type that could have been seen only in

evidence that in Minoan times groundwater in the areas adjacent to the palace was plentiful and of very high quality.

9.3.2 Historical times

The end of Minoan civilization (around 1100 BC) was followed by the Dark Ages (*ca.* 1100–500 BC), a period characterised by a decrease in the Cretan population and very limited technological achievements. Historical times in Crete extended from early *ca.* 1000 BC (early Iron Ages) and comprise the Classical, Hellenistic and Roman periods.

9.3.2.1 The Hellenistic period (ca. 323–67 BC)

Hellenistic Crete did not stand in the cultural foreground during this time, nor did it play an influential role in international politics. The history of this long period, which spans the three centuries from Alexander the Great's death in 323 BC to the beginning of Roman period in 67 BC, is a record of incessant strife among the Cretan cities, in which external powers were often involved, especially Macedon and the Ptolemies. Crete was torn by internal warfare. Major and minor wars among the strongest cities, such as Knossos, Gortys, Lyttos, and their allies led sometimes to catastrophic results (Davaras, 1976). Significant developments relevant to water supply and to hygienic lifestyle were achieved in Crete this period, probably as a continuation of those developed during the Minoan Era. However, there is no evidence suggesting the connection between the Minoan era and the Classical and Hellenistic periods regarding the water systems. One could argue that all the hydraulic engineers' knowledge was lost with the Minoans and was re-gained only much later in the Hellenistic and then the Roman period, but this idea may be far from the truth (Panagiotakis, 2006). During Classical Antiquity many Greek cities (on mainland Greece) provide evidence of water management, which seems to have been closely related to the process of urbanisation (Fahlbusch, 2010). As Crouch (1993) says, "development of water supply, waste removal, and drainage made dense settlement possible".

Several Cretan cities of the Hellenistic and Roman periods, such as Lato, Dreros, Eleutherna, and Polyrrhenia were dependent on rainwater storage, the former using a large number of small domestic cisterns, while the latter utilised two major cisterns with capacities of some hundreds of m^3 each. Several ancient cities were exclusively dependent on rainwater storage and thus, became self-sufficient and ensured their survival in case of a siege. Such was not the fate of Carthage in Northern Africa, which was provided with an aqueduct a century after its destruction by the Romans in 146 BC.

In the castle areas of Classical and Hellenistic Crete on the top of the hills there was neither a spring nor a deep well. In order to guarantee the water supply for the inhabitants, especially in the case of a siege, cisterns had been constructed to collect rainwater during the rainy winter season. The Greeks improved on the cisterns technology of the Minoans by building not only circular cross-section shaped cisterns, but also ones with rectangular cross-sections (e.g. Lato and Dreros). Also, in rocky castle areas casters were hewn into rocks. For example in Polyrrhenia city in western Crete, which was built on top of a high hill (more than 400 m elevation); a location which offers excellent views to all the surrounding areas (from Crete to the Libyan sea), flourished during the Classic times. It was a powerful political centre and had two excellent harbours, Kissamos and Falasarna. Several small scale cisterns, which are hewn into the rocks, still exist in Polyrrhenia. They were mostly pear shaped. At least one layer of hydraulic plaster prevented water losses through the bottom and the walls. The estimated size of those cisterns is 10 m^3. In addition to the small scale cisterns much bigger cisterns were excavated in the rocky castles. In ancient Eleutherna a cistern of about 1000 $m^3$was constructed, including the tunnel outside of the cistern (Figure 9.9a, b).

worth noting that aqueducts were constructed using technology similar to that of qanat tunnels (e.g. Polyrrhenia).

9.3.2.2 The Roman period (ca. 67 BC–330 AD)

During the roman occupation Crete was incorporated with Cyrene (the African Libya) into a single Roman province. At that time the large town of Gortys in the Messara valley became the provincial capital. The island was exploited by the conquerors mainly for its grain. On the other hand, with the ending of the wars and incessant strife between the cities, under the Pax Romana, Crete entered a period of prosperity reminiscent of the old Minoan time.

Greek civilization continued to exist and flourish, and several cities destroyed during the Roman campaign were rebuilt. Emperor Hadrian was a great patron of some of the Cretan cities. Several cities were established and public buildings, often with fine mosaics, toilets, and other hydraulic works were constructed. Examples are the capital, Gortys as well as Hierapytna, Aptera, Lyttos, and Lebena. In addition, public engineering works like roads and large scale aqueducts and cisterns were constructed. As Pendlebury wrote, Crete was fortunate in being in the backwater of history and in lying well away from the storm where emperors rose and fell (Davaras, 1976).

The Romans built "mega water supply systems" including many magnificent structures. The advanced water and wastewater technologies developed in Minoan and Hellenistic Crete were expanded and improved upon during the Roman domination of the Greek world. The achievements of this era, which met the hygienic and functional requirements of ancient cities, were so advanced that they could only be compared to the modern urban water systems which developed in Europe and North America in the second half of the 19th century (Mays *et al.* 2007). Moreover, it should be noted that hydraulic technologies in ancient Greece were not limited to urban water and wastewater systems. The progress in the supply of urban water was even more astonishing, as numerous aqueducts, cisterns, wells, and other water facilities indicate (De Feo *et al.* 2010).

In Roman Crete there were several aqueducts (Table 9.2) such as those in Axos, Chersonessos, Minoa, Kissamos, Gortys (Figure 9.10a) and Falassarna (Figure 9.10b). Also, a very impressive Roman aqueduct of total 22 km length has been referred to in ancient Lyttos (Kelly, 2006b: Oikonomaki, 1984 and others). Its water source was located at the west flank of the present Oropedio Nissimou mountain (its summit is 1148 m high), at a location called Kournias, a few kilometres south of Krassi located at an altitude over 600 m.

In that aqueduct the channel was 0.35–0.40 m wide. This is a region with many springs which is underlined by the present springs and the spring building in the village of Krassi. The aqueduct continued along the NE and NW sides of Louloudaki, where remains are still visible. Close to Kastamonitsa village the channel made a sharp turn to the north heading for Lyttos. Remnants of the aqueduct in the area of Kastamonitsa are shown in Figure 9.11a; the most visible ones can be seen between Kastamonitsa and Tichos. Notice that tichos means wall. It is the substructure and partly the aqueduct bridge is composed of stretches of several dozens of meters of stone wall, at some places 10m high and at others less than one meter. At ground level the wall is over 2 m wide. Unfortunately, no visible remains have yet been identified from this point up to Lyttos. The presence of several stone pipe blocks in the area, most of which were used locally as building material, is quite unique. The one on Figure 9.11b with dimensions of 0.52×0.52 m^2 and 0.66 m long, was found in the garden of a villager of Kastamonitsa, and had a bore of 21 cm in diameter without any sign of connection (no male or female side or rings). These stone pipes can serve as a conduit under pressure and have been used to build an inverted siphon on top of the wall (tichos), as was also stated by Kelly (2006a) and Harrison (1993) who adds 'including lead lines'.

Table 9.2 Characteristics of major Hellenistic and Roman Aqueducts in Crete, Greece.

Aqueduct name	Site	Construction	Length (km)	Reconstruction
1. Aghia Pelagia	Aghia Pelagia	Roman	na	na
2. Axos	Axos	(Hellenistic 631BC) Roman	2.5	na
3. Archadia (Ini)	Ini	Classical	na	na
4. Archanes	Archanes	Turkish	na	na
5. Chersonissos	Chersonissos	First half of the 2nd century AD	13	na
6. Eleutherna	Eleutherna	Hellenistic	3	Roman
7. Elyros	Rodovani, Chania	Roman	2	na
8. Falassarna	Falassarna	Roman	1.5	na
9. Fountana (Knossos)	Fountana, Skalani	Roman	11.10 plus tunnel 1.15	Egyptian
10. Gavdos	Gavdos	Roman	1.0	na
11. Gortys	Zaros, Gortys	1st century AD	15	Byzantine
12. Hierapytna	Agios Ioannis, Hierapytna	Roman	na	na
13. Karidaki	Karidaki, Iraklion	Venetian	na	na
14. Kastellian	Kastellian	Roman	na	na
15. Kydonia	Kydonia, Chania	Roman	na	Venetian (1551 and 1554)
16. Kissamos	Kissamos, Chania	Roman	7.57	na
17. Lappa	Argiroupolis	Roman	6	na
18. Lassaia	Lassaia	Roman	na	na
19. Lefki[a] (Kouphonisi)	Kouphonisi	Roman	na	na
20. Lebena (Lendas)	Lentas	Hellenistic	na	Roman
21. Lisos	Agios Kirilos, Lisos	Hellenistic	na	Roman
22. Lyktos or Lyttos	Agios Panteleimon, Lyctos	Roman	22	na
23. Minoa	Marathi, Chania	Roman	2.77	na
24. Mastampa	Mastampa, Iraklion	Venetian	na	na

(*Continued*)

to low N. 3199/03, Greece is divided in 14 river basin districts (RBDs). The Ministry of Environment, Energy, and Climate Change considers all Greek islands (included Crete) as one single RBD.

Overall the RBDs of Crete island is characterised by high water availability, which is necessary to sustain agricultural production. In modern Crete, irrigation is responsible for more than 83.3% of the water consumption (while domestic use including tourism is 15.6% and industrial use 1%) and, to provide this quantity several large hydraulic works have been built. The demand for irrigation water is high, while at the same time only 31% of the available agricultural land is irrigated. The main issue with water resources management on the island is focused on the uneven geographical distribution of water resources in relation to the water demand hotspots on the island (Donta *et al.* 2005).

The contribution of surface water to the potential water resources of Crete is about 35%. The real contribution though is about 5%, which means that almost all the water quantity used on Crete comes from subterranean sources (springs, wells and boreholes) (Donta *et al.* 2005). In Crete there are no perennial streams and the total mean annual runoff is estimated to be 740 Mm^3 (RGC, 2002).

The contribution of groundwater to the potential renewable water resources in Crete is about 65%. Almost all the water quantity used comes from subterranean sources. It is estimated that approximately 20% of the groundwater resources of the island, totalling approximately 2100 Mm^3 are associated with Neogene-Quaternary aquifers while the remaining 80% are associated with their deep karstic counterparts (RGC, 2002). It is estimated that 40 to 55% of the mean annual precipitation infiltrates into the ground in karstic aquifers, creating a renewable groundwater potential of approximately 1800 Mm^3. Of that groundwater potential 80% is discharged from karstic springs the majority of which are unfortunately located along the coast, thus being contaminated by intruding seawater. Shallow aquifers hosted by Quaternary alluvial deposits have a renewable groundwater potential of 300 Mm^3.

Several boreholes and pumping wells exist on the island where the Regional Governor of Crete and other public services measure groundwater levels and chloride concentrations. In overexploited aquifers, such as the major Messara valley, a drop in the groundwater level has been observed. This is mainly due to over pumping of the aquifers for irrigation purposes. In the Iraklion area there is an intense exploitation of the existing aquifers (Donta *et al.* 2005).

It should be noted that water availability in average terms is not the limiting factor. Much more important are the significant regional and seasonal variations in water availability and demand. To overcome future water shortage, several measures should be taken for the conservation of water resources, decentralised management, promotion of water reuse and protection of the environment. Today 40% of the total population of the island live in Iraklion, Chania, Rethymnon and Agios Nikolaos (the four bigger cities). Providing the water supply details of these cities gives a general idea of how water is supplied at present times on Crete. Smaller towns and villages usually have individual drilling wells and/or use any surface water available. The majority of the settlements in the high elevations enjoy excess of superb quality water. Water supply in the city of Chania mainly depends on spring water (e.g. Agyia, located at a distance of 15 km south of the city) and secondarily to groundwater. On the other hand, the water supply of Iraklion and Agios Nikolaos cities depends on groundwater. Finally, water supply in Rethymnon depends on spring water and on the storage dam located at a distance of 30 km south of the city.

9.4 DISCUSSION AND CONCLUSIONS

From a global point of view, ancient water supply technologies significantly contributed to the development of human culture. Considering the evolution of the water technologies on Crete, Greece, the development of science and engineering is not linear but is often characterised by discontinuities and regressions during the so called dark times. "Bridges" from the past to the future are always present, although they are often invisible to those who cross them! Thus, in addition to many ancient constructions that have been continuously or intermittently in operation to date, substantial information from ancient Crete written sources has also been preserved including such specifics as detailed contracts between the public and constructors of hydraulic works (Koutsoyiannis & Angelakis, 2003).

The climatic and geo-hydrological conditions in the island are highly differentiated. Thus water availability is spatially and temporally uneven and it is clear that water technologies, developed through the centuries, such as aqueducts, cisterns, wells, and other systems were created and adapted to the local conditions and present an ecological sustainable system with a considerable potential in water supply. Cretan civilizations as many others, which were great centres of power and culture, were built in locations that could support the respective populations (Mays, 2010). Today, we find ourselves in similar situations in many places around the world. How do we balance the mega water projects with the methods of traditional knowledge? Traditional knowledge allowed ancient societies to keep ecosystems in balance, and carry out outstanding technical, artistic, and architectural work that has been universally admired (Laureano, 2006). The use of traditional knowledge has been able to renew and adapt itself. Traditional knowledge incorporates innovation in a dynamic fashion, subject to the test of a long term, achieving local and environmental sustainability (Mays, 2007). The ancients for the most part, lived in harmony with nature and their environment. Those that did not failed. The Cretan civilizations are a cogent paradigm. Their actions should be a warning to us. The legacies and lessons on water supply evolution in Crete since the Minoan era are summarised as follows:

(a) The definition of sustainability in modern times should be re-examined in light of ancient public works and management practices. Technological developments based on sound engineering principles can have extended useful lives.
(b) In water-short areas, development of cost-effective decentralised water and wastewater management is essential.
(c) Ancient Cretan knowledge could play an important role for sustainable water supply in the future cities.
(d) Cretan ancient (particularly Minoan) water technologies should be considered not as historical artefacts but as potential paradigms for sustainable water technologies. Development of effective water supply management projects, in short-water areas could be based on traditional knowledge.

In conclusion, this Chapter provides valuable insights into the water supply technologies developed in ancient Crete which are the underpinning of modern achievements in water engineering and management practices. It is the best proof that "*the past is the key for the future*". Further research is necessary to recover the forgotten subsurface water ways in order to get a closer insight to their local or even regional water supply potential.

Acknowledgements

The authors greatly appreciate the critical comments of Mr. Y. Christodoulakos, Ministry of Culture, 25th Dept. of Antiquities, 73134 Chania, Greece, gchkeepka@yahoo.gr.

REFERENCES

Alexiou S. (1964). Minoan Civilization. S. Alexiou and Sons, Iraklion, Greece (in Greek).
Angelakis A. N. and Spyridakis S. V. (1996). The status of water resources in Minoan times: A Preliminary study. In: Diachronic Climatic Impacts on Water Resources with Emphasis on Mediterranean Region, A. N. Angelakis and A. S. Issar (eds), Springer–Verlag, Heidelberg, Germany, Ch. 8: 161–191.

Angelakis A. N. and Koutsoyiannis D. (2003). Urban water resources management in Ancient Greek times. In: The Encycl. of Water Sci., Markel Dekker Inc., B. A. Stewartand and T. Howell (eds), Madison Ave, New York, N.Y., USA, pp. 999–1008.

Angelakis A. N. and Spyridakis D. S. (2010). Water supply and wastewater management aspects in Ancient Greece. *Water Sci. and Tech. Water Supply*, **10**(4), 618–628.

Angelakis A. N., Koutsoyiannis D. and Tchobanoglous G. (2005). Water resources technologies in the Ancient Greece. *Water Resources*, **39**(1), 210–220.

Angelakis A. N., Savvakis Y. M. and Charalampakis G. (2007). Aqueducts during the Minoan Era. *Water Sci. and Tech, Water Sci and Tech. Supply*, **7**(1), 95–102.

Antoniou A. and Angelakis A. N. (2011). Evolution of sanitary and purgatory structures in Ancient Greece. *Journal of Early Medicine Issue on Latrines, Hygiene and Health* (in press).

Buffet B. and Evrard R. (1950). The Potable Water Through the Centuries. Editions Soledi, Rue de la Province, 37, Liege, Belgium (in French).

Cadogan G. (1978). Pyrgos, Crete, (1970–1977) Archaeological Reports, Archaeopress, Cordon House, Oxford, UK, 24 (1977–78), 70–84.

Cadogan G. (2007). Water management in Minoan Crete, Greece: The two cisterns of one middle Bronze age settlement. *Water Sci. and Tech. Water Supply*, **7**(1), 103–112.

Crouch D. P. (1993). Water Management in Ancient Greek Cities. Oxford University Press, New York, USA and Oxford, UK.

Davaras C. (1976) Guide to Cretan Antiquities. Noyes Press, Noyes Building, Park Ridge, New Jersey 07656, USA.

De Feo G., Mays L. and Angelakis A. N. (2011). Water and wastewater management technologies in Ancient Greek and Roman civilizations. In: Treatise on Water Science, P. Wilderer (ed.), vol. **1**, Academic Press, Oxford, UK, pp. 3–22.

Despotakis V. K. (1999). Water Resources Management for the new Kapodistrias Municipalities using Arcview GIS. Paper presented at the 9th Arc/Info Users Meeting, Athens, Greece, November 2 & 3, 1999.

Dialynas M. G., Lyritztis A. G. and Angelakis A. N. (2006). Historical development of water supply of the city of Iraklion, Greece. In: Proc. of 1st IWA International Symposium on Water and Wastewater Technologies in Ancient Civilizations, 28–30 October 2006, Iraklion, Greece, pp. 671–676.

Donta A., Lange A. M. and Herrmann A. (2006). Water on Mediterranean Islands: Current Conditions and Prospects for Sustainable Management, Published by Centre for Environment Research (CER) University of Muenste, vol. **5**, MEDIS program.

EU (European Communities) (2000). Framework for Community Action in the Field of Water Policy. The Official Journal of the European Communities L 327/1, 2000/60/EC. Brussels, Belgium.

Evans S. A. (1921–1935). The Palace of Minos at Knossos: A Comparative Account of the Successive Stages of the Early Cretan Civilization as Illustrated by the Discoveries. Vols. **I-IV**, Macmillan and Co., London, UK (Reprinted by Biblo and Tannen, New York, USA).

Driessen J. M. and MacDonald C. F. (1997). The troubled Island. Minoan Crete before and after the Santorini eruption. *Aegaeum*, **17**, 64–80.

Fahlbusch H. (2010). Introduction: political situation, attitude to the water and necessity of transfers. In: Water engineering and Management in Classic Civilizations. Water Engineering and Management through Time – Learning from History, E. Cabrera and F. Arregui (eds), CRC Press, Taylor and Francis Group, New York, Leiden, London, UK, pp. 77–115.

Fahlbusch H. (2008). Municipal Water Supply in Antiquity. http://www.romanaqueducts.info/webteksten/waterinantiquity.htm (accessed 10 December 2011).

Garbrecht G. (2001). Altertumer von Pergamon (Antiquities of Pergamon), I, 4, Berlin, Germany.

Gikas P., Christodulakos Y., Gikas V. and Angelakis A. N. (2009). Water supply in the Roman city of Aptera, Crete, Greece. In: Proc. of the 2nd IWA International Symposium on Water and Wastewater Technologies in Ancient Civilizations, May 28–30, 2009, Bari, Italy.

Gorokhovich Y., Mays L. and Ullmann L. (2011). A survey of water technologies of the Ancient Minoans and traditional knowledge. *Water Sci. and Tech., Water Supply*, **11**(4), 388–399.

Graham J. W. (1987). The Palaces of Crete. Princeton University Press, Princeton, USA.

Hazzidakis J. (1934). The Minoan Villas of Tylissos. P. Geuthner, Paris, France (in French).

Haberey W. (1972). The Roman Water conduit of Colony. Rheinland–Verlag, Bonn, Germany (in German).

Hunchinson R. W. (1950). Prehistoric Town Planning in Crete. Town Planning Review, **21**, pp. 199–220.

Kelly A. (2006a). The impact of constructing aqueducts on the settlement patterns of roman crete. Cura Aquarum in Ephesus. In: The 12th International Conference on the History of Water Management and Hydraulic Engineering in the Mediterranean Region. Ephesus, October 2–10, 2004, Leuven, Belgium, Volume **II**: 303–310.

Kelly A. (2006b). Distributions of cretan aqueducts: a window onto romanization. In: Cretologic Conference, **A5**, 391–402.

Koutsoyiannis D. and Angelakis A. N. (2004). Agricultural hydraulic works in Ancient Greece. In: The Encycl. of Water Science, B. A. Stewart and T. Howell (eds), Markel Dekker New York, N.Y., USA, pp. 1–4.

Koutsoyiannis D. and Angelakis A. N. (2003). Hydrologic and hydraulic sciences and technologies in Ancient Greek Times. In: The Encycl. of Water Sci., Markel Dekker Inc., B. A. Stewart and T. Howell (eds), Madison Ave. New York, N.Y., USA, pp. 415–418.

Koutsoyiannis D., Mamassis N. and Tegos A. (2007). Logical and illogical exegeses of Hydro meteorological phenomena in ancient Greece. *Water Sci. and Tech, Water Supply*, **7**(1), 13–22.

Koutsoyiannis D., Zarkadoulas N., Angelakis A. N. and Tchobanoglous G. (2008) Urban water management in Ancient Greece: legacies and lessons. *ASCE, Journal of Water Res. Planning & Manag.*, **134** (1), 45–54.

Laureano P. (2006). Ancient water techniques for proper management of Mediterranean ecosystems. In: Proceedings of the 1st IWA International Symposium on Water and Wastewater Technologies in Ancient Civilizations, A. N. Angelakis D. Koutsoyriannis (ed.), National Agricultural Research Foundation, Iraklion, Greece, 28–30 Oct. 2006, pp. 209–218.

Lolos Y. A. (1997). The Hadrianic Aqueduct of Corinth. *Hesperia*, **66**(2), 271–314. Published by American School of Classical Studies at Athens, Athens, Greece.

Lyrintzis A. G. and Angelakis A. N. (2006). Is the labyrinth a water catchment technique? people and water management in Minoan Crete. In: Proc. of 1st IWA International Symposium on Water and Wastewater Technologies in Ancient Civilizations. Iraklion, Greece, 28–30 October 2006, pp. 163–174.

Mays L. W. (ed.) (2007). Water Resources Sustainability. McGraw–Hill, New York, USA.

Mays L. W. (2008). A very brief history of hydraulic technology during antiquity. *Envir. Fluid Mechanics*, **8**(5), 471–484.

Mays L. W. (ed.) (2010). Ancient Water Technology. Springer, Dordrecht Heidelberg, London, UK and New York, USA.

Mays L. W., Koutsoyiannis D. and Angelakis A. N. (2007). A brief history of urban water supply in antiquity. *Water Sci. and Techn, Water Supply*, **7**(1), 1–12.

Oikonomaki N. E. (1984). The Aqueduct of Lyttos. Λύκτος **1**, 66–69 (In Greek).

Panagiotakis N. (2006). Water management in the pediada region in central crete through time. In: Proc. of 1st IWA International Symposium on Water and Wastewater Technologies in Ancient Civilizations. Iraklion, Greece, 28–30 October 2006, pp. 271–280.

Paranychianakis N. V., Kritsotakis M. and Angelakis A. N. (2011). A Sustainable Water Resources and Wastewater Future for Crete, Greece, Problems and Perspectives (in preparation).

Platon N. (1974). Zakros: The New Minoan Palace. The Archaeological Society at Athens, Athens, Greece.

Platon E. (2001). Minoan terracotta water spouts. In: Proc. of the 7th Intern. Congress of Cretan Studies. Rethymno, Greece, **A2**, 767–775 (in Greek).

Perlès C. (2001). The Early Neolithic in Greece: The First Farming Communities in Europe. Cambridge University Press, Cambridge, UK and New York, USA.

RGC (Regional Governor of Crete) (2002). Integrated Water Resources Management on Crete. Study Made Under the Department of Water Resources Management. Iraklion, Greece.

Sakellarakis Y. and Sapouna-Sakellaraki E. (1997). Archanes: Minoan Crete in a New Light. Ammos/Eleni Nakou Foundation. Athens, Greece.

Savvas D. (2008). Turkey's Fountains. Iraklion Patris Newspaper, 18/08/08. Iraklion, Greece (in Greek).

Spanakis S. (1981). The Water Supply of Iraklion, 828–1939. The Technical Chamber of Greece, Iraklion, Greece (in Greek).

Strataridakis A. I., Chalkiadakis E. G. and Gigourtakis N. M. (2009). The significance of the foundana spring aqueduct for water supply of Heraklion City (Crete) through the ages. In: Proc. (in DVD) of the 2nd IWA International Symposium on Water and Wastewater Technologies in Ancient Civilizations, May 28–30, 2009, Bari, Italy.

Tassios T. P. (2007). Water supply of Ancient Greek cities. *Water Sci. and Techn., Water Supply*, **7**(1), 165–191.

Tsonis A. A., Swanson K. L., Sugihara G. and Tsonis P. A. (2010). Climate change and the demise of Minoan civilization. *Clim. Past Discuss.*, **6**, 801–815.

Tzombanaki Ch. (2005). The Architecture in Crete. Period of Modern Times, 19th Century and Period of Autonomy. Vol. A1, Iraklion, Greece.

Webster L. and Hughes R. (2010). The Mystery of Minoan Tapered Pipes. Department of Civil and Environmental Engineering, The University of Melbourne, Victoria, Australia, pp. 1–4.

Chapter 10

A brief history of urban water management in ancient Greece

N. Zarkadoulas, D. Koutsoyiannis, N. Mamassis and A. N. Angelakis

10.1 INTRODUCTION

It is commonly accepted that technologies and infrastructures related to use of water were developed as early as the sixth millennium BC in Egypt and Mesopotamia. In Greece, such technologies first appeared in the Minoan Crete and the Mycenaean Mainland Greece during the third and second millennium BC, respectively. The end of the Mycenaean civilization (around 1100 BC) is followed be the Dark Ages (*ca.* 1100–700 BC) a period characterised by a decrease in the population and the disruption of the cities. Under a general lack of archaeological data for this period, no evidence for hydraulic works has been found so far. At about 700 BC a new period, known as archaic Greek antiquity, starts, which is characterised by urbanisation, spread of the colonisation, and the emerging of city-states. The first urban water infrastructures are developed under a political system known as tyranny.

Evolution of water systems continued in later periods, in the Classical and Hellenistic antiquity (*ca.* 500–67 BC). Exceptional examples of advanced water infrastructures and management practices of these periods are found in Athens, the Aegean islands and Asia Minor. The infrastructures impress not only for their scale, but for their integration even by today's standards. Some of the infrastructures are still in use, demonstrating how far sustainability can go.

Although the archaeological and technological documentation on ancient Greek water systems and management practices is extensive (mainly following an object-oriented approach), the following aspects need further investigation:

(a) What are the conditions that led societies or city states to construct and operate complex and costly infrastructures?
(b) How the developments in the form, the structure and the scale of the city of the past interrelate with the contemporaneous developments on water resource technology and management?
(c) How water management and legislation interrelate with the technology of urban water systems?

Previous works (Koutsoyiannis *et al.* 2008) shed light on the management frame of urban water management in the city of Athens. The basic conclusion of this analysis was that water infrastructures were combined by non-structural measures, including support by legislation (e.g. Solon's laws). Here we further explore the above questions using several examples. The interested reader can find more information on these examples at the *Web Based Information System for the Inspection of the Hydraulic Works in Ancient Greece* (http://itia.ntua.gr/ahw/works/; see also Mamassis and Koutsoyiannis, 2010), which contains data for almost 100 important hydraulic works from the Minoan era up to the Roman period.

10.2 PREHISTORIC GREECE

The hydraulic technological frame of ancient Greece was formed very early, during the Minoan/Cycladic and Mycenaean eras. There is a difference of scale between Minoan and Mycenaean water systems. The former were mainly focused on domestic water use at the palatial scale whereas the latter targeted agricultural water use (irrigation and drainage) and were larger-scale systems.

The technologies and practices developed in Minoan Crete, at the scale of the palace, include (Angelakis *et al.* 2005):

- *potable water systems*, like the network of terracotta piping located beneath the palace floors in Knossos;
- *wastewater and storm water sewage systems*, like those found at Knossos, Zakros and Phaistos palaces;
- *hygienic water use systems*, like bathrooms with flushing toilets.

The study of those elaborate infrastructures has revealed a degree of development that presupposes a good understanding of water flow. Similar urban sewer systems have also been found in the island of Thera (Santorini) and other prehistoric sites of the Aegean civilization (*ca.* 3200–1100 BC).

Mycenaean civilization depended on agricultural production. In order to cover the increased water needs for agriculture (even in modern Greece, about 85% of total water consumption is used for irrigation), Mycenaeans chose closed river basins for their settlements and developed large scale flood control and drainage infrastructures, for the first time in Greek territory (Koutsoyiannis *et al.* 2012). The prosperity of areas like the Arcadian and Boeotian Orchomenos is directly connected to the successful operation of these projects.

10.3 ARCHAIC GREECE: THE FORMATION OF THE CITY-STATE AND THE IMPACT OF TYRANNY

10.3.1 Urbanisation: The model of the city–state

In archaic Greek antiquity, a new political and societal model started to form: the city-state. Additionally, spread of the colonisation occurs. An explanation of the emerging city-state model has been given by Doxiadis (1964):

> *'Greece is divided by mountains into small plains. It is in these plains that the major part of land cultivation takes place and we can roughly say that these do not exceed 20% of the ancient Greek peninsula. The physical boundaries of the small plains form the boundaries of the city-state. These areas range from fairly small states with an area of 100 km^2 such as the state of Aegina, to fairly large states, such as the states of Arcadia and Laconia which spread over an area of about 5000 km^2. Diagrammatically, we can thus visualise the ancient Greek states as squares of 10 by 10 km, which could be crossed from end to end in 2 hours or so, to squares of 70 by 70 km, which need 14 hours to cross on foot.'*

During that period, cities depended on trade, rather than on agricultural production, which triggered new criteria for city siting. The locations of the new cities tended to appear in dry places, at a distance from rivers or lakes, but attention was paid to water adequacy in the event of war. In addition, the core of the city moved from the palace or the Acropolis to the Agora. As Agora gradually became the centre of political, social and commercial activities, hygienic technologies were gradually implemented on a larger scale, the scale of the Agora. The cities grew considerably and the first large-scale urban water infrastructures were developed. At the political level, institutional progress led to tyranny.

10.3.2 City siting

Observation of the most important cities of the Greek antiquity leads to the conclusion that they are mainly located in dry climates (with annual precipitation of about 500 mm/year or less). City siting in dry places, must have been primarily driven by the laws of the natural selection with the populations established in dry climates having larger probability to survive, as they were protected from waterborne diseases. In the centuries that followed, Greeks must have progressively assimilated the fact that dry climates are generally more convenient to live and healthier. As a result, most major Greek cities during the several phases of the Greek civilization that lasted for millennia were established in those areas that had the minimal rainfall across the continental and insular Greece.

As the model of city-state starts to dominate, the economy moves from being agriculture- to trade-based. This change of the societal model gradually reduces the dependence of the economy on water, allowing the relocation of the centre of human activity at a distance from large water bodies (rivers and lakes). Thus, contrary to Mycenaeans who used to settle close to flood plains of rivers, that is, the most fertile areas, no major city in flood-prone areas existed in the Classical Greek antiquity. The importance of city siting can be inferred from Aristotle's Politics (VII, X, 4):

> '*The site of the city itself we must pray that fortune itself may place on sloping ground, having regard to four considerations: first, as a thing essential, the consideration of health (for cities whose sites slopes east or towards the breezes that blow from the sunrise are more healthy, and in the second degree those that face away from the north wind, for these are milder in winter); and among the remaining considerations, a sloping site is favourable both for political and for military purposes*' (translation by Doxiadis, 1964, pp. 346–364).

However, the above criteria have not been applied for cultural and places, many of which (e.g. Delphi, Olympia, and Dodone) are situated in areas with adequate water resources, for example, close to springs or inside groves, and if this was not practical, great efforts were made to transfer water and plant the area (Camp, 2004, p. 113).

10.3.3 Safety of water supply

Safety in the event of war imposed city development around a natural rock (e.g., the Acropolis of Athens or the Acrocorinth). Even if there was enough water to the wider area around the town, additionally, there should be enough water within the walls of the acropolis, in case of siege. According to Aristotle the driving force behind the development of small-scale constructions (such as cisterns) was ensuring adequacy of water in the event of war:

> '*There should be a natural abundance of springs and fountains in the town, or, if there is a deficiency of them, great reservoirs may be established for the collection of rainwater, such as will not fail when the inhabitants are cut off from the country by war*' (Aristotle, Politics, 7, XI, translation by Benjamin Jowett; http://classics.mit.edu/Aristotle/politics.7.seven.html).

Thus, water adequacy was primarily ensured by natural springs and/or wells. In the case of the Acropolis of Athens, the Klepshydra spring, which is believed to have been discovered in the second half of the 13th century BC during the fortification works, ensured water adequacy (Figure 10.1). In addition beneath the

10.3.5.2 Samos

In Samos during the rule of tyrant Polykrates (second half of the sixth century BC), impressive civil infrastructures were constructed. The advancement of urban water technology and management is illustrated through the extraordinary example of the water supply of the island of Samos. The most amazing part of the water supply system of ancient Samos (located at the site of the modern-day village of Pythagoreio) is the "Eupalinean digging," more widely known as the Tunnel of Eupalinos, named after the engineer from Megara who designed and constructed it. The aqueduct includes the 1036 m long tunnel and two additional parts, so that its total length exceeds 2800 m. Its construction started in 530 BC and lasted 10 years. From Herodotus (484–425 BC), the "Father of History", we learn:

> '*I have dwelt the longer on the affairs of the Samians, because three of the greatest works in all Greece were made by them. One is a tunnel, under a hill one hundred and fifty fathoms high, carried entirely through the base of the hill, with a mouth at either end. The length of the cutting is seven furlongs- the height and width are each eight feet. Along the whole course there is a second cutting, twenty cubits deep and three feet broad, whereby water is brought, through pipes, from an abundant source into the city. The architect of this tunnel was Eupalinus, son of Naustrophus, a Megarian. Such is the first of their great works; the second is a mole in the sea, which goes all round the harbour, near twenty fathoms deep, and in length above two furlongs. The third is a temple; the largest of all the temples known to us, whereof Rhoecus, son of Phileus, a Samian, was first architect. Because of these works I have dwelt the longer on the affairs of Samos*' (Herodotus, Histories, III, translation by George Rawlinson; http://classics.mit.edu/Herodotus/history.3.iii.html).

The tunnel was in operation until the fifth century AD. It is certain that the tunnel Eupalinos constructed was not the only solution to the problem of conveying water to Samos. A simple alternative solution could have been the construction of a chain of open channels and tunnels at shallow depths with shafts, following a route around the mountain. This solution, already well known (cf. Peisistratean aqueduct), would certainly have been easier technically, faster, and less expensive. The reasons that led to the construction of such a costly structure are not obvious. Probably Eupalinos and the tyrant Polycrates preferred this unorthodox and breakthrough solution because they wished to build a monument of technology rather than simply solving a specific water transportation problem.

10.3.5.3 Naxos

In a recent excavation, the remnants of an ancient aqueduct conducting water from Melanes, a fertile place in inland Naxos to the littoral ancient town of Naxos were investigated (V. Lambrinoudakis, personal communication). The original phase of the aqueduct construction is dated to the late sixth century BC, either during the tyranny of Lygdamis or during the succeeding brief interval of democracy on the island. The aqueduct ran over 11 km on hillsides at the upper limit of the fertile land and consisted of socket-jointed clay pipes of a diameter of about 0.30 m buried in a ditch of a depth of about 1 m (Figure 10.3). Its slope varied from 1 to 4%. The aqueduct contained a tunnel, hewn in the heights to the north of the Melanes basin, which allowed the aqueduct on the one hand to pass by another spring and to secure a supplementary resource, and on the other to irrigate, before entering the plain, a small fertile valley named Cambones. This tunnel was 220 m long, 1.60 m high and 0.80 m wide. Its entrance and its exit, well preserved behind reconstructions of the Roman Period, were recently investigated at lengths 3 to 5 m inside. Parts of the series of pipes are sometimes well preserved and other times blocked, as indicated by a by-pass series of pipes found along the route of the aqueduct. The original phase of the aqueduct is of the same type as the Peisistratian and Eupalinos aqueducts. Sometime in the Roman imperial period the aqueduct was reconstructed on the traces of the ancient pipeline. The Roman aqueduct ended in the ancient city of Naxos in a vaulted fountain, whose interior is totally preserved

had the disease myself, and watched its operation in the case of others' (Thucydides, The History of the Peloponnesian War, 2, VII, translation by Richard Crawley; http://classics.mit.edu/Thucydides/pelopwar.2.second.html).

Small scale infrastructures, such as wells and cisterns are regarded more resilient in war conditions. Most of the cisterns found in the Athenian Agora date from the fourth to the first centuries BC. The storm-water cisterns, whose maintenance in ancient Athens must have been obligatory for citizens, is an appropriate provision for the maximisation of system stability and safety. In addition to providing a source of water for private use and enhancing the security of the overall system, cisterns, still in wide use today in anhydrous Greek islands, also reduce the amount of storm water to be discharged (Koutsoyiannis *et al.* 2008).

10.4.1.1 Potable and sub-potable water quality

Rainwater, collected from the roofs and stored in cisterns was normally used for washing whereas well water was used for drinking (Lang, 1968). The distinction between potable and non-potable water is also reflected by Aristotle:

'*Special care should be taken of the health of the inhabitants, which will depend chiefly on the healthiness of the locality and of the quarter to which they are exposed, and secondly, on the use of pure water; this latter point is by no means a secondary consideration. For the elements which we use most and oftenest for the support of the body contribute most to health, and among these are water and air. Wherefore, in all wise states, if there is a want of pure water, and the supply is not all equally good, the drinking water ought to be separated from that which is used for other purposes*' (Aristotle, Politics, 7, XI, translation by Benjamin Jowett; http://classics.mit.edu/Aristotle/politics.7.seven.html).

10.4.1.2 Water administrators

From Plutarch's Life of Themistocles (XXXI,I), it is known that there was at least one public official concerned with waterworks even in the early fifth century BC, named "κρουνῶν επιμελητής" (superintendent of fountains). He was appointed to operate and maintain the city water system, to monitor enforcement of the regulation, and to ensure the fair distribution of water (Koutsoyiannis *et al.* 2008). The superintendent of fountains was one of the most important public officials. According to Aristotle:

'*All the magistrates that are concerned with the ordinary routine of administration are elected by lot, except the Military Treasurer, the Commissioners of the Theoric fund, and the Superintendent of Springs. These are elected by vote, and hold office from one Panathenaic festival to the next*' (Aristotle, The Athenian Constitution, Section 2, Part 43, translation by Sir Frederic G. Kenyon; http://classics.mit.edu/Aristotle/athenian_const.2.2.html).

10.4.1.3 Public and private works

Water resource management of the city of Athens has not always been based on public financing. In periods where democracy had troubles in finding its own financial resources, development was driven by the private sector. In the fifth century BC hydraulic benefactors of Athens included two statesmen, Pericles and Kimon, and an astronomer, Meton. Specifically, according to Lang (1968) Meton provided a fountain on Kolonos Agoraios and Pericles (495–429 BC) generously offered to restore a springhouse. Also, according to Plutarch, Kimon made a grove by bringing in water in the Academy:

'*And the place where they built them being soft and marshy ground, they were forced to sink great weights of stone and rubble to secure the foundation, and did all this out of the money Kimon supplied them with. It was he, likewise, who first embellished the upper city with those fine and ornamental places of exercise and resort, which they afterwards so much frequented and delighted in. He set the market-place with plane-trees;*

and the Academy, which was before a bare, dry, and dirty spot, he converted into a well-watered grove, with shady alleys to walk in, and open courses for races' (Plutarch, Kimon, XIII, 8, translation by John Dryden; http://classics.mit.edu/Plurarch/cimon.html).

10.4.2 The contribution of Hippodamos on city planning and its effects on water infrastructures

Ancient Greek cities are divided in two categories: those formed through natural growth (e.g. like Athens or Corinth) and those created on the Hippodameian system, named after Hippodamos the Milesian (498–408 BC). The choice of building a city on the Hippodameian system is based on the benefits of the parallel streets forming a grid, and is dictated by functional reasons. According to Aristotle, Hippodamos was the first architect to design a city in the form of a grid:

> '*His system was for a city with a population of ten thousands, divided into three classes; for he made one class of artisans, one of farmers and the third the class that fought for the state in war and was the armed class. He divided the land into three parts, one sacred, one public and one private: sacred land to supply the customary offerings to the gods, common land to provide the warrior class with food, and private land to be owned by the farmers*' (*Aristotle*, Politics, II, V, 2, translation by Doxiadis, 1964, pp. 346–364).

Despite their differences, the concept hidden behind both building processes was the same: To take advantage of the natural landscape and to create both public and private spaces according to rational and functional considerations with man at the centre (Doxiadis, 1968).

The Piraeus peninsula was uninhabited when Themistocles decided that the port of Athens had to be relocated from Phalero to a more appropriate and safer area. The Piraeus port was designed from scratch by Hippodamos in 470–460 BC and is considered to be the first city designed under the Hippodameian system. Hippodamos introduced a series of innovations that established Piraeus as a design standard throughout antiquity. More specifically Zissimou (2007) reported the following:

(a) He separated the function of each part of the city, creating three separate zones: public, private and sacred.
(b) He divided the city in equal building blocks, each of which contained the same number of houses. Each block had dimensions of 140×160 ft^2 (40.37×47.40 m^2) and contained eight properties of 40×70 ft^2 (11.87×20.31 m$^2 = 241$ m^2).
(c) He enrolled the Pythagorean theory of numbers in the design of roads creating a harmony that affected citizens. There were three categories of roads: the *culs de sac* in simple blocks with a width of 14 ft (about 5 m), roads connecting the settlements with a width of 20 ft (about 8.20 m) and "wide roads", connecting the ports, the market and the entrances of the city with a width of 45 ft (about 15 m).

The most famous example of a Hippodameian city is Priene in Western Anatolia (Figure 10.4).

The consequences for urban management from the "pre-designed" urban space are:

(a) The scale of the city changes, requiring bigger infrastructures.
(b) The location of all public infrastructures (agora, temples, public baths and so on.) is pre-defined

The benefits for water management from the "pre-designed" urban space are:

(a) The organised city-planning and the regularity of the grid allow engineers to optimise design and construction of hydraulic infrastructures, similar to modern ones.
(b) The construction and maintenance of infrastructures are easier

smaller line, carefully jointed with lids in place. It seems that at that time period, the overflow water was not just disposed of but kept for some secondary purpose. In the second period, the original line was given up in favour of larger pipes. It seems that the new aqueduct not only brought in more water, which required larger overflow pipes but also that the increased supply of water in the Agora may has obviated the need for the overflow. At the same time, private installations like wells and cisterns tended to be abandoned (Lang, 1968).

In the Hellenistic world, the prevalence of the Hippodameian system resulted in reduced importance of security in city planning. This had been predicted and cautioned by Aristotle, who made the following comment:

> *'The arrangement of the private dwellings is thought to be more agreeable and more convenient for general purposes if they are laid out in straight streets, after the modern fashion, that is, the one introduced by Hippodamus; but is more suitable for security in war if it is on the contrary plan, as cites used to be in ancient times; for that arrangement is difficult for foreign troops to enter and to find their way about in when attacking. Hence, it is well to combine the advantages of both plans, and not to lay out the whole city in straight streets, but only certain parts and districts, for in this way it will combine security with beauty'* (Aristotle, Politics, VII, X, 4, translation from Doxiadis, 1964, p. 346–364).

Technological innovations of this period include the construction of pipelines under pressure. A prominent example is the water supply of the Hellenistic city of Pergamon in Western Anatolia. In about 200 BC, the system of cisterns could not meet the demands of the growing population. One of the three aqueducts constructed transferred water from Madragad Mountain passing a valley by an inverted siphon of length exceeding 3 km and with a maximum pressure head of about 180 m. The inverted siphon was made of lead and anchored with big stone constructions (Koutsoyiannis *et al.* 2008).

In the centuries that followed, the ancient Greek city-state was replaced by the powerful Roman Empire. The enforcement of "Roman Peace" throughout the Mediterranean reduced the importance for the protection of water in the event of war and ensured the reliability of water supply through the massive construction of large scale aqueducts. Paradoxically, the Roman technology made little use of the flow under pressure, replacing siphons with water bridges, which became a characteristic mark of the Roman aqueducts. The explanation for the preference of free surface flow over pressurised flow is explained by Vitruvius (VIII, 6, 10):

> *'Water supply by earthenware pipes has these advantages. First, if any fault occurs in the work, anybody can repair it. Again, water is much more wholesome from earthenware pipes than from lead pipes. For it seems to be made injurious by lead, because white lead is produced by it; and this is said to be harmful to the human body'* (translation adapted from Lang, 1968).

This is correct, of course, but on the other hand it is a technological regression. Without pressurised pipes, the modern urban water systems would be impossible. But these had to wait for the industrial revolution, to replace the lead pipes with iron (cast iron, steel) pipes (and more recently with plastic pipes) which are not harmful to health.

10.6 DISCUSSION AND CONCLUSIONS

Sound know-how about urban water systems existed in Greece from the Minoan times. The strong technological base of the Minoan and Mycenaean period has been the base of the advanced technological progress and water management of the centuries that followed.

The stagnancy during the Dark Ages was followed by significant progress in the Archaic Greek antiquity. During that period, city-states were formed and urbanisation occurred. As cities depended on trade and not on agricultural production, they tended to be located at dry places, at a distance from rivers or lakes. Under tyranny, cities grew significantly and the first large scale urban water infrastructures were developed.

The period of democracy that followed, with its small-scale structures and its non-structural measures is a lesson of sustainable management and marks the importance of the institutional progress in water management.

During the Hellenistic period, urban city planning acquired a new dimension in the form of the Hippodameian city, which implemented a different design philosophy. The evolution of the "designed city" is mainly reflected on the scale of the projects, which resulted in water adequacy and more widespread hygienic water use.

Obviously, the scale of the city today is much greater than in antiquity. This is also reflected in the scale of water infrastructures. As a result, a direct comparison is not possible. Nevertheless, we can assume that the following elements of ancient Greek water management should be re-considered:

(a) City planning has to include urban water criteria; protection from floods should be a major consideration.
(b) The use of small-scale infrastructures, in parallel to the large-scale ones, is a big step towards sustainability and resilience. The principles and practices of sustainable water use should not be forgotten even in periods of water adequacy.
(c) Safety and security of water supply in emergency situations, including turbulent and war periods, should be kept in mind in our designs of urban water systems.

Acknowledgements

Special thanks are due to Dr. V. Lambrinoudakis, Emeritus Professor at the University of Athens, Greece for the information he provided on the ancient aqueduct in Naxos.

REFERENCES

Angelakis A. N., Koutsoyiannis D. and Tchobanoglous G. (2005). Urban wastewater and stormwater technologies in ancient Greece. *Water Research*, **39**(1), 210–220.

Crouch D. P. (1993). Water Management in Ancient Greek Cities. Oxford University Press, New York–Oxford, U.K.

Camp J. M. (2004). The Athenian Agora, Excavations in the Heart of Classical Athens. National Bank of Greece Cultural Foundation, Athens, Greece (in Greek).

Doxiadis C. (1964). The ancient Greek city and the city of the present. *Ekistics*, , **18**(108), 346–364.

Hydria Project (2009). The Springs and Fountains of the Acropolis hill. http://www.hydriaproject.net/en/cases/athens/acropolis_hill/water_works.html (accessed 8 september 2011).

Koutsoyiannis D., Mamassis N., Efstratiadis A., Zarkadoulas N. and Markonis I. (2012). Floods in Greece, changes in flood risk in Europe. In: Z. W. Kundzewicz (ed.), IAHS Special Publication no. 10, IAHS Press, Wallingford, UK.

Koutsoyiannis D., Zarkadoulas N., Angelakis A. N. and Tchobanoglous G. (2008). Urban Water Management in Ancient Greece: Legacies and Lessons. *Journal of Water Resources Planning and Management, American Society of Civil Engineers*, **134**(1), 45–54.

Lang M. (1968). Waterworks in the Athenian Agora: Excavations of the Athenian Agora. Picture Book No.11, American School of Classical Studies at Athens, Princeton, NJ, USA.

Lezine-Velissaropoulou V. (2002). Architecture and Politics in Ancient Greece, The Experience of Archaic Tyranny. Kardamitsa Publications, Athens, Greece (in Greek).

Mamassis N. and Koutsoyiannis D. (2010). A web based information system for the inspection of the hydraulic works in Ancient Greece. In: Ancient Water Technologies, L. W. Mays (ed.), Springer, Dordrecht, Germany, pp. 103–114.

Zissimou T. (2007). Peiraeus, a walk on antiquity. The Port of Agony Citizen's Movement. Athens, Greece (in Greek).

Chapter 11

Sustainable water supply in pre-Columbian civilizations in Ancient Peru and South America

A. Reyes-Knoche

11.1 INTRODUCTION

In South America, especially in the Western Andean region, the catchment and secure provision of water has been, still is, and will remain a very demanding challenge due to specific geographic, hydrologic and climatic conditions. Ancient pre-Columbian civilizations achieved significant water supply and sanitation solutions and ensured sustainability in the same difficult natural habitats of today without the aid of modern resources of our times.

During the 4500 years of currently known South American pre-Columbian history several highly developed civilizations emerged mainly in the Western Coast and Andean regions. Around the year 1530 almost the total of their former territories, people and knowledge formed part of the Inca Empire "Tahuantinsuyo", the pre-Hispanic state with the largest extension in America, with a population of about 15 million inhabitants of different cultures and languages, a surface area of about 900,000 km^2, and a length of more than 5000 km, encompassing territories of the coast and the Andes of actual Peru, Bolivia, Ecuador, Colombia, Chile and Argentina.

But besides their unique administration, notable knowledge and remarkable achievements in many disciplines, probably the most outstanding success of the Incas of Peru was the eradication of hunger.

Along with presenting an overview of the most remarkable water supply and sanitation solutions achieved by ancient Peruvians under different geographic, hydrologic and climatic conditions, the present Chapter also depicts the framework and organisation under which pre-Columbian civilizations and especially the Incas succeeded in ensuring the subsistence of their whole population within the same living environments of today.

The present Chapter summarises information, impressions and experiences gathered during the last 25 years, from several journeys to the different archaeological sites cited, numerous visits to museums, revisions of diverse historical and archaeological literature, especially primary sources, discussions with archaeologists and other experts, and takes into account results of professional practice in water projects in developed and developing countries. It is the product of fascination and interest for pre-Columbian cultures and the Incas ever since youth combined with the conviction that many solutions opted for in the past are still applicable in the present.

11.2.2 Three drainage basins

The Peruvian rivers originate mainly in the Andes (Figure 11.2). They drain into three basins, the Pacific Ocean basin (283,600 km^2, 22%), the Atlantic Ocean basin (952,800 km^2, 74%) and the Lake Titicaca basin (48,800 km^2, 4%).

The rivers of the Pacific Ocean basin, 53 in total, are steep and short. Most of them carry water only during the rainy season and remain dry the rest of the year. Only a few of them keep significant flow rates along the year, such as the rivers Tumbes, Santa, Cañete and Majes.

The rivers of the Atlantic Ocean basin, 44 in total, are much longer, have a much larger flow, and are less steep once they exit the "*sierra*" (highlands). The main rivers of this basin are the Urubamba, Mantaro, Apurímac, Ucayali, Huallaga, Marañón, Napo, the Amazonas River, the largest river in the world by volume and length, and the Madre de Dios, Putumayo and Yavarí rivers.

The rivers of the Titicaca Lake basin, 9 in total, are generally short and non navigable, with small gradients and sparse flow variations. The main rivers of this basin are the Ramis, Illave and Coata rivers.

11.2.3 Climate

The Peruvian territory has a great climatic diversity due to the influence of the Andes and the cold Humboldt Current (Von Humboldt). The *costa* (coast) is characterised by moderate temperatures, low precipitations, and high humidity, except for its warmer, moist northern region. In the *sierra* (highlands), rain is frequent during summer months, and temperature and humidity diminishes with altitude up to the frozen peaks of the Andes. The *selva* (amazon forest) features heavy rainfall and high temperatures, except for its southernmost part, which has cold winters and seasonal rainfall. Table 11.1 shows meteorological data for representative cities.

Table 11.1 Altitude, average mean temperature and precipitation of cities in the Peruvian coast, Sierra and Selva. Own compilation based on data of SENAHMI, Lima, Peru.

Region	City	Altitude [m above sea level]	Average mean temperature [°C]	Average mean precipitation [mm]
Coast	Piura	55	24.4	72
	Chiclayo	34	21.0	62
	Trujillo	26	19.2	12
	Lima	30	19.2	15
Sierra	Arequipa	2508	15.4	99
	Cuzco	3249	12.5	736
	Chachapoyas	2435	15.3	796
Selva	Tingo María	665	24.8	3302
	Yurimaguas	184	26.9	2047
	Iquitos	126	26.2	2853

11.2.4 El Niño phenomenon

In Peru, the global coupled ocean-atmosphere phenomenon "*El Niño*", also known as "ENSO" (El Niño Southern Oscillation) exerts its influence especially in the northern coast.

Every year around December, the warm Equatorial Current descends towards the northern coast of Peru, replacing the cold Humboldt Current and originating thermal variations (sea water temperature and air temperature in the coastal zone rise, atmospheric pressure falls, winds become weak and the sea level augments). The more the warm current advances towards the south and the more time the warm current stays, the stronger are the effects experienced. In exceptional years this phenomenon causes heavy rains in the northern coast, generating river overflows and flooding, intense droughts in the southern sierra (particularly in the *altiplano*), plagues and diseases for certain crops, epidemics, and alters the marine and coastal ecosystems. The major economic damages occur in transportation (roads and bridges) and in agriculture (irrigation and drainage infrastructure, crop losses). However some positive effects are also experienced, such as the consumption of fish and seafood normally available only in the northern region, the presence of vegetation in the arid coast, the increment of water reservoir volumes, and the increase of groundwater levels.

The strength and consequences of the *El Niño* phenomenon are extremely variable. It is very difficult to establish clear patterns that allow reliable forecasts, similar to earthquakes. However, based on historical data the phenomenon can be classified into three categories: The "normal", the "very strong" and the "mega" *Niño*. The "normal" *Niño* appears on average every 3 to 4 years. The "very strong" *Niño* with strong temperature anomalies and heavy rains could appear on average every 9 to 12 years. The two last strong ones appeared in 1982–83 and 1997–98. The example of the River Piura illustrates the magnitude of *El Niño* effects especially in the northern coast of Peru. During the *El Niño* of 1998 the Piura river reached a flow rate of 4424 m^3/sec compared to the average annual flow rate of 29.1 m^3/s (SENAMHI). The "mega" *Niños* are believed to appear every 500 or 1000 years. "Mega" *Niños* could have been the reason for the sudden fall or rise of some pre-Columbian civilizations, such as the Mochica, Nazca, Tiahuanaco, Huari and/or the Chavín cultures.

Several attempts have been made and are currently been made in order to reconstruct climate changes and environment history and to somehow understand and predict climate behaviour as well as to find reasons for the rise and fall of Pre-Columbian civilizations. The scenarios are often linked to *El Niño* occurrences of different magnitudes. The investigations are supported by records of events in primary and secondary sources and/or by evidences of environmental changes such as glacial ice, alluvial sediments, presence of molluscs, and other climate-proxies (Caviedes, 2005; García-Herrera *et al.* 2008; Sandweiss *et al.* 1996, 2009; Ortloff, 2009; Eitel *et al.* 2009; Reindel *et al.* 2009, Mächtle *et al.* 2006, 2010).

11.2.5 Other natural occurrences

Peru is particularly exposed to earthquakes as the country is located in a seismic zone. The interface between the Nazca and the South American tectonic plates is located near the Peruvian coast. Minor earthquakes occur constantly. The most recent severe earthquakes took place in 1946, 1970, 2001, 2005, and 2007 with different epicentres (IGP).

Another very common natural hazard in the Andean Region is the so called "*huayco*" (debris avalanche/flash flood) and landslides caused by earthquakes and/or heavy rains occurring in the Andes heights. The heaviest contemporary "*huayco*" known occurred directly after the earthquake of 1970, when a huge glacial ice mass atop the Huascarán glacier was loosened, slid down and created an enormous avalanche burying the towns of Yungay and Ranrahirca, and causing major damage in other nearby localities.

11.3 INDICATIVE TIMELINE OF SELECTED SOUTH AMERICAN PRE-COLUMBIAN CIVILIZATIONS

An indicative timeline of the pre-Columbian civilizations mentioned in the present chapter is shown in Figure 11.3. The indicative timeline is intended to serve as a summarised quick reference guide. This timeline has been prepared and updated taking into account several timeline versions and/or chronology attempts from various archaeologists indicating differing dates. Due to the lack of written evidence the exact dates are mostly unknown. The timeline includes also actual discoveries, such as the Caral-Supe complex, Mochica and Wari culture archaeological sites. The shaded segments in Figure 11.3 indicate the most probable periods for each civilization (matching of indications in the various sources consulted). The dashed lines indicate further possible periods of existence (diverging indications in the various sources consulted). The Inca Empire "Tahuantinsuyo" existed only the last 94 years of the *ca.* 4000 years of the known pre-Columbian Peruvian history.

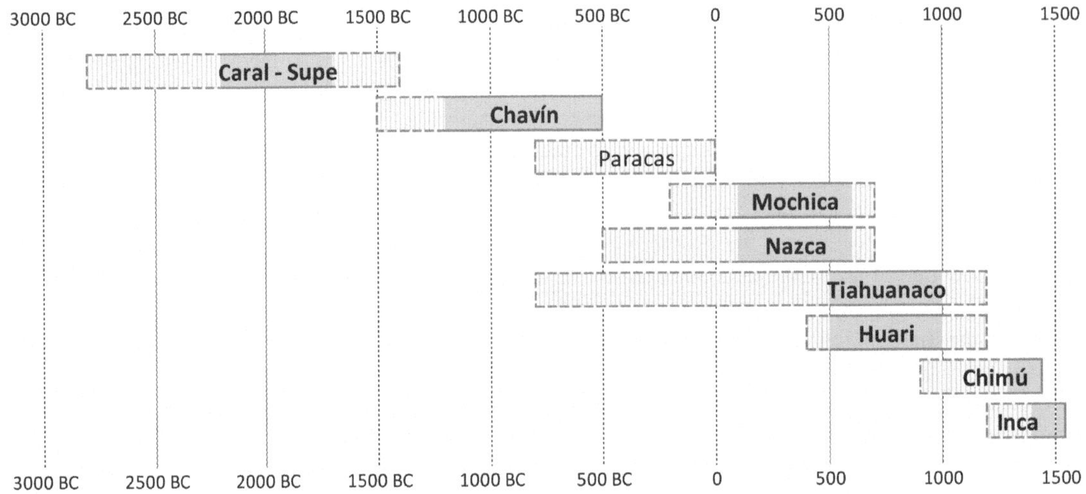

Figure 11.3 Indicative timeline representing the chronological appearance of the mentioned pre-Columbian civilizations within *ca.* 4500 years of known pre-Columbian history. The shaded segments indicate the most probable periods for each civilization. The dashed lines indicate further possible periods of existence

It is worth mentioning that several important discoveries have been made in recent times in Peru, such as the archaeological sites "*Caral-Supe sacred city*", "*Señor de Sipán*", "*El Brujo*", "*Cerro Pátapo*" and most recently, in February 2011, the "*Señor de Huari*" archaeological site. The results of the systematic investigations of these new archaeological finds are contributing to increase, update and sometimes even rectify the actual knowledge of the pre-Columbian civilizations.

11.4 THE INCA EMPIRE "TAHUANTINSUYO"

11.4.1 Extension, origin and accomplishments

The Inca Empire "*Tahuantinsuyo*" began its formation in about 1438, more than 500 years ago. At the arrival of the Spanish "conquistadores" in 1532 the Incas administered the pre-Hispanic state with largest extension in America, with a population of about 15 million inhabitants of different cultures and languages, a surface area of about 900,000 km^2, and a length of more than 5000 km, which is longer than the distance between the North Cape in Norway and Sicily. The northern limit was the river Ancasmayo, located north of today's city of San Juan de Pasto in Colombia (4° N), while the southern border was the River Maule in today's

Chile (36° S). It encompassed territories of the coast and the Andes of actual Colombia, Ecuador, Peru, Bolivia, Argentina and Chile (Figure 11.2).

Tahuantinsuyo means "the four regions" in Quechua, the official language of the Incas. The empire was divided into four regions (suyos), the *Chinchaysuyo* (NW), the *Antisuyo* (NE), the *Contisuyo* (SW) and the *Collasuyo* (SE). They met at the city of Cuzco, the administrative, political and military centre of the empire. Cuzco means "the navel of the world".

The origin of the Incas is dated around 1200 (see timeline in Figure 11.3), when, according to a legend Manco Cápac and Mama Occlo emerged from the Lake Titicaca to found the Inca dynasty. Until the creation of the "*Tahuantinsuyo*" in the middle of the 15th century the Incas were a regional chiefdom around Cuzco. Then the Incas accomplished their rapid expansion and consolidation.

Almost every contemporary civilization in the Andes and adjacent coastal regions was annexed or conquered within a little less than 100 years. The Incas assimilated all knowledge they encountered including the progresses in "water engineering", maintained and/or further developed the water and sanitation infrastructure of the conquered or annexed civilizations and built up new infrastructure with admirable and unique precision, some of them still operating today. The characteristic style the Incas implemented all over their territories was already noticed by the German naturalist and explorer Alexander von Humboldt, who wrote 'you could think that this huge amount of buildings must have been built by only one architect' (Von Humboldt). Some of the most significant and representative samples of Pre-Columbian water resources heritage of almost 4500 years of today's known development are duly presented further down.

11.4.1.1 Eradication of hunger

But besides all legacies in arts, sciences and engineering, the most prominent achievement of the Incas was the eradication of hunger, despite often adverse and challenging conditions (Section 11.2). This crucial success as well as all the progress and sustainable development achieved was only possible due to a certain all-embracing order implemented consequently with severe discipline.

For this reason, before the description of the ingenious water infrastructure and solutions conceived (Section 11.5), an outline of the institutional framework and organisation of the Inca Empire is being presented below. Main sources consulted were preferably primary written sources from chroniclers of the beginning of colonial times (Guaman Poma de Ayala, 1615; Garcilaso de la Vega, 1609, 1616; Cieza de León, 1553) as well as several contemporary literature based on the same primary sources (Cáceres Macedo, 1989; Kauffmann Doig, 1970; Lumbreras, 1988, 2003; Moseley, 1992; Palma, 1894; Rostworowski, 1988, 2009).

11.4.2 Institutional framework and organisation

11.4.2.1 Economy based on agriculture

The economy was based on agriculture. Hence water catchment, water distribution and the technical use of water was of utmost importance for the subsistence of the population, and for the stability and power of the empire. Despite the irregular distribution of water and the limited availability of arable land especially in the Andes the ancient Peruvians accomplished different adaptive measures. They assimilated all knowledge developed by their ancestors including the progresses in "water engineering". They maintained and/or further developed the water infrastructure of the conquered or annexed civilizations and also built up new infrastructure with admirable and unique precision, some of them still operating today. The performance, effectiveness, duration, accuracy, and mode of operation of their water infrastructure are now, even after hundreds of years, motive for admiration (Reyes-Knoche, 2009).

also in military campaigns. As long as enough males remained at home to farm the fields, the state was free to determine the numbers of conscripts to be mobilised and the length of service. Other types of work also existed. The "ayni" was a work form of mutual support during the agricultural labours. The "minca" was a work type realised in the lands of the sun, consisting mainly of road maintenance and repair and water channel cleaning. The mincas concluded with celebrations sponsored by the clergy, during which the food and beverages were taken out of the clergy storage.

11.4.2.5 Communitarian organisation

The families were organised in an "*ayllu*", a communitarian organisation unit supposing blood relationship of patri-lineal descendants. The state looked for an economic and political organisation permitting an effective control over the population and guaranteeing the work tribute and the mandatory works. This was accomplished by each "*purec*" (family chief) and each "*pachaca kamayoc*" also called "*curaca*" (chief of 100 families), who was practically the one responsible for one *ayllu*. This responsible role was elected every year during a festivity called "*camachicu*". The curacas were the intermediaries between the imperial hierarchy and the populace. Each five curacas had to report to a "*pisca pachac camayoc*" (500 purec), each two of those to a "*huranca camayoc*" (1,000 *purec*), and these again had to report to a "*unu camayoc*" (chief of 10,000 families), the governors of a province.

Four of those governors had a military chief called "*wamaní*", and all those belonging to a "*suyo*" (region) reported directly to a leader called "*suyuyoc apu*", the head of that *suyo*. As mentioned before, the *Tahuantinsuyo* consisted of four *suyos* (regions). Another Inca official was the "*tucuy ricuy*" (the one that sees and hears everything), the secret general inspector, who was delegated to the annexed regions.

11.4.2.6 Pragmatic justice system

The Incas applied the principle of reciprocity, "give and take" in modern terms, but with clear rules, disciplined implementation, and severe penalties.

Justice was based on three principles, "*ama sua*" (don't steal), "*ama qella*" (don't be lazy), and "*ama llulla*" (don't lie). Sanctions were executed drastically and without delay. Murder, robbery, adultery and any rebelliousness towards the authorities were also severely punished. On the other hand, the disabled or anyone not able to work were supported by the state.

11.4.2.7 Expansion policy

The expansion policy of the Incas envisaged first a peaceful attempt. An "ambassador" was sent to the monarch with generous presents and the offer to become part of the empire under the rules and protection of the Inca. The monarch and his relatives would become part of the Inca nobility. Only if this pacific mission failed were military actions with well organised and strong disciplined soldiers taken. The leaders of the conquered would then become part of the Inca populace.

In both cases, expansion via peaceful persuasion or by military enforcement, the Incas introduced their institutional framework and organisation in the new territories, leaving the functioning local structures intact. They assimilated all knowledge they encountered including the progresses in "water engineering". They maintained and/or further developed the water infrastructure of the annexed civilizations, such as the "*andenes*" (terraces) for agriculture.

11.4.2.8 Road network and communications

The Incas maintained a road network of at least 23,000 km over some of the earth's most uneven terrains, connecting territories of today's Peru, Bolivia, Ecuador, Colombia, Chile and Argentina ("*Qhapac Ñam*",

Great Inka Trail, inscription process in UNESCO's World Heritage List currently under way). The "*Qhapac Ñam*" served for transport, communication and administration. Many Inca roads are intact and in excellent condition. Some reach heights of 5000 m above sea level. They include rope suspension bridges and tunnels to overcome natural obstacles. The roads were mostly wide in the coast and often narrow and steep in the Andes. Every 2 to 15 km, at key points along the road, shelter and rest houses ("*tambos*") with food and water were located. The "*tambos*" served also as relay stations for the "*chasqui*", relay messenger runners positioned all over the empire. The "*chasqui*" carried a spoken message, a "*pututo*" (horn made of a conch), a "*quipu*" (string and knot recording device) and a "*qipi*" (backpack) to transport packages (Figure 11.5). Chasquis would start at one "*tambo*", and run to the next "*tambo*" where a rested *chasqui* was waiting to carry the message to the next tambo. A message could be delivered from Cuzco to Quito within a week (Guamán Poma de Ayala, 1587–1615; Garcilaso de la Vega, 1609, 1616; Lumbreras, 2003; Hyslop, 1990).

Despite several assertions in many archaeological writings it is difficult to believe that the Incas and other ancient Peruvians didn't know the wheel. They knew the sphere and the circle, since both can be found in many artefacts, sculptures, ceramics and jewellery. Most likely the Incas deliberately didn't apply the wheel by reason of respect towards the "*Inti*", the sun goddess, for whom the circular form was reserved.

It is believed that the Incas didn't have a scripture, since no written Inca document containing phonetic symbols has been found. The Inca recording device known is the "*quipu*" (Figure 11.5). Quipus contain within their manifold coloured strings and knots mostly coded numbers in a decimal system. However, the colour, strings and knots codes have not been completely deciphered yet. The *quipus* could only be read by trained persons with good memory and intellect, the "*quipu camayoc*" and the Inca nobility. The investigation of the *quipus* is still under way.

11.5 THE PRE-COLUMBIAN WATER RESOURCES HERITAGE

The present section contains a brief description of noteworthy samples of Pre-Columbian water solutions, sorted by civilizations and following the chronology presented in Section 11.3. Each and every civilization developed special solutions for their particular environments and needs themselves. The reader will find key data obtained from sources cited and from personal on-site observations, as well as carefully prepared and selected illustrations. Additionally, a quick spatial overview of the sites named can be obtained with the indicated geographical coordinates and the aid of online geographical tools such as Google Earth.

11.5.1 Caral-Supe

Caral-Supe is the oldest known civilization in the Americas (*ca.* 5000 years, see timeline in Figure 11.3). The archaeological complex of Caral-Supe, included in 2009 in UNESCO's World Heritage List is situated in the Supe valley north of Lima about 16 km away from the coastline (10°53′28.82″S 77° 31′04.32″W), and consists of several urban settlements along the valley. The settlements include pyramids, public buildings, circular plazas, residential compounds and domiciles.

The first to identify and document settlements in the Supe Valley was the German archaeologist Max Uhle, who in 1905 explored the vestiges in Áspero, about 20 km from the Caral-Supe archaeological complex. The Peruvian archaeologist Julio C. Tello, explored the same place by the end of the 1930s. Later on, the North-American researcher Paul Kosok identified and documented a series of sites in the middle Supe valley including a "ceremonial complex" located close to the Hacienda Chupa Cigarro Grande (Kosok, 1965, pp. 222–223). This site happened to be what is today known as "Sacred City of Caral" which has been systematically excavated and studied together with many other sites identified in the Supe-Valley under the direction of the Peruvian archaeologist Ruth Shady Solís.

The aqueduct has a total length of 9 km and consists of three sections. The first two sections are situated in the occidental side of the Andes. The aqueduct starts with the water catchment located at 3513 m altitude. In this first section the aqueduct is completely carved in volcanic rock and has a length of 850 m. The second section is 2600 m long and is built along the mountain side. It ends at the watershed at about 3350 m altitude. The third section is situated in the oriental side of the Andes and has a length of 5650 m, ending at a reservoir at the foot of the Santa Apolonia Mountain in the city of Cajamarca.

About 25 years ago, during an expedition along the first section of the aqueduct carved in volcanic rock, the author of the present section could find very close to one narrow border of the channel section, hidden behind dense vegetation, one almost vertical cavity of enough size to "swallow" a full-grown youngster, by slipping involuntarily into it and disappearing almost instantaneously and silently for the surprise of himself and the other members of the expedition.

11.5.3 Mochica

The Mochica or Moche culture developed around several valleys on the north coast of Peru (see timeline in Figure 11.3). During the last 20 years, since the discovery of the "*Señor de Sipán*" at "*Huaca Rajada*" (6°47′42″S 79°36′04″W), several discoveries and systematic archaeological on-site investigations have been and still are being made. The results help to continuously increase the knowledge about the Mochica culture. The main legacy of the Mochica is their ceramics detail depicting objects, plants, animals, human portraits and erotic scenes, their filigree gold and silver ornaments, monumental pyramidal constructions of adobe bricks, and their irrigation channels and regional water resource management schemes. Monumental adobe brick constructions of the Mochica worth mentioning here are the pyramids of Túcume (6°30′57″S 79°50′39″W), first identified and documented by the German engineer Brüning (Brüning; Raddatz, 1990), the Huaca del Sol and Huaca de la Luna, as well as the Batán Grande, El Brujo and Huaca Rajada archaeological sites, among many other sites along the northern coast of Peru.

The Mochica culture was based on agriculture. It is believed that the Mochica cultivated about 35% more land than is cultivated nowadays in their former territories. The main agricultural products of the Mochica were corn (*maíz*), chirimoya, guanábana, avocado, beans, peanuts, coca, cacao, manioc (*yuca*), papaya, lúcuma, sweet potato, Peruvian chilli (*ají*), tomato, potato, pepino, caigua, squash, loche (pre-Columbian pumpkin), among others, all depicted in ceramics, pictographies, sculptures and/or textiles (Larco Hoyle, 1938, 1940, 1978, and 2001).

11.5.3.1 Irrigation systems

In order to assure the supply of water needed for irrigation the Mochica diverted river waters into vast networks of irrigation channels. Examples of such hydraulic channel works are the Raca Rumi and the Taymi channels that divert water from the Chancay-Lambayeque River, irrigating agricultural areas of the Lambayeque valley and the Leche valley. Both channels are still in use, rehabilitated, slightly modified or built-up following the original channel traces. They are part of the major irrigation system Tinajones in operation since 1968 and co-financed by KfW Entwicklungsbank, the German Development Bank (Figure 11.8). The Taymi channel also supplied water to several settlements such as the ancient Moche pyramid city of Túcume (Reyes Carrasco, 2009) (6°40′43″S 79°49′29″W). The city of Chiclayo, the only major Peruvian city which was not founded by the Spanish conquistadores, is still supplied today with water taken from the Taymi channel. The Talambo-Zaña channel, a component of the major irrigation scheme Jequetepeque-Zaña, co-financed as well by the German KfW Entwicklungsbank, also follows the routes and/or sections of a main pre-Columbian Mochica channel.

culture, and their beautiful polychrome pottery are stupendous. They reclaimed the desert with irrigation channels and underground aqueducts, some of which are still in operation today. The main agricultural products were maize, squash, sweet potatoes and manioc. The Nazca people were also the creators of the Nazca lines, included in 1994 in UNESCO's World Heritage List, a series of enigmatic figures scratched in the ground, and visible as a whole only from the air (Figure 11.11).

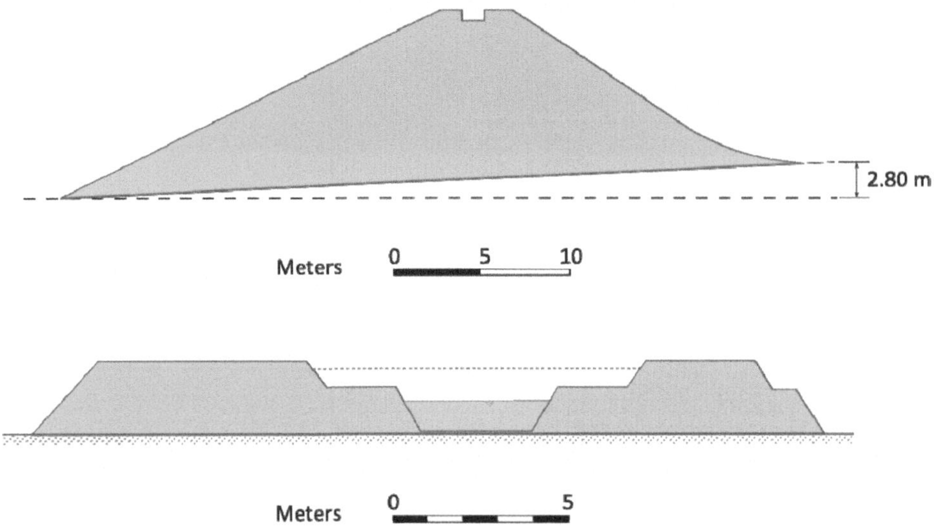

Figure 11.9 Representative cross section of a Mochica aqueduct (Aqueduct of Áscope, top, nearly rectangular cross-section) and of a Mochica channel (Vichansao Channel, bottom, double trapezoidal compound channel cross-section). Sketches drawn based on Larco Hoyle (1938, 1940, and 2001)

11.5.4.1 Underground aqueducts

Underground galleries ("*puquios*"), constructed by the Ancient Nazcas covered with stones, flagstones and/or Huarango-tree trunks collect and conduct underground water to the arid zones of the Aja, Tierras Blancas and Nazca valleys. Openings located along the conduction line permit the access into the deeper underground galleries for maintenance and cleaning works through spiral formed descending ramps, serving also as wells (Figure 11.10). The distances between the openings depend on the terrain characteristics and vary from 10–20 m to about 180 m. The length of the underground aqueducts varies from about 500 m to about 1500 m, with depths in the order of 10–20 m below ground. The tunnels are quite narrow with an average cross-section of approximately one square metre. The lower aqueduct sections are mostly uncovered trapezoidal channels with steep curvilinear sidewalls of cobbled stones. The water arrives mostly in reservoirs ("cochas") from which the water is distributed via open irrigation channels mainly for agricultural purposes. The Nazca constructed several similar underground aqueduct systems. From the 46 aqueducts known today 32 are currently in operation (Schreiber & Lancho, 2006). The most commonly known and one of the best conserved aqueduct is the "puqio" of "Cantalloc" (Figure 11.10).

Although the exact origins of the aqueducts remain unknown, a legend narrated by the Peruvian writer Ricardo Palma known as "La Achirana del Inca" ("achirana" meaning what neatly flows towards the beautiful) gives a possible hint around the construction of the aqueducts in the Nazca Region (Palma, 1894).

11.5.4.2 Nazca lines

Even though several colleagues have studied and still are studying the Nazca lines (14°41′18″S 75°7′23″W), the real purpose remains a mystery. However, the lines are somehow linked with the movement of the orbs. For example, one line points to the rising sun on the summer solstice. The Nazca, being mainly an agricultural society, had the necessity of a precise calendar, especially because of their harsh habitat. The Nazca lines (Figure 11.11) could have served as a calendar to determine the seasons, the right time to plough, seed and irrigate (Reiche, 1947). The archaeological site of Cahuachi, considered to have been a major ceremonial centre of the Nazca culture, is located close to the Nazca lines (14°49′7″S 75°7′0″W).

11.5.5 Pre-Columbian surveying instrument

In order to be able to build paths, aqueducts and irrigation channels with the precision stated, the pre-Columbian master builders must have had the aid of a levelling instrument. The pre-Columbian surveyor's instrument identified and described by Reyes Carrasco (1980) is considered to have served similar purposes as today's levelling instruments and theodolites, particularly for tracing slopes and for on-site alignments. Even though the instrument in question has been previously catalogued as a musical instrument the true purpose of the apparatus is quite clear for the hydraulic-engineering-versed colleague.

The ceramic device consists of two fundamental parts, the upper component "A", similar to a soup bowl, followed by a hollow cylindrical thin walled body "B" fixed over another component similar to "A" (Figure 11.12). The part "B" shows on one side two thin perpendicular incisions 1-1 and 2-2, and diametrically opposed to them two incisions of the same characteristics. Both crosses are located in such way, that they coincide perfectly from one side to the other. The straight lines 2-2 and 2′-2′ are contained in the same plane "C", and the straight lines 1-1 and 1′-1′ are contained in the plane "D", perpendicular to the plane "C".

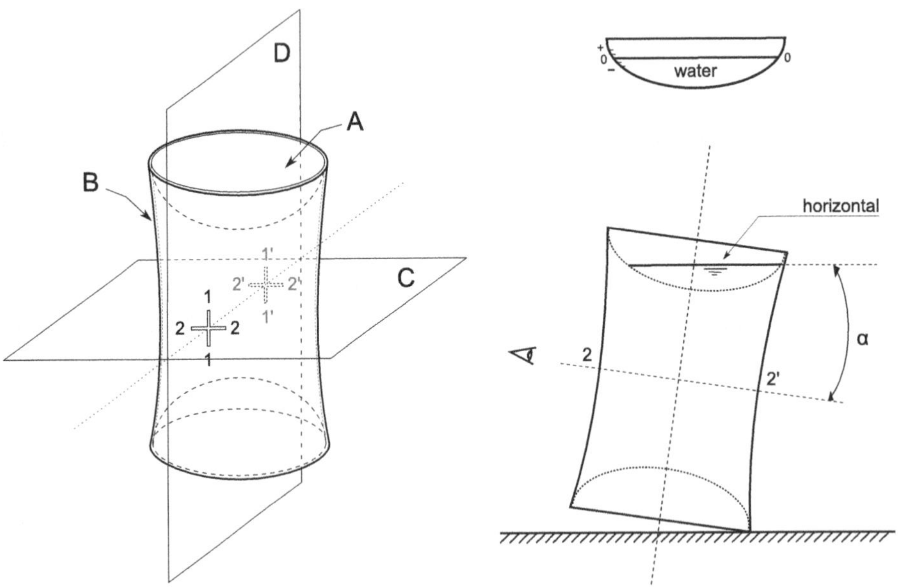

Figure 11.12 Pre-Columbian surveying instrument, precursor of today's surveyor levels and theodolites (Drawn after Reyes Carrasco, 1980, 2005)

waru" (Figure 11.13). The raised fields surrounded by ditches filled with water retain the sun's heat during frost and cold nights keeping the crops from freezing, leach out salts and catch organic material, silt and algae that can be used as fertiliser. Today's local farmers still continue using the "*waru waru*" technique. The raised fields are mostly 2–10 m wide by 10–20 m long and 0.4–0.80 m high. These dimensions can vary depending on the terrain slope and groundwater level (Canahua Murillo, 1992).

11.5.6.2 Huari, terraces

During their period of prosperity the also highly developed Huari state was twice as large as the Tiahuanaco state, covering a territory over 1500 km in length ranging from the southern Andes to the northern Coast and Andes of Peru. The centre of the Huari state was the city of Wari, situated close to today's city of Ayacucho. The Huari established administrative centres, connected cities and military tags with roads, developed new regions through resettlements, and built terraces ("*andenes*") for agriculture. It is believed that a great portion of roads and terraces used during the Inca epoch and attributed to the Incas were already constructed by their Huari predecessors.

It is presumed that a long period of drought related to strong ENSO occurrences broke the supremacy of the Tiahuanaco and Huari states (Section 11.2.4). The Huari are also believed to have been the strong warrior nation of the "Chanka" mentioned by several chroniclers as archrivals of the Incas during their early days.

Vestiges of the Huari civilization can be found all along their former territories. Worth mentioning are the archaeological sites of "*Cerro Pátapo*", 23 km from the city of Chiclayo in the northern coast of Peru (6°43′17″S 79°38′41″W), the archaeological site of "*Pikillacta*", 20 km east of Cuzco in the southern Andes (13°37′00″S 71°42′53″W), and also the archaeological site of "*Espiritu Pampa*" located in the upper region of the Amazon forest of the department of Cuzco (12°54′08″S 73°12′27″W), home of the recently discovered "*Señor de Huari*".

11.5.7 Chimú

The origin of the Chimú culture lies most probably in the Mochica culture. The Chimú occupied the same territory the Mochica inhabited centuries before (see timeline in Figure 11.3). Their territory ranged 1300 km along the mostly arid coast from today's Tumbes in the north to the city of Lima, the capital of Peru. Their capital was the city of Chan Chan, the biggest abode brick city of the Americas. The Chimú were also versed gold and silver smiths, weavers, and fishermen. Like the Moche, in order to satisfy their water demand, they designed, built and operated suitable hydraulic infrastructure for irrigation and water supply. Some of the ancient irrigation channels built by the Mochica and Chimú are still in use or are incorporated into regional irrigation schemes. Prominent examples are among others the irrigation scheme of Tinajones (Section 11.5.4) and the Chavimochic irrigation scheme in the Department of La Libertad, connecting the valleys of the Chao, Virú, Moche and Chicama Rivers with a main channel transporting water from the Santa River, a similar concept already applied with success by the Mochica and Chimú master builders.

11.5.7.1 Chan Chan

The ancient city of Chan Chan, included in 1986 in the UNESCO's World Heritage in Danger List, is located close to the Pacific Ocean, 4 km west of the city of Trujillo, on the way to the airport. Chan Chan covered an area of 18–20 km^2 and accommodated about 100,000 inhabitants. Today's archaeological site (8°6′31″S 79°4′30″W) encompasses nine walled palaces, each supposed to be owned by a king ("cacique"). A plan view and cross-section of a water reservoir ("*wachaque*") located inside the Tschudi palace is shown in Figure 11.14. The Tschudi palace measures 480 × 455 m. The adobe outer walls are up to 7 m high. Chan Chan's water sources were groundwater and water transported by open channels.

11.5.9.1 Terraces

The "andenes" (terraces) were constructed on hillsides of different slopes, from the bottom to the top, becoming smaller on the top of the mountains. The "*anden*" (terrace) consists of three stone walls, one frontal and two laterals. The width and height depend from the gradient of the mountain. In steep gradients the andenes have a width of 9.80 to 1.70 m. The normal width is about 1.80 m to 4.50 m, and sometimes the width can be between 15 and 60 m.

The terraces were separated by gangways to permit the access, and drainage channels that drained the water from the top of the mountain to the bottom without washing out the soils. The *andenes* were filled up with soils of good quality carried in baskets by workers. Flagstones were fixed on accessible front-sides of the walls as steps arranged on an angle of about 30°.

With the terraces a notable increase in the area of cultivable land was achieved. In ancient Peru several hundred thousand hectares of cultivable land were made available with terraces. The andenes prevented erosion. They also permitted the growth of different types of crops at the appropriate altitude and climatic conditions. A partial view to the andenes of Pisac (13°24′28.86″S 71°50′33.02″W) in the sacred valley of the Incas nearby Cuzco, and a detailed view of the andenes in Machu Picchu are shown in Figure 11.16. Also, a panoramic view of the archaeological site of Machu Picchu (13°9′48″S 72°32′46″W, UNESCO's World Heritage List inclusion in 1983), including the *andenes* that permitted the self-sustenance of the inhabitants is presented in Figure 11.17. More andenes can be found all along the Andes, in the Colca Valley (15°41′42.05″S 72°06′15.37″W), in Ollantaytambo (13°15′24.55″S 72°15′55.89″W), Moray (13° 19′48.87″S 72°11′48.02″W), Tipón (13°33′34.17″S 71°48′13.67″W), Andamarca (14°23′28.33″S 73° 57′24.43″W), Laraos (12°21′23.51″S 75°47′28.32″W), and other places in the Departments of Arequipa, Apurímac, Cusco, Ica, Lima, Moquegua, Puno and Tacna. According to an inventory (INRENA, 1996) the area covered by *andenes* within the Peruvian territory totals 256,945 ha. The actual project "Terraces Recuperation in the Andes", co-financed by the Inter-American Development Bank, aims to reconstruct pre-Columbian terraces as a practical way to execute adaptation projects in the field, in the scope of Andean indigenous communities impacted by Global Warming (IDB, 2010). The "Plan MERISS Inka" is another example of bilateral cooperation efforts aiming to rehabilitate and improve irrigation practices in the Andean region. The program has been co-financed by the German KfW Entwicklungsbank since 1980 and has also been executed with the support of the German Technical Cooperation GTZ (today's GIZ). World Bank's recent "Sierra Irrigation Project" (P104760) development objective is to contribute to increasing agricultural production and productivity in targeted areas of the Sierra (World Bank, 2010).

11.5.10 Water cult/culture and mythology

From the very beginning ancient Peruvian worshiped the "God of Water". Given that climatic changes were recurrent and negatively influenced the production of foodstuffs, the God of Water developed into the most eminent supernatural entity within the magic-religious structure of ancient Peru. Up to the present time it has been depicted in the iconography and is the permanent object of worship (Kauffmann Doig, 2002).

Two samples for today's rituals and celebrations around water are the "*Pachamama Raymi*" in Cuzco and the "*Yaku Raymi*" in Ayacucho. Both have their origin in ancient Inca rituals and celebrations. During the "*Pachamama Raymi*" a special homage is rendered to the "*pachamama*" (mother earth), the "*pagapu*", "payment/offering to the earth". This rite marks the beginning of the Andean year. The "*Yaku Raymi*" (celebration of the water) consists of the cleaning of the irrigation channels to invoke the rain and in the ceremonies of "*pagapu*" (payment/offering to the earth), carried out in between songs and dancing performances.

11.6 CONCLUSIONS AND FINAL COMMENTS

Pre-Columbian water resources heritage of Peru and South America is the result of *ca.* 4500 years of sustainable development and successful adaptation to challenging environments.

The catchment and secure provision of water in Peru and South America, especially in the western coast and Andean region, has been, still is, and will continue to be an outstanding challenge. The ancient Pre-Columbians knew how to adapt to the specific and different conditions of their living spaces. They achieved extraordinary water supply and sanitation solutions, ensuring sustainability in the same difficult natural habitat of today and without the aid of the modern resources of our times.

The Inca Empire "*Tahuantinsuyo*" existed for only 100 years, but they inherited and applied the knowledge gained by almost all other pre-Columbian cultures during *ca.* 4500 years of known development. The level of civilization, eradication of hunger and sustainability achieved was only possible with the consequent implementation of their institutional framework and organisation, strong leadership and skilled experts and workers.

The "master builders" of ancient Peru must have done integrated strategic planning and consequent implementation, both after an integral assessment of challenges and opportunities.

The architecture of the Incas seems to melt with the geographic surrounding, and is well integrated into the natural water cycle.

In today's Peru the challenge to achieve the right balance between water availability and water demand is practically the same, since despite the effects of global warming the manifold living environments are practically almost the same.

Today's Peruvian water sector is characterised by increasing water demand and competing interests especially in the highly populated and mostly arid coastal region, endangered water quality (e.g. from the mining industry), poorly performing irrigation, deficient institutional capacity, insufficient water supply and hygiene, and vulnerability to climate change including extreme climatic conditions and glacier melting. In addition to that, it is foreseeable that the Millennium Development Goal related to eradication of hunger will not be achieved by the due date in 2015.

Nevertheless, the water supply and sanitation solutions implemented in the past can be considered more than just starting points for new solutions. It is difficult to find reasons why not to apply "best practice" and proved systems in the same environment where they were already developed and successfully implemented.

REFERENCES

ANA, Autoridad Nacional del Agua (2011). Water resources management data, www.ana.gob.pe, Lima, Perú (last accessed 20 September 2011) (Website contents in Spanish).

Brüning, H. H. and Raddatz C. (eds) (1990). *Fotodoukmente aus Nordperu von Hans Heinrich Brüning (1848–1928)* (Photographic documents of Northern Peru, Hans Heinrich Brüning, 1848–1928) Hamburgisches Völkermuseum, Hamburg, Germany (in German).

Burger, R. L. (1992, 1995). Chavin and the origins of Andean civilization. Thames and Hudson Ltd, London, United Kingdom, p. 248.

Cáceres, Macedo J. (1989). *Las Culturas Prehispanicas del Perú* (The Pre-Hispanic Cultures of Peru), Concytec, Lima, Perú, pp. 116.

Canahua Murillo, A. and Larico Mamani, L. (1992). *Manual técnico de Waru Waru para profesionales y técnicos* (Technical Handbook of Waru Waru for professionals and technicians), PELT/INADE-IC/COTESU, Puno, Perú (in Spanish).

Caviedes, C. N. (2005). *El Niño. Klima macht Geschichte* (El Niño in History. Storming through the Ages), Wissenschaftliche Buchgesellschaft, Germany, pp. 167 (in German).

Cieza de León (1553). Crónica del Perú, *El Señorío de los Incas* (Chronicle of Peru, the chiefdom of the Incas), Sevilla, Spain / Ed. Fundación Biblioteca Ayacucho, 2005, Caracas, Venezuela, pp. 544 (in Spanish).

Eitel, B. and Mächtle, B. (2009). Man and environment in the eastern Atacama Desert (Southern Peru): Holocene climate changes and their impact on Pre-Columbian cultures. In: New Technologies for Archaeology, M. Reindel and G. A. Wagner (Hrsg.) Springer Verlag, Berlin, Heidelberg, pp. 17–37.

Firenze Mostre S.p.A. (2003). *Perù, tremila anni di capolavori* (Peru, three thousand years of master works), Mondadori Electa S.p.A. Milano, Italy, p. 319 (in Italian).

García-Herrera R., Díaz H. F., García R. R., Prieto M. R., Barriopedro D., Moyano R. and Hernández E. (2008). A chronology of El Niño events from primary documentary sources in Northern Peru. *Journal of Climate*, American Meteorological Society, **21**, 1948–1962.

Garcilaso de la Vega I. (1609, 1616). *Comentarios Reales de los Incas* (Commentaries of the Incas), Lisboa, Portugal (in Spanish).

Graves, C. (ed.) (2001) *La papa, Tesoro de los Andes* (The Potato, Treasure of the Andes), International Potato Center, Lima, Peru, pp. 212.

Grewenig, M. M. (2004). *Inka Gold, 3000 Jahre Hochkulturen, Meisterwerke aus dem Larco Museum Peru.* (3000 years of remarkable Cultures, master pieces of the Larco Museum), Kehrer Verlag Heidelberg, Germany, pp. 224 (in German).

Guamán Poma de Ayala, F. (1587–1615). *La Nueva Corónica y buen Gobierno* (The new chronicle and good governance), pp. 1189 (in Spanish).

Guidoni, E. and Magni, R. (1972, 1987). *Monumente großer Kulturen – Inka* (Monuments of great cultures – Inca), Karl Müller Verlag, Erlangen, Germany (in German).

Heyerdahl, T. (1995). Pyramids of Túcume: The Quest for Peru's Forgotten City, Thames & Hudson Ltd, New York, USA, pp. 240.

Hyslop, J. (1984). The Inka Road System, Institute of Andean Research, USA. Academic Press, Inc., New York.

IDB, Inter-American Development Bank (2010). Terraces Recuperation in the Andes (PE-T1165), Convenio de Cooperación Técnica (Technical Cooperation Agreement), Washington, USA (in Spanish).

IGN, Instituto Geográfico Nacional (1989). *Atlas del Perú* (Atlas of Peru). Instituto Geográfico Nacional, Lima, Perú (in Spanish).

IGP, Instituto Geofísico del Perú. Seismological and Geophysical Data, Lima, Perú, www.igp.gob.pe (last accessed 20, September 2011) (Website contents in Spanish).

INRENA, Instituto Nacional de Recursos Naturales (1996). *Informe técnico del estudio de inventario y evaluación de Andenes* (Report on the inventory and evaluation of andenes), Ministerio de Agricultura, Lima, Peru (in Spanish).

INRENA, Instituto Nacional de Recursos Naturales (2009). *Política y Estrategia Nacional de Recursos Hídricos del Perú* (National water resources policy and strategy of Peru), Comisión Técnica Multisectorial, Lima, Perú (in Spanish).

Kauffmann Doig, F. (1970). *Arqueología Peruana, Visión Integral* (Peruvian archaeology, integral vision), Promoción Editorial Inca, S. A., Lima, Perú, pp. 641 (in Spanish).

Kauffmann Doig, F. (2002). *Los Dioses Andinos: Dioses del sustento* (Andean gods: Gods of Sustenance), Precolombart, Boletín de la Asociación de Amigos del Museo Barbier-Mueller, Barcelona, España, 4–5 (in Spanish and English).

Kosok, Paul (1965). Life, Land and Water in Ancient Peru, Long Island University Press, Brooklin, New York.

KfW Entwicklungsbank (2009). Schlüsselzahlen Wasser &Klima 2009. Key figures for water and climate, Kompetenzzentrum Wasser und Abfallwirtschaft, KfW Entwicklungsbank, Frankfurt/Main, Germany (in Germany).

KfW Entwicklungsbank (2009). *10. Bericht über die Evaluierung der Projekte and Programme in Entwicklung sländern 2006–2008* (10th report on evaluation of projects and programs in developing countries), KfW Entwicklungsbank, Frankfurt/Main, Germany (in German).

Larco Hoyle, R. (2001). *Los Mochicas I & II.* Revised version of the original editions of 1938 and 1940 supplemented with unpublished manuscripts of the author, Museo Arqueológico Larco Herrera, Lima, Peru (in Spanish).

Larco Hoyle, R. (1938, 1940). *Los Mochicas I & II.* Lima, Peru (in Spanish).

Larco Hoyle, R. (1978). *Peru Checan – Studie über die erotischen Darstellungen in der peruanischen Kunst* (Peru Checan – Study of erotic representations in the Peruvian Art), Wilhelm Heyne Verlag, München (in German).

Lumbreras, L. G. (1988). De los orígenes de la civilización en el Perú (About the origins of the civilization in Peru), Peisa, Lima, Perú, pp. 143 (in Spanish).

Lumbreras, L. G. (2003). *Qhapaq Ñam – El camino de los Incas* (Qhapaq Ñam – The path of the Incas), Chasqui, Año 1 número 1, Boletín Cultural del Ministerio de Relaciones Exteriores, Lima, Peru (in Spanish).

Mächtle, B., Eitel, B., Kadereit, A. and Unkel I. (2006). Holocene environmental changes in the northern Atacama desert, southern Peru (14°30'S) and their impact on the rise and fall of pre-Columbian cultures. In: Eitel, B. (ed.) Holocene landscape development and geoarchaeological research, Gebrüder Borntraeger, Berlin, Germany, pp. 47–62

Mächtle B., Unkel I., Eitel B., Kromer B. and Schiegl S. (2010). Molluscs as evidence for a Late Pleistocene and Early Holocene humid period in the northern Atacama desert, southern Peru (14.5°S), Quaternary Research, **73**, 39–47.

Moseley, M. E. (1992). The Incas and their ancestors, the archaeology of Peru, Thames and Hudson Ltd., London, UK, pp. 272.

Museo Oro del Perú (1980). *Oro del Perú* (Gold of Peru), Museo de Oro, Lima, Perú, p. 107 (in Spanish).

Oficina Nacional de Evaluación de Recursos Naturales (ONERN, today INRENA) (1978). *Guía de Información Cartográfica y de Recursos Naturales del Perú* (Cartographic and natural resources information guide), ONERN, Lima, Perú (in Spanish).

Ortloff C. R. (ed.) (2009). Water engineering in the ancient world. Archaeological and climate perspectives on societies of ancient South America, the Middle East and South-East Asia, Oxford University Press, New York, USA, pp. 440.

Palma, R. (1894). Tradiciones Peruanas - Tercera serie, La achirana del Inca (Peruvian Traditions – Third series, The "achirana" of the Inca), Montaner y Simón, Barcelona, Spain (in Spanish).

Plate, E. J. and Merz B. (eds) (2001). *Naturkatastrophen. Ursachen, Auswirkung, Vorsorge* (Natural disasters. Causes, impacts, provision), E. Schweizerbart'sche Verlagsbuchhandlung, Stuttgart, Germany, pp. 475 (in German).

Petersen, G. (1967). *Cumbemayo: Acueducto arqueológico que cruza la divisoria continental* (Cumbemayo: Archaeological aqueduct crossing the continental watershed), Tecnia, n. 3, pp. 112–139, UNI, Lima, Perú (in Spanish).

Regal, A. (1970). *Los trabajos hidráulicos del Inca en el antiguo Perú* (The Hydraulic Works of the Inca in Ancient Peru). Lima, Perú, p. 142 (in Spanish).

Reiche, M. (1947). *Geheimnis der Wüste* (Secrets of the desert). Nazca, Peru, p. 93 (in German, Spanish and English).

Reindel, M. and Wagner, G. A. (eds) (2009). New Technologies for Archaeology, Multidisciplinary Investigations in Palpa and Nasca, Peru, Springer Verlag Berlin Heidelberg, p. 512.

Reyes Carrasco, L. V. (1973). *Análisis hidráulico del sitio arqueológico de Chavín de Huantar* (Hydraulic analysis of the archaeological site of Chavín de Huantar), Universidad Nacional de Ingeniería, Lima, Perú (in Spanish).

Reyes Carrasco, L. V. (1980). *Reflexiones sobre un dispositivo precolombino, precursor de los niveles y teodolitos actuales* (Pre–Columbian device, precursor of actual levelling instruments and theodolites), III Congreso Nacional de Ingeniería Civil, Cuzco, Perú (in Spanish).

Reyes Carrasco, L. V. (2005). *Mecánica de Fluidos, Hidráulica y Diseños Hidráulicos, Tomo I & Tomo II* (Fluid Mechanics, Hydraulics and Hydraulic Designs, Volume I and II). Chiclayo, Perú, p. 410 and 413 (in Spanish).

Reyes Carrasco, L. V. (2009). *Sobre la alimentación de agua de las poblaciones que habitaron las pirámides de Túcume* (About the water supply of the former populations around the Pyramids of Túcume), XVII Congreso Nacional de Ingeniería Civil, Chiclayo, Perú (in Spanish).

Reyes-Knoche A. (1991). *Traditionelle Bewässerung am Beispiel des alten Perus* (Traditional Irrigation in Ancient Peru), Seminar "Ökologische und soziale Folgen von Stauanlagen", Institut für Hydromechanik, University of Karlsruhe (today's KIT Karlsruhe Institute of Technology), Germany.

Reyes-Knoche A. (2009). Water supply and sanitation in ancient Peru: Pre-Columbian cultures and the Incas, 2nd IWA symposium on Water Technologies in Ancient Civilizations, Bari, Italy.

Reyes-Knoche A. (2009). Water solutions in ancient Peru – Overview and options to solve today's and future water challenges, 1st IWA International Development Congress, Mexico City, Mexico.

Reyes-Knoche A. (2011). *Nachhaltige Wasserwirtschaft und Anpassung an den Klimawandel im Präkolumbischen und heutigen Peru – Überblick und Rekonstruktion als Beitrag zur Lösung heutiger und zukünftiger Herausforderungen am Beispiel des Siedlungsraumes der Mochica in Nordperu* (Sustainable water management and adaptation to climate change in Precolumbian and today's Peru). Synopsis of Ph.D. Thesis, University of Heidelberg, Germany (in German).

Rostworowski de Diez Canseco, M. (1988, 2009) *Historia del Tahuantinsuyo* (History of the Tahuantinsuyo), IEP Instituto de Estudios Peruanos, Lima, Peru, pp. 359 (in Spanish).

Sandweiss, D. H., Richardson, J. B., Reitz, E. J., Rollins, H. B. and Maasch, K. A. (1996). Geoarchaeological Evidence from Peru for a 5000 Years B.P. Onset of El Niño, *Science*, **273**, 1531–1533.

Sandweiss, D. H. and Quilter, J. (2008). *El niño, catastrophism, and culture change in ancient America*, Dumbarton Oaks Trustees for Harvard University, Harvard University Press, USA, pp. 290.

Sandweiss, D. H., Shady Solís, R., Moseley, M. E., Keefer, D. K. and Ortloff, C. R. (2009). Environmental change and economic development in coastal Peru between 5,800 and 3,600 years ago. *Proc. Natl. Acad. Sci. USA*, **106**, 1359–1363

Schreiber, K. and Lancho Rojas, J. (2006). *Aguas en el desierto: Los Puqios de Nasca* (Water in the desert: The Puquios of Nazca), Fondo Editorial de la Pontificia Universidad Católica del Perú, Lima, Perú (in Spanish).

SENAMHI, *Servicio Nacional de Meteorología e Hidrología* (Meteorological and hydrological historical data), www.senamhi.gob.pe, Lima, Peru (last accessed 20 September 2011) (Website contents in Spanish).

Shady Solís, R. (2005). *Caral-Supe y su entorno natural y social en los orígenes de la civilización* (Caral-Supe and the natural and social environment within the origin of civilization), Investigaciones Sociales, Año IX, N° 14, UNMSM/IIHS, Lima, Perú, pp. 89–120 (in Spanish).

Shady Solís, R. (2007). *Los valores sociales y culturales de Caral-Supe, la civilización más antigua del Perú y América y su rol en el desarrollo integral y sostenible* (Social and cultural values of Caral-Supe, the oldest civilization in Peru and America, and their role within sustainable development), Proyecto Especial Arqueológico Caral–Supe/INC, Lima, Perú, p. 69 (in Spanish).

Shimada, I. (1994). Pampa Grande and the Mochica Culture, University of Texas Press, Austin, USA, p. 323.

Silverman, H. and Isbell, W. H. (eds) (2008). The Handbook of South American Archaeology, Springer Science+Business Media, LLC, New York.

Squier, E. G. (1877). *Peru – Incidents of travel and exploration in the land of the Incas*, Macmillan & Co., London, United Kingdom, pp. 599.

Tello, J. C. (1943). *Discovery of the Chavin culture in Peru*. American Antiquity, **9**(1), 135–160.

Uhle, M. and Wurster, W. W. (ed.) (1999). *Max Uhle (1856–1944) – Pläne archäologischer Städten im Andengebiet* (Max Uhle – Plan of archaeological sites in the Andean region), Ibero–Amerikanisches Institut, Berlin, Germany, p. 199 (in German).

UNSTATS, *United Nations Statistics Division, Millennium Development Goals Indicators*, unstats.un.org (last accessed 20 September 2011).

Vargas Llosa, M. (2005). *Diccionario del Amante de América Latina, Chavín de Huántar* (Dictionary of the Lover of Latin America, Chavín de Huantar), Ediciones Paidós Ibérica, S. A., Barcelona, Spain, pp. 106–109 (in Spanish).

Villa Hügel e.V. and Deutsch-Peruanische Gesellschaft e.V. (1984). *Peru durch die Jahrtausende, Kunst und Kultur im Lande der Inka* (Peru through the Milennia, Art and culture in the land of the Incas), Verlag Aurel Bongers KG, Recklinghausen, Germany, p. 471 (in German).

Von Humboldt A. (1810). *Vues des cordillères, et monumens des peuples indigènes de l'Amérique* (Views of the Cordilleras, and monuments of the indigene people of America). F. Schoell, Paris, France (in French)

Von Humboldt A. (1819–1825). *Voyages aux régions équinoxales du Nouveau Continent, fait en 1799, 1800, 1801, 1803 et 1804 avec Aimé Bonpland* (Travels to the equinox regions of the New Continent done in 1799, 1800, 1801, 1803 and 1804 with Aimé Bonpland), Paris, France (in French).

Wagner, Günther A. (ed.) (2007). *Einführung in die Archäometrie* (Introduction to Archaeometry). Springer Verlag Berlin Heidelberg, Germany, p. 374 (in German).

The World Bank (2010). Project Appraisal Document on a proposed Loan to the Republic of Peru for a Sierra Irrigation Project (P104760), Sustainable Development Department, Washington, D.C., USA.

Chapter 12

Historical development of water supply in Cyprus

C. A. Kambanellas

12.1 INTRODUCTION

Cyprus, like Crete, has one of the world's oldest recorded histories. It was during ancient times that the Greek language and culture of the island was permanently established.

Geography has been perhaps the major determining factor in the development of the island throughout its history, and is both a blessing and a curse. Strategically, located at the crossroads of the three continents (Africa, Asia and Europe), and of major civilizations, Cyprus was conquered by powers that dominated the eastern Mediterranean at various periods (Mallinson, 2010).

The first signs of civilization go back 10,000 years. The discovery of copper on the island in the third millennium BC brought wealth and trade to the island. Settled by Mycenaean Greeks in the 13th century BC, Cyprus evolved into a flourishing centre of Greek civilization. Because of its strategic position and natural wealth, it was conquered by various nations. Hyskos, the Peoples of the Sea, Phoenicians, Assyrians, Egyptians, Persians, Romans, Byzantines, Crusaders, Franks, Venetians, Turks and British passed through Cyprus from 1200 BC up to 1960 AD (PIO, 1985).

Like the history of many other semi-arid lands, the history of this island reflects the story of its water supplies. Periods of prosperity and progress like the Archaic and Classical period (750–325 BC), the Hellenistic period (325–58 BC) and the Roman period (58 BC–330 AD), brought with them the need for water, both for crop irrigation and for the amenities of civilization, and as the population grew so did the demand for water. In addition, the location, the prevailing climate conditions and the hydrology of Cyprus are briefly discussed.

12.2 PHYSICAL SETTING

12.2.1 Location

Cyprus is the third largest island in the Mediterranean after Sicily and Sardinia, with an area of 9251 km^2. It is situated at the north-east corner of the Mediterranean, at a distance of 300 km north of Egypt, 105 km west of Syria and 75 km south of Turkey. Greece lies 380 km to the north-west. Cyprus lies at a latitude of 34°33′–35°34′ North and longitude 32°16′–34°37′ East. A map of Cyprus is shown in Figure 12.1.

12.2.3 Hydrology

12.2.3.1 Precipitation

The average annual total precipitation increases up the south-western slopes from 450 mm to nearly 1100 mm at the top of the central massif. On the leeward slopes amounts decrease steadily northwards and eastwards to between 300 and 350 mm in the central plain and the flat south-eastern parts of the island. Rivers are seasonal and only flow after heavy rain. The narrow ridge of the Pentadaktylos range, stretching 160 km from west to east along the extreme north of the island, produces a relatively small increase in rainfall to nearly 550 mm along its ridge at about 1000 m.

Rainfall in the warmer months contributes little or nothing to water resources and agriculture. The small amounts which fall are rapidly absorbed by the very dry soil and soon evaporated in high temperatures and low humidities. Autumn and winter rainfall, on which agriculture and water supply generally depend, is somewhat variable. The average rainfall for the year as a whole is about 480 mm (covers the period 1951–1980). Statistical analysis of rainfall in Cyprus reveals a decreasing trend of rainfall amounts in the last 30 years (Hadjioannou, 1997).

12.2.3.2 Air temperature

Cyprus has a hot summer and mild winter but this generalisation must be modified by consideration of altitude, which lowers temperatures by about 5°C per 1000 m and of marine influences which give cooler summers and warmer winters near most of the coastline and especially on the west coast. The seasonal difference between mid-summer and mid-winter temperatures is quite large at 18°C inland and about 14°C on the coast.

Differences between day maximum and night minimum temperatures are also quite large especially inland in summer. These differences are in winter 8 to 10°C on the lowlands and 5 to 6°C on the mountains increasing in summer to 16°C on the central plain and 9 to 12°C elsewhere. In July and August the mean daily temperature ranges between 29°C on the central plain and 22°C on the Troodos mountains, while the average maximum temperature for these months ranges between 36°C and 27°C respectively. In January the mean daily temperature is 10°C on the central plain and 3°C on the higher parts of Troodos mountains with an average minimum temperature of 5°C and 0°C respectively (Hadjioannou, 1997).

12.2.3.3 Winds

Over the eastern Mediterranean generally, surface winds are mostly westerly or south-westerly in winter and north-westerly or northerly in summer. Usually of light or moderate strength, they rarely reach gale force.

Over the island of Cyprus, however, winds are quite variable in direction with orography and local heating effects playing a large part in determination of local wind direction and strength. Differences in temperature between sea and land, which are built up daily during predominant periods of clear skies in summer, cause considerable sea and land breezes. Whilst these are most marked near the coast they regularly penetrate far inland in summer reaching the capital, Lefkosia (Nicosia), and often bringing a welcome reduction of temperature and also an increase in humidity (Hadjioannou, 1997).

12.2.3.4 Humidity

Elevation above mean sea level and distance from the coast also have considerable effects on the relative humidity which to a large extent is a reflection of temperature differences. Humidity may be described as

12.3.11 Present time – Republic of Cyprus (1960–2011)

In 1960 Cyprus gained its independence and a great emphasis was attached by the government to water resources and development. The main aim for domestic water supply was the supply, in sufficient quantities, of piped water in villages and towns for domestic and industrial use. Since 1967 all towns and villages of Cyprus have been serviced with a piped water supply (Konteatis, 1978).

The water from the various sources is conveyed through a trunk main to a service reservoir and/or storage tank near each town or village community. From the service reservoir or storage tank the water is distributed to each house through a distribution network. The pipes used for the distribution of the domestic water are asbestos cement (AC) pipes, unplasticised poly-vinyl-chlorides (uPVC) pipes and galvanised iron pipes. The service reservoirs or storage tanks are made of reinforced concrete.

Each house has its own water meter which records the consumption. At the end of each water consumption period, which is usually two or three months, a water bill is sent to each consumer by the water board, municipality, improvement board or village water committee, which is responsible for the collection of money. The water consumption is recorded on a special form by the meter readers. The water tariffs of each town cover all costs for the production and distribution of water, that is both the capital, and the operation and maintenance costs. For the villages, full operation and maintenance costs and a proportion of capital costs are covered by the water tariffs.

The government finances a proportion of the capital cost of a water supply scheme for a village, which ranges from 50% to 85% in most cases; the remainder is borne by the community itself. There are cases in which the government finances 100% of the water supply scheme. An example of this is the Lymbia water supply for which the government funded the whole scheme because this village is on the *green line*, that is, the separating line between the government controlled part of the island and the position of the Turkish troops occupying the north part of the island since 1974. On the other hand, the town water supply works are paid for in full by the respective authorities, for example, the water boards. For the towns for which the responsible authority is the municipality, such as Paphos, the government contribution is one third of the cost (Kambanellas, 1991).

12.3.11.1 The water resources situation

Water resources in Cyprus are very limited and expensive to develop. Water is required both for domestic and agricultural purposes which constitute the main uses. Industrial and other uses are limited. Out of the total water demand now used (2010), 37% (48.4 Mm^3) are for irrigation and the remaining 63% (82.1 Mm^3) are for domestic, industrial and tourist uses.

Until 1974 only groundwater was used to supply urban areas with potable water. Coupled with the much greater demands placed on groundwater by irrigation needs, this resulted in over-exploitation of groundwater. The depletion of aquifers and the intrusion of the sea into several coastal aquifers were problems that had to be tackled by turning to the development of surface water sources.

Thus from 1960 to 1996 dam storage was increased 50-fold from just 6 to 300 Mm^3, and the proportion of surface water used until 1996, for the urban water supply sector has increased from nil to 60%.

A reverse osmosis sea-water desalination plant came on line at the beginning of April 1997. Today the proportion of the domestic water supply is 10% ground water, 26% surface water and the remaining 64% is desalinated water. Almost all towns and villages in Cyprus are served by piped house-to-house water supply systems, covering almost 100% of the population. The supply to each consumer is metered and charged on the basis of a rising block tariff designed to discourage waste. Different tariffs apply to domestic and industrial consumers (Kambanellas, 2003).

12.3.11.2 Conservation of drinking water

In arid and semi-arid areas, water supply is a permanent problem for a number of reasons, mainly the following: low precipitation, high evaporation, over pumping of aquifers, increase of population and tourism, rise in the standard of living, per capita consumption, and so on. Unfortunately Cyprus is no exception. This becomes evident if one takes a look at the water development projects from 1960 to the present day. Despite huge expenditure for the construction of several dams and other hydraulic projects, the island's water problem not only has not been solved but, on the contrary, has been aggravated. From 1997 onwards the government of the Republic of Cyprus has decided, in parallel with new projects, the implementation of water conservation measures at household level (Kambanellas, 2005 & 2011).

Potable water used in households and industry is normally taken directly from the drinking water system and discharged into a central sewerage system or into an onsite wastewater system. Water, suitable for potable use, is therefore taken from the supply system and used for other purposes, before discharge into the wastewater system. It is quite obvious that water of this quality is not needed for many domestic and industrial applications. To meet these non-potable water demands with an appropriate quality of water, two solutions are currently available:

(a) The installation of a separate processed water system,
(b) The decentralised recycling of at least a suitable part of the discharge water for reuse as processed water.

Only about half of the average annual supply of domestic water in Cyprus needs to be of drinking water quality. Over 50% of the demand for water could be met by water of a lower grade quality, such as processed water. Processed water is usually defined as water which meets the requirements – in terms of dissolved components and other essential parameters – for various systems such as instruments, machinery or living organisms.

Conservation of drinking water has been initiated as a practical means of assisting water demand management where, for instance, capital expenditure on water resource development (new dams, main conveyors, water treatment, etc.) might be reduced or deferred. 'Water saved is exactly the same as water supplied' and 'One person's reduction in water use makes water available for someone else to use'.

As a practical measure, to save drinking water, a scheme has been put into practice during the last 14 years, for subsidising the recycling of "grey water" for watering the garden and/or for the operation of the WC's in the individual households or buildings (Kambanellas, 2011).

12.4 CONCLUSIONS

The history of Cyprus reflects the history of its water supplies. Periods of prosperity and progress bring with them the need of water, both for crop irrigation and for the amenities of civilization and as the population grows, so does the demand for water.

To satisfy the water demand, water supply technologies such as aqueducts, cisterns, water conduits, chains-of-wells and other systems were used.

These water supply technologies were developed through the centuries. The trade with Egypt, Crete, etc., the settlements of Achaean Greeks and Mycenaeans, and the various conquerors of Cyprus, had positive influences on the development of the water supply technologies.

These water supply technologies were adapted to the local conditions so that the supply of water was based on sustainable development.

Even though Cyprus is located in a semi-arid region of the world, sustainable water supply systems were used for many years (since the Neolithic Age up to the Modern Times). Unfortunately, during the present time (1960–2011) the rapid and inconsiderate development of water supply systems has brought the depletion of the aquifers and the intrusion of the sea into several coastal aquifers.

Despite the huge expenditure for the construction of several water projects at the present time, the island's water problem has been aggravated. This is why implementing means of water conservation is an essential and beneficial solution in dealing with water scarcity. Therefore, development of cost-effective decentralised water and waste water systems is essential.

Acknowledgements

The author appreciates very much the critical comments of Dr. P. Flourentzos, Director of the Department of Antiquities for the period 2004–2009.

REFERENCES

Aupert P. (2000). Guide to Amathus, The Bank of Cyprus Cultural Foundation in collaboration with the Department of Antiquities, Nicosia, Cyprus.

Cyprus Tourism Organization (CTO). (2001). Cyprus. Cyprus Tourism Organization, Nicosia, Cyprus.

Flourentzos P. (2004), Excavations at Amathus. Department of Antiquities, Nicosia, Cyprus (in Greek).

Hadjioannou K. (1971–1983). The Ancient Cyprus from Hellenistic Sources. Holy Archbishopric of Cyprus, Nicosia, Cyprus (in Greek).

Hadjioannou L. (1997). The Climate of Cyprus. Meteorological Service, Nicosia, Cyprus.

Kambanellas C. A. (1991). Water Consumption and Recycling of Grey Water in Cyprus. POW, UK.

Kambanellas C. A. (2003). Best Practices on Non-Conventional Water Resources Management in Cyprus. In: 3rd World Water Forum, Kyoto, Japan.

Kambanellas C. A. (2005). Recycling of Grey Water in Cyprus. World Water Week in Stockholm, SIWI, Stockholm, Sweden.

Kambanellas C. A. (2011). CYPROBELL Grey Water Recycling Systems. www.hydranos.org, (accessed 10 March 2011), Nicosia, Cyprus.

Kambanellas C. A. and Kambanellas S. C. (2011). Recycling of Grey Water in Cyprus. In: The 6th IWA Specialist Conference on Efficient Use and Management of Water, Dead Sea, Jordan.

Knauss J. (2006). *Prahistorische Grundwasser brunnen rund um die Agais* (Prehistoric Wells in the Aegean Area). Association for the study and Spread of Hellenic history, Munich, Germany.

Konstantinou G., Panayides J., Xenofondos C., Afrodises S., Michailides P. and Kramvis S. (2002). The Geology of Cyprus, Geological Survey Department, Nicosia, Cyprus (in Greek).

Konteatis C. A. C. (1978). Department of Water Development, Annual Reports for 1964 and 1966 to 1978. Published by the Press and Information Office of the Government of Cyprus, Nicosia, Cyprus.

Last J. S. (1975). Kourion: The Ancient Water Supply. In: Proceedings of the American Philosophical Society, **119**(1), 39–72.

Lyssiotis P. (2009). About Cyprus, Press and Information Office, Cyprus.

Mallinson W. (2010). Cyprus a Historical Overview. Press and Information Office, Nicosia, Cyprus.

Pavlides A. (1991). A Brief History of Cyprus and its Greek Origin. Sponsored by G. Paraskevaides, Nicosia, Cyprus.

Raeburn C. (1947). Water Supply and Irrigation Department, Water Supply in Cyprus. Printed at the Cyprus Government Printing Office, MP 284/48, Nicosia, Cyprus.

Press and Information Office (PIO), (1985). Cyprus, No Man is an Island. Published by the Press and Information Office of the Government of Cyprus, Nicosia, Cyprus.

Ward I. L. (1956). Department of Water Development, Annual Reports for 1948 to 1956. Printed at the Cyprus Government Printing Office, MP 284/49, Nicosia, Cyprus.

Ward I. L. (1959). Department of Water Development, Cyprus Water Development. Printed at the Cyprus Government Printing Office, Nicosia, Cyprus.

Chapter 13

Water supply in the Middle East during Roman and Byzantine periods

Benoît Haut and Didier Viviers

13.1 INTRODUCTION

In this chapter, the focus is given to water supply systems in the Middle East during the Roman period and the beginning of the Byzantine period.

The geographical area covered includes Asia Minor (or Anatolia), Syria and Jordan. The time period covered extends from the conquest of the Middle East by the Romans (first century BC) to the end of the seventh century AD.

The numerical data that are not referenced were acquired on the sites by the authors.

13.1.1 Some elements of Middle East history

Alexander the Great died in Babylon in 323 BC. His Empire, covering the entire Middle East, was then divided into three and inherited by his generals, Seleucus, Antigonus and Ptolemy son of Lagos. Subsequently, Hellenistic civilization was dominant in the Middle East (Figure 13.1a).

In the West, Rome had won the Punic Wars against Carthage. These three wars (264–241 BC, 218–201 BC, 149–146 BC) resulted in the destruction of Carthage and its annexation by Rome. Consequently, the Roman Republic emerged as a powerful military power whose hegemony in the region was firmly established. Following these victories, Rome proceeded to conquer the entire Middle East in less than three centuries. Thus, Syria was occupied in 64 BC and Petra, the capital of the Nabataean Kingdom, was taken over in 106 AD. To ensure the domination of such a large territory, the Republic was reformed and transformed into an Empire which reached the apogee of its power during the second century before its decline in the chaotic years of the third century AD (Figure 13.1b).

In the fourth century, the gravity centre of the Roman Empire slowly moved to the east. As a consequence, Constantine, the first Christian Emperor, brought his capital from Rome to Byzantium in 330 AD and renamed it Constantinople. The Roman armies suffered heavy defeats, notably at Adrianople by the Visigoths in 378 AD. Consequently, after the death of Theodosius I, in 395 AD, the Roman Empire was permanently divided into the Western and Eastern Roman Empires, the latter also known as the Byzantine Empire. The Western Roman Empire ended officially with the abdication of Romulus Augustus on September 4, 476 AD (Figure 13.1c). In the Middle East, the so-called Roman

On the other hand, it has been attested by much evidence that, in former Nabataean cities, the Romans usually operated already existing water systems, as the engineering knowledge of the Nabataeans was high (Dentzer *et al.* 2001; Mouton & Al-Dbiyat, 2009; Ortloff, 2009).

The Romans used water for the irrigation of cultivated fields and for urban needs. In the cities, fountains flourished; large baths and latrines were present. This led to a high need for water; daily consumptions between 200 and 500 l per capita are reported (Bailhache, 1979; Hodge, 2008; Kessener & Piras, 2008). According to Vitruvius, water not only satisfies the needs of the people, but also their pleasure (Morgan, 1960).

It is commonly admitted that the Romans had a remarkable engineering knowledge of water supply (Hodge, 2008; Ortloff, 2009; Viollet, 2004). Water was carried to the Roman cities through aqueducts that could reach more than 100 km in length. Within the cities, water was distributed through a complex combination of water towers and pipelines. Wastewater was evacuated from the cities by drainage systems. Only a few Roman writings on this engineering practice have been preserved. However, archaeology offers, especially in the Middle East where numerous Roman water systems are very well preserved, some precise illustrations of their techniques. The surviving written records of Frontinus (Evans, 1994; Herschel, 1973) and Vitruvius (Morgan, 1960) provide some understanding of water supply systems in the Roman period. While these works give insight into the design methodology of water supply systems of that period, they reflect pre-scientific views of hydraulic principles (Ortloff, 2003). For instance, in the work of Frontinus, such a concept as the flow rate is not known. However, recent analyses of Roman water systems using fluid mechanics have demonstrated their high level of quality and their smartness (Chanson, 2001; Haut & Viviers, 2007; Ortloff, 2001; Ortloff, 2003).

During the second, third and fourth centuries, water use in the Middle East provides an excellent picture of the best technology available at this period. In Byzantine times, a decline in the quality of the water systems is commonly admitted. In many cities, like Constantinople, the aqueducts were abandoned and the Romans focused on the construction of cisterns, often very large, intended to collect rain water (Viollet, 2004). For instance, in Smyrna, beautiful large cisterns, extremely well preserved, were constructed at the end of the Roman period under the western stoa (columned gallery) of the agora (Lancaster, 2007; Levi, 2003). However, some counterexamples exist, like those of Apamea where channels and pipelines of an astonishing quality were constructed in the Byzantine period, in the frame of the re-organisation of the area of the city where the aqueduct goes into the town (Haut & Viviers, 2007).

13.1.3 Structure of this chapter

In this chapter, the path of water from its source to its use is followed. Some general comments on the different components of the water systems in the Roman and Byzantine periods are presented. These comments are illustrated by numerous examples observed in cities with remarkable hydraulic remains: Smyrna, Ephesus, Perge, Aspendos, Apamea, Philippopolis, Bosra, Gerasa and Petra. These nine cities are located on the maps of Figure 13.1. Some historical elements regarding four of these nine cities are given in the next section.

The following components of water systems are considered in the next sections:

(a) aqueducts for water transport outside the city limits,
(b) entrance of water in the city and distribution systems,
(c) water transport inside the city limits, and
(d) urban water use (latrines, baths, fountains and cisterns).

A section of this chapter is dedicated to each of these items. Wastewater collection systems are not considered due to the lack of archaeological evidence on these systems in the cities studied by the authors.

13.2 INFORMATION ABOUT SOME IMPORTANT CITIES IN THE MIDDLE EAST DURING ROMAN AND BYZANTINE PERIODS

In this section, some information about four important cities in the Middle East during Roman and Byzantine periods are given. Numerous examples from the water system of these cities are used to illustrate the Roman techniques of water management.

13.2.1 Ephesus

Ephesus is located in Turkey, 50 km south of the city of Izmir, near the coastline. A simplified map of the actual remains is given in Figure 13.2.

The earliest occupation of the region of Ephesus as a Mycenaean trading post (1699–1400 BC) is located north of the archaeological site. The nearby Sanctuary of Artemis drew numerous pilgrims in the area from the eighth century BC and it became a huge temple complex in the sixth century BC. The pre-archaic settlements gave place to a small town in the sixth century BC. Later on, at the beginning of the third century BC, Lysimachus, a successor to Alexander the Great, moved the city southwards, to the actual position of the remains. He built the surrounding walls (Ortloff, 2009).

When Ephesus passed under Roman rule (around 130 BC), numerous new buildings were erected in the city centre. At its heyday in the second century AD, during the Roman Imperial period, Ephesus had approximately 250,000 inhabitants and was the second largest city in the Roman Empire. The residences, the six bathing complexes, the numerous fountains and the latrines (see Figure 13.2) needed a large amount of water. Four aqueducts were operated at this time (Ortloff, 2001; Wiplinger, 2010).

By the fourth century AD, the shoreline had changed drastically. The efforts to maintain a port at Ephesus were abandoned in the seventh century AD and Ephesus' importance as a commercial and religious centre started to decline.

13.2.2 Perge

Perge is located 11 km from the Mediterranean coast, near Antalya (Turkey). A simplified map of the actual remains is given in Figure 13.2.

The site has been occupied since the third millennium BC. After the arrival of Alexander the Great in 334 BC, it passed under the control of various Hellenistic kingdoms. Later on, Perge gained a genuine prosperity in the second and third centuries AD, during the Roman Imperial period. Most of what can be seen today at the site comes from these centuries, such as the theatre, the stadium, the agora and the baths. A fountain at the bottom of the acropolis delivered water in a canal at the centre of the colonnaded main street (Figure 13.2). There was a last period of prosperity during the fifth and sixth centuries AD. The city walls were extended southwards and the city was decorated with many churches (Abbasoglu, 2009; Tunçer, 1992).

Later on, in the eighth century AD, the raids by mountain tribes and the Arabs, coupled with the development of Antalya, led Perge to decline. Perge was finally ruined during the Seljuk and Arab raids in the twelfth century AD and was deserted by the people (Tunçer, 1992).

13.2.3 Aspendos

Aspendos is situated 50 km east of Antalya, 12 km north of the coastline. A simplified map of the actual remains is given in Figure 13.2.

Aspendos was initially founded as a Hellenic colony of the Argives (1000 BC). In 333 BC, Alexander the Great took the city off the Persians. Later on, in 133 BC, the city came under Roman rule and an extensive urban development plan was established. In the Roman period, the importance of the city resulted from its position on commercial roads and on the Eurymedon River, with a direct and easy access to the sea (Ortloff, 2001).

The acropolis of the city, sketched in Figure 13.2, lies 60 m above the sea level. Today, it is dominated by the remains of a monumental fountain (2nd–3rd century AD), thought to have ended the aqueduct supplying water to the town, and the entrance of the basilica (Kessener & Piras, 2008).

Toward the end of the Roman period the city began a decline that continued throughout the Byzantine period.

13.2.4 Apamea

The archaeological site of Apamea in Syria, on the right bank of the Orontes, between Hama and Alep, has a continuous human activity that can be traced back to the Middle Palaeolithic. After the conquest of Alexander the Great, a Greek city was established there in 300/299 BC by Seleucus Nicator, the King of Syria. Apamea became one of the main cities of the Seleucid Empire, with Antioch, Seleucia and Laodiceia. Apamea served as an administrative and military centre of the north of Syria.

Later on, Apamea was attached by Pompeii to the Roman Empire (64 BC). Because of several violent earthquakes, the city saw large reconstructions, which sometimes profited from liberalities of Roman Emperors.

As the capital of the Roman province of Syria Secunda, with a dense population, Apamea reached genuine prosperity during the fifth and sixth centuries AD, in the Byzantine period. The Persian Wars combined with dramatic earthquakes subsequently weakened the city and made it vulnerable to the Arabic conquest in 638 AD.

Since the beginning of the second century AD, a large street with porticoes offered to the city a monumental artery by which the town was intersected from north to south. This Cardo Maximus (see Figure 13.2), approximately 37 m wide and 1850 m long, was also used for wheeled transport. A second feature of urbanism was the large wall that surrounded the city. The foundations of this, about 7 km long, fortification were set in Hellenistic times (see Figure 13.2). The only known water supply of the city is an aqueduct (see Figure 13.2), used from 47 AD (Balty, 2000) until at least the seventh century. It was bringing water into the town from a spring located about 80 km away from Apamea (Balty, 1987).

Excavations in the area of the baths and in the north-east corner of the city, where the aqueduct enters the town (see Figure 13.2), were performed from 2002 by a team of archaeologists from the Université Libre de Bruxelles (Belgium) (Viviers, 2008).

13.3 WATER TRANSPORT OUTSIDE THE CITY LIMITS: THE AQUEDUCTS

Springs were, by far, the most common sources of water for aqueducts. Water sources for the Roman systems also included reservoirs that were developed by the building of dams (Mays, 2010).

Aqueducts were used to transport water from the source to the locations where water was needed, either for irrigation or for urban supplies.

Roman aqueducts included various components such as channels with an open surface flow following the surface of the land, aqueduct bridges built with arches and inverted siphons.

Three considerations have an influence on the aqueduct construction: the need for a regular slope, the obligation to sometimes cross valleys, and the need to deliver water in the city into a water tower constructed as high as possible, in order to benefit from a high potential energy to distribute the flow in the city. This is why inverted siphons and aqueduct bridges are often observed near to cities.

The maximum flow rate of water that could be transported by this system can be calculated approximately. A schematic lateral view of the overflow of the water from the pool to the channel is proposed in Figure 13.21. The staircase structure on this representation, with six 14 cm high steps, can be observed in Figure 13.20.

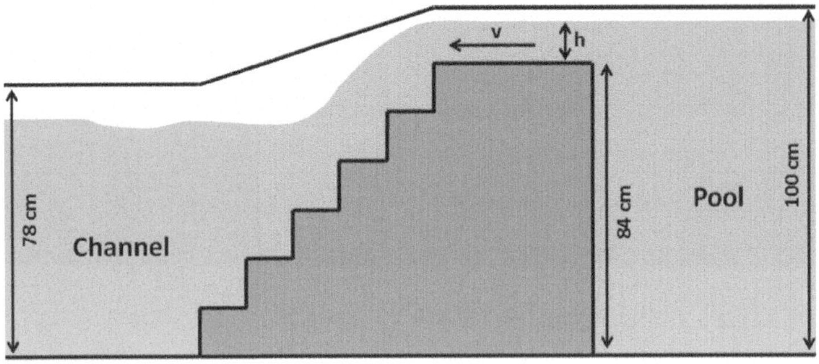

Figure 13.21 Lateral view of the overflow of water from the pool to the channel

The largest possible value of h, the height of water above the staircase structure (see Figure 13.21), is 16 cm. Indeed, if h was larger than 16 cm, the water would have flown over the lateral walls of the pool; but no calcareous concretions are observed on these walls.

It can be assumed that the flow on the top of the staircase structure was critical, that is, characterised by a Froude number (Fr) equal to unity (Lencastre, 1995):

$$\mathrm{Fr} = \frac{v}{\sqrt{gh}} = 1$$

where v is the velocity of the water on the top of the staircase structure (see Figure 13.21) and g is the acceleration of gravity.

Therefore, Q, the volumetric flow rate of water in the channel, can be calculated as follows:

$$Q = hLv = L\sqrt{g}h^{\frac{3}{2}}$$

where L (= 1.8 m) is the width of the cross section of the flow on the top of the staircase structure. $Q =$ 31,000 m^3/day is calculated.

The inner width of the channel is equal to 2.4 m. This large value, compared to usual values of channel inner width in Roman engineering, might have been chosen for aesthetic reasons, as the channel is in the middle of the main street. The channel has an average slope of 1.8 cm/m between points 1 and 3 in Figure 13.18. As the obstacles in the channel are separated by 7 m, the height difference between two successive obstacles is 13 cm (between points 1 and 3 in Figure 13.18).

If there were no obstacles in the channel, the Manning equation, presented previously, could be used to calculate the height of water that would have been observed in the channel. Using a friction coefficient $n = 0.014\ m^{1/6}$, a channel slope $\theta = 1.8$ cm/m, a channel inner width $b = 2.4$ m and a volumetric flow rate $Q = 31{,}000\ m^3/day$, a value of 8 cm is obtained for the water height in the channel between points 1 and 3 in Figure 13.18. This small value of the water height in the channel is obviously a consequence of the large width of the channel. Such a small value of the water height in the channel would have led to a small pressure at the beginning of the derivations starting from the channel, and hence to difficulties in operating such derivations.

The flow in the channel was of course considerably influenced by the presence of the obstacles. It can be assumed that the flow of the water over the obstacles in the channel was critical, that is, characterised by a

REFERENCES

Abbasoglu H. (2009). Perge 2008. *ANMED, News of Archaeology from Anatolia's Mediterranean Areas*, **8**, 61–70.

Bailhache M. (1979). Evolution of the flow rate of Roman aqueducts. *Dossiers de l'Archéologie*, **38**, 74–81 (in French).

Balty J. C. (1987). Water in Apamea, Syria. In: L'homme et l'eau en Méditerranée et au Proche-Orient. IV. L'eau dans l'agriculture, Lyon, France, (in French).

Balty J. C. (2000). Claudia Apamea. New data on the history and the topography of Apamea. *Comptes rendus de l'Académie des Inscriptions et Belles – Lettres*, **1**, 459–481 (in French).

Chanson H. (2001). Conception Hydraulique des Aqueducs Romains (Hydraulic conception of the Roman aqueducts). *La Houille Blanche*, **6/7**, 43–57.

Darrous N. and Rohmer J. (2004). Chahba-Philippopolis (Hauran): essai de synthèse archéologique et historique (Chahba-Philippopolis (Hauran): archaeological and historical synthesis). *Syria*, **81**, 5–41.

Dentzer J. M., Blanc P. M. and Fournet T. (2001). Le développement urbain de Bosra de l'époque nabatéenne à l'époque byzantine: bilan des recherches françaises 1981–2002 (Urban development of Bosra from the Nabatean period to the Byzantine period. French archaeological research 1981–2002). *Syria*, **79**, 75–154.

Erdemgil S. (2010). Ephesus. Net Turistik Yayinlar A.S., Istanbul, Turkey.

Evans H. B. (1994). Water Distribution in Ancient Rome: The Evidence of Frontinus. The University of Michigan Press. Ann Arbor, Michigan, USA.

Herschel C. (1973). Frontinus. The Two Books on the Water Supply of the City of Rome of Sextus Julius Frontinus. New England Water Company Publication, Cambridge, UK.

Haut B. and Viviers D. (2007). Analysis of the water supply system of the city of Apamea, using Computational Fluid Dynamics. Hydraulic system in the north-eastern area of the city, in the Byzantine period. *Journal of Archaeological Science*, **34**, 415–427.

Hodge A. T. (2008). Roman Aqueducts and Water Supply, 2nd edn, Duckworth, London, UK.

Kessener P. and Piras S. (2008). The pressure line of the Aspendos Aqueduct. *Adalya*, **II**, 160–187.

Lancaster L. C. (2007). Early Examples of So-Called Pitched Brick Barrel Vaulting in Roman Greece and Asia Minor: A Question of Origin and Intention. In: The Bautechnik imantiken und vorantiken kleinasien, Internationale Konferenz, 13–16 June 2007 in Istanbul, Turkey.

Lencastre A. (1995). General Hydraulics. Eyrolles, Paris, France (in French).

Levi E. A. (2003). The Agora of İzmir and Cultural Tourism, The International Committee for Documentation of Cultural Heritage (CIPA), Antalya Symposium.

Mays L. W. (2010). Ancient Water Technologies. Springer-Verlag, Heidelberg, Germany.

Morgan M. H. (1960). Vitruvius. Ten Books on Architecture. Dover Publications, New York, USA.

Mouton M. and Al-Dbiyat M. (2009). Stratégies d'acquisition de l'eau et société au moyen orient depuis l'antiquité (Water systems in Middle-East from antiquity). Institut Français du Proche Orient, Beyrouth, Lebanon.

Ortloff C. R. and Crouch P. D. (2001). The urban water supply and distribution system of the Ionian city of Ephesos. *Journal of Archaeological Science*, **28**, 843–860.

Ortloff C. R. (2003). Computational fluid dynamics investigation of the hydraulic behaviour of the Roman inverted siphon system at Aspendos, Turkey. *Journal of Archaeological Science*, **30**, 417–428.

Ortloff C. R. (2009). Water Engineering in the ancient World. Archaeological and Climate Perspectives on Societies of Ancient South America, the Middle-East and South-East Asia. Oxford University Press, New York, USA.

Tunçer M. (1992). Perge Conservation Plan Report. Perge Conservation Plan Research Report, Akman Project Co.

Viollet P. L. (2000). Hydraulics in Ancient Civilizations. Presse de l'Ecole des Ponts et Chaussées, Paris, France (in French).

Viviers D. and Vokaer A. (2007). Travaux de la Mission archéologique belge à Apamée de Syrie XLème campagne (2006) (Belgian archaeological mission in Apamea. Report of the 2006 campaign). *Revue belge de philosophie et d'histoire*, **85**, 125–156.

Viviers D. (2008). Recent research in Apamea. *Studia Orontica*, 109–115.

Wiplinger G. (2010). The Aqueducts of Ephesos, Austrian Archeological Institute (OeAI). Website, http://www.oeai.at/index.php/water-supply.html (last accessed August 11, 2011).

Chapter 14

Water supply management technologies in the Ancient Greek and Roman civilizations

G. De Feo, P. Laureano, L. W. Mays and A. N. Angelakis

14.1 PROLEGOMENA

'.... *αλλά Θαλής μεν ο της τοιαύτης αρχηγός φιλοσοφίας ύδωρ είναι φησίν είναι δι'ό και την γήν εφ' ύδατος αποφαίνεται είναι, λαβών ίσως την υπόληψιν ταύτην εκ του πάντων οράν την τροφήν υγράν ούσαν ..... και διά το πάντων τα σπέρματα την φύσιν υγράν έχειν, το δέ ύδωρ αρχήν της φύσεως είναι ...*' (Aristotle, Metaphysics, 983 b, 20–30).

According to "*Thales, the Founder of this philosophy, water is the most important natural element and thus the earth rests on water, perhaps from observing that the nutriment of all things contains moisture ... and that the seeds of all things possess a moist nature, so water is the origin of nature ...*" (transl. by the authors) (Aristotle, Metaphysics, 983 b, 20–30).

During the Hellenistic and Roman periods, the progress in urban water supply management technologies was admirable, as witnessed by several aqueducts, cisterns, wells, and other water facilities discovered (Koutsoyiannis *et al.* 2008). As a matter of fact, numerous hydraulic works which were developed during the two mentioned periods are still standing such as those described in this chapter.

In particular, in this chapter a rather synoptic description of the main concepts of water supply management during the Greek and Roman civilizations is presented. Water supply technologies in the ancient Greek and Roman civilizations, aqueducts, cisterns, water distribution systems, fountains, thermae and baths are considered. The main principles and challenges are also discussed.

14.2 WATER SUPPLY SOURCES, DAMS, AND RESERVOIRS

A systematic evolution of water supply management in ancient Greece began in Crete during the early Bronze Age, that is, the Early Minoan period (*ca.* 3500–2150 BC). Water supply sources for the settlements and palaces of the early Greek civilizations included rivers, springs, wells, rainwater harvesting, underground cisterns, and the bringing of water from higher altitudes using aqueducts. A summary of the sources of water sources for settlements and palaces of the early Minoan civilizations (4000–1100 BC) is presented in Table 14.1.

Ancient dams were built in order to develop a reservoir or to divert water from a river. There are remains of a number of dams built by the Mycenaeans around 1200 BC. The Mycenaeans (named from the site of Mycenae in north-eastern Argolis in Peloponnese of southern Greece) flourished between 1600 BC and *ca.* 1100 BC. The Mycenaeans dams were essentially long, low dykes in Peloponnese and Beotia as reported by Knauss (1991). In Peloponnese these included: Pheneos (2.5 m high, 2500 m long); Stymphalos (2.5 m high, 1900 m long); Orchomenos (2 m high, 2100 m long); Mantinea (3 m high, 300 m long); and Taka (2 m high, 900 m long). In Beotia these included: Boedria (2 m high, 1250 m long); Thisbe I (2.5 m high, 1200 m long); and Thisbe II (4 m high, 200 m long). The Tiryns Dam (located 4 km east of Tiryns near the modern village of Ayios Adrianos) and a 1.5 km long diversion canal were built to divert flood flows, and thus prevent them from entering the town, in the small river that flowed through it to the sea (Zangger, 1994). This diversion canal diverted the flood flows to another river that also flowed to the sea. Tiryns Dam was 10 m high and 100 m long constructed with an earth core and two masonry walls. The width of the dam was 103 m on the right bank and 57 m on the left bank.

Table 14.1 Summary of water sources for settlements and palaces of the early Minoan civilizations (4000–1100 BC).

Source of water	Palaces and settlements
Rainwater harvesting systems	Agia Triadha, Chamaizi, Knossos, Myrtos-Pygros, Phaistos, Kato Zakros
Wells	Palaikastro, Knossos, Kato Zakros, Kommos
Aqueducts from sources (e.g. springs)	Knossos*, Mallia*, Tylissos, Pylos, Thebes
Underground cisterns w/ steps	Zakros, Tylissos
Springs	Knossos, Tylissos, Syme

*Probable.

The later ancient Greeks did not accomplish a great deal in the construction of dams for developing reservoirs. One example is the ancient Greek dam built in the Mytikas Valley between the mid fourth and third centuries (Murray, 1984). The Mytikas Valley is located halfway up the western coast of Akarnania, north of Kalamos (ancient Karnos) Island. The dam was approximately 11 m in height constructed of blocks (varying is size with many approximately $1.5 \times 1.0 \times 1.0$ m) set in stepped courses.

Water sources for the Roman systems included not only springs, percolation wells, and weirs on streams, but also reservoirs that were developed by building dams. The source of water for the aqueduct in Segovia, Spain was a weir across the Rio Acebeda. Springs were, by far, the most common sources of water for aqueducts. Dams were also built in many regions of the Roman Empire to develop water supply reservoirs. Figure 14.1 shows four Roman dams used to create reservoirs in Spain. One of the Roman water systems that has the sources well preserved was the system at ancient *Augusta Emerita*, present day Merida, Spain. This system included two reservoirs, now known as the Cornalvo dam and the Proserpina dam. The Proserpina dam is an earthen dam, approximately 427 m long and 12 m high, shown in Figure 14.1a. The Cornalvo dam is an earthen dam, approximately 194 m long and 20 m high with an 8 m dam crest width, shown in Figure 14.1b. Both of these dams are still in use today, obviously with modifications over the years. Schnitter (1994) and Smith (1971) provide additional information on the history of dams.

14.3 WATER TRANSMISSION: AQUEDUCTS

Aqueducts are man-made conduits for carrying water across hollows or valleys (from the Latin *aqua*, "water," and *ducere*, "to lead"). In modern engineering, "aqueduct" refers to a system of pipes, ditches, canals, tunnels, and supporting structures used to convey water from its source to its main distribution points (Fonder & Xanthoulis, 2007).

The third aqueduct, the Madradag aqueduct (see Figure 14.7), was built to the acropolis and was apparently built by Eumenes II (197–159 BC). This 50 km (Falbusch, 2010) long aqueduct was constructed with a triple terracotta pipeline of approximately 0.18 m inner diameter with a wall thickness of 3.5 cm and 50 to 60 cm pipe segments. Ozis (1996) provides a length of the three terracotta pipes as 44 km. The last stretch of 3.5 km of aqueduct was a lead pipe siphon under a maximum pressure of 200 m. No sign of the lead pipes has survived, however soil samples close to the pipeline location proved to have 50 times higher lead-content than soil samples further away (Fahlbush, 2010). The three pipelines ran parallel to each other until they reached a header tank at an elevation of 376 m on the southern slope of Mount Hagios Georgios, which is north of Pergamon. The header tank consisted of two adjacent chambers of 3.62 × 1.21 m each (Fahlbusch, 2010). The pipes of the triple pipeline aqueduct merged into the north-eastern corner of the first chamber. Water flowed through the first chamber then through the second one into a pressure pipeline from the south-eastern corner of the second chamber. The enlarged cross sections of the chambers compared to the pipe diameters allowed them to function as settling tanks. From the second chamber an inverted siphon was used to transport water from the header tank to the acropolis with a maximum pressure height of approximately 190 m water column (Fahlbusch, 2010). The exact point where the double siphon terminated on the acropolis is not known.

Table 14.2 lists the characteristics of major Greek aqueducts, discussed so far, and the Roman aqueducts, discussed in the next paragraph.

14.3.3 The Roman aqueduct systems

The first major Roman aqueducts were built to supply waters to the city of Rome. The water supply system of Rome is considered one of the marvels of the ancient world (Hodge, 2002; De Feo & Napoli, 2007). Although sewer and water pipes were not inventions of the Romans, since they were already present in other eastern civilizations as well as in Greek civilization, they were certainly perfected by the Romans. The Romans resumed the engineering works of the Assyrians, and turned their concepts into major infrastructure to serve all the citizens (Lofrano & Brown, 2010). In fact, the Romans were "urban people" and consumed enormous quantities of water in order to supply baths, public and decorative fountains, residences, garden irrigation, flour mills, aquatic shows and swimming pools (Hodge, 2002; Tolle-Kastenbein, 2005; De Feo & Napoli, 2007; Mavromati & Chryssaidis, 2007). As a matter of fact, on the whole the 11 Imperial Age aqueducts serving Rome had a total flow rate of $1.13 \times 10^6\ m^3 d^{-1}$, and since the population of Rome at the end of the first century AD was about 500,000 inhabitants (Bono & Boni, 1996), an extraordinary mean specific consumption of around 2 m^3 $inhabitant^{-1}$ d^{-1} was produced. This value is surprising if compared with the current specific water use of 200–300 l $inhabitant^{-1}$ d^{-1} (Chanson, 2008; De Feo *et al.* 2011).

construction and management of a Roman aqueduct was not different from the modern practice, with several modern technologies coming from Roman engineering. The building of an aqueduct started with the search for a source. Water was collected after permeating through vaults and walls of draining channels and settled. From the source, water flowed into an open channel flow and air was present over the water surface (Monteleone *et al.* 2007). The water in the aqueducts descended quietly through concrete channels. During the route, there were multi-tiered viaducts, inverted siphons and tunnels to exceed valleys or steep points. At the end of its course, the channel entered into a so called "*piscina limaria*", a sedimentation tank to settle particulate impurities. Then, the channel flowed into a partitioning tank called "*castellum divisorium*" (or "*castellum dividiculum*") where there were some walls and weirs to regulate the water flowing into the urban pressure pipes (De Feo & Napoli, 2007; Monteleone *et al.* 2007).

Table 14.2 Characteristics of major Hellenistic and Roman aqueducts.

Name	Location	Period	Construction	Length (km)	Flowrate ($m^3\ d^{-1}$)
Eupalinos	Greece	Archaic	530–520 BC	28	na
Peisistratean (Athens)	Greece	Archaic	546–527 BC	na	na
Paggaio Mountain 1 (Amphipolis)	Greece	Classical	4th century BC	14	na
Paggaio Mountain 2	Greece	Classical	4th century BC	20	na
Attalos	Turkey	Hellenistic	na	15	na
Demophon	Turkey	Hellenistic	na	na	1500
Madradag (Pergamon)	Turkey	Hellenistic	200–190 BC	44	40,000
Cherchell	Algeria	Roman	na	>45	40,000/6600
Cuicul	Algeria	Roman	na	5–6	na
Gunugu	Algeria	Roman	na	na	na
Arles	France	Roman	1st century AD	48.0	8000
Beaulieu (Aix-en-P.)	France	Roman	na	na	na
Brévenne (Lyon)	France	Roman	*ca.* 1–50 AD	70.0	10,000
Gier (Lyon)	France	Roman	50 AD	86.0	15,000
Gorze (Metz)	France	Roman	100/200 AD	22.3	na
Mons (Fréjus)	France	Roman	100/200 AD	39.4	na
Mont d'Or (Lyon)	France	Roman	40 BC	26.0	2000–6000
Montjeu (Autun)	France	Roman	na	na	na
Nîmes	France	Roman	41–54 AD	49.8–50.0	35,000
Yzeron-Craponne (Lyon)	France	Roman	*ca.* 20–10 BC	40.0	13,000
Köln	Germany	Roman	na	95.4	na
Athens Hadrianean	Greece	Roman	117–138 AD	25.7	na
Chersonisos	Greece	Roman	First half of the 2nd century AD	13	na
Corinth	Greece	Roman	125/160 AD	85.0	80,000
Elyros (Crete)	Greece	Roman	na	2	na
Gortys (Crete)	Greece	Roman	Late Roman	15	na
Lyttos	Greece	Roman	33 BC to 14 AD	22	na
Minoa (Crete)	Greece	Roman	na	2.77	na

(*Continued*)

Table 14.2 Characteristics of major Hellenistic and Roman aqueducts (*Continued*).

Name	Location	Period	Construction	Length (km)	Flowrate ($m^3 d^{-1}$)
Moria (Lesvos)	Greece	Roman	End of 2nd begging of 3rd century AD	26	na
Naxos	Greece	Roman	6th century BC[a]	11.0	na
Nikopolis	Greece	Roman	33 BC–138 AD	70.0	na
Olympia	Greece	Roman	160 AD	na	na
Rhodos	Greece	Roman	na	na	na
Samos (Eupalineon)	Greece	Roman	530 BC	2.8	na
Anio Novus (Rome)	Italy	Roman	52 AD	86.876	189,520
Anio Vetus (Rome)	Italy	Roman	273 BC	63.640	175,920
Aqua Alexandrina (Rome)	Italy	Roman	226 AD	22,000	21,025
Aqua Alsietina (Rome)	Italy	Roman	2 BC	32.882	15,680
Aqua Appia (Rome)	Italy	Roman	312 BC	16.561	73,000
Aqua Augusta (Serino)	Italy	Roman	33–12 BC	≈140.0	47,520
Aqua Claudia (Rome)	Italy	Roman	52 AD	68.977	184,280
Aqua Julia (Rome)	Italy	Roman	33 BC	22.830	48,240
Aqua Marcia (Rome)	Italy	Roman	144 BC	91.331	187,600
Aqua Tepula (Rome)	Italy	Roman	127 BC	17.800	17,800
Aqua Trajana (Rome)	Italy	Roman	109 AD	58.000	113,100
Aqua Virgo (Rome)	Italy	Roman	19 BC	20.875	100,160
Italica (Hadrian)	Spain	Roman	2nd century AD	na	na
Italica (Trajan)	Spain	Roman	1st century AD	na	na
Carthage	Tunisia	Roman	160 AD	132.0	17,280
Dougga	Tunisia	Roman	na	12	na
Patara	Turkey	Roman	100 AD	25	na
Valens	Turkey	Roman	4th century AD	250	1,000,000

na: not available

Source: Ozis, 1996; Adam, 1988; Chanson, 2000; Chanson, 2008; Hodge, 2002; Rodgers, 2004; De Feo & Napoli, 2007; Angelakis *et al.* 2007; Nikolic, 2008; Fahlbush, 2010; De Feo *et al.* 2011.

Minoan, Greek, and Roman water public works are characterised by simplicity, robustness of operation, and the absence of complex controls.

The meaning of sustainability in modern times should be re-evaluated in light of Minoan, Greek, and Roman hydraulic works and water and wastewater management practices.

Technological developments based on sound engineering principles can have extended useful lives.

In areas of water shortage, development of a cost-effective and environmental friendly water resources management practice, based on Minoan, Greek, and Roman civilizations principles, is essential.

The use of traditional knowledge does not directly apply techniques of the past but instead, attempts 'to understand the logic of this model of knowledge' (Laureano, 2007). Traditional knowledge allowed ancient societies to keep eco-systems in balance and carry out outstanding technical, artistic, and architectural work that has been universally admired. The use of traditional knowledge has been able to renew and adapt itself. Traditional knowledge incorporates innovation in a dynamic fashion, subject to the test of a long term, achieving local and environmental sustainability. An important subject for the sustainability in developing nations of the world is to research the implementation of methods of traditional knowledge for water supply. Many of these techniques may prove to be very valuable over the more conventional (more sophisticated) ones.

The ancients for the most part lived in harmony with nature and their environment, those that did not failed. Their actions should be warnings to us. In other words the ancients have warned us. Today we do not live in harmony with nature and the environment.

Usually we define "ancient civilizations" as those confined far away into the past and, therefore, dated as "very old". However, compared to the timescale, they were the dawn of civilization; their being ancient is more properly referred to as being "young civilizations". If we relate the evolution of civilization using the human life as the timescale, rather than centuries, it would be more appropriate to recognise the "ancient civilizations" as "young" and the "modern civilization" as "old". It is well known that young people have a greater risk attitude, compared to the elderly and thus the "first civilizations" were more genuine, spontaneous and instinctive as well as having greater risk attitudes which led them toward the construction of wonderful and fantastic works, and a better understanding of the human needs and wishes. In the light of the water and wastewater technologies perspective, this is particularly true because water is the beginning of life as stated by Aristotle (Metaphysics, 983 b).

REFERENCES

Adam J. P. (1988). L'Arte di Costruire presso i Romani. Materiali e Tecniche (Roman building: materials and techniques), Longanesi, Milan, Italy.

Angelakis A. N. and Spyridakis D. S. (2010). Water supply and wastewater management aspects in ancient Greece. *Water Science and Technology: Water Supply*, **10**(4), 618–628.

Angelakis A. N., Koutsoyiannis D. and Tchobanoglous G. (2005). Urban wastewater and stormwater technologies in the ancient Greece. *Water Research*, **39**(1), 210–220.

Angelakis A. N., Savvakis Y. M. and Charalampakis G. (2007). Aqueducts during the Minoan era. *Water Science and Technology: Water Supply*, **7**(1), 95–101.

Angelakis A. N. and Koutsoyiannis D. (2003). Urban Water Resources Management in Ancient Greek Times. In: The Encycl. of Water Sci., B. A. Stewart and T. Howell (eds), Markel Dekker Inc., Madison Ave. New York, N.Y., USA, 999–1007.

Antoniou A. and Angelakis A. N. (2011). Evolution of sanitary and purgatory structures in ancient Greece. *Journal of Early Medicine Issue on Latrines, Hygiene and Health* (in press).

Athena Review (2004). History and exploration. *Journal of Archaeology*, **1**(4), 56.

Bono P. and Boni C. (1996). Water supply of Rome in antiquity and today. *Environmental Geology*, **27**, 126–134.

Chanson H. (2000). Hydraulics of Roman aqueducts: steep chutes, cascades and dropshafts. *American Journal of Archaeology*, **104**(1), 47–72.

Chanson H. (2008). The hydraulics of Roman aqueducts: What do we know? Why should we learn? Proceedings of world environmental and water resources congress 2008: Ahupua'a. Roger Babcock Jr. & Raymond Walton (Editors), Honolulu, Hawaii.

Crouch D. P. (1993). Water Management in Ancient Greek Cities, Oxford University Press, New York, USA.

De Feo G. and Napoli R. M. A. (2007). Historical development of the Augustan Aqueduct in Southern Italy: twenty centuries of works from Serino to Naples. *Water Science and Technology: Water Supply*, **7**(1), 131–138.

De Feo G., Laureano P., Drusiani R. and Angelakis A. N. (2010a). Water and wastewater management technologies through the centuries. *Water Science and Technology: Water Supply*, **10**(3), 337–349.

De Feo G., De Gisi S., Malvano C. and De Biase O. (2010b). The greatest water reservoirs in the ancient Roman world and the "Piscina Mirabilis" in Misenum. *Water Science and Technology: Water Supply*, **10**(3), 350–358.

De Feo G., De Gisi S., Malvano C., Tortora A., Del Prete S., Maurano F. and Tropeano E. (2009). The Roman Aqueduct and the Helvius' Fountain in Sant'Egidio del Monte Albino, in Southern Italy: a Historical and Morphological Approach. Proceedings of 2nd IWA International Symposium on Water and Wastewater Technologies in Ancient Technologies, Bari, Italy.

De Feo G., Mays L. W. and Angelakis A. N. (2011). Water and Wastewater Management Technologies in the Ancient Greek and Roman Civilizations. In: Treatise on Water Science, Peter Wilderer (ed.), Oxford: Academic Press, Vol. 4, 3–22.

Fahlbusch H. (2010). Water engineering and management in the classic civilizations. In: Water Engineering and Management through Time – Learning form History, E. Cabrera and F. Arregui, (eds), CRC Press, Taylor and Francis Group, Leiden, The Netherlands, pp. 75–115.

Figueiredo M. O., Silva T. P. and Veiga J. P. (2001). Contributions of ceramics chemistry to the dating of restoration works at the Aqueduct of Carthage, Tunisia (in Portuguese). In: Proceedings of 5th Conf. Inorg. Chem., Portuguese Chem. Soc., Monte-Real, Portugal.

Fonder N. and Xanthoulis S. (2007). Roman aqueduct and hydraulic engineering: case of Nîmes aqueduct and its Pont du Gard bridge. *Water Science and Technology: Water Supply,* **7**(1), 121–129.

Gikas P., Christodulakos Y., Gikas V. and Angelakis A. N. (2009). Water Supply in the Roman City of Aptera, Crete, Greece. In: Proceedings of 2nd IWA International Symposium on Water and Wastewater Technologies in Ancient Technologies, Bari, Italy.

Hodge A. T. (2002). Roman aqueducts & water supply, 2nd edition. Gerald Duckworth & Co. Ltd, London, UK.

Jones D. E. (1967). Urban hydrology: A Redirection. *Civ. Eng.*, **37**, 58–62.

Knauss J. (1991). Arkadian and Boiotan, centres of Mycenaean hydraulic engineering. *Irrigation and Drainage,* **5**, 363–381.

Koutsoyiannis D., Zarkadoulas N., Angelakis A. N. and Tchobanoglous G. (2008). Urban water management in ancient Greece: legacies and lessons. *ASCE J. Water Resour. Plann. Manag.*, **134**(1), 45–54.

Laureano P. (2007). Ancient water techniques for proper management of Mediterranean ecosystems. *Water Sciences and Technology, Water Supply*, **7**(1), 237–244.

Lofrano G. and Brown J. (2010). Wastewater management through the ages: a history of mankind. *Sci. Total Environ.*, **408**(22), 5254–64.

Mamassis N. and Koutsoyiannis D. (2010). A Web Based Information System for the Inspection of the Hydraulic Works. In: Ancient Water Technologies, L. W. Mays (ed.), Springer Dordrecht Heidelberg, London, UK and New York, USA, pp. 103–114.

Martini P. and Drusiani R. (2009). History of the water supply of Rome as a paradigm of water services development in Italic Peninsula. In: Proceedings of 2nd IWA International Symposium on Water and Wastewater Technologies in Ancient Technologies, Bari, Italy.

Mavromati E. and Chryssaidis L. (2007). Aqueducts in the Hellenic area during the Roman period. *Water Science and Technology: Water Supply*, **7**(1), 139–145.

Mays L. W., Koutsoyiannis D. and Angelakis A. N. (2007). A brief history of water in antiquity. *Water Science and Technology: Water Supply*, **7**(1), 1–12.

Monteleone M. C. (2009). Fountains in the streets of Pompeii: the evaluation of the outlet velocity and of the expected flowrate from the features of fifteen fountains. Proceedings of 2nd IWA International Symposium on Water and Wastewater Technologies in Ancient Technologies, Bari, Italy.

Monteleone M. C., Yeung H. and Smith R. (2007). A review of ancient Roman water supply exploring techniques of pressure reduction. *Water Science and Technology: Water Supply*, **7**(1), 113–120.

Murray W. M. (1984). The ancient dam of the Mytikas Valley. *American Journal of Archaeology*, **88**(2), 195–203.

Nikolic M. (2008). Cross-Disciplinary Investigation of Ancient Long Distance Water Pipelines, Ph.D. Dissertation, University of Victoria, Canada.

Niniou-Kindeli V. and Christdoulakos Y. (2004). Roman Aptera – An Initial Approximation. Proceedings of International Conference: Creta Romana e Protobizantina, Heraclion, Greece, Bottega d'Erasmo, Padova, Italy.

Ozis U. (1996). Historical water schemes in Turkey. *Water Resources Development*, **12**(3), 347–383.

Rodgers R. H. (2004). Sextus Iulius Frontinus. On the Water-Management of the City of Rome. De Aquaeductu Urbis Romae, Cambridge University Press, Cambridge, UK.

Schnitter N. J. (1994). A history of dams, the useful pyramids. A. A. Balkema, Rotterdam, Netherlands.

Smith N. (1971). A history of dams. Peter Davies, London, UK.

Tolle-Kastenbein R. (2005). Archeologia dell'acqua (Water archaeology), Longanesi–Collana Biblioteca di archeologia, Milan, Italy.

Vikatou O. (2006). Olympia – The Archaeological Site and the Museums (Archaeological Guides) Ekdotike Athenon, Athens, Greece.

Zangger E. (1994). Landscape changes around Tiryns, *American Journal of Archaeology*, **98**, 189–212.

In modern day terms Mays (2007) has defined, 'Water resources sustainability is the ability to use water in sufficient quantities and quality from the local to the global scale to meet the needs of humans and ecosystems for the present and the future to sustain life, and to protect humans from the damages brought about by natural and human-caused disasters that affect sustaining life.' The water resources sustainability issues faced by the ancients, in many ways, are not so different than what we face today.

A five-point framework for the collapse of societies was proposed by Diamond (2005): (a) damage that people inadvertently inflict on their environment; (b) climate change; (c) society's responses to its problems; (d) hostile neighbours; and (e) decreased support by friendly neighbours. Essentially the first three relate, directly or indirectly, to water resources. As pointed out by Kolbert (2006, 2009) climate shifts more than politics and war may have undermined several ancient civilizations, which left them vulnerable to internal instability and external conquest. Other studies (Hodell, *et al.*, 1995; Polyak and Asmeron, 2001; Haug, *et al.*, 2003) have also confirmed the timing of climate shifts with the timing of the failure of some ancient civilizations.

15.2 AMERICAN SOUTH-WEST

Three major cultures – the Ancestral Puebloans (Chaco Anasazi), the Hohokam, and the Mogollon – existed in the American south-west during the late pre-contact period (see Figure 15.2). The concept of prehistoric regional systems has been used to describe these cultures (Crown and Judge, 1991). The Hohokam and Chaco Anasazi regional systems have received particular attention as two of the most important. The extent of the Hohokam regional system has been defined by ball courts and material culture, and the Chaco regional system has been defined by roads and other architectural criteria. The American southwest is a difficult and fragile environment consisting of arid and semi-arid lands. The Chaco and Hohokam systems evolved in quite different environments as will be explained. The Hohokam system had the benefits of the Sonoran desert and the Gila River. The Chaco system was located in the harsh and uncertain environment of the San Juan Basin. Each of these occupied a distinctive ecological niche within the south-western environment, and as a consequence their infrastructures significantly differed.

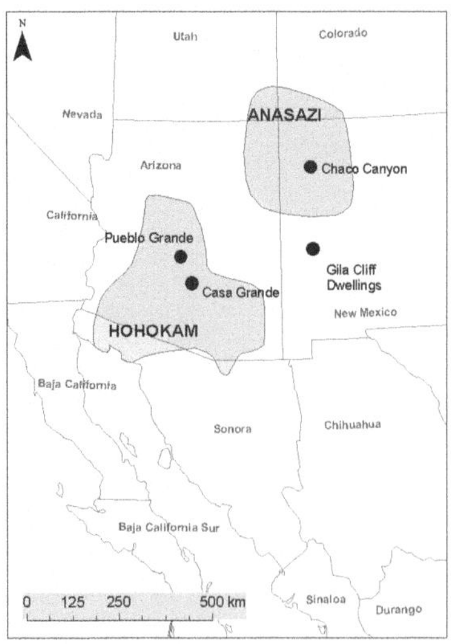

Figure 15.2 Ancient cultures in the American Southwest (with permission of Yuri Gorokhovich)

were very deeply incised causing the water to be at depths well below any usable level for corn and beans, but would have been ideal to water the deep tap roots of the mesquite.

15.2.1.2 Casa Grande and the Gila River

Casa Grande (Casa Grande Ruins National Monument) is south of Phoenix, Arizona, located in the Sonoran Desert in the drainage area of the McClellan Wash which is a tributary of the Gila River. Casa Grande (or Great House) was in the heartland of an extensive irrigation (or agricultural) society. Adjustments of the Hohokam society were tied to the preservation of their irrigation based society.

During the Pioneer period, 300–700 AD, the Hohokam lived as simple farmers in small villages along the middle Gila River. Location of the small villages depended on good arable land, a suitable location so that irrigation water could be obtained from the Gila River, and a shallow aquifer so that wells for domestic water could be dug. Between 300 and 500 AD the Hohokam acquired cultivated plants that first included cotton and tepary beans from Mexico, followed by sieva and jack beans, green-striped cushaw squash, warty squash and pigweed. In addition the Hohokam augmented their food supply with wild plants and they hunted jackrabbits, cottontails, and mule deer, along with some fish and clams from the Gila River.

Later during the Colonial period (950–1175 AD) ball courts were constructed, serving as a focal point for games and ceremonies. During the Sedentary period (950–1175 AD) the population continued to increase resulting in changes. The society was altered as a more complex and expanded canal system was built that served more than one village. This in turn required the leadership to coordinate water distribution. During the Classic period organisational changes began to occur as the population stabilised and shifted into fewer and larger villages.

15.2.1.3 Water sustainability issues of the Hohokam

In 899 AD a flood caused decentralisation and widespread population movement of the Hohokams from the Salt-Gila River Basin to areas where they had to rely upon dry farming. The dry farming provided a more secure subsistence base. Eventual collapse of the Hohokam regional system resulted from a combination of several factors. These included flooding in the 1080s, hydrologic degradation in the early 1100s, and larger communities forcibly recruiting labour or levying tribute from surrounding populations (Crown and Judge, 1991). In 1358 a major flood ultimately destroyed the canal networks, resulting in the depopulation of the Hohokam area. Culturally drained the Hohokam faced obliteration in about 1450 AD. Parts of the irrigation system had been in service for almost 1500 years and most likely were in severe disrepair: canals silted requiring extensive maintenance, and problems with salt. See Haury (1978), Hunt *et al.* (2005), Masse (1981), and Woodbury (1960) for further information.

Basically nature ultimately robbed the Hohokam society of their lifestyle. The high river flows caused the deepening of the river channel such that the canal intakes along the river could no longer divert sufficient water for irrigation. Intakes had to be moved further upstream and thus it became a struggle to continue farming. The consolidation of canal systems combined with the extension of political and religious control was one approach to circumvent the problems. The Casa Grande canal was ultimately consolidated and extended to obtain water from a location several miles upstream on the Gila River. As more catastrophic flooding occurred failure of the managerial/religious system to deal with the situation more than likely resulted in a slow societal collapse. Sometime between 1355 and 1450 AD the Hohokam abandoned their large, central settlements. The social system became decentralised as groups moved into the desert or established small villages along the Gila River. Later these tribes were encountered by the Spanish at the end of the 17th century.

15.2.2 Ancestral Puebloans (600 to 1200 AD)

15.2.2.1 Chaco Canyon

In the high deserts of the Colorado Plateau (see Figure 15.2), the Anasazi (a Dine' (Navajo) word meaning "enemy ancestors"), also called the "ancient ones", had their homeland. "Anasazi" is no longer favoured and now is referred to as "Ancestral Puebloan" to indicate these people were ancestors of the modern Pueblo people of the South-west (Mesa Verde Museum Association, 2000). When the first people arrived in Chaco Canyon, there were abundant trees, a high groundwater table, and level floodplains without arroyos. This was, most likely, an ideal environment (conditions) for agriculture in this area. Chaco is beautiful with four distant mountain ranges: the San Juan Mountains to the north; the Jemez Mountains to the east; the Chuska Mountains to the west; and the Zuni Mountains to the south.

The first Anasazi settlers, also called basket makers, arrived in Mesa Verdi around 600 AD. They entered the early Pueblo phase (700–900 AD) which was the time they transitioned from pit houses to surface dwellings, evidenced by their dramatic adobe dwellings, or pueblos. Chaco Canyon was the centre of Anasazi civilization, with many large pueblos probably serving as administrative and ceremonial centres for a widespread population of the Chaco regional system. Also of particular note is the extensive road system, built by a people who did not rely on either wheeled vehicles or draft animals. The longest and best-defined roads (constructed between 1075 and 1140 AD) extended over 80.5 km in length. The rise and fall of the Chacoan civilization was from 600 to 1200 AD, with the peak decade being 1110 to 1120 AD.

Chaco Canyon is situated in the San Juan Basin in north-western New Mexico. The basin has limited surface water, most of which is discharged from ephemeral washes and arroyos. The water, collected from the side canyon that drained from the top of the upper mesa, was diverted into canals by either an earthen or a masonry dam near the mouth of the side canyon (Vivian, 1990). These canals averaged 4.5 m in width and 1.4 m in depth; some were lined with stone slabs and others were bordered by masonry walls. The canals ended at a masonry head gate, where water was then diverted to the fields in small ditches or to overflow ponds and small reservoirs.

A three tiered multiple headgate design could have been used at Chaco Canyon for adding waste material to irrigation water to produce soluble nitrates for the field of fertiliser dehydration basins (Richard Fisher). This closely resembles the "gate and pool" system found at Lefthand Canyon by Prof. Neely. Contrasting this observation, R. Gwenn Vivian proposed that the three boxes were built at different time periods as canyon terrace levels changed.

15.2.2.2 Mesa Verde

The Ancestral Puebloan people made what is now known as Mesa Verde (Mesa Verde National Park) their home for over 700 years, from 600 to 1300 AD. Today, Mesa Verde National Park preserves this ancient culture with over 4000 known archaeological sites including cliff dwellings (see Figure 15.5a) and the mesa top sites of pithouses (Figure 15.5b), pueblos, masonry towers, and farming structures. Four reservoirs have been identified in Mesa Verde National Park: Far View Reservoir (950–1180 AD); Morefield Reservoir (750–1100 AD); Sagebrush Reservoir (950–1100 AD); and Box Elder Reservoir (800–950 AD) (Wright, 2006).

Far View Reservoir (also known as Mummy Lake), located in Chapin Mesa, was built during two different periods and was used to store water for domestic uses. Reservoir dimensions are about 90 feet (27 m) in diameter and 12 feet (3.6 m) deep (with a depth of water storage of about 4.6 ft). This reservoir structure (see Figure 15.6) contains masonry work, a diversion ditch (inlet structure), and channels. The restored inlet structure is shown in Figure 15.6. The masonry walls suggest two construction periods – one between *ca.* 900 and 1100 AD and another between *ca.*1100 and 1300 AD – based

deforested completely to build roofs, and even today the area remains deforested. Between *ca.* 1125 and 1180 AD, very little rain fell in the region. After 1180, rainfall briefly returned to normal. Another drought occurred from 1270 to 1274, followed by a period of normal rainfall. In 1275, yet another drought began which lasted 14 years.

15.3 MESOAMERICA

Mesoamerica includes Mexico and the northern Central America. The earliest Mesoamerican civilization, the Olmecs, evolved sometime before 1000 BC along the Gulf of Mexico. After about 800 BC the Mesoamerican civilization exerted social and religious influence in an area extending from the Valley of Mexico to modern El Salvador. Many Mesoamerican civilizations developed and failed for various reasons. The period or era from about 150 to 900 AD (called the Classic) was the most remarkable in the development of Mesoamerica (Coe, 1994). During the Classic period the people of Mexico and the Mayan area (see Figures 15.1 and 15.8) built civilizations comparable with advanced civilizations in other parts of the world. In Mesoamerica the ancient urban civilizations developed in arid highlands where irrigation (hydraulic) agriculture allowed high population densities. In the tropical lowlands, however, there was a dependence on slash-and-burn (milpa) agriculture which kept the bulk of the population scattered in small hamlets. Sanders and Price (1968) suggest that the non-urban lowland civilization resulted from responses to pressures set up by the hydraulic, urban civilization. Teotihuacan (City of the Gods) in Mexico is the earliest example of highland urbanism.

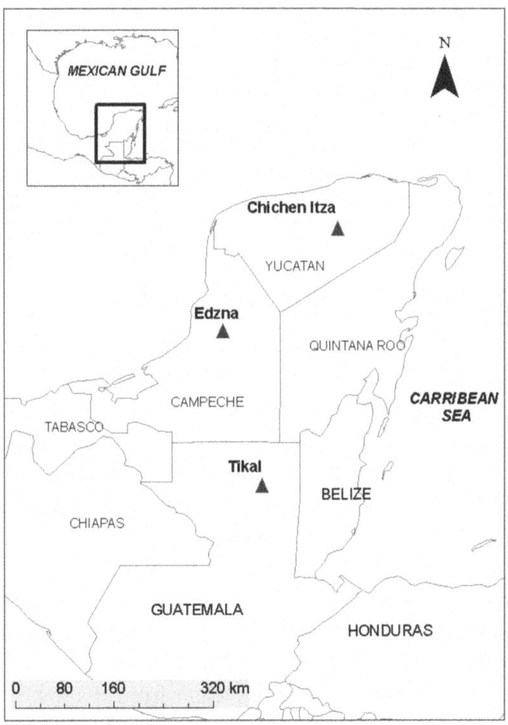

Figure 15.8 Map of Maya Area (copyright with Yuri Gorokhovich)

15.3.1 Water technologies in Mesoamerica

The regional chronology and dates of developments in various aspects of canal irrigation technology is presented in Table 15.1. In particular the technological developments for the Valley of Mexico and Oaxaca are presented. Other technologies that were not developed as a part of canal irrigation included dams: cut, fill and mortared stone construction; core walls; buttress dams; and storage dams on perennial streams (Doolittle, 1990). Other water control devices that were developed included: broad-type head and sluice gates; arched aqueducts; flow control boxes; drop structures; chutes; pipes; and inverted siphons (Doolittle, 1990).

Table 15.1 Approximate dates of developments in various aspects of canal irrigation technology (dates from Doolittle, 1990).

Approximate date	Water related technology
1400 AD	Relocation of rivers, large masonry aqueducts
800	Advanced relocation of ephemeral streams
400	Rock diversion dams, trellised aqueducts
300 AD	Relocation of ephemeral streams
200 BC	Use of valley bottoms, advanced channelisation, furrows
300	Use of perennial streams, earth and brush diversion dams, Head gates, small earthen aqueducts
350	Use of permanent springs, sluice gates
550	Masonry dams (arch) with floodgates, chiselled canal
750	Earthen dams
900	Incipient channelisation, weirs, water spreaders
1200 BC	Use of ephemeral streams, site drainage, small canals, Rock storage dams (gravity) with spillway

15.3.2 Teotihuacan Empire (300–600 AD)

Teotihuacan was a very impressive civilization which evolved about twenty-five miles north of Mexico City (see Figure 15.1). Prior to 300 BC, Teotihuacan valley had a small population spread over the valley and was the dominant urban centre in Mesoamerica throughout the classic period. By *ca.* 100 AD Teotihuacan covered an area of 12 km^2 which has been linked to the development of so-called hydraulic agriculture (Haviland, 1970). The urban area expanded in size, there was an increased socio-economic diversity, and an expanding political influence. At its height, around 600 AD, Teotihuacan was fully urban with a population of approximately 85,000 people and covering an area of 19 km^2 (Haviland, 1970). Others (such as Millon, 1993) have estimated that the maximum population was approximately 125,000 during the Xolalpan phase. Teotihuacan was the largest urban centre of the time in Mesoamerica.

Around 300 BC the use of canals for irrigation rapidly spread throughout the central highland basin, the location of Teotihuacan (Doolittle, 1990). South of Teotihuacan, near Amanalco, Texcoco east of the Basin of Mexico, irrigation would have consisted of diverting water from shallow spring-fed streams into simple irrigation canals and then onto fields only a few metres away. Flood water systems were also used. North-east of Teotihuacan, south of Otumba, a series of ancient irrigation canals (dating between 300 and 100 BC) were excavated. Other canals in the same area date to *ca.* 900 to 1600 AD. Evidence also exists of canals that were built between *ca.* 200 and 800 AD near Teotihuacan. One of these was the first confirmed relocation of a natural stream. The reader can refer to Doolittle (1990) for further information on these canals.

Teotihuacan (city of the Gods), Mexico was abandoned mysteriously around *ca.* 600 to 700 AD. Teotihuacan was a very impressive civilization which evolved about 40 km north of Mexico City at about the same time as Rome. During this time the collapse of civilized life occurred in most of central Mexico. One possible cause was the erosion and desiccation of the region resulting from the destruction of the surrounding forests that were used for the burning of the lime that went into the building of Teotihuacan. The increasing aridity of the climate in Mexico may have been a related factor. The entire edifice of the Teotihuacan state may have perished from the loss of agriculture. Even though the city had no outer defensive walls, it was not an open city easy for hostile outsiders to attack. The collapse of Teotihuacan opened civilized Mexico to nomadic tribes from the north. Human malnourishment has been indicated by skeletal remains.

Of the five-factor framework for social collapse suggested by Diamond (2005), the only factor that did not play a role in the collapse of the Anasazi was hostile neighbours. Water resources sustainability was affected by the deforestation, the erosion (cutting) of the arroyos from the diversion of water resulting in lowering of the groundwater levels and the supply source to the irrigated fields, and finally, the repeated periods of drought caused the final collapse.

15.3.3 Xochicalco (650–900 AD)

After the disintegration of Teotihuacan's empire in the seventh century AD, foreigners from the Gulf Coast lowlands and the Yucatan Peninsula appeared in central Mexico. Cacaxtla and Xochicalco, both of Mayan influence, are two regional centres that became important with the disappearance of Teotihuacan. Xochicalco (in the place of the house of flowers), was located on a hill top approximately 38 km from modern day Cuernavaca, Mexico, and became one of the great Mesoamerican cities in the late classic period (650–900 AD). Despite the Mayan influences, the predominant style and architecture is that of Teotihuacan. There were no rivers or streams or wells to obtain water, so rainwater harvesting was the source of water. Rainwater was collected in the large plaza area and conveyed using drainage structures (see Figure 15.9a) and drainage ditches (Figure 15.9b) into cisterns such as the one shown in Figure 15.9c. From the cisterns water was conveyed to other areas of the city using pipes as shown in Figure 15.9d. The collapse/abandonment of Xochicalco, most likely, resulted from drought, warfare, and internal political struggles.

15.3.4 The Maya

The ancient Maya lived in a vast area covering parts of present-day Guatemala, Mexico, Belize, and the western areas of Honduras and El Salvador. Mayans settled in the last millennium BC and their civilization flourished until around 870 AD. The environment that the Mayans lived in was less fragile than that of the semi-arid lands where the Ancestral Puebloans and Hohokam lived. Tikal was one of the largest lowland Maya centres, located some 300 km north of present day Guatemala City. The city was located in a rain forest setting with a present day average annual rainfall of 135 cm. The urbanism of Tikal was not because of irrigated (hydraulic) agriculture. A number of artificial reservoirs were built in Tikal, which became more and more important as the population increased.

the hydraulic system are not known; however, it is safe to say that the rainwater was used for drinking. The modern Maya use the water in the ancient channels for drinking and bathing until about February when the water level is low and the channels are full of aquatic life and unfit for drinking (Matheny, 1976). Normal dry season is from January through May. The Maya of the late Classic period chose not to dig wells and only 12 chultans have been found at the site (Matheny, 1976).

It is possible that the water system of Edzna is an example of how the Maya people of the lowlands controlled and used water. The typical Mayan chultan were bottled shaped in cross-section with a narrow restricted neck and a large globular-shaped chamber below. The chultan at Edzna is slightly more than 5 m deep. Frequently the chultans were lined with plaster to prevent seepage. The average capacity was 7500 gallons, enough to supply around 25 people for one year (Matheny *et al.* 1983).

Rainfall varies significantly from the north (46 cm/year) to the south (254 cm/year) of the Yucatan Peninsula. The soils are also deeper in the southern part resulting in more productive agriculture and consequently supported more people. Rainfall was very unpredictable, resulting in droughts that destroyed crops. Ironically though, the water problems were more severe in the wetter southern part. Ground elevations increased from the north to the south causing the depths down to the water table to be greater in the south.

15.3.4.2 Water resources sustainability of the Maya

Centuries before the Spanish arrived, the collapse of many other great Mayan cities occurred within a fairly short time period. Several reasons have emerged as to why these cities collapsed, including overpopulation and the consequential exhaustion of land resources possibly coupled with a prolonged drought. A drought from 125 AD until 250 AD caused the pre-classic collapse at El Mirador and other locations. A drought around 600 AD caused a decline at Tikal and other locations. Around 760 AD a drought started that resulted in the Mayan classic collapse in different locations from 760 to 910 AD.

The soil of the rain forest is actually poor in nutrients so that crops could be grown for only two or three years, then to go fallow for up to 18 years. This required ever-increasing destruction of the rain forest (and animal habitat) to feed a growing population. Other secondary reasons for the collapse include increased warfare, a bloated ruling class requiring more and more support from the working classes, increased sacrifices extending to the lower classes, and possible epidemics. The Maya collapsed as a result of four of the five factors in Diamond's (2005) framework. Trade or cessation of trade with friendly societies was not a factor for the Maya. Water resources sustainability was most likely a factor in the collapse of the Maya.

15.3.5 The Aztec Empire (1150–1519 AD)

Starting in the 12th century the nucleus of Aztec Empire was the Valley (or Basin) of Mexico. Approximate boundaries of the Aztec Empire are shown in Figure 15.1. Surrounded by enemies in their beginning, they built a city in the middle of a lake. The city of Tenochtitlan (see Figure 15.11), located on a reclaimed island in the saline Lake Texcoco, was the capital of the Aztec Empire. The Aztecs built aqueducts as well as flood control works in addition to a new form of irrigation. The Aztec overcame the lack of conventional farming land by building floating fields, *chinampas*, in Lake Texcoco (Woolf, 2005). In shallow areas of the lake (see Figure 15.11a) the *chinampa* were built up using earth and plants. These artificial fields were held in place using wooden poles driven into the lake bed. In deeper areas of the lake reed beds, anchored to the lake bottom, were filled using earth. Obviously such a system was subject to flooding and could be surrounded easily by enemies. The chinampa cultivation may have originated with the Teotihuacan in the early Classic period (Coe, 1994).

By 1465 AD a new aqueduct was rebuilt to replace the first one with a chain of islands connected with hollowed out logs. It's possible that remains of the old islands were used. The new islands were much larger than the original ones, with widths of 10 to 12 m and lengths of tens of metres, and were higher above the water level (Doolittle, 1990). The new aqueduct super structures were masonry with a lime-based mortar and dimensions of 1.6 m high and 2.5 to 3.0 m top width, with a wider width at the bottom. The trapezoidal shaped superstructures were built on a foundation of mixed lime and small stones and the foundation was also supported using pilings. The new aqueduct consisted of two conduits, which were mortared as the rest of the superstructure. Each conduit was trapezoidal with a 75 centimetre (cm) depth, and 30 cm width at the bottom and 60 cm width at the top.

Another important aqueduct was built to obtain water from Acuecuexcatl Springs, located in the southern part of the valley in the neighbouring kingdom of Coyoacan. After the Acuecuexcatl aqueduct was completed around 1900 AD and water flowed through the aqueduct, large uncontrollable quantities of flow caused flooding in the city of Tenochtitlán.

Flood control works were built in the early part of the Aztec Empire (Raynal-Villasenor, 1987). This was the first comprehensive flood control works in Mesoamerica. Roads served an important role as structural measures for flood control. The Tacuba road was the first built by the Aztecs around 1418 AD Later the Tlaltelolco–Atzcapotzalco road and the Coyoacan–Xochmilco roads were built. With these three roads the fresh waters of the west and south-west parts of the valley were confined. The most important flood control project of the Aztecs was the Netzahualcoyotl's dyke. This dyke was 16 km long and 20 m wide. The dyke was completed along with the construction of the Cuitlahuac road, which divided the Chalco and Xochmilco lakes, and the construction of Mexicaltzingo road which divided the Mexico and Xochimilco lakes.

Mesoamericans had a large number of fertility gods and goddesses, of which the rain god was among the most senior. The Aztecs were faithful worshippers of the rain god, Tlaloc (see Chapter 1), whose cult dates back as far as the Olmec civilization (Woolf, 2005). Tlaloc was honoured with sacrifices in the form of blood and other offerings. He stored rainwater in four huge jars, which he kept in the north, south, east, and west, and from the eastern jar he sent life giving rains, and from others storms and droughts (Woolf, 2005).

15.4 ADVANCEMENT OF WATER TECHNOLOGY AFTER THE SPANISH CONQUEST

The Valley of Mexico, *ca.* 1519 on the eve of the Spanish conquest of Mexico is shown in Figure 15.11a. The siege of Tenochtitlán occurred in 1521 by the Spanish conquistador Cortez. Water was not only the major contributing factor to the development of the Aztec civilization, but was also its downfall during its final days as Cortez destroyed the aqueducts that brought fresh water from the mainland (Back, 1981). After the Aztec Empire was conquered, the Spaniards rebuilt the aqueducts and continued to use the spring water until the mid 1850s.

From the destruction and rubble the Spaniards built the City of Mexico and continued to be plagued with the same water management problems of the Aztecs. The last large remnant of the lakes that exists today is Lake Texcoco, which has always contained salt and still causes hydrologic and environmental problems, as it has for the last seven centuries (Back, 1981).

The Spanish built several aqueducts including the Chapultepec aqueduct (Figures 15.12 and 15.13), the Queretaro Aqueduct (Figures 15.14, 15.15 and 15.16), the Morilia aqueduct (Figure 15.17), and the Tepotzotlán aqueduct (Figure 15.18). What did the Spanish accomplish based upon what the indigenous people had accomplished and what did they learn from the Romans? The Spanish had obviously learned a tremendous amount about water management and aqueduct building from the many Roman water structures and aqueducts built in Spain. Also they must have gained a lot of knowledge from the

driest of the middle to late Holocene, which coincided with the collapse of the Classic Mayan civilization. Such a continuous climate proxy record provides evidence that climate deterioration occurred in the Mayan region during the terminal Classic period.

According to Haug *et al.* (2003) 'a seasonally resolved record of titanium shows that the collapse of Maya civilization in the Terminal Classic Period occurred during an extended regional dry period, punctuated by more intense multiyear droughts centred at approximately 810, 860, and 910 AD'. These investigators studied the bulk-titanium content of undisturbed sediment in the riverine environment in the Cariaco Basin of the southern Caribbean and the hydrologic cycle over the northern tropical South America. A century-scale decline in rainfall put a general strain on resources in the region, which was then exacerbated by abrupt drought events. These events would have obviously contributed to the social stresses that led to the failure or collapse of the Maya.

The Ancestral Puebloans of the American south-west achieved perhaps the most advanced civilization of any Native American group. Researchers have surmised that the rise and fall of these people was related to climate shifts. Polyak and Asmeron (2001) used stalagmites from the Carlsbad Cavern and Hidden Cave in the Guadalupe Mountains, New Mexico to develop a late Holocene climate reconstruction for the south-western United States. They used annual banding, hiatuses, and high-precision uranium-series dating to develop a 4000-year annually resolved climate history. The researchers reported that the stalagmite record correlates well with the archaeological record of changes in cultural activities of indigenous people. From 4000 to 3000 years ago a present-day-like climate existed and a much wetter and cooler period occurred from 3000 to 800 years ago. This was followed by the present-day-like conditions with a slightly wetter period from 440 to 290 years ago. Climate change was not the sole reason, but is certainly an important underlying explanation for the cultural shifts and evolution in the arid south-western United States during the late Holocene. Refer to Kolbert (2006, 2009) to study the role of climate shifts on the ancient civilizations.

What relevance does the collapse or decline of ancient civilizations have upon modern societies? Learning from the past and discovering the reasons for the success and failure of other societies seems very logical. We certainly are a much more advanced society than those of the ancient societies, but will we be able to overcome the obstacles to survival before us? The collapse of some civilizations may have been the result of the very processes that had been responsible for their success (e.g. the Mayans and Romans and others).

REFERENCES

Ackerly N., Howard J. and McGuire R. H. (1987). La Ciudad Canals: A Study of Hohokam Irrigation Systems at the Community Level. Arizona State University, Anthropological Field Studies 17, Department of Anthropology, Arizona State University, Tempe, AZ, USA.

Andrews J. P. and Bostwick T. W. (2000). Desert Farmers at the River's Edge: The Hohokam and Pueblo Grande. 2nd Edn, Pueblo Grande Museum and Archaeological Park, City of Phoenix, Phoenix, AZ. USA.

Back W. (1981). Hydromythology and ethnohydrology in the new world. *Water Resources Research*, **17**(2), 257–287.

Bauer B. (1998). The Sacred Landscape of the Inca: The Cuzco Ceque System. University of Texas Press, Austin, TX, USA.

Bauer B. (1922). The Development of the Inca State. The University of Texas Press, Austin, TX, USA.

Bribiesca-Castrejon J. L. (1958). Drinking water in the Mexican Republic: Water Supplies in Pre-Hispanic Times. *Ingeniera Hidraulica en Mexico (Hydraulic Engineering in Mexico)*, **12.2**, 69–82 (in Spanish).

Coe M. M. (1993). The Maya. 5th edition. Thames and Hudson, Inc., New York, USA.

Coe M. M. (1994). Mexico: From the Olmecs to the Aztecs. 4th edn. Thames and Hudson, Inc., New York, USA.

Crown P. L. and Judge W. J. (eds) (1991). Chaco and Hohokam: Prehistoric Regional Systems in the American Southwest. School of American Research Press, Santa Fe, NM, USA.

D'Altroy T. (2003). The Incas. Blackwell Publishing. Hoboken, NY, USA.

de Menocal P. B. (2001). Cultural responses to climate change during the late Holocene. *Science*, **292**(5517), 667–673.

Dahlin B. H. (1983). Climate change and prehistory on the Yucatan peninsula. *Climate Change*, **5**, 245–263.

Dart A. (1989). Prehistoric Irrigation Arizona: A Context for Canals and Related Cultural Resources. Center for Desert Archaeology, Technical Report 89–7, Prepared for Arizona State Historic Preservation Office, Arizona State Parks Board, Phoenix, AZ, USA.

Diamond J. (2005). Collapse: How Societies Choose to Fall or Succeed, Viking, New York, USA.

Doolittle W. E. (1990). Canal irrigation in prehistoric Mexico: The sequence of technological change. University of Texas Press, Austin, TX, USA.

Fairley J. P. (2003). Geologic water storage in pre Columbian Peru. *Latin American Antiquity*, **14**(2), 193–206.

Falkenmach M. and Lindh G. (1993). Water and economic development. In: Water in crisis: Guide to the World's Freshwater Resources. P. H. Gleick (ed.), Oxford University Press, Inc., New York, USA, pp. 81–91.

Folan W. J., Gunn J., Eaton J. D. and Patch R. W. (1983). Paleoclimatological patterning in southern Mesoamerica. *Journal of Field Archaeology*, **10**, 453–468.

Folan W. J. and Hyde B. H. (1985). Climatic forecasting and recording among ancient and historic Maya. In: An Ethnohistorical Approach to Epistemological and Paleoclimatological Patterning, in Contributions to the Archaeology and Ethnohistory of Greater Mesoamerica, W. J. Folan (ed.), Southern Illinois University, Carbondale, IL, USA.

Garraty J. A. and Gay P. (eds) (1972). The Columbia history of the world. Harper and Row Publishers, Inc., New York, USA.

Gunn J. and Adams R. E. W. (1981). Climate change, culture and civilization in North America. *World Archaeology*, **13**, 87–100.

Haviland W. A. (1970). Tikal, Guatemala and Mesoamerican urbanism. *World Archaeology*, **2**(2), 186–198.

Haug G. H., Gunther D., Peterson L. C., Sigman D. M., Hughen K. A. and Aeschlimann B. (2003). Climate and collapse of Maya civilization, *Science*, **299**(5613), 1731–1735.

Haury E. W. (1978). The Hohokam: Desert farmers and craftsmen. University of Arizona Press, Tucson, AZ, USA.

Hemming J. (1982). Monuments of the Incas. Little, Brown, Boston, USA.

Hodell D. A., Curtis J. H. and Brenner M. (1995). Possible role of climate in the collapse of Classic Maya civilization. Nature, **375**, 391–394.

Hunt R. C., Guillet D., Abbott D. R., Bayman J., Fish P., Fish S., Kintigh K. and Neely J. A. (2005). Plausible ethnographic analogies for the social organization of Hohokam canal irrigation. *American antiquity*, **70**(3), 433–456.

Hyslop J. (1990). Inka settlement planning. University of Texas Press, Austin, TX, USA.

Kolbert E. (2006). Field Notes from a Catastrophe: Man, Nature, and Climate Change. Bloomsbury, New York, NY, USA.

Kolbert E. (2009). Outlook: Extreme. *National Geographic*, pp. 60–65, April.

Labrador D. (2001). Climate changes coincide with cultural shifts in ancient American southwest. *Scientific American*, October 5.

Lekson S. H., Windes T. C., Stein J. R. and Judge W. J. (1988). The Chaco canyon community. *Scientific American*, **259** (1), 100–109.

Lister R. H. and Lister F. C. (1983). Those who came before. Southwestern Parks and Monuments Association. University of Arizona Press, Tuscon, AZ, USA.

Masse B. (1981). Prehistoric irrigation systems in the Salt River Valley, Arizona. *Science*, **214**(3), 408–415.

Masse B. (1991). The quest for subsistence, sufficiency and civilization in the Sonora Desert. In: Prehistoric Regional Systems in the American Southwest, P. L. Crown and W. J. Judge (eds). Chaco & Hohokam., School of American Research Press, Santa Fe, New Mexico, USA, pp. 195–224.

Matheny R. T. (1976). Maya lowland hydraulic systems. *Science*, **193**(4254), August 10.

Matheny R. T., Gurr D. L., Forsyth D. W. and Hauck F. R. (1983). Investigations of Edzna, Campeche, Mexico 1(1). The Hydraulic System Papers, 46. Brigham Young University. New World Archeological Foundation, Provo, UT, USA.

Mays L. W. (2007a). Water Resources Sustainability. McGraw–Hill, New York, USA.

Mays L. W. (2007b). Water sustainability of ancient civilizations in Mesoamerica and the American southwest. *Water science and technology: water supply*, **7**(1), 229–236.

Mays L. W. (2010). Ancient Water Technologies, Springer Science and Business Media. B.V., Dordrecht. The Netherlands.

Millon R. (1983). The place where time began: an archaeologist's interpretation of what happened in Teotihuacan history. In: Teotihuacan: Art from the City of the Gods, K. Berrin and E. Pasztory (eds). The Fine Arts Museums of San Francisco, San Francisco, USA.

Mesa Museum Association, in cooperation with Mesa Verde National Park, Far View: An Ancient Farming Community, 2000.

Messenger L. C., Jr. (1990). Ancient winds of change: Climate settings and prehistoric social complexity in Mesoamerica. *Ancient Mesoamerica*, **1**, 21–40.

Patrick H. R. (1903). The Ancient Canal System and Pueblos of the Salt River Valley, Arizona. Bulletin No. 1, Phoenix Free Museum, Phoenix, AZ, USA.

Polyak V. J. and Asmeron Y. (2001). Late Holocene climate and cultural changes in the southwestern United States. *Science*, **294**, 148–151.

Raynal-Villasenor J. A. (1987). The remarkable hydrological works of the Axtec civilization. *Water for the Future:* Hydrology in Perspective Proceedings of the Rome Symposium, IAHS Publication No. 164.

Sanders W. T. (1981). Ecological adaptation in the basin of Mexico: 23,000 BC to the present. In: Supplement to the handbook of middle American Indians, V. Riefler Bricker and J. A. Sabloff (eds), Vol. I, Archaeology, University of Texas Press, Austin, USA, 147–197

Sanders W. T. and Price B. J. (1968). Mesoamerica: The Evolution of a Civilization. Random House, New York, USA.

Schimmelmann A., Lange C. B. and Meggers B. J. (2003). Palaeoclimate and archaeological evidence for a 200-yr recurrence of floods and droughts linking California, Mesoamerica and South America over the past 2000 years. *The Holocene*, **13**, 763–778.

Thomas D. H. (1994). Exploring Ancient Native America: An Archaeological Guide. Macmillan, New York, NY.

Turney O. (1929). Prehistoric irrigation in Arizona. *Arizona Historical Review*, **2**(5), Phoenix, AZ, USA.

Vivian R. G. (1974). Conservation and diversion: Water control systems in the Anasazi southwest. Irrigation Impact on Society, Anthropological Paper. In: T. Downing and M. Gibson (eds), No. 25, University of Arizona, Tucson, AZ, USA, pp. 95–112.

Vivian R. G. (1990). The Chacoan prehistory of the San Juan basin. Academic Press, San Diego, CA, USA.

Woodbury R. B. (1960). The Hohokam canals at Pueblo Grande, Arizona. *American Antiquity*, **26**(2), 267–270.

Woolf G. (General Editor) (2005). Ancient Civilizations. Duncan Baird Publishers, London, UK.

Wright K. R. (2006). The Water Mysteries of Mesa Verde, Johnson Books, Boulder, CO, USA.

Wright K. R. and Valencia Zegarra A. (1999). Ancient Machu Picchu drainage engineering. *Journal of Irrigation and Drainage, ASCE*, **125**(6), 360–369.

Wright K. R. (2006). Tipon: Water Engineering Masterpiece of the Inca Empire. ASCE Press, Reston, Virginia, USA.

Wright K. R. and Valencia-Zegarra A. (2000). Machu Picchu: A Civil Engineering Marvel. American Society of Civil engineers ASCE Press, Reston, Virginia, USA.

Wright K. R., Kelly J. M. and Valencia-Zegarra A. (1997). Machu Picchu: Ancient hydraulic engineering. *Journal of Hydraulic Engineering, ASCE*, **123**(10), 838–843.

Chapter 16

Water supply of Athens in the antiquity

E. D. Chiotis and L. E. Chioti[1]

Gilgamesh is awesome to perfection. It was he who opened the mountain passes, who dug wells on the flank of the mountain.

(The Epic of Gilgamesh, Tablet 1, Kovacs)

16.1 INTRODUCTION

This chapter focuses on the efforts for water supply in ancient Athens, under rather adverse conditions, over a period of 12 centuries from the sixth century BC up to the sixth century AD. The exploitation of local and distant water resources is examined on the basis of archaeological data in a geological context.

The historical background, the climate and the hydrogeological conditions are examined in this introduction. The solutions adapted or invented for the water supply in the ancient city of Athens were certainly shaped to a great degree by the above critical factors. Particular emphasis was given to the development of the underground water resources, because of the inadequate capacity of springs and rivers in the vicinity of Athens.

16.1.1 Historical introduction

The development of the city-states in ancient Greece was based to some degree on the necessary public infrastructure of hydraulic and drainage works. Engineering achievements such as the Eupalinean tunnel on the island of Samos as well as the Peisistratean aqueduct and the Great Drain in Athens made possible the expansion of the cities in antiquity.

Polycrates, the tyrant of Samos, engaged engineer Eupalinos in the sixth century BC to carry out various public works; among them the most impressive is the "Eupalinean" tunnel for the transfer of water in terracotta pipes from a spring across a mountain to the city. The tunnel, 1036 m long, was dug horizontal and begun from both sides of the mountain by two teams; they advanced in a straight line and met, with a remarkable accuracy, in the middle of the tunnel.

[1]Co-author of Section 6.

A common feature of these ancient works is their sustainability; although they were forgotten for a long period, they were accidentally rediscovered and put in operation in modern times. It was only recently realised that the National Garden in Athens has been constantly irrigated with a daily rate of about 1200 m^3 from an ancient aqueduct. This happened when the foundation of an underground parking station in 2004 cut off the ancient aqueduct from the Hymettos Mountain at a depth of 12 m.

The Hadrianic aqueduct, constructed in the middle of the second century AD, was identified in 1847 during the cleaning works of a supposed "spring", which actually was the water overflow from a collapsed section of the ancient underground aqueduct. The Hadrianic aqueduct was then laboriously cleaned over a length of 20 kilometres and remained the main source of water of Athens till 1935, when the Marathon reservoir was constructed. By then all efforts for water supply were mainly limited to the resources of the Athenian hydrological basin (Chiotis, 2008).

The Great Drain, the stone-built channel for the drainage of the Athenian Agora, was made early in the fifth century BC. This relieved the low lying Agora from the frequent flooding during heavy rainstorms and made possible the architectural growth in the flat part of the Agora. It was during the excavations in 1932 that heavy rainstorms eroded the ground and disclosed the Great Drain which still drains the Agora today. No doubt, these works are not only subject to admiration, but also models of a skilful design (Chiotis, 2011).

Political power in Athens from the eighth to the sixth century BC was in the hands of several large aristocratic families which controlled large areas of Attica, in the territory around Athens. As happened in many other Greek cities in the sixth century BC, a tyrant seized power in Athens in 546 BC. The first great hydraulic work of Athens is the Peisistratean aqueduct, constructed in the period of the tyrant Peisistratos.

Cleisthenes, the founder of the Athenian Democracy, instituted a crucial reform in 508 BC in his attempt to break up the aristocratic power structure. The city-state of the ancient Athens played since then an important role in the shaping of Western civilization. Statesmen, poets, writers, artists and philosophers flourished here during the fifth and fourth centuries BC, when Athens was the leading city-state in Greece. Two more aqueducts were constructed in that period, the Hymettos and the Acharnian one.

The influence of Rome in Athens starts in 86 BC, when the Romans besieged the city. Athens recovered soon and prospered through the second century AD; it is during this period when the Roman Emperor Hadrian initiated the Hadrianic aqueduct.

Hard times began for Athens in the third century, when the city was damaged by northern invaders, the Herulians, in 267 AD. More barbarian incursions occurred in the following centuries. The aqueducts and their distribution networks were among the first victims of the barbarian hostilities. However, a new aqueduct was constructed in the Late Roman period in the fifth century AD which also supplied a perfectly maintained water mill in Athens.

It is difficult to estimate the population of Athens over time in antiquity; as an indirect measure of population, the extent of the city is mentioned hereafter. The boundaries of the city are delineated by the remnants of the walls which after the Persian invasion in 490 BC protected an area of about 460 acres (1.85 km^2). The Athenian *Agora*, the political and cultural centre of the city, occupied the central part of the fortification. The city fortification was extended much later, in a period of political turmoil and instability, a few years before the Herulian invasion in 267 AD. This fortification surrounded an additional area of about 100 acres (0.4 km^2), which comprised the expansion of Athens during the Roman period. After the Herulian invasion, the city was rebuilt and the new fortification, the Post-Herulian wall, enclosed a reduced area of about only 30 acres (0.115 km^2); even the old Agora was left outside the new fortified circuit.

As noted by Judeich (1931 p. 47) Plato's model is more than a faint game of imagination. In harmony with Plato, Judeich (p. 46) considers that a continuous cover of blue-grey Cretaceous limestone extended westwards from Anchesmos (Tourkovounia) to Aigaleo; furthermore, he assumes that the limestone cover collapsed due to erosion of the supporting underlying layers and that it is only maintained on the hilltops.

Marinos *et al.* (1971) performed a detailed stratigraphic study of the formation of the Athenian schists and settled the conflicting views supported in the past; the stratigraphic sequence is schematically presented in Figure 16.2, in the middle. They reached the conclusion that the formation of the Athenian schists with the marls, the limestones of Acropolis etc. are all part of the same stratigraphic unit of Upper Cretaceous age, mainly Senonian. The upper part of the formation is composed of a thick series of marls, platy limestones and sandstones, crowned in the hills by the Upper Cretaceous limestones. Shales and phyllites participate in the lower part of the formation.

Based on the comparative evaluation of the geological maps and the geologic literature it is summarised that the formation of the Athenian schists represents flyschoid sediments composed of an upper unit of sediments such as marls, sandstones etc., dated to the Maastrichtian, and a lower unit of shales and phyllites; gradual transition between them is locally observed. The Upper Cretaceous limestones on the top of the hills in the Athenian basin are masses locally enclosed in the upper unit of the Athenian schists; it has been, therefore, suggested that these limestones were emplaced as large olistholiths into the upper unit of the Athenian schists (IGME, Kifissia sheet).

The above geotectonic views are presented in both, the simplified geological map in Figure 16.3 and the corresponding geological sections in Figure 16.4. The geological formations of the map in Figure 16.3 are classed into groups of similar hydrogeological behaviour, which is the particular point of interest of this presentation. The original data were compiled from the 1:50,000 geological map of Greece of the Hellenic Institute of Geology, namely the sheets of Eleusis, Kifissia, Koropi and Peiraias. The geotectonic model introduced in the Kifissia sheet is extended over the whole Athenian basin in Figure 16.3. The older routes of the rivers Kifissos, Ilissos and Eridanos are depicted according to the geological sketch of Judeich (1931, his Fig. 6). It is noted that the Ilissos River was technically diverted eastwards, to flow directly into the sea in 1930s, in order to reduce the frequent flooding of the Kifissos valley.

Attica is structured by two major geotectonic units, the Pelagonian zone comprising the mountains of Parnitha and Aigaleo, and the Attic-Cycladic mass, to which Penteli and Hymettos belong. The Pelagonian zone is overthrust over the autochthonous Attic members along a linear "suture" zone of Alpine formations, as depicted in the sections of the Figure 16.4.

The Mesozoic members of the Pelagonian zone comprise non-metamorphosed carbonate sediments resting on Middle Triassic to Palaeozoic phyllites and schists. The numerous springs at the foothills of Parnitha are mostly connected with this setting of carbonate rocks overlying schists. The Mesozoic carbonate members of Penteli and Hymettos are, by contrast, metamorphosed into marbles. The carbonate rocks of the surrounding mountains, either limestones or marbles, are the main water reservoirs of the Athenian basin. In addition, the limestones in the vicinity of the ancient Athens on top of the hills of Acropolis, Areopagus and Nymphs are the most significant water reservoirs in the city, especially because they are underlain by less permeable layers (Figure 16.2, bottom).

Thus, the formation of the Athenian schists is presented in Figure 16.4 as a unique tectonic unit delimited between the overthrust fronts of the Figure 16.3. It is intensively folded and emplaced eastwards over the Neohellenic Tectonic Nappe Unit (NHTN) and underlain below the Pelagonian zone; it is tentatively named the Aphidnae-Acropolis tectonic unit.

16.2 MANAGEMENT OF LOCAL WATER RESOURCES IN ANCIENT ATHENS

Local water resources from springs, wells and other complex underground works have never been abandoned, despite their low yield, even when water was available from aqueducts. Furthermore, they were the exclusive solution for areas not supplied by the aqueducts. On the other hand, some deep wells were, in a few cases, so productive that continuous pumping was necessary during their excavation. In addition to the wells, cisterns for the storage of rainwater and sophisticated systems of wells and galleries were very common.

The concern and the efforts for independent private water accumulated very good knowledge of the geological rules controlling underground water as well as experience for the stability of rocks in underground excavations, cisterns in particular. The experience for the exploitation of local water resources was developed progressively over a long period of time.

16.2.1 Springs

Water springs, the most convenient source of water, were very few in ancient Athens; a couple of them existed near the bed of the Ilissos River and four minor ones around the Acropolis. The modest limestone rock of the Acropolis has always been the principal point of Athens. The area around the hill was settled already in the Neolithic times, thanks to a favourable combination of geomorphologic and geological conditions. The steep hill and the caves in the limestone provided a safe shelter for the control of the broader fertile area, as well as water springs at the contact of the limestone cliff with the underlying marly unit.

The Klepsydra, located at the NW slope of the Acropolis and the Asclepieion spring at the SW side were the most significant ones of the four springs shown in Figure 16.6. Klepsydra is considered by Parsons (1943) as the most copious of the series of small springs which girdle the Acropolis; its hardness led people to describe it, ever since antiquity, as brackish and unfit for drinking.

Klepsydra itself was though seasonally inadequate even for the Neolithic people; within a short radius of the spring no less than 20 shallow Neolithic wells or pits have been excavated, the purpose of which was clearly to "tap the underground vein of water" (Parsons, 1943). During the Turkish siege of Acropolis in 1826, the Klepsydra was protected by the Greeks by a bastion to secure the supply of about two cubic metres of water daily (Cordellas's, 1879, 69)

On the other hand Cordellas quotation that the spring provided water to the tower of winds is not sufficiently documented. As noted by Parsons (1943, 223) 'it was not until Turkish times that the water of Klepsydra was piped off in the direction of the Market of Caesar and Augustus; and it is unlikely that it ever served to run the water-clock in the Horologion of Andronicus; throughout antiquity it was carried in a north-westerly direction, clearly toward the ancient Agora'.

A meaningful remark for the fountain Klepsydra is mentioned in the comments related to Aristophanes' comedy Lysystrata: 'the fountain was named Klepsydra for the reason that in some instances it overflows, in others it is poor' (Parsons, 1943, 265). This comment makes sense when taking into account that a couple of days after a good rainfall the fountain's rate of flow is significantly increased; in dry periods the water flow is reduced and seeps into the ground along the numerous faults and fractures (Figure 16.2, bottom).

Greater importance is attributed in recent studies to the Asclepieion spring than to Klepsydra; it has been revealed that the Post-Herulian wall also extended in the south side of the Acropolis in order to encircle the Propylaia, the monumental entrance to the citadel, as well as the Asclepieion spring which was vital for the water supply of the Acropolis. By contrast, the flow of the Klepsydra is considered to have been intermittent, as suggested by its name (Gioles, 2005, 26).

The existence of another spring can be possibly deduced from Strabo's quotation of an earlier Hellenistic author, Kallimachos, referring to "the junction of the rivers" of whom one only is referred by name, i.e., Eridanos. In his Collection of the Rivers, Kallimachos says that 'it makes him laugh if anyone makes bold to write that the Athenian virgins "draw pure liquid from the Eridanos," from which even cattle would hold aloof' (Shear, 1997).

Apparently, this passage has to do with marriage rites. On the wedding day the bribe was washed with water from a spring or a fountain house. Confusion perhaps has been caused from the fact that the name Eridanos was used in antiquity not only for the river originating from Lycabettus but also for a tributary stream of the Ilissos River. This tributary is labelled as Eridanos in Kaupert's map of Attica and its junction with Ilissos is about 1.5 km upstream of the Kallirrhoe spring (close to Athens's Hilton Hotel, at the crossing point of Michlakopoulou and S. Merkouti Streets); the same stream is mentioned by Gräber (1905) as pseudo-Eridanos (p. 58). With reference to the particular spring, Kallimachos mentions further that 'its sources are indeed existent now, with pure and potable water, as they say, outside the so-called Gates of Diochares, near the Lykeion', a description that matches with the above mentioned junction.

16.2.2 Wells

The Neolithic wells near the Klepsydra spring, a couple of metres deep, were the forerunners of more than 400 deeper wells found and carefully studied in the Agora excavations (Figure 16.6). Digging wells was a standard practice all over the city of Athens, due to the lack of springs and rivers (Camp, 1977). Early in the sixth century BC, Solon was appointed as a lawgiver; among others he passed a law to regulate and encourage the use of wells. According to it 'wherever there was a public well within half a mile this was to be used; where it was farther away, one could dig one's own: but if, having dug to a depth of 60 feet, one did not find water one was permitted to fill a five-gallon jar twice a day from one's neighbour's well' (Travlos, 1971, 11). After passing his legislation, Solon went into voluntary exile to avoid being pressured into amending his legislation.

Most of the wells in the Agora and especially the deeper ones were dug in lenses of sandstones and platy limestones of the marly unit; they were mostly concentrated in the hill slopes north of Areopagus and along the N-S trending hill of Agoraios Kolonos, shown in Figure 16.5. Few and rather shallow wells were also dug in the alluvial sediments in the flat part of the Agora.

The earliest wells were usually unlined; a lining was sometimes built in the sixth century BC with masonry of small stones, and from the fourth century drums consisting of strong and well made terracotta sections were used. Some wells proved abortive and were at once abandoned; some were in use for only a few years, others for several centuries (Camp, 1977).

Deep wells of high capacity were dug in the so called industrial district and near the "old Kallirrhoe spring" (No 1 in Figure 16.6). The latter one yielded so much water that it was necessary to pump it in order to control the water inflow. A conduit was found at a depth of 22 m directed eastwards, which was impossible to follow because of the small dimensions. According to Gräber (1905, 14) it could not have been but a pipe of an aqueduct; it was at a depth of about 60 m above sea level and its mouth could be north of Areopagus in the Agora.

In the excavations of the industrial district SW of the Agora a deep well was dug, the so called great shaft, to a depth of 24.85 m without finding bottom; the flow of water was so great that it seemed hopeless to try to bail it out to greater depth. The industrial district was self-sufficient in water, supplied from many wells; the area was a favourable location for digging wells from the Prehistoric till the Byzantine times. There was sufficient water not only for domestic use but also for workshops and baths; the central part of the area

by the authors that another function was performed improving the efficiency of the system. The tunnels captured additional rainwater, which infiltrated directly into the system through the ground above the tunnels.

This combination of complex nets, including cisterns, wells and tunnels was also applied in the north slopes of the Areopagus hill, as well as in the labyrinth of the underground works revealed by Dörpfeld in the excavations west of the Odeion of Herodes Atticus.

16.2.3.2 The Areopagus slopes

The underground water potential in the north slopes of the Areopagus was known and intensively exploited by the use of underground works from the fifth century BC till the Late Roman times. The sloping topography and the stability of the rocks in the area were also favourable for the construction of cisterns.

An early and very simple installation of the fifth century BC is described by Thompson (1958). A small rectangular collecting basin (0.50 × 0.80 m) made of rubble stone work was set near the bottom of the hill slope, apparently at the site of a water seepage. From this basin the water was carried northward in a conduit made of reused, round terracotta pipes of the archaic type.

A source of water maintained in use for long was a well on the slopes of the Areopagus (well N 19:2, House A) which was first tapped in the first century AD and was reused in the fifth or possibly even the sixth century AD. For its re-use the well itself was closed above by means of a vaulted spring house accessible through a door and a short flight of steps. The water flowed from the well to the House A northward, over a horizontal distance of 40 m and an elevation difference of 7–8 m in a brick-vaulted tunnel (Thompson, 1958, 147).

An impressive underground system described by Shear was found in the Areopagus houses west of the Panathenaic way. A large brick-lined vaulted tunnel leads into the hillside sloping from north to south at a grade of about 3 to 1 for a distance of 29 m, at which point the floor becomes flat and continues horizontal for another 7 m. At that point the tunnel opens into a tile-lined well. The date of the construction of the tunnel is in the early Roman period, and it seems probable that this elaborate engineering achievement was part of the water-supply system (Shear, 1939).

It is surprising that similar ancient shafts were widespread all over the city and reused in houses of the 19th century AD. The shafts were lined with terracotta sections; they were equipped at the bottom with chambers and conduits or galleries with a vaulted roof made of bricks, as described by Cordellas (1879, 70).

An elaborate and still usable installation of the third century AD afforded the Areopagus House C an ample water supply. Springs farther up the slope were tapped by a large rock-cut shaft from which the water was led off through a channel protected by a vaulted tunnel cut through bedrock and paved with rectangular tiles. The channel continued under the floor of the springhouse and emptied into the pool of a Nymphaeum (Franz *et al.* 1988).

16.2.3.3 Dörpfeld's Excavation

Dörpfeld excavated the area in the valley between the hills of Acropolis and Nymphs along the road of Apostle Paul from 1892 to 1898; among others he revealed a real labyrinth of underground hydraulic works, consisting of a net of interconnected twisting galleries, many wells and cisterns (Figure 16.9). The unprecedented complexity of this net, the diligence of the excavations and the detailed topographical surveying compose indeed a great contribution to the understanding of the elaborate underground water systems developed in ancient Athens. Dörpfeld invited Gräber, a specialist on hydraulics, to evaluate the excavation findings; it is thanks to his study that we possess a detailed description of the fascinating underground works demonstrating the great struggle for the supply of water in the dry environment of Athens.

Garden. Ziller's sketch (Figure 16.11 and 28 in Ziller's plate VIII) depicts a typical Roman tunnel, similar to the outlet of Hadrian's aqueduct in the Agora. Given its location and that the floor is about 8.4 m below the ground, the elevation of the floor is estimated at about 81.5 m, below the level of the "Old Kallirrhoe Spring". Ziller attempted to correlate the channel with the Peisistratean aqueduct on the reasoning that it is a repair of an older aqueduct which was damaged during works for the extension of the theatre's scene.

A critical point for the course of this hypothetical branch or any other aqueduct in the area is the geomorphological saddle between Acropolis and Lycabettus at the elevation of 83.2 m after Gräber (83.9 m on modern 5k map), at the crossing of the Kydhathenaion and Hadrian's streets. For the tunnel to reach 83.85 m at the endpoint near the "Old Kallirrhoe Spring", Gräber estimated (p. 22) that the aqueduct should pass at least 4 m above the ancient ground at the saddle point.

It is argued by the first author that the twisting galleries described, of both the upper and the lower level, constitute a complex local system, designed for capturing both rain and underground water, irrelevant to the Peisistratean aqueduct. The capturing function of the underground works is indicated also from the stalactites found during the excavations; they have been formed in the galleries of the upper level, at a depth of about 10 m from the surface, under the rock of Acropolis. The water capturing function of the upper gallery is also accepted by Gräber in his remark that the floor of the upper tunnel at the eastern end is shaped in the form of a basin which is maintained full of water before overflow starts westwards (Figure 16.10). The infiltration of water into the marly unit is facilitated by the dense faulting shown at the bottom section in Figure 16.2.

It the author's opinion the archaeological findings do not justify the assumption of a southern branch of the Peisistratean aqueduct extending from the National Park to the Theatre of Dionysus and further east to the Odeion of Herodes Atticus; however, an aqueduct branch of similar trend at a lower elevation cannot be excluded. Furthermore, the ancient aqueduct in the National Garden does not belong to the Peisistratean aqueduct, as explained below on the basis of the recent findings in the excavations for the Metropolitan Railway which have revealed the actual Peisistratean branch in the area.

16.3 AQUEDUCTS SUPPLYING THE AGORA

At least four aqueducts supplied water to the Agora over the period from the sixth century BC to the sixth century AD, namely the Peisistratean, the Hemyttos, the Hadrianic and the Late Roman aqueduct. The drainage net and the water lines in the Agora follow a pattern imposed by the topography. All water supply lines were brought into the Agora from the SE side, at relatively high elevation, whereas the drainage lines converged to the same point at the Eridanos River (Figure 16.12).

The Peisistratean pipeline in the Agora brought water to one of the earliest public buildings, the Southeast Fountain House (Figure 16.13). It was perhaps the most effective motive for the attraction of the public interest to an area of past cemeteries, usually flooded during the rainstorms. The water was distributed in Athens through terracotta pipelines, first in the late sixth century BC. The conduit was made of round terracotta pipes which underlay the east-to-west road at the southern side of the Agora. The terracotta pipes, commonly described as pipes of Peisistratean or Archaic type, vary in size in different branches. They are in general about 0.25 m in diameter, slightly wider near the centre and about 0.65 m in length. Each pipe section has a large oval access hole near the upstream end, designed to facilitate sealing of the joints on the inside and to permit subsequent cleaning of the line if necessary. Pipes of this type have been in use for some centuries, mainly for drinking water, but occasionally for drainage too.

Another section of a pipeline with similar features was found at P2 over a length of 15 m, again in a carved trench; it is also dated from the late sixth century BC (Zachariadou, 2000b).

A similar terracotta pipeline of significant length was found in front of the Greek Parliament (location K in Figure 16.14), in the Syntagma Square, in three sections with a total length of about 160 m. The pipes are of the Peisistratean type; however, they are dated in the first half of the fifth century BC (Zachariadou, 2000a). They constitute therefore a branch later than the Peisistratean pipeline mentioned earlier. The pipe sections are connected with lead to avoid the leakage of water; the oval access holes are again placed near the upstream end, as indicated by the flow direction of fused lead, in the exhibition of archaeological findings at the Syntagma Station.

We correlate this branch with the Kimonian pipeline found in the Agora, dated in the second quarter of the fifth century BC, although the Syntagma pipeline might be a bit earlier; we use the term Kimonian net for both of them to indicate that they belong to the same distribution net, running north of the Eridanos River, not necessarily the same branch. Evidence of a local branch departing from the main line was also found in the excavation at Syntagma Square (Zachariadou, 2002).

In our opinion, to the Kimonian net should also be allocated not only the nearby findings in Kerameikos as suggested by Shear (1984), but also the discovery of the pipeline along the ancient road to the Academy by Karagiorga-Stathakopoulou (1978).

A corridor of aqueducts (Aq in Figure 16.14) was disclosed at the southern end of the same excavation, at the crossing with a local stream of the Eridanos River. Fortunately, the findings have been relocated and are exhibited at the University campus in Zographou; the aqueducts are dated to the Roman Age, in the very instructive posters of the Exhibition. In addition, close and parallel to the above hydraulic works, a late Roman stone-built straight channel was revealed over a length of 47 m, which is considered as a drainage channel (Zachariadou, 2000a, 158); however, in the author's opinion, it has the features of an aqueduct, which might have been used for drainage at a later stage. This site is a typical key point through which aqueducts pass consequently over various periods, apparently due to restrictions imposed by the topography. We believe that it is a candidate location for the route of both the Hadrianic as well as for a Late Roman aqueduct for the supply of water in a NE–SW direction.

Another important finding, identified as a section of the Hadrianic aqueduct, was excavated at the location H in Figure 16.14 (Spathari & Chatzioti, 1983), at the north-east corner of the National Garden over a total length of about 50 m. The aqueduct was built in a trench 0.5 m deep in the natural rock (Athenian schists); the external width was 2.05 m and the walls were 0.55–0.60 m thick, allowing for an inside width of 1 m.

The lateral walls were maintained to a height of 1 to 1.5 m and they ended in a vault. The direction of the aqueduct was N16°W and the flow direction was from north to south.

An additional section of exactly the same construction was excavated later over a length of 26 m at the extension of the previous one. It was described in this case as an early Roman conduit, similar to that revealed on the Amalias Avenue and many other sites in Athens. It was considered as a section of a central drainage system, on the basis of the fact that six pi-shaped terra cotta pipes ended in its interior (Zachariadou, 2000b). In our opinion, the interpretation of this bold channel in the National Garden as a branch of the Hadrianic aqueduct for the supply of water to the Hadrianic Athens is plausible.

Another important aqueduct, described by Ziller (112–113) is the ancient aqueduct of the Royal Garden; it is composed of a tunnel 12–14 m below the surface and wells along the route of the tunnel. Ziller followed this aqueduct from the western side of the National Garden up to Rizareios, near the point P1 in Figure 16.14. This aqueduct has been misidentified as the Peisistratean aqueduct by Judeich and Tölle-Kastenbein. However, the route of the Peisistratean pipeline in this area was recently uncovered by Lygouri-Tolia, as already mentioned. There is no evidence suggesting that these dissimilar hydraulic works are of the same

The Peisistratean aqueduct is only partially known from the Peisistratean and the Kimonian pipelines shown in Figure 16.15. It has been the forerunner of the Hemyttos aqueduct with which it should have a similar path, along the Ilissos River; because of that it is commonly confused with the apparently later Hymettos aqueduct. Unfortunately, its course from Athens to the Hemyttos Mountain remains unknown. It is most likely that it has been found, so far, at one location only, at Goudi, not far from the tunnel of the Hymettos aqueduct. The tunnel floor there was 7.65 m below the surface and its section was 1.2 m high by 0.8 m wide; a well was also found, 1 m in diameter, connected with the tunnel (Alexandri, 1975).

The Hymettos aqueduct, as named by Ziller, is a combination of lateral channels collecting water from springs and underground capturing works into a tunnel 12–14 m below the surface. The tunnel follows the upper part of the Ilissos riverbed and then it turns westwards towards Athens. It continues currently to capture underground water from the riverbed, the overlying sediments and perhaps from unknown lateral branches; it continues to supply about 1200 m^3 daily to the National Garden, although there is no current contribution from springs.

The stone built channel in the Agora, constructed in the fourth century BC and the Agia Trias aqueduct described by Ziller were most likely branches of the Hymettos aqueduct, in the author's opinion. During the recent tunnelling works for the Metropolitan Railway a tunnel was discovered at the northern side of the Metropolitan Church, passing through a broad chamber. It is hardly a coincidence that it is found along the route of the Agia Trias aqueduct after Ziller's Plate VII. In any case, the Hemyttos aqueduct seems to be an expansion and amplification of the Peisistratean aqueduct undertaken not later than the early fourth century BC.

The Acharnian aqueduct is a hydraulic work of the early fourth century BC (Platonos-Yiota, 2004) that brought water from springs in the foothills of the Parnitha Mountain. Large symmetric half-sections of terracotta, were joined together to form an elliptical conduit for the flow of water; the conduit was buried in the earth, some metres below the surface. The upper part of the route of the Acharnian aqueduct, up to the Kifissos River, is known from excavations of the aqueduct as well as from ancient inscriptions, describing the terms of agreement between the commissioners of the aqueduct and the landowners.

According to the inscriptions, as evaluated by Vanderpool (1965), the commissioners of the aqueduct are permitted to dig underground collecting galleries in the water-bearing strata near the springs at any place and to any depth that they see fit. This combination of water resources is a typical feature of all the ancient aqueducts in the dry land of Attica; they usually collected water from springs where available, but in addition they captured underground water which was actually the main supply of the aqueducts. It is guessed that a similar practice of digging wells on the flanks of the mountains is suggested in the lyrics of the Epic of Gilgamesh quoted in the introduction.

There are two alternative assumptions for the continuation of the Acharnian aqueduct southwards, shown in Figure 16.16. Vanderpool's interpretation that the aqueduct brought water to Athens is quite feasible topographically. The alternative interpretation (Petropoulakou & Pentazos, 1973) assumes that the aqueduct served for the irrigation of the area of the Academy north of Athens.

It seems that the drought in the fourth century BC, inferred on the basis of archaeological evidence by Camp (1982), was faced in Athens by the construction of both the Hymettos and the Acharnian aqueducts. Another period of drought might have triggered the construction of the Hadrianic aqueduct. Among the monuments on the terrace of the north side of Parthenon, Pausanias described the statue of the Goddess Earth, beseeching Zeus for rain. Pausanias suggests that the Athenians needed rain or a drought had plagued the whole Greece, whereas Stevens (1946) suggests that the statue dates to the Hadrianic period.

The Hadrianic aqueduct collected water from springs on the slopes of the Parnitha Mountain in the north through two channels, which finally merged into one at the point A (Figure 16.17). However, most of the water of the aqueduct was of underground origin. The tunnel which was over most of the route below the

deformation of the tunnel locally. The tunnel inclination from the bottom of the first well at 211.9 m in the northern end is about 2.5‰ over a length of 1.5 km; then comes the steepest section up to the Kifissos River with an average dip of 13‰ over a path of 2.5 km and finally the inclination is maintained more or less constant at 2.2‰ for the rest 16 km beyond the Kifissos River. The Hadrianic tunnel ended at the ancient cistern on the Lycabettus hill at an elevation of 133.37 m on the floor.

The construction of the aqueduct started in 125 AD in the period of the Roman Emperor Hadrian and was completed after his death in 140 AD. During the operation in the past century, the flow rate of the Hadrianic aqueduct itself in the summer season is estimated to 7000 m^3 daily. It is expected though that the flow rate might be higher in antiquity, if there was water input from the river itself, as commonly was the practice with the Roman aqueducts.

The Late Roman aqueduct is known from the excavations in the Agora where it supplied water for the operation of water mills; a proposal for the route of this aqueduct is presented in the next section.

16.6 WATERMILLS OF THE ATHENIAN AGORA

Among the ancient watermills shown in Figure 16.13, the only one that has been confidently identified and dated is the "central" one, m2, which was annexed to the Post-Herulian Wall; it was discovered in 1933 and remains one of the best preserved ancient watermills. It was in operation from the third quarter of the fifth century AD up to the last quarter of the sixth century (Parsons, 1936, 88; Thompson, 1959, 69–70; Thompson & Wycherley, 1972, 214; Frantz *et al.* 1988, 81).

As phrased by Spain (2008), Parsons (1936, Figure 16.10) produced an unusual and stimulating interpretation of the evidence skilfully comparing and illustrating his analysis and findings with the well-known text of Vitruvius (Vitruvius, De architectura, X.4.1, X.5.1, X.5.2). Spain reconsidered and improved some technical elements of Parsons' interpretation (Spain, 1987, 335).

Both the wheel-race, A, and the mill room, B, with the gear pit were excavated and found in good condition (Figure 16.19). The wheel-race, some 5.5 × 1.1 × 4.2 m, was dug 3 m deep into the bedrock to create a high head. The diameter of the water-wheel is measured as 3.24 m from the scoring on the lime deposits on the wheel-pit wall. A water channel (head-race, Figure 16.19 DC) with an internal section 0.42 × 0.42 m approaches from the south taking a straight course directly towards the wheel over the last 20 m. The wheel was vertical overshot, it received the impulsion of the water on its upper perimeter and turned in a counter-clockwise direction as seen from the west. The head-race delivered water at a height of 1.4 m above the top of the wheel (Parsons, 1936, 76–82).

A slot, measuring internally 0.80 × 6.40 m, was excavated in 1959 by Thompson (Thompson, 1960, 349, Figure 16.1, Frantz *et al.* 80–81) in the area of the pronaos of the Southeast Temple and was interpreted as a pit for a mill-wheel (Figure 16.20). Some slight vestiges, possibly of the mill-room, were observed to the west. The mills' aqueduct came from the east, across the Post-Herulian Wall (aq in Figure 16.20). It is worth noting that a small part of it survived above the east foundation of the Southeast Temple (mr in Figure 16.20). Thompson suggested that the water was further carried northward to drive the central mill and thereafter the precious water was carried once more northward in an arched aqueduct that implies the existence of a third mill (Thompson, 1960, 349; Frantz *et al.*1988, 81).

The necessity of a system for the diversion of water, before feeding the central mill, was plausibly supported by both Reynolds (Reynolds 1983, 42) and Spain (Spain, 1987, 336; Spain, 2008, 56–57); however, nothing of the kind was located. Fortunately, the careful study of the rich archive of excavation photos of the ASCSA (American School of Classical Studies at Athens) revealed that such a diversion channel existed at the supposed wheel-race of the south mill, m1. One can observe the diversion channel (dc) in Figure 16.21 (Image 1997.15.0014, http://www.agathe.gr) along with the wheel-race (wr) of the

It is argued that the Late Roman aqueduct comprised the Perissos and Kapodistriou water bridges which are dated to the fifth century AD (Leigh, 1998, 62). It is furthermore supported that the Late Roman aqueduct comprised also the Kifissia (Figure 16.16, 8.1) and Herakleion (Figure 16.16, 8.2) aqueduct branches which after passing over the above mentioned water bridges, were joined into a single channel according to Ziller (1877). The aqueduct was explicitly described by Ziller up to a point in the vicinity of Aghia Glykeria at Galatsi, where Ziller lost its continuation. The final part of the aqueduct is unknown but it can be approximately traced on the topographic map between the two extreme alternatives shown on Figure 16.16, as 8.3 and 8.4.

It is supported by the authors that this extension is topographically quite feasible, both from the Kapodistriou water bridge at an elevation of *ca.* 140 m and even from Perissos, elevated at about 123 m. In respect with the second one it is noted that over a length of about 10 km, from the Perissos water bridge to the National Garden of Athens at an elevation of 100 m, the average inclination would be 2.3‰; this is comparable to the inclination of 2.2‰ for the Hadrianic aqueduct over a length of 16 km, from the crossing point with the Kifissos river up to the ancient water cistern on the Lycabettus hill (Chiotis, 2008).

It is noted that the Kifissia branch passes along three Roman cisterns (Skilardi, 2005, 53–60). It is likely therefore that the Late Roman aqueduct, feeding the Agora watermills, would be part of a complex water management system that supplied water from springs at Kifissia and Herakleion. By means of the cisterns the surplus of water from local consumption would be supplied to Athens (Chiotis & Chioti, 2011).

16.7 CALCITE DEPOSITION IN PIPES AND AQUEDUCTS

The original water quality is often and arbitrarily associated with the deposition of calcite either in the terracotta distribution pipes, as in the case of the Eupalinean aqueduct, or in aqueduct tunnels or channels, as in the case of the Hadrianic aqueduct. However, calcite formation might not necessarily be related to the original quality of water. Calcium carbonate sintering is also a serious problem in the drainage of modern tunnels, the study of which has elucidated the calcification mechanism (Gamisch & Girmscheid, 2003; Dietzel *et al.* 2008).

It has been clarified in modern tunnels, that the formation of calcite deposits is influenced by biological and chemical processes in the upper layers of the soil. The rain water seeps in the water-unsaturated soil whose pores are filled with subsurface air. This air contains less oxygen and considerably more carbon dioxide because of the soil respiration and the restricted adjustment with the atmosphere. Thus the seepage water contains considerable amounts of carbon dioxide and mineral nutrients and absorbs numerous further substances such as humic acid, ammonium compounds, nitrite, etc. from the soil. The calcium hydroxide [$Ca(OH)_2$] in the concrete in the tunnels, and also the sodium aluminate [$NaAl(OH)_4$] of alkali sprayed concrete catalysts raise the pH-value of the seepage water when it comes into contact with the building materials. A shift of the carbonic acid balance reduces the solubility of calcium-ions and results in the precipitation of the solved calcium-ions in the form of calcium carbonate. To reduce the calcium carbonate precipitation the amount of soluble calcium compounds has to be minimised. This can be achieved in the tunnels by using concrete with puzzolanic additives which do not contain any lime (Gamisch & Girmscheid, 2003).

It is suggested therefore that the lime of concrete in the walls of the Roman aqueducts is a source of calcium, which is dissolved during the flow of water to the destination site. The original calcium ions in the water would have been depleted early before reaching the destination, if the ions were not replenished by additional dissolution. Contrary to the Eupalinean aqueduct, calcite deposition was not a problem with the Peisistratean or the Kimonian terracotta pipelines.

Epilogue

In the dry land of Athens water supply was a difficult task; it was approached through the systematic exploitation of both, the local water resources in the city as well as the surface and underground water from the distant aquifers of the surrounding mountains in Attica.

The search for water created a good geological experience adequate for the institution by Solon of rules for the wise management of underground water, as well as Plato's assumption on the geological past of the limestone hills surrounding the ancient city, in his dialogue Kritias. The importance given by the Athenians to the correct management of water and the drainage of rains is reflected in Plato's Laws (books 6 and 8) which appear to have as a prototype the drainage and water supply nets of the Athenian Agora.

The common characteristic of the ancient aqueducts was the exploitation of underground water, in addition of course to the use of springs; however, the major contribution was that of the phreatic zone. This is clearly attested in the case of the Hemyttos aqueduct; although there is currently no contribution from springs, the Hemyttos aqueduct still supplies about 1200 m^3 daily, sufficient for the irrigation of the National Garden. Furthermore, Ziller had correctly noticed for the ancient Long Wall aqueduct, shown in Figure 16.16, that it drew the water from the Ilissos riverbed along a tunnel below the riverbed, accessible through wells, a few metres apart laterally. The Hadrianic tunnel was also a continuous system of underground water capture, as correctly observed by Ephie Nestoridou (oral communication).

Thanks to the recent excavations in the centre of Athens two distribution lines of the Peisistratean aqueduct can be traced, although the Peisistratean aqueduct itself remains unknown. The stone-built channel of the fourth century BC and the Agia Trias aqueduct are correlated with the Hymettos aqueduct. The commonly believed extension of the Peisistratean aqueduct from the National Garden south of the Acropolis, towards the Theatre of Herodes Atticus is considered as an unjustified assumption. The hydraulic works W-SW of the theatre are considered by the authors as a complex net of works designed to capture underground water.

The Roman water bridges at Perissos and Kapodistriou are topographically lower than the Hadrianic tunnel and cannot not be therefore related to the Hadrianic aqueduct, as sometimes considered. They are assigned to the Late Roman aqueduct which finally supplied water for the operation of the watermills in the Athenian Agora.

It remains an open question whether the Acharnian aqueduct supplied water to the city of Athens, it is clear though that this was topographically possible.

Drainage was a critical prerequisite for city planning due to the uncommon but heavy rainstorms which could cause sudden floods in the ancient Agora. The understanding of the historical evolution of the drainage and hydraulic works provides the necessary background to follow up the planning of Athens and Agora in particular.

It is impressive that the solutions applied in antiquity were practical and sustainable and that the technical works for water supply and drainage are operational today.

Acknowledgements

We greatly benefited from the important contribution on the subject by Cordellas, Ziller and Gräber in a period when direct observations on ancient hydraulic works were possible. The American School of Classical Studies at Athens is the most prolific source of published and publicised data on the Athenian Agora since the 1930s, thanks to which our evaluation of the water supply conditions in Athens could be more specific. The findings of the recent archaeological excavations by the Greek Archaeological authorities, under particularly pressing conditions, are acknowledged as the most precious component for the integration of previous data.

The permission of the American School of Classical Studies at Athens for the publication of the photos and drawings in Figures 16.8, 16.19, 16.20, 16.21 and 16.22 is gratefully acknowledged.

The long involvement of the first author with this subject, at the expense of other obligations or alternatives, would not be possible without the support of his family.

Thanks are also addressed to Prof. P. Marinos at the Technical University of Athens, for the suggested essential improvements to this work, as well as to Dr. A. Angelakis.

REFERENCES

Alexandri O. (1975). Arch. Deltion. Ministry of Culture, Athens, Greece, 30, 27.

American School of Classical Studies at Athens, Athenian Agora Excavations, http://www.agathe.gr, 13/01/12.

Andronopoulos B. and Koukis G. (1976). Engineering geology study in the Acropolis area, Athens, Engineering geology Investigations No 1, IGME, Athens, Greece.

Camp J.McK., II (1977). The water supply of ancient Athens from 3000 to 86 BC, Ph.D. Thesis, Princeton Univ., New Jersey, USA.

Camp J.McK., II (1979). A drought in the late eighth century BC. *Hesperia*, **48**, 397–411.

Camp J.McK., II (1982). Drought and famine in the 4th century BC, *Hesperia Supplements*, 20, Studies in Athenian Architecture, Sculpture and Topography, pp. 9–17.

Camp J.McK., II (1996). Excavations in the Athenian Agora: 1994 and 1995, *Hesperia*, **65**, 231–261.

Chiotis E. (2008). Revision of views on the aqueduct of Hadrian in Athens and its relationship with previous hydraulic works, 5th Symposium of the Hellenic Society of Archaeometry, Athens, Greece.

Chiotis E. and Chioti L. (2011). Industrial activities in the ancient Agora of Athens in the Late Roman times, In: The Agora in the Mediterranean from Homeric to Roman time, A. Giannikouri (ed.). Ministry of Culture, Athens, Greece, pp. 181–196.

Chiotis E. (2011). Aqueducts and drainage works in the ancient Agora of Athens. In: *The Agora in the Mediterranean from Homeric to Roman times*, A. Giannikouri (ed.), Ministry of Culture, Athens, Greece, pp. 165–180.

Cordellas A. (1879). Αθήναι εξεταζόμεναι υπό υδραυλικήν έποψιν, (Athens examined from the point of view of hydraulics) , Cordellas (Finokalias press), Athens, Greece, 172.

Dietzel M., Rinder T., Leis A., Reichl P., Sellner P., Draschitz C., Plank G., Klammer D. and Schöfer H. (2008). Koralm Tunnel as a Case Study for Sinter Formation in Drainage Systems–Precipitation Mechanisms and Retaliatory Action. *Geomechanics and Tunnelling*, **1**, 271–278.

Frantz A., Thompson H. A and Travlos J. (1988). Late Antiquity: AD 267–700. *The Athenian Agora*, **24**, 1–156.

Gamisch T. and Girmscheid G. (2003). Life-cycle maintenance management of traffic tunnels–strategy assessment to develop new calcification reduction methods in tunnel drainage system. In: System-based Vision for Strategic and Creative Design, F. Bontempi (ed.). A. A. Balkema, Amsterdam, The Netherlands. 1253–1258.

Gräber W. (1905). The Enneakrounos., Mitteilungen Deutsch. *Arch. Inst., Athenische Abteilung*, **30**, 1–64.

Gioles N. (2005). The Athens in the first Christian centuries., G. Gelbesis, Athens, Greece, p. 142.

IGME (Institute of Geology and Mineral Exploration). Geological Map of Greece 1:50,000, Sheets Athens-Eleusis, Kifissia, Koropi-Plaka, Athens-Peiraias.

Judeich W. (1931). Topography of Athens. C. H. Beck, Munich, Germany, p. 473.

Karagiorga-Stathakopoulou Th. (1978). *Arch. Deltion*. Ministry of Culture, Athens, Greece, **33**, 21–22.

Kovacs M. G. (1989). The Epic of Gilgamesh, **4**, Stanford University Press, Stanford, California, USA.

Marinos G. P. and Petrascheck W. E. (1956). Laurium, Inst. Geology, Athens, Greece, 247.

Marinos G. P., Katsikatsos G., Georgiades-Dikaioulia E. and Mirkou R. (1971). The Athens' Schists Formation I: Stratigraphy and Structure, *Annales Géologiques des Pays Helléniques*, Athens, Greece, **23**, 183–216.

Leigh S. (1998). The Aqueduct of Hadrian and the water supply of Roman Athens. Ph. D. Thesis, University of Pennsylvania, USA, p. 365.

Lygouri-Tolia E. (2000). Evangelismos Station, In: The City under the City, L. Parlama and N. Stambolodis (Eds.). Museum of Cycladic Art, Athens, Greece, pp. 208–217.

Parsons A. W. (1936). A Roman water-mill in the Athenian Agora, *Hesperia*, **5**, 70–90.

Parsons A. W. (1943). Klepsydra and the Paved Court of the Pythion, *Hesperia*, **12**, 191–267.

Petropoulakou M. and Pentazos E. (1973). ATTICA: Residential elements- first report, Ancient Greek Cities, **21**. Center of Ekistics, Athens, Greece.

Platonos-Yiota M. (2004). ACHARNAI, Historical, and topographic overview of Ancient Acharnai, of the neighbouring Municipalities and the Parnitha fortifications, Municipality of Acharmai, Athens, Greece, p. 539.

Reynolds T. S. (1983). Stronger than a Hundred Men: A History of the Vertical Water Wheel, Johns Hopkins University Press, Baltimore, Maryland.

Shear T. L. (1939). Excavations in the Athenian Agora: Sixteenth Report, *Hesperia*, **8**, 201–246.

Shear T. L., Jr (1984). The Athenian Agora: Excavations of 1980–1982, *Hesperia*, **53**, 1–57.

Shear T. L, Jr (1997). The Athenian Agora: Excavations of 1989–1993. *Hesperia*, **66**, 495–548.

Skilardi D. (2005). Attic Road 2004: Excavations, Findings, New Museums. Activities of the B′ Ephorate of Classical and Prehistoric Antiquities, Athens, Greece, pp. 53–60.

Spathari E. and Chatzioti M. (1983). Hadrianian Aqueduct, Arch. Deltion. Ministry of Culture, Athens, Greece, **38**, 24–25.

Spain R. J. (1987). The Roman Watermill in the Athenian Agora: A New View of the Evidence, *Hesperia*, **56**, 335–353.

Spain R. (2008). The power and performance of Roman water-mills: hydro-mechanical analysis of vertical-wheeled-water-mills, BAR International Series 1786. John and Erica Hodges Ltd., Oxford, UK.

Stevens G. P. (1946). The northeast corner of the Parthenon, *Hesperia*, **15**, 1–26.

Thompson H. A. (1940). The Tholos of Athens and its predecessors, *Hesperia Supplements*, **4**.

Thompson H. A. (1958). Activities in the Athenian Agora: 1957, *Hesperia*, **27**, 145–160.

Thompson H. A. (1960). Activities in the Athenian Agora: 1959, *Hesperia*, **29**, 327–368.

Thompson H. A. (1968). Activity in the Athenian Agora: 1966–1967, *Hesperia*, **37**, 36–72.

Thompson H. A. and Wycherley R. E. (1972). *The Agora of Athens: The history, shape and uses of an ancient city center*. The American School of Classical Studies at Athens, Princton, New Jersey, USA.

Tölle-Kastenbein R. (1994). The archaic net of water conduits for Athens and its later construction phases. P. Von Zabern and Mainz am Rhein, Germany, 120.

Travlos J. (1971). Pictorial dictionary of Ancient Athens, Praeger Publishers, New York and Washington DC, USA, 361.

Young R. S. (1951). An industrial district of ancient Athens, *Hesperia*, **20**, 135–288.

Vanderpool E. (1965). The Acharnian Aqueduct, *Χαριστήριον εις Α.Κ. Ορλάνδον* (Honorary Volume in Memory of A. K. Orlandos), **1**, 166–174.

Zachariadou O. (2000a). Syntagma Station, In: *The City under the City*, L. Parlama and N. Stambolodis (Eds.) Museum of Cycladic Art, Athens, Greece, pp. 148–161.

Zachariadou O. (2000b). Herodes Atticus's Shaft, In: *The City under the City*, L. Parlama and N. Stambolodis(Ed.). Museum of Cycladic Art, Athens, Greece, pp. 190–195.

Zachariadou O. (2002). Syntagma Square: relocation of ancient findings, Arch. Deltion. Ministry of Culture, Athens, Greece, **57**, 236.

Ziller E. (1877). Exploration of the ancient aqueducts of Athens. *Mitteil. Deutsch. Arch. Inst., Athen, Abteilung*, **2**, 107–131.

17.2.2.1 Aqueducts from the Alban Hills

The water resources from this area (Latium Volcano) are of excellent quality. They include the following *Aquae: Appia* (described earlier), *Tepula, lulia, Virgo and Alexandrina.*

Aqua Tepula was tapped by Censors Servilius Caepius and Cassius Longinus (125 BC) with an aqueduct of about 18 km. Later, the *Aqua Iulia* was collected by Marcus Vipsanius Agrippa (33 BC) into an aqueduct 23 km long. The sources of the Aqua Tepula are located near the 27 km of the Via Latina and those of the Iulia, near the present town of Grottaferrata. The name "Tepula" derives from the "tepid" temperature (16–17°C) of the water conveyed.

In the year 19 BC the same Marcus Vipsanius Agrippa channelled the *Aqua Virgo* to Rome, to provide water for the Baths built by him near the Pantheon. This aqueduct ran almost entirely underground, and the flow rate was 1200 L/sec.

In the year 226 AD, Emperor Alexander Severus brought the water from the springs in the Pantano estate, located in the district of Montepulciano, to supply the baths built under his direction, near the *Campus Martius*. This aqueduct, *Aqua Alexandrina*, was about 22 km long and delivered some 260 L/sec.

17.2.2.2 Aqueducts from the Aniene River Valley

This group comprises the most important Roman aqueducts, the *Anio Vetus*, the *Aqua Marcia*, the *Aqua Claudia* and the *Anio Novus*. Censors M. *Curius Dentatus* and *Lucius Papirius* built an aqueduct to draw water from the Aniene River, above Tivoli. The water carried by this aqueduct, called *Anio Vetus*, was judged to be of poor quality and was used exclusively for cleaning and irrigation. The spoils won in battle from *Pyrrhus*, King of Epirus after his defeat by the Roman Army in the *Maleventum* battle, furnished the pay of the workmen engaged in the building operation.

In the middle of the second century BC the two existing aqueducts (*Appia* and *Anio Vetus*) feeding Rome had fallen into decay by neglect, and had been damaged by private people drawing off the water at different parts of their course. The Senate commissioned the praetor Quintus Marcius to repair the old aqueducts, and to build a third, which was called *Aqua Marcia* in his honour. This water works carried water from the *Serenae* springs into the Aniene River Valley near Arsoli and to the city; the quality was excellent with a large flow rate (about 2250 L/sec).

By the year 38 AD, Rome already possessed seven aqueducts, but Emperor Caligula decided that the water supply was inadequate and ordered the construction of two new aqueducts, which were completed under his successor Claudius, an emperor quite knowledgeable about hydraulic engineering.

The spring sources of the *Aqua Claudia* were located near to but higher than those of the *Aqua Marcia*, and possessed the same qualities of purity. This aqueduct required the building of an arched aqueduct of considerable height (approx. 18 m). It was one of the most important and strategically situated aqueducts supplying the capital. One of its secondary branches was later to supply the Palatine Hill, the centre of political power in Rome.

The *Anio Novus* was built at the same time, tapping the Aniene River and other spring waters upstream from the *Anio Vetus*. This aqueduct runs very close to the *Aqua Claudia*. In fact in certain sections their conduits were superimposed on the same arches, which are still visible in the gate of Porta Maggiore (Figures 17.9 and 17.10). Because most of the aqueducts from the Aniene River entered Rome at Porta Maggiore, this entrance gate, higher than the other part of the city, was considered a "water hub" for the water feeding Rome.

17.2.2.3 Aqueducts from the Sabatine Volcano

This group comprises the *Aqua Alsietina* and the *Aqua Traiana*, both drawing water from the cone of the Sabatine Volcano. The first water to be drawn, in the year 2 BC, was that from Lake *Alsietinus* (now Martignano), by order of Emperor Augustus, to supply his *naumachia* under the *Janiculus* Hill.

Inside the city of Rome there were 247 *Castella*, heads or reservoirs constructed of masonry, in which the water was stored, and out of which the supply-pipes for the various regions of Rome were taken. Normally each area served by a *Castellum* was independent from the others, in this way the collapse of the whole network was avoided if a pipe break happened in a single section.

Roman used pressured pipes in only a few situations, like the inverted siphon (made of lead) to cross over a deep valley. Although it is extremely likely that the siphon systems were used in the Rome water supply system (especially in view of the shape of the terrain), no archaeological evidence has yet been found to support this hypothesis (Staccioli, 2002).

For most of their length the early aqueducts were simply channels excavated through the rock. The depth of the channel below the ground varied in order to have a slight constant gradient (normally less than 1/200) throughout the length of the aqueduct; while, vertical shafts were excavated at regular intervals to provide ventilation and access.

The tunnelling technique was the same used for the construction of the qanat system still present today in the Middle East and North Africa. Normally, the aqueduct channel was raised on arches in the final stretches in order to give a sufficient head for distribution of the water within the city.

From the above-mentioned Castellum, water could be carried to public collecting tanks. With the increasing number of aqueducts, it was possible to foresee "private" supplies for the people of the elites. While, the private supply of water to the richest had become a permanent rule starting from the end of the 19th century.

A popular thesis correlated the fall of the Roman Empire with the effects of a largely diffused lead-poisoning of the population. Carbonatic deposits on the internal side of the pipes, depending on the quality of supplied waters and the typical uninterrupted water-flow which determines low residence time, do not sustain this theory. It is important to highlight that the Romans were conscious about the unhealthiness of the lead pipe workmanships. The workers involved in lead pipe manufacture were exposed to vapours from the molten lead. *Vitruvius (De Architectura Liber VIII)* described precisely the symptoms of lead poisoning diseases, by writing: '*non est dubium quin ipsum quoque non sit salubre exemplar autem ab artificibus plumbariis possumus accipere, quod palloribus occupatos habent corporis colores*' *or:* 'there can be no doubt that itself cannot be a wholesome body. This may be verified by observing the workers in lead, who are of a pallid colour.'

The Romans were also aware of the importance of the correct use of water, relating quality characteristics to the final uses. As far as quality was concerned, Aqua Marcia was deemed to be the best, closely followed by Aqua Claudia. Frontinus (De Aqueductu Urbis Romae, Chap. XIII) with reference to the latter, wrote that: 'Haec bonitatis proximae est Marciae' ('the quality of which is close to that of Aqua Marcia'). While, Vitruvius (De Archicectura, Chap. VIII) suggests the careful observation of the health and quality of life of people living near the sources from which water was derived.

Normally, public uses (street and sewer washing) and industrial activities (potter factory, tannery, etc.) were supplied by fountain spill-over (aqua caduca), while the final use of the Aqua Traiana were hydraulic mills. As reported earlier, in Rome there were some aqueducts (Anius Novus and Alsietina) reserved for non-drinking uses.

Below, some information about materials used will be given. Stones and bricks were primarily used by the Romans as building blocks, while the early cement was made by a mixture of sand, water and lime, with some stone fragments, pottery, or tile. The so called "Pozzolana" was largely used and appreciated in waterworks. It was a cement with volcanic ash from Vesuvius, which proved a very effective and strong material due to its capacities of resistance and impermeability in dry conditions.

The Romans also used machines in order to raise water. These devices of Greek origin (Archimedean screw, Ctesibica pump and Noria) are described in Vitruvius' books. These devices were not used in water distribution but only for special purposes such as bilge pumps in the ships, mine draining machinery, in the internal scheme of water in public baths, etc.

In order to pass over river crossings and gorges, spectacular arched bridges were built. In particular, the size of some of these structures is astounding. Due to the large above ground structures, as well as the amount of tunnelling that was required to complete aqueducts, these projects usually took many years and involved a lot of workers. The enormous labour force consisted not only of slaves, or prisoners of war, but also engineers and specialised workers.

In the Colonies, far from Rome, soldiers were often employed for the building of aqueducts (and also for the streets). We have to consider that some typical aqueduct works, like tunnelling, had a great importance from the point of view of military arts, like in the operation of conquering the besieged cities. For these reasons the Curator Aquarum (people responsible of aqueducts management) were often Generals of the Roman Army.

17.2.4 Ancient Roman culture, expertise and administration in Roman aqueducts

Most of what is currently known about the Roman aqueducts has been deduced from books obviously concerning water but also architecture, natural sciences and history.

Most important authors of books and their texts are indicated in Table 17.2:

Table 17.2 Ancient Roman authors of books concerning water and aqueducts.

	Name of author	Years of life	Book title
1	Marcus Vitruvius Pollio	First century BC	De Architectura
2	Plinius Cecilius Secundus*	23–79 AD	Naturalis Historia
3	Sestus Julius Frontinus	30–103 AD	De aquaeductu urbis Romae
4	Magnum Aurelius Cassiodorus	490–550 AD	Variarum libri

*know also as "Plinius the Elder"

The most important Roman writer, in the water field, was Frontinus, who served as Water's Master (Curator Aquarum) in Rome from 97 AD to 104 AD. After the army experience as General of the Legion Secunda Adjiutrix in Germany, he was appointed to this office by Nerva Emperor in order to restore the water system neglected and abused for a long time. Frontinus was very systematic in restoring the aqueducts, doing away with the corruption and fraudulent practices.

Frontinus was the head of a staff of engineers, surveyors and clerks, and a crew of 700 governmental slaves. These people worked as inspectors, foremen, masons, plumbers, and plasterers. Frontinus kept very accurate records, taking inventory of the entire system and detailing the technical aspects of how such a system had to be managed. After correcting all problems occurred, Frontinus began making improvements to the system in order to make it more efficient, and to ensure that the work he had done would be continued when he left office. He set also a standard to be followed, that remains famous (De aquaeductu urbis Romae Cap 39–63) and it is now considered the first example of technical rules. *Tabula fistularum* is a table containing the physical dimensions and hydraulic characteristics of 25 kinds of lead pipes (diameter from 1 to 10 inch) that constituted the greatest part of the distribution network in Rome. *Frontinus also* wrote about the history of the water system of capital, the state of the system in his time, and the problems of maintenance.

An important aspect of the Roman water administration was the tariff. The water users had to pay for the amount of water arriving into their homes or businesses, while the access to fountains was free of charge. At those times, every user such as public baths, artisans, and private houses had calibrated (diameter and length) nozzles made of bronze, and the water was paid for with reference to the dimensions of these nozzles. Frontinus has described also some possible illegal forms of water uses and the way to discover them.

The revenue derived from the application of tariffs was not enough to cover the cost of managing and investment, and thus public financing was necessary. Private capital was involved as well, but there was not a form of investment comparable with the modern practices. The rich families normally used these forms of sponsorship in order to obtain benevolence of civil society (and political advantages). In general, the approach to the water-supply issue was essentially based on the principle that water management should be limited to public utility, for which the State was responsible. The building of aqueducts was directed and supervised first by the magistrates of the republic and then by emperors, but ad usum populi (for the benefit of the community). In the Republican age, water management was the responsibility of the Censors in charge of finance and public works. They built and operated the water supply system. Under the reforms introduced by Augustus, these responsibilities were turned over to the Curatores Aquarum, who were fully in charge of water management. The Curatores were assisted by an engineering staff. They supervised the so-called familiae aquariae, those distinguished by water property; the familia publica, in charge of public waters, and the familia Caesaris, responsible for the Imperial waters.

Marcus Vipsanius Agrippa (Augustus friend and son-in-law) was the first to be appointed by the emperor to perform this task. He was responsible for the realisation of a series of monumental architectural works for the embellishment of the city of Rome, such as the Thermal Baths and above all, the Pantheon. He commanded a large work force composed by technicians, architects and engineers, administrative clerks and, obviously, slaves.

Other than the above mentioned text of Frontinus, the famous edict (Johnson *et al.* 1961) of Venafro (town located in south Italy), enacted by Emperor Augustus between 18 and 11 BC also illustrated the organisational rules, including the applicable rates and administrative procedures, that ruled such water supply service during the Roman empire. It was written: 'Quaeque aqua in oppidum Venafranorum it fluit ducitur, eam aquam distribuere discribere uendundi causa, aut ei rei uectigal inponere constiltuere' meaning, 'It is right that the flowing water taken by the town of Venafria is distributed and assigned through sale, or that a tax is fixed and imposed by the colony's judges through decree of the decurions majority…'. Most parts of a modern administration of the water services (e.g. cost recovery, public control, assembly rules) were included in this edict. A "water code" was ultimately laid down in eight articles that Frontinus describes in full (De aquaeductu urbis Romae Cap 94–130); it included the administrative penalties (fines and confiscations) that could be as high as 100,000 sesterces (about $150,000 in today's currency).

17.2.5 The fall of the Roman Empire and the end of the Roman aqueduct systems

Just as all great things must come to an end, so did the Romans system of aqueducts. As was pointed out previously, with a water distribution system of this scale, a continuous maintenance effort was required. The water infrastructures followed the decline of the Romans' economy and policy organisation. The final destruction of the Romans aqueducts system occurred during 537 AD as a consequence of the siege of the city by the Goths, leaded by Vitigis.

A few days after the inauguration (September 20, 1870), the temporal power of the Popes came to an end, by effect of entrance into the City of the Piedmont Army led by the General Raffaele Cadorna. A few months later (December 5, 1870) Rome was declared the capital of the kingdom of Italy.

A new era of water services of Rome began.

17.3.2 The beginning of the new era in Roman water services

It may be confidently claimed that the last aqueduct built by the Popes (*Aqua Pia*) ideally bridges the gap between the traditional concept that we find in *Frontinus* and *Vitruvius* and the current aqueduct systems. A large part of this renewed aqueduct used the most recent technologies; from Tivoli to Rome the water was carried, for the first time, in a cast-iron pipe under high pressure that permits the gradual abandonment of the ancient design. The height of the source of the aqueduct situated in the Aniene valley at an altitude of 329 m asl. allowed the higher areas of the city, as far as Monte Mario and peripheral areas as far as Ostia, to be served without the need for pumping.

This new system can be regarded as the first modern aqueduct, architecturally and also from an administrative standpoint, because of the type of management entrusted to an industrial company.

The entrepreneurial conception (AA.VV. 1997) was an initial idea of James Shepherd, an Englishman living in Rome where he directed (it was not a case) a factory producing and distributing gas for lighting. The plans for the new aqueduct received the approval of the Pontifical State and on November 15, 1865 the "Anglo-Roman Water Company" was set up in London. The equity was mainly foreign owned (Belgian and above all British) and the company was granted a 99 year concession (until 1964) of the aqueduct in conditions of absolute monopoly. Later, before the aqueduct actually became operational, the company's name was changed to "Società Acqua Pia Antica Marcia" and the foreign shareholding was reduced.

It is not surprising to learn at this point that during the 19th century in Rome, as in other huge Italian cities, the development of public utilities, such as gas lighting and water services, drew foreign capital and technology by granting concession contracts for the construction and management. Unlike the rapidly developing countries of Central-Northern Europe, Italy offered little in the way of mineral ores and fuels nor could the country rely on any colonial possessions like other nations.

The weak domestic financial market was also run by bankers who were reluctant to venture beyond the safe waters of low risk investments or traditional productive activities such as agriculture and the textile industry, particularly in a period of rapid and often unpredictable changes in the power centres of a country moving towards national unity. For the above reasons, capital and technical knowhow of English, French, Swiss and other origin characterised the first modern aqueduct systems required to support the rapid ongoing urban development. Other cities that relied on foreign capital during the 19th century in order to boost or build their water supply systems were Genoa, Bologna, Florence and Naples.

17.3.3 Water services in present-day Rome

In the early 20th century, an important discussion (Montemartini, 1902) began in the wake of what had emerged in several northern European countries (such as the political platform of the "Fabian Society" founded in England in 1884) by different schools of thought calling for the intervention of the local authorities in public utility management. The municipalisation law introduced by the Giolitti Cabinet on April 11, 1902 marked the beginning of a new era in Italy as far as the organisation of public utilities, including water services was concerned. Whereas in Rome water services management remained the

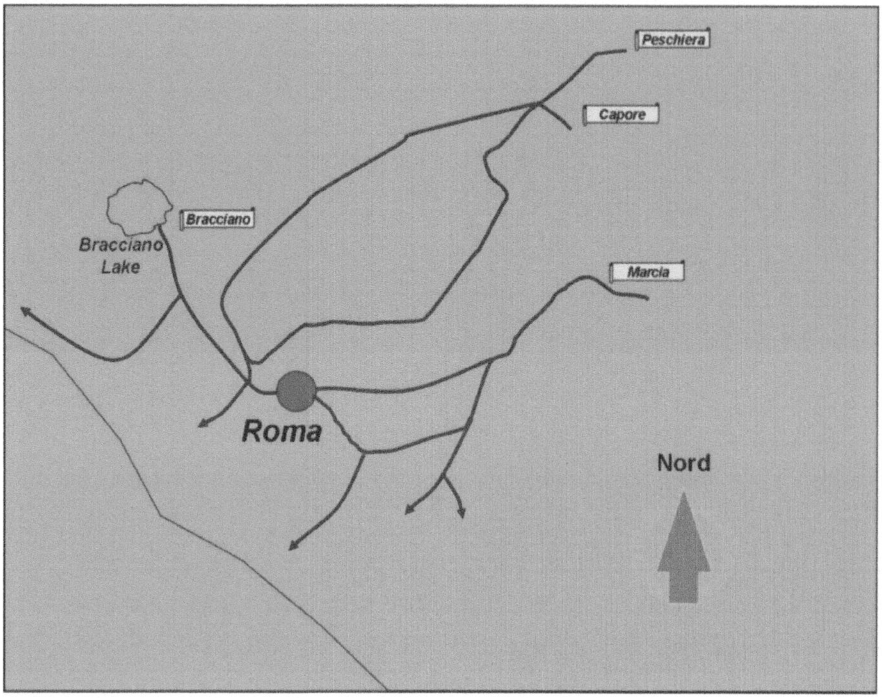

Figure 17.20 Current system of aqueducts supply

The Peschiera-Capore aqueduct, with its overall capacity of 14,000 litres per second, is the largest water supply system in Italy and its main source (Peschiera) is fed from numerous springs supplying the conveyance line to Rome, together with additional carbonatic springs from Capore. The overall length of the system is 130 km and the flow, due to elevation of its departure point, is entirely by gravity and permits the generation of 40 MW of hydroelectric power in the Salisano station. Such production exceeds the needs of the water supply system of Rome and the surplus feeds the electrical grid of the city. In 2009 some 464 million cubic metres of drinking water were supplied to about 3 million inhabitants in Rome and Fiumicino, as well as to 64 additional municipalities in Lazio. In continuation of the tradition, first of the Romans and then of the Popes, a commemorative fountain has been built for this aqueduct (Figure 17.21).

Other sources are used for supplies of non drinking water for fountains and municipal sprinkling systems. In particular, the ancient Vergine and Paolo-Traiano aqueducts are used for this purpose.

The entire intake and distribution network is 6445 and 400 km long for drinking and non drinking water, respectively, while the respective storage capacities are 480,904 and 5962 m^3 respectively.

Starting in 1998, ACEA has expanded its activity to the water and sewage services of the other municipalities of Rome metropolitan area, serving an overall population of about 3.7 million.

In 1985 the new "statutory management of water services" was defined in Rome, Italy, which defines the management of the water cycle, including wastewater, in addition to traditional drinking water activity. The ACEA has taken over the wastewater treatment services, thus laying the foundations for the integrated management of the entire water cycle coherently with the new lines indicated by the European Water Framework Directive European Directive (n. 60/2000).

In the 1990s, as a result of the EU Maastricht Treaty coming into force (November 1, 1993), the political and institutional framework of public utilities began to change again. The EU indicate to the member countries (including Italy) economic policy guidelines aimed at opening up the markets and at doing away with monopolies, also of a public nature. Italian national legislation thus undertook to follow this new path. As usual, energy services, like gas and electricity led this race.

This means that the larger Italian municipal utilities (starting with that of Genoa in 1996), providing public services such as water supply, began to change the composition of their equity capital, which had hitherto been completely public.

In July 1999 the ACEA became a public-private company placing 49% of its equity on the stock market. Today it is the largest Italian company resulting from the conversion of former municipal utilities, with over 6000 employees distributed among the various operating companies belonging to the group. In the year 2009 its annual consolidated turnover was almost 3 billion Euro.

In 2009 the ACEA celebrated 100 years of activity and a new cycle began. The company consolidated its position in the fields of water and electricity supply and the environment (since 2006 it has been active in the urban waste treatment and waste-to-energy sectors), thus establishing itself as a multi-service operator at the national and international level, providing water services to about 8 million inhabitants in Italy and as much as that overseas.

17.4 CONCLUSIONS

It is known that the ancient Romans, taking advantage from the experience of different peoples, developed an original system not only in waterworks building but also in legislation and organisation of the water services. For these reasons, the collapse of the Roman Empire did not coincide with the collapse of the "Roman way" of construction and operation of aqueducts, whose effects were strongly present until the 19th century. This chapter shows the development of water services in Rome, form the city's origins until today.

The historical development of water supply in Rome was not a local affair and greatly exceeded the town dimensions. By the effect of the spreading and consolidation over many centuries of culture and organisation of the Roman Empire in the large multi-continents area, this development has involved a part of the history of human civilisation related to the water management resources.

This important heritage is a proof not only of the power of Roman Empire but also of the great passion of the men who managed the water services in that period. It is a passion and a desire that continues today in, not only a national but also, a European dimension with similar rules and a large vision which involves all countries in a similar fashion to what happened in the time of Roman Empire.

Remembering the first words of the *Frontinus* book: 'Inasmuch as every task assigned by the Emperor demands especial attention; and inasmuch as I am incited, not merely to diligence, but also to devotion, when any matter is entrusted to me, be it as a consequence of my natural sense of responsibility or of my fidelity'. Or in latin text: '*Cum omnis res ab imperatore delegata intentiorem exigat curam, et me seu naturalis sollicitudo seu fides sedula non ad diligentiam modo verum ad amorem quoque commissae rei instigent...*'. These characteristics can still be observed in today's societies.

Acknowledgements

We wish to thank Ing. Andrea Mangano of ACEA SpA of Rome for information and data on the current plant and management situation regarding the water supply in the city of Rome.

REFERENCES

AA.VV. (1986). The Triumph of the Water: Water and Aqueducts of Rome. Paleani Editrice, Roma, Italy (in Italian).

AA.VV. (1991). Etruscans Masters of Hydraulics. Electa Editori Umbri, Perugia, Italy (in Italian).

AA.VV. (1997). Rebirth of a Myth. Marsilio Editore, Venezia, Italy (in Italian).

Barreca F. (1988). Phoenician-Punic Civilization. Carlo Delfini Editore, Sassari, Italy (in Italian).

Blackman D. R. and Hodge A. T. (2001). Frontinus' Legacy: Essays on Frontinus' de acquis urbis Romae. Ann Arbor Univ. of Michigan Press, Ann Arbor, Michigan, USA.

Carcopino J. (1939). Daily Life in Rome During the Empire Era. Libr. Hachette, Paris, France (in French).

Drusiani R., Bersani P. and Penta P. (2008). The Ancient Lake Albano Tunnel: Origins and Considerations Regarding the Hydraulic Regulation Achieved. *Proc. of the 1st IWA International Symposium on Water and Wastewater Technologies in Ancient Civilizations*, Iraklion, Greece, October 2008.

Drusiani R. and Martini P. (2009). History of the water supply of Rome as a paradigm of water services development in Italic Peninsula. *Proc. (in DVD) of 2nd IWA International Symposium on Water and Wastewater Technologies in Ancient Civilizations*, Bari, Italy, May 2009.

Hooper J. (2010). Secrets of Roman Aqueduct Lie in Chapel Say UK Film-makers. Guardian.co.uk 24 January 2010.

Johnson A. C., Coleman-Norton P. R. and Bourne F. C. (1961). Ancient Roman Statutes. Lawbook Exchange Ltd. Clark, New Jersey, USA, pp. 114–115.

Lanciani R. (1893–1901). Forma Urbis Romae. Milano (rist. Roma), Italy (in Italian), 1988.

Leoncini L. and Richter M. (2007). Discover Rome: the Fountains. Polo Books, Roma, Italy (in Italian).

Montemartini G. (1902). Municipalization of Public Services. Soc. Editrice Libraria, Milano, Italy (in Italian).

Staccioli R. A. (2002). Aqueducts, Fountains and Baths of Ancient Rome. Newton & Compton, Roma, Iyaly (in Italian).

Sorcinelli P. (1979). Poverty and Disease in the Nineteenth Century. Franco Angeli Editore, Milano, Italy (in Italian).

Chapter 18

The historical development of water supply to Iraklion, Crete, Greece from antiquity to the present

A. I. Strataridaki, E. G. Chalkiadakis and N. M. Gigourtakis

18.1 INTRODUCTION

Iraklion, the most populated city of the island of Crete (Figure 18.1) in the eastern Mediterranean, has always had the problem of water supply. At the end of the Neolithic period, the climate on the island was warmer and wetter than it is today. After the Neolithic Age, the climate became more or less like that of today (http://crete.classics.ox.ac.uk/U1S4/U1S4L2.html). Today, the climate is mild with relatively warm winters (mid 10s °C) and hot (mid-high 20s, low 30s °C) summers. In general, the Mediterranean climatic zone influences a large part of Crete and makes spring and autumn short in comparison to the longer summer and winter seasons (Hellenic National Meteorological Service). Iraklion, following its initial function as a small port in early antiquity and its gradual development from a settlement to a growing city, fell to various conquerors throughout its long history. In the course of time, the water supply to Iraklion has remained a vital issue. In the present study, an attempt is made to trace the history of the water supply to Iraklion through the city's major historical phases from the Minoan times to the present.

18.1.1 The Minoan age to the Greco-Roman period (*ca.* second millennium – first century BC)

Iraklion was one of the two seaports that ancient Knossos had (Figure 18.2). According to Strabo (*Geogr.* 10.476.7–8), these ports were Amnissos and Iraklion. The former was the main Knossian port during the Minoan times (Coldstream 2004, 64), whereas Iraklion must have been in use at least in the Roman age (Rigsby 1976). Strabo is the earliest literary source (first century BC) citing the toponym 'Ηράκλειον (Iraklion). As a port, Iraklion was located at the mouth of the Kairatos river (modern Katsambas), which flowed by Knossos (Guarducci 1935).

In order for Iraklion (Pliny the Elder, *Hist. Nat.* 4.12.59) to have functioned as a port, it must have met the criteria for the existence of all ancient ports, i.e., a naturally protected bay, the flow of drinkable water, and a broad, sandy coast (Ioannidou-Karetsou 2008, 56). As one of these criteria was the presence of potable water, a question is raised as to the water source of Iraklion.

The evidence of wells is an indication of limited water resources. Some big underground cisterns have been found, but none of them has been excavated systematically.[4] The absence of any relevant publications that could have presented the finds of the surface survey in the old town of Iraklion deprives us of further knowledge about the town in the Byzantine era.

Furthermore, in the last 20 years the excavations that have taken place in Iraklion have brought to light two big baths (*hypocausta*), but as no subsequent publication exists, there is no evidence about a main water supply system. Thus, only assumptions can be formulated, for instance, that there was no main aqueduct that supplied water to the town of Iraklion during this period. The old aqueduct, which in the Late Roman times supplied water to *Knossos* – and perhaps the broader Iraklion area – was destroyed, therefore it is not possible to form a complete Figure of the water supply to Iraklion at that time. (Strataridaki *et al.* 2009).

18.1.4 The Venetian presence (1204–1669)

18.1.4.1 The early concern for water

Under the Venetian rule the town was renamed *Candia.* In the early years *Candia* retained the same infrastructure as that in the Second Byzantine period. In time and especially during the last decades of the 13th century the population was growing and started spreading outside the walls of the town. Naturally, the water demand increased then. Lack of water had already been a problem in the early centuries of the Venetian dominion, as there was no running water. Our information on this comes from Cristoforo Buondelmonti (*Perigraphi tis Nisou Kritis*, 2002, 44), who, during his visits to Candia in 1400–1420, realised a minimum water quantity in the town. He observed, however, that the available water was sold and came from wells in the Katsambas area. The water quality was poor, because it often contained substances brought into the well by the violent current of the adjacent river *Kairatos.* Another, but anonymous traveller wrote that in 1612 Candia 'suffered from lack of water; the available water from water cisterns and from house wells was not adequate and of bad quality; these were the reasons that led people to fetch water from Katsambas' (Spanakis 1981, 38).

In 1403 the Venetian Senate[5] issued a Decree by which the immediate repair of the cistern at the Duke's Palace[6] and its filling up with water were ordered. This was mandatory, as there was no other water source at the town centre, which would have provided those living in the Palace with adequate water for their needs (*Nea Eleftheria*, 1903).

In the meantime, the Turkish aggression against Crete had alarmed Venice, which in 1474 sent an official document to the Cretan Government with instructions – due to the water supply problem to *Candia* – to take urgent action in building three large cisterns, which would have had to be always filled with rain water or with water from the Katsambas wells. Furthermore, all inhabitants were encouraged to repair their own cisterns (Spanakis 1981, 40)[7] and fill them up with water within 15 days. That was the time when the increased needs for water in the developing town led the authorities of Candia to take urgent steps to

[4]Three cisterns are known today: a big underground cistern close to St. Titus church, another underneath the building of the Municipal Library, and a third one underneath the 'Theotokopoulos Park'.

[5]This Senate, known as 'Senato Mar', was in charge exclusively of the islands that belonged to Venice.

[6]The Duke's Palace was located at the modern *Lions' Square*. It was the same palace in which the Byzantine Duke of Crete lived, and there were wells and a cistern in it.

[7]They were surface cisterns, which were filled up with either rain water or pumped water from wells.

18.1.5 The Ottoman occupation (1669–1830)

When the Turks conquered *Candia* (*Candiye* in Turkish), their concern was to attend to the direct problem of the town water supply. Specifically, Kioprulu Fazil Ahmet Pasha, the conqueror of *Candia*, ordered the repair of the Morosini aqueduct, as the Turks had successfully cut off the water flow from the springs of the Archanes area including the Youktas mountain, during the siege (Karantzikou 2004/5, 197) (Figure 18.2).

Furthermore, the Turks renamed the Morosini aqueduct as *Kemer Soyiou* aqueduct (Turkish Archive of Iraklion, 1688, 1698), meaning "water of the arch", because the central water conduit passed through the arch of the south gate of the town, the *Jesus* Gate. Water flowed to Morosini's fountain through Morosini's aqueduct. According to Evligia Çelebi, a Turkish traveller, when Ottomans conquered the town, the oversized statue of Poseidon still stood in Morosini's fountain. Then, the square where the fountain existed was renamed *Havouz Meidani* due to the presence of a water cistern there. Today, no information is available, as to what happened to the statue. Only assumptions can be made, and one is that it was probably removed from the fountain by Kioprulu Fazil Ahmet Pasha, as the Moslem religion does not espouse any iconic representation of God. The statue could also have been destroyed by an earthquake. At any rate, Ahmet Pasha constructed an iron rail around the fountain, and ordered that it be painted red (Dimitriadis 1993, 217–218).

Çelebi noted characteristically that during the first period of the conquest of *Candia*, there were 70 drinking fountains in the town (Dimitriadis 1993, 217–218). Water had a direct connection with the Ottomans' faith, as is the case with most – if not all – religions; in every mosque there was a fountain for the religious needs of the Moslems (Spanakis 1981, 89). The Ottomans tried to secure the water supply to the mosques and fountains. Later on, the water was distributed into the public fountains and to the rich citizens. Moreover, the Ottomans used the water from other springs as well. To satisfy the needs of the Moslems and citizens in general, the Turks built fountains next to every mosque (Karantzikou 2004/5, 200–201). Every fountain was named after the Pasha who built it (Spanakis 1981, 90).

During the first period of the Ottoman conquest, all fountains existed in Turkish neighbourhoods. However, when Christians and Moslems lived side by side, the water was not enough; on the other hand, only a few houses had running water or cisterns and they usually belonged to the Ottoman officials (Karantzikou 2004/5, 203).

One third of the water flow was channelled to the aqueduct *Kemer Soyiou*, and belonged to the public. In 1671 the Turks utilised the springs at *Karydaki*, near Archanes. In 1688, the spring water in the area *Mihanilski*, which is most likely the place named *Assomatos* today, was also channelled to the aqueduct *Kemer Soyiou* (Spanakis 1981, 91).

Generally speaking, during the period of the Turkish conquest there were charitable drinking fountains in the town, known as 'sebils' (Ottoman Public fountains). People built 'sebils' on purpose, in order to quench one's thirst, as there was the belief that a person's sins would be forgiven if that person offered water to others for quenching their thirst. Almost all the charitable drinking fountains were destroyed, except for the charitable fountain which is still located at the Kornarou square today. This fountain was built by Hadji Hibrahim Aga, who had sent for snow from the Idi mountain, so the fountain water was cold in the summer (Spanakis 1981, 95–98).

Furthermore, water was treated like private property and in some cases it was inherited. That was a privilege that few people and especially Moslems had. In the end, Moslem religious foundations 'Evkafs' sold water to rich Moslems and that was a real burden for poor Christians and Moslems alike (Spanakis 1981, 107).

Nevertheless, the Ottomans could not face effectively the serious problem of water shortage for supplying water to *Candia*. The deficient conservation of the whole aqueduct, which the Venetians had

springs in Archanes (*Assomatos*, *Grammata*, *Skalounakia*, *Rossiki Ekklesia*, *Karydaki*, *Paradeisos*) and in *Fundana*, which watered the Iraklion aqueduct, before he wrote his report (Spanakis 1981, 98–100).

In 1887, a new cistern of 480 m^3 capacity was built next to the *Jesus* Gate (Figure 18.13). A 1892 document of the Turkish Archives of Iraklion (Turkish Archive of Heraklion, 1892) refers to the number of the springs that supplied Iraklion with water. The main springs supplying water to Iraklion aqueduct and the water flow rate in *massoures* are shown in Table 18.1.

Table 18.1 Springs of Iraklion in 1892 and their water flow rate.

Springs	Water flow rate (in *massoures*)
Fundana	200
Assomatos	75
Grammata	80
Skalounakia	40
Paradeisos	87
Rossiki Ekklesia	35

According to this document, the *Fundana* spring was the only one among all other springs that continued to supply Iraklion with the largest water flow rate during the 19th century. The enormous water flow rate of the *Fundana* reflects its great significance as a water source, this being the reason for its dedication to the Sultan Abdul Metzit. In fact, the Turkish tradition prescribed that every water spring had to be dedicated to an eminent person.

In the period of the Cretans' struggles against the Ottomans, when the former's aim was to be united with Greece (this being known as the 'Cretan Question'), the *Fundana* spring became significant again as a water source for Iraklion. During the Revolution of 1897, all springs supplying Iraklion with water, as well as a large part of the aqueduct (from the *Fundana* to Skalani), remained under the Cretan revolutionaries' control, whose centre of operations was the Archanes town. According to the revolutionaries, if the water supply to Iraklion stopped, this would be a measure for exercising pressure against the Turks. In fact, a bridge aqueduct near Skalani – north-east of Archanes – was blown up by the Cretans *(The Historical Archive, 1897)* who in turn accused the Turks of this act. At any rate, the destruction of the aqueduct meant the shutting down of the *Fundana* water supply to Iraklion. To face this problem, the revolutionaries sent a committee to the spot of the destruction, in order to inspect the damage to the aqueduct. The committee noted that it was urgent to repair the destroyed duct. We can only assume that the duct was repaired, although there is no relevant evidence (Psaroulaki 1997; Strataridaki *et al.* 2009).

The defeat of Greece during the Greek-Turkish War in 1897 had an effect on the development of the Cretan Question. The Greek army withdrew from Crete and the Cretans accepted the proposal of autonomy temporarily. In 1898 the Great Powers decided that the areas which had been occupied by the revolutionaries should have been administered by the Committee which was appointed by the General Assembly, while towns and areas held by the Turks should have fallen under the supervision of the Powers. British troops were responsible for the District of Iraklion.

On August 25, 1898, the Turks refused to hand over the tax office in Iraklion to the officials of the temporary government called 'Executive Committee'. Serious riots broke out in the town by the Turkish mob. During these events tens of Christian Cretans were murdered by the Turks; in addition, 17 British

Table 18.2 Springs of Iraklion in 1913 and their water flow rate (Spanakis 1981, 116).

Springs	Water flow rate (in L/sec)
Astrakoi	40
Miliara	23
Keramoutsi	23
Morosini New springs	8
Morosini	4
Fundana	2
Karydaki	2

18.1.9 From the Union of Crete with Greece to Modern Times (1913–today)

The Union of Crete with Greece was officially declared on December 1, 1913, following the victory of Greece during the Balkan Wars, under the leadership of the Cretan politician Eleftherios Venizelos (Chalkiadakis 2007, 438–456). The period of 1910–1927 was the worst in the water supply of Iraklion, for the town had expanded and was transforming into a city (Banasakis 1969, 149). After the union of Crete with Greece, the authorities of the island invited the Hydraulics engineer I. Libritis from Egypt, who would use the *Fundana* aqueduct to channel water from Astrakoi to Iraklion (14.5 km distance). In a report Libritis noted that, on one hand, the *Fundana* spring constituted a most valuable spring for supplying water to Iraklion but on the other hand, the springs in Astrakoi could solve the problem of the water supply to the city. For this purpose, Libritis suggested that water should be channelled via pipes or a built aqueduct – different from that at *Fundana* – due to large elevation difference (Spanakis 1981, 109, 110, 112–117).

In 1922 the Greek army's expedition to Asia Minor was followed by their total defeat by the Turks, and the Greek population living in the Ottoman Empire were forced to abandon their homes. In January 1923 Greece and Turkey signed the Population Exchange Accord. Thus, Greeks living on Turkish land and Moslems living in Greece were mutually obliged to move, the former to Greece and the latter to Turkey. Moslems left Crete, and Greek refugees from Asia Minor arrived at Iraklion, particularly in the area now called New Halikarnassos (named after the ancient Halikarnassos in Asia Minor). The arrival of the refugees and the subsequent growth of the population increased the need for water supply (Spanakis 1981, 117–120). In 1927, when mayor of Iraklion was I. Voyiatzakis, work began for the construction of a new aqueduct starting at the Astrakoi springs in a cave, named *Neraïdospilios* (Banasakis 1969, 149) (Figure 18.22).

In 1933 the stone conduit of the *Fundana* aqueduct was substituted with iron pipes and, as a result, the old Roman tunnel went out of use (Banasakis 1969, 149–151). In the following years new springs were added to the city's aqueduct: those were the *Miliara* Springs (1934), and *Kryes Piges* ('Cold Springs') (1936), east of the *Miliara* Springs, in the agricultural district of the Myrtia village (Spanakis 1981, 128–132). In the following years, the need for new springs was intense, but without any result (Archive of the Municipality of Heraklion, Records of the Municipal Council, 7-9-1936).

It is worth mentioning that constructions related to the water supply of Iraklion were destroyed either by accident or due to the authorities' negligence. Specifically, in 1938, the storage cistern of the *Bembo*

In 1967, at the beginning of dictatorship in Greece, there were four different aqueducts: 1) the *aqueduct of Malia*; here, water was pumped up from six bores in the area of Malia (it was constructed under the financial support of the US, when Krassadakis was Mayor) and was channelled to the storage cistern in Hagios Ioannis at 70 m altitude; its storage capacity was 6000 m^3; 2) the *aqueduct of Astrakoi/Miliara*; water was collected initially in the *Miliara* storage cistern and later in the cistern of Hagios Ioannis; 3) the *Archanes' aqueduct*, in which the water flowed from *Morosini's springs* and *Karydaki spring*, and was channelled to a small storage cistern, in Fortezza, a suburb of Iraklion; and 4) the *Fundana aqueduct*, in which the water was channelled to a small cistern in Hagios Ioannis. The water flow rate and the length of the exterior conduit of each aqueduct are shown in the Table 18.3 (*Patris*, 1967a, b, c, d).

Table 18.3 Aqueducts of Iraklion in 1967.

Aqueducts	Water flow rate (m^3/d)	Length of conduits (m)
Malia	8000	31,000
Astrakoi	4200	15,000
Archanes	720	13,950
Fundana	480	8855

The drought which occurred in the following years forced the municipal authorities in 1970 to supply water to Iraklion every other day, due to the low water reserves (*Patris*, 1970). Although the authorities considered again utilising the Almyros river water, their attempt was unsuccessful, as the water was saline (*Patris*, 1972, *Ethniki Phoni*, 1973a). Then they took temporary measures and utilised the water from the bores at the villages of Kastelli (*Ethniki Phoni*, 1973b) and Thrapsanos (*Patris*, 1974; *Ethniki Phoni*, 1974c).

In the following years, the Municipality of Iraklion essentially managed the water supply problem of the city by transferring water from Malia and Thrapsanos (*Ethniki Phoni*, 1976d). In 1976, the opening ceremony of the Thrapsanos conduit took place; water was channelled to the cistern at Fortezza (*Demokratis*, 1976) thus meeting the water needs of the city. According to Manolis Karellis, Mayor of Iraklion, 'there was enough water for the water supply of the city for the future' (*Allagi*, 1979).

Today, the aqueduct supplying water to Iraklion is comprised of the nets of *Papa-Titos* storage cistern (to which water is channelled via conduits from Malia, Skalani, Aïtania and Tylissos), the nets of the cisterns called "D2/D1" (this net is composed mainly of the nets at Hagios Myron and Voutes), and the nets at Fortezza (Figure 18.27). These three nets are interconnected. It is worth saying that today the Iraklion aqueduct includes the old springs of Astrakoi (*Miliara, Kryes Vryses, Fragma* (the small dam), *Neraïdospilios*), *Fundana* and *Karydaki* (Municipal Water Supply and Sewerage Service of Iraklion, 2003, 8–9, no. of plans 1,6 and 7). Furthermore, some experts support the view that the construction of the Aposelemis dam, east of Iraklion, and the utilisation and desalination of the Almyros river, west of the city, could ensure water for the future needs of the city (*Ethniki Phoni*, 1976d; *Mesogeios*, 1959a; *Patris*, 1959a). The Aposelemis dam is under construction today, despite the reactions of environmental organisations against destructive effects on nature (*Patris*, 2010; Kavvadas *et al.* 2000).

In general, the water supply to Iraklion is closely connected to the city's history and monuments, and, at the same time, it reflects the human struggle to satisfy basic needs, like water. Such needs have been intensified on the one hand by the climatic conditions, which have not changed dramatically through the last centuries; and on the other hand, by Iraklion population increase, which at times has pointed to the investigation of new water sources and subsequently to the implementation of more effective water supply projects.

REFERENCES

Primary Sources

Buondelmonti Cr. (2002). Perigraphi tis Nisou Kritis (Description of the Island of Crete), Styl. Alexiou (eds.), Mikros Nautilos, Iraklion, Greece.

Heraklion Municipality Archive (1936). Records of the Municipal Council, 7.9.1936, v. 41. Iraklion, Greece.

M.W.S.S.S (Municipal Water Supply and Sewerage Service of Iraklion) (2003). Meleti Diachirisis Pedion Ekmetalleusis DEYAH (Management Study of Distribution Areas), February 2003, Plan Nos. 1, 6 & 7. M.W.S. S.S, Iraklion, Greece.

Leo Diaconis, Historia, 2.8.16.

Pliny the Elder, Historia Naturalis, 4.12.59.

Stavrinidis N. S. (1986). Metaphraseis tourkikon eggrafon aphoronton eis tin Historian tis Kritis, II, eggrafa periodou 1672-1694, Eti Hegiras 1083-1105, Vikelaia Dimotiki Vivliothiki Herakliou (Translations of Turkish Documents regarding the History of Crete, II, Documents on the years 1672–1694, Hegira years 1083-1105), Vikelaia Municipal Library of Iraklion, Iraklion, Greece.

Strabo, Geographica, 10.476.7-8.

The Historical Archive (1897). The Historical Archive of Archanes, Record No. 196–Letter of Major-General Sir H. Chermside to Antonios Triphopoulos and others, Heraklion, Greece 3/15 April 1897.

Turkish Archive of Heraklion (1688). No. 950, 5th of January 1688, Vikelaia Municipal Library of Iraklion, Iraklion, Greece.

Turkish Archive of Heraklion (1698). no. 1472, 31st of October 1698, Vikelaia Municipal Library of Iraklion, Iraklion, Greece.

Turkish Archive of Heraklion (1892). Translation Office Z8, No. 4356, Catalogue of Springs of the year 1892, Vikelaia Municipal Library of Iraklion, Iraklion, Greece.

Turkey (1898). Turkey No. 7, Further correspondence respecting the affaires of Crete, presented by both houses of parliament by command of her Majesty, London 1898, printed for Her Majesty's Stationery Office by Harrison and sons, St. Martin's Lane, London, UK.

Iraklion Press

Allagi (1965). 24th of September 1965, Iraklion, Greece.

Allagi (1979). 24th of March 1965, Iraklion, Greece.

Demokratis (1976). 22nd of July 1976, Iraklion, Greece.

Drasis (1959). 2nd of October 1959, Iraklion, Greece.

Ethniki Phoni (1973a). 2nd of April 1973, Iraklion, Greece.

Ethniki Phoni (1973b). 17th of September 1973, Iraklion, Greece.

Ethniki Phoni (1973c). 14th of January 1974, Iraklion, Greece.

Ethniki Phoni (1973d). 24th of May 1976, Iraklion, Greece.

Mesogeios (1959a). 18th of August 1959, Iraklion, Greece.

Mesogeios (1959b). 21st of August 1959, Iraklion, Greece.

Nea Eleftheria (1903). 14th of September 1903, Iraklion, Greece.

Patris (1959a). 29th of August 1959, Iraklion, Greece.

Patris (1959b). 2nd of October 1959, Iraklion, Greece.

Patris (1962). 31st of March 1962, Iraklion, Greece.
Patris (1963a). 9th of March 1963, Iraklion, Greece.
Patris (1963b). 28th of April 1963, Iraklion, Greece.
Patris (1967a). 13th of May 1967, Iraklion, Greece.
Patris (1967b). 14th of May 1967, Iraklion, Greece.
Patris (1967c). 16th of May 1967, Iraklion, Greece.
Patris (1967d). 19th of May 1967, Iraklion, Greece.
Patris (1969). 4th of May 1969, Iraklion, Greece.
Patris (1970). 3rd of June 1970, Iraklion, Greece.
Patris (1972). 28th of June 1972, Iraklion, Greece.
Patris (1974). 12th of March 1974, Iraklion, Greece.
Patris (2010). 10th of March 2010, Iraklion, Greece.

Secondary Sources

Angelakis A. N. and Koutsoyiannis D. (2003). Urban water resources management in ancient greek times. In: The Encycl. of Water Sci., B.A. Stewart and T. Howell (eds.), Markel Dekker Inc., Madison Ave. New York, N.Y., USA, pp. 999–1007.

Banasakis G. (1960). Cultural Data on Heraklion, Crete, Iraklion, Greece (in Greek).

Chalkiadakis E. G. (2002). Ta dimosia erga tis Kritikis Politeias 1898–1913: Heraklion (The Public Works of the Cretan State1898–1913: Heraklion), Ellotia, 10, Municipality of Chania, Greece, pp. 149–169.

Chalkiadakis E. G. (2007). He politiki tou Eleutheriou Venizelou sto Zitima tis Enosis tis Kritis me tin Ellada: 1910–1913. In: *90 chronia apo tin Enosi tis Kritis me tin eleutheri Ellada* (Eleutherios Venizelos' Policy on the Question of the Union of Crete with Greece: 1910–1913), (90 Years after the Union of Crete with Greece) (*Symposium Proceedings*, 5–7 December 2003). Rethymnon, Greece, pp. 443–463.

Coldstream N. J. (2004). The early greek period: geometric and archaic (*ca* 1050–480 BC) In: Heraklion and Its Area: A Journey through Time, N. M. Gigourtakis (ed)., Iraklion, Greece, pp. 59–76.

De Feo G., Mays L. W. and Angelakis A. N. (2011). Water and wastewater management technologies in the ancient greek and roman civilizations. In: Treatise on Water Science, vol. 1, P. Wilderer (ed.)., Oxford, UK, pp. 3–22.

Detorakis Th. (1990). Historia tis Kritis (History of Crete), Typokreta (ed)., Iraklion, Greece.

Dimitriadis B. (1993). Mnimia tou Heraklíou kata ton evliya çelebi (Monuments of heraklion according to evliya çelebi). *Ariadni.*, **6**, 217–218.

Gerola G. (1905). Monumenti Veneti di Creta, v. IV, Venice, Italy.

Guarducci M. (1935). Inscriptiones Creticae. I, XI. Roma, Italy.

Hellenic National Meteorological Service: http://www.hnms.gr/hnms/english/climatology/climatology_region_diagrams_html?dr_city=Heraklion&dr_region=ClimCrete (accessed 04 April 2011).

Ioannidou-Karetsou A. (2008). To epineio ton ellenistikon-romaikon chronon. In: Heraklio: He agnosti historia tis archaias polis, (The Unknown History of the Ancient City). A. Ioannidou-Karetsou (ed.), Iraklion, Greece, pp. 54–74.

Karantzikou H. (2004–05). He Hydreusi tou Chandaka kata ta prota chronia tis Tourkokratias (The Water Supply of *Chandax* during the first years of the Turkish occupation). *Palimpsiston*, 19/20, Vikelaia Municipal Library of Iraklion.

Kavvadas M., Kotsonis A., Somakos L., Giolas A. and Lazaridou S. (no date). Geotechnika themata tou fragmatos Aposelemi Kritis – Technika stoicheia tou ergou (Geotechnical aspects of Aposelemiis dam in Crete island – Project technical data). http://portal.tee.gr/portal/page/portal/teelar/EKDILWSEIS/damConference/eisigiseis/1.5.pdf (accessed 5 February 2011).

Markoulaki St. (2008). Psifidota apo to choro tes voreias epektases tou Archaiologikou Mouseiou Herakleiou kai to perivallon tous. In: Heraklio: He agnosti historia tis archaias polis, (The Unknown History of the Ancient City). A. Ioannidou-Karetsou (ed.), Iraklion, Greece, pp. 109–147.

Psaroulaki A. (1997). The water supply to Heraklion (1897) and the armies of the International Forces. In: Archanes 1897. In: A Scientific Symposium Proceedings [30–31 August 1997]), 3rd edn, Iraklion. Municipality of Arhanes, Arhanes, Iraklion, Greece, pp. 226–230 (in Greek).

Rashed and Zenap Ismat Dr. (1978). *Crete under the Egyptian Rule 1830–1840* (transl. from the Arabic by Dr. Eug. Michailidis). St. Spanakis (ed). Iraklion, Greece (in Greek).

Rigsby K. J. (1976). Cnossus and Capua, Transactions of the American Philological Association, **106**, pp. 313–330.

Sanders I. F. (1982). *Roman Crete*. Warminster, Wilts., UK.

Spanakis St. (1953). *Mnimia tis Kritikis Historias*, **vol. III** (Monuments of Cretan History), Private Edition, Iraklion, Greece.

Spanakis St. (1969a). *Monuments of Cretan History*. Private Edition, **vol. IV**, Iraklion, Greece (in Greek).

Spanakis St. (1969b). *Monuments of Cretan History*. Private Edition, **vol. V**, Iraklion, Greece (in Greek).

Spanakis St. (1981). *The Water supply to Heraklion 829–1939*, Technical Chamber of Greece, Dept. of Eastern Crete, Iraklion, Greece (in Greek).

Spanakis St. (1990). *The Heraklion through the Centuries*, Municipality of Iraklion, Iraklion, Greece (in Greek).

Spanakis St. (1981). The *Crete I, Central-Eastern, Tourism-History-Archaeology*. Sphakianos, 3rd edn, Iraklion, Greece (in Greek).

Strataridaki A. I., Chalkiadakis E. G. and Gigourtakis N. M. (2009). The history of the Fundana spring aqueduct and its significance for the water supply of Heraklion city (Crete) through the ages. In: Water and Wastewater Technologies in Ancient Civilizations, In: Proc. (in DVD) of 2nd IWA International Symposium. Bari, Italy, 28–30 May 2009, pp. 1–8.

Xanthoudides St. (1916). *Archaeologikon Deltion*. 2, 23 [Reprint: *Meletimata*, Heraklion: Vikelaia Municipal Library, 1980, pp. 41–42].

Xanthoudides St. (1980). *Meletimata (Studies)*. N. Panayiotakis and Th. Detorakis (eds.), Iraklion, Greece.

Xanthoudides St. (1927). Chandax–Heraklion. Historical Notes, Son of S. Alxiou Publ., Iraklion, Greece (in Greek).

Xyritakis D. (2008).Nikolaos Krassadakis. The Life, Adventures and Work of an Eminent Politician), Philopolis Publ., Iraklion, Greece, (in Greek).

Websites

http://www.report24.gr/periferiarxis-kritis.htm (accessed 04 April 2011)

http://www.alpha-omegaonline.com/road/route_20b.htm (accessed 04 April 2011)

Chapter 19

The historical development of water supply technologies in Barcelona, Spain

M. Salgot and A. N. Angelakis

19.1 INTRODUCTION

The Iberia Peninsula has a millenary tradition of the exploitation of water resources due to the fact that water is temporarily scarce in most of the areas of the Mediterranean catchments of the present Spain and Portugal. Greek and Roman influences, as well as Arabic ones, could still be detected in the context of water resources management to ensure cities' supply along the entire year. In this context, one of the oldest Spanish towns, Barcelona, shows interesting features throughout the centuries. This chapter studies the history of Barcelona's water supply since the Roman times of this colony, featuring the difficulties encountered to deliver sufficient amounts of good quality water to the different users and the evolution of the supply in terms of quantity and quality up to the beginning of the 20th century.

19.2 PHYSICAL SETTINGS

Barcelona is a Spanish town in the north-east of the Iberian Peninsula, around 150 km south of the French border and lying on the Mediterranean shore. Its origins are discussed, but for certain the town existed in Roman times as a colony, which was then less important than other neighbouring towns, like Tarraco or Baetulo. The town gained importance throughout the centuries finally becoming the capital town of Catalonia (Soldevila & Coll, 1979). Its development throughout the more than 2 millennia since its foundation has been uneven but its major increase in inhabitants was stated during the 20th century.

It is considered that in terms of water supply and sanitation, the worst times arrived when the walls of the city constrained its physical expansion and created an insane environment due to the high population density and the presence of factories and cattle inside the walls (Gran Enciclopèdia Catalana, 1971).

For comparison purposes, it is worthwhile noting that the Greater Barcelona has at present around 5 million inhabitants living in a surface of 636 km^2, with a population density of 1542 inhabitants per km^2. The central nucleus, called Area Metropolitana (Metropolitan Area) has a population of 3.2 million inhabitants and a density of 5060 inhabitants per km^2. The strict Barcelona, the municipality, has a population of 1.6 million inhabitants (INE, 2011) in an area of 101 km^2.

streams. Not far away, two rivers drain water from the mountains. The Llobregat in the south flows all year round, while the Besòs in the north has a more temporary pattern with periods without any water but that in the associated aquifer.

A relatively high number of temporary streams and gullies crossed the area where the town is presently located, and their courses coincide mainly with the big drainage infrastructures now in service (Arandes, 1998). The groundwater resources were heavily used over the centuries, but due to health reasons were progressively abandoned in the last century causing a present elevation of the piezometric levels.

19.2.3.1 Precipitation

The annual average precipitation in Barcelona is around 600 mm, but in 2010 was 694 mm (Idescat, 2011). There are recurrent droughts where the annual rains are reduced to nearly half of their usual figure; for example, 321 mm during 2007. There are two maxima, autumn (primary maximum) and springtime (secondary). October is the rainiest month with around 95 mm and July the driest with 27 mm on average. The average of the rainy days is 90 mm. Sometimes heavy rainfalls can occur pouring more than 140 mm in half a day (http://www.bcn.es/publicacions/b_informacio/bi_58/bi_58_4.htm) or more than 20 mm in 2 hours (http://www.fabra.cat/meteo/episodis/EpisodiMARC2011.pdf).

19.2.3.2 Humidity and potential evapotranspiration

Humidity levels in Barcelona remain high all year round, with percentages from 60 to more than 70 even in the winter. The levels can get uncomfortable, especially during August and September with peaks near 90%.

The potential evapotranspiration, is around 839 L/year in Barcelona city although small changes exist (http://www.aemet.es/es/ and http://www.bcn.es/estadistica/castella/dades/anuari/cap01/C0102030.htm) which depend on the specific location of the observatory.

19.3 BARCELONA WATER SUPPLY BEFORE THE ROMAN PERIOD AND HISTORICAL FACTS

The initial location of the human settlement in the area was southwards from where it is presently, besides the Montjuïc Hill and used to have a small harbour, but the sandy sediments brought by the Llobregat River made the harbour unusable for long periods of time which caused the change of the village to a new place not so menaced by the river flows. This new location seems to coincide with the present one.

The origins of the town are a little bit confused and before the Rome domain, the area preserved remnants from the end of the Neolithic and beginnings of the Chalcolithic. Afterwards, the Laietani (Iberian people) culture was developed in the area and certain commercial activity existed. During the second Punic war (218–202 BC) Cartago occupied the site and in many papers, this is established as the data of the foundation of the town.

The first structure archaeologically documented in relation to the artificial water supply of the area is a well-cistern localised in one of the silos of Montjuïc Harbour dated in the fourth-third centuries BC (Asensio *et al.* 2008, in Miró & Orengo, 2010). Underground water from the aquifers located in the area has been used since the ancient times via wells and fountains either public or private. The wells were used throughout the Middle Age and several street names remind us of this (Voltes, 1967). In relation with this, there is a theory which relates the name of the town with a possible Carthaginian origin; *Bar-kino* signifying well of the bay in Punic (Miró & Orengo, 2010).

Although there are discussions on the date of the foundation of Barcelona, there is a consensus on the Roman settlement named "Colonia Iulia Augusta Paterna Faventia Barcino" in the time of the Roman

With the new resources the Municipality was able to supply various public fountains. Institutions and members of the nobility of Barcelona such as the Cathedral, the government of the city or the Consell de Cent (the council formed by 100 members), and various manors and convents also received a continuous water supply (Voltes, 1967).

In the 14th century, the network inside the city walls was subject to numerous provisions of the town council, in order to control the purity and the hygiene in the public fountains and the houses of institutions and individuals. It is noted that the breaks and the scaling of the pipes frequently hindered the supply. Moreover, the long periods of drought, typical of the Mediterranean climate of Barcelona, prevented a real regularity of supply, relying partly on the surface water flows of the Besòs River.

The Canal was also a cause for concern in the 14th century. Its last section entered the walls and the water quality generated a continuous risk for health. Various municipal and county's reforms intended to increase the volume available and prevent the disposal of domestic, animal, industrial and agricultural waste. Fear of the spread of the plague, which wreaked havoc in the second half of the century, and other epidemics, led the public authorities of the time to maximise the health controls of the Rec Comtal.

From the 15th century to the beginning of the 18th there were no major transformations in the water supply to the Plain of Barcelona. The stability and the lack of economic and demographic growth were certainly the determinant factors. The ageing or abandonment of the pipelines and the persistence of long periods of drought led at certain times to a great reduction in the water reaching the city. The town council then performed emergency repairs on the damaged network or increased the supply from the Collserola sources (Voltes, 1967).

When the water available for the city reached intolerable minimums a search began for new alternative resources outside the territorial sphere of the town. At least in the years 1551, 1627, 1632 and 1633 the town council unsuccessfully undertook negotiations with the viceroy to bring water to Barcelona from the Llobregat River, south of town. At the end of the 16th century there was even an attempt to divert water from the Ter River, 60 km north, with the aim of channelling it to the Besòs basin and thus increasing the flow of the Comtal canal (Conillera *et al.* 1986).

The Consell de Cent concentrated a good part of its resources on improving the control of the scarce flows reaching the city from the Collserola galleries. The water concessions for institutions and members of the high society of Barcelona and for public consumption were administered in a restrictive manner, as the lack of water encouraged infractions by individuals. In 1650 the most well-known master of the fountains from the 17th century, Francesc Socies, wrote *The Book of the Fountains*, a manuscript in which he described the city's water supply network in detail. Socies considered it to be insufficient as a result of municipal negligence and private usurpations (Conillera *et al.* 1986).

The Consell de Cent granted the water on behalf of the eminent domain of the monarchy and reserved the beneficial ownership of use. The concessions for the political, religious, cultural or military institutions of the city and for important elements of the nobility and the upper middle class were granted in accordance with the prestige and the merits of the applicant. In situations of scarcity, the municipal authorities could suspend the concession without further explanations. There were frequent lawsuits and, at the time of the master Socies, numerous disputes concerning the ownership of the water used by individuals are documented.

The Rec Comtal continued to serve the needs of the irrigated land and the flour mills and factories. However, the Batllia General (the institution in charge of administering the Royal Assets in Catalonia) promoted reforms in the canal with the aim of conserving and maintaining it. To guarantee the hydraulic force necessary for the mills and sufficient water to irrigate the lands vegetable gardens outside the city walls were absolutely necessary (Conillera *et al.* 1986).

19.6 THE NEW SUPPLY NEEDS RELATED TO THE ECONOMIC AND DEMOGRAPHIC RECOVERY OF THE 18TH CENTURY

Throughout the 18th century Catalonia experienced an intense transformation compared to the previous centuries. There are two key factors for this change. The first one was the disappearance of the Catalan political, legal and cultural system imposed by the Reorganisation Decree. This was passed in 1716 by the Spanish Bourbon monarchy, Philip V reigning, after the end of the War of Spanish Succession. Second, the economic and demographic impulse experienced by Catalonia as the century advanced, with a clear repercussion on the urban expansion of Barcelona. Both events had a crucial influence on the evolution of Barcelona's water supply.

The Crown became the holder of the whole of Catalonia's water domain. The Intendència (old name for province and its administration) could grant concessions to individuals. The latter, in exchange for an initial payment and an annual fee, took on the beneficial ownership of the water granted. Barcelona's City Council, which arose from the new Bourbon legislation, which abolished the Consell de Cent, conserved exclusive powers over the water supply service, and maintained its beneficial ownership over the waters which served the city. Despite this, the municipality could not grant concessions to individuals in Barcelona under the regime of emphyteutic lease (a type of real estate contract), a power reserved for the Intendència, and, therefore, the city council did not receive any revenues on these concessions. The municipality of Barcelona restricted the access of individuals and institutions to the domestic water service, and thus was deprived of an important source of financing for the modernisation of the supply infrastructures. Moreover, the scarcity of municipal financial resources was chronic throughout the century. After the abolition of the tax fees that the city received under the Catalan legal system, the so-called Municipal Financing Charter of 1718 granted the city a volume of income which did not exceed 25% of that of the previous era, and this meagre amount, at the same time, depended to a large extent on the direct approval of the Intendència. Consequently, Barcelona City Council was not in a position to allocate significant budgetary items to the improvement of the water supply system (Voltes, 1967).

The new conditions of productive and commercial development and of population growth which were taking place around Catalonia led, in Barcelona, to a progressive increase in the population density and to an industrial occupation of the free spaces still available within the city walls. The Raval area (a neighbourhood) in the southern half of the city, which still had large bare spaces, was gradually filled in as the 17th century progressed. The Barceloneta neighbourhood, in the north, was also built.

While the urban and demographic growth, factories, and irrigated agriculture demanded new water resources, the municipal financial and political weakness made it very difficult to improve the water supply system necessary for the 18th century city.

The city of Barcelona did not have sufficient water resources to meet the increasing urban demand. This was coupled with the poor condition of the supply network due in part to the quality of waters – highly calcareous – which generated precipitations (scaling) in the pipelines, reducing their capacity to transport water. Given Barcelona City Council's paucity of financial resources, the majority of the works planned were aimed at maintaining this old hydrological infrastructure. There were frequent complaints by the population of Barcelona, and at times of special need there were disputes between the neighbours to fill the jugs at the public fountains.

Among the different alternatives that it was possible to consider in the 18th century, only the Comtal canal offered a minimum potential for increasing the resources available. Other projects which were of interest to bring water from the rivers Llobregat or the Ter were not possible in the technological and political context of Barcelona at that time. The Rec Comtal was still bringing a considerable volume of supposedly clean water from the River Besòs to Barcelona for drinking purposes. Advantage had already been taken of this fact in 1703 by the Consell de Cent when it agreed to build a closed canal from a gate of the canal to the city. The initial objective was watering the trees of the Rambla (one of the main streets in downtown) and animals.

This was the embryo of the supply service that the municipality set up in the new Raval neighbourhood at the height of urban and demographic growth. Starting from 1709, and in different phases of expansion, a distribution network was built from the wall to meet the needs of the population of the Raval. By 1754, in addition to the irrigation of the trees, this network already served three public fountains, the charitable institutions, the convents of the Raval and the artillery foundry near the seaside (Conillera *et al.* 1986; Voltes, 1967).

Although this new network allowed the population of the new districts to receive a supply which the former distribution system could not in any way offer, the water from the Comtal canal did not reach the Raval with sufficient quality for drinking purposes. The decision to hide the new conduit and to build a covered canal was taken in 1703 precisely for health reasons. It is noted that the Canal was converted into a sewer due to the disposal of wastewater from the sewerage and industries, such as textiles and leather (Voltes, 1967).

Despite these precautions, the water caught did not offer health guarantees. Between the origin and the entrance into the town precinct, the Rec Comtal crossed various sources of contamination. Among others, hemp was soaked upwards in the river, and the canal itself received waste, wastewater, ink and remains from washing wool, clothes and animal waste, and herds drank in it. Health hazards increased during the 18th century with the new industrial uses to which the waters of the canal were initially destined. Thus, in the middle of the century, the canal provided a water service to calico factories and calico meadows situated outside the wall. Some of these establishments took water and discharged waste before the origin of the pipe serving the population and the institutions of the Raval.

Under these conditions, in the second half of the 18th century the supply was already a much more pronounced element of social differentiation than it had been until the 17th century. Traditionally, a privileged minority, the ecclesiastic and political leaders, the nobility and the trading bourgeoisie, enjoyed home water service through a concession. The majority of the population only had the insufficient urban fountains, supplied by the Collserola pipeline, and the numerous wells distributed everywhere, frequently hazardous due to the proximity of the sewers or the domestic latrines. Starting from 1709, with the beginning of the urban supply service through a pipeline, the population had to use the dubious waters from the Rec Comtal. Within the Raval, the differentiation for reasons of merit and social prestige occurred again. Home direct access to the water from the canal was reserved for the concessions granted by Barcelona City Council to charitable or military institutions or the convents of the area. The working-class sectors of the neighbourhood only had access to the three new and sole public fountains. If it is considered that this new neighbourhood did not have sewers, in the 18th century it was a clear situation of health marginalisation. In the north-eastern extreme of the city two districts, hardly reached by the Collserola distribution network, were also in a disadvantageous situation. The vicinity of the most hazardous section of the Rec Comtal limited the supply of safe water to the domestic wells of the area.

Barcelona City Council managed to extend the distribution of urban water to the city thanks to the dual contribution of the Collserola-Rovira catchments and the Clot-Canaletes pipe. The system was, however, extremely precarious. The volume of the supposedly abundant flows of the canal was, in turn, conditioned by various factors.

First, the Besòs regime was subject to considerable seasonal irregularity. Like most of the Mediterranean streams, this river suffered from severe low water levels all year round and especially in summer, and dangerous floods in the autumn which wreaked havoc in the Portal Nou and intramural area of the city, close to its mouth.

Second, the canal supplied the water resources needed by the textile and flour producers of Barcelona at the height of protoindustrial evolution. The majority of the first calico factories and meadows of the town date from the central decades of the 18th century. The facilities for the production and printing of textile pieces had enough water availability as an essential condition for their location and success. This resource could be provided in the area by the Rec Comtal although some industries began to use groundwater for the same purpose. A significant proportion of the cotton weaving factories installed inside the city were located near the path of the final section of the canal. However, the economic development of the 18th century in Catalonia was linked to the operation of the milling activities and certain "mill fever" can even be described at that time. The 10 flour mills powered with hydraulic energy of the water from the Rec Comtal would have experienced a considerable increase in production.

Third, the agricultural development of Barcelona Plain during the 18th century accelerated during its last decades. This development was characterised by the increase in the cultivated land, an improvement in agricultural techniques, the specialisation in crops commercialised in the city, an increase in the income obtained and the presence of the trading and manufacturing bourgeoisie of Barcelona in the social composition of agricultural ownership. The biggest transformation took place in the sphere of irrigated agriculture. The surface irrigated with water from the Rec Comtal increased throughout the century. The number of concessions granted by the Intendència for the use of canal water for irrigation is by far the highest in Catalonia in the 18th century and the value of these irrigated lands, in accordance with the intensification of the harvests and the income obtained, easily exceeded the value of irrigated areas in other places of the country. The stimulus provided by the proximity to Barcelona, with a constantly growing economy and population and with social minorities with great purchasing power, was the decisive factor for this outstanding development of the production of vegetable crops associated with the Comtal canal.

As a consequence, well into the second half of the 18th century the water offer in the area reached a peak due to the process of urban, demographic and productive transformation and was not able to fill the demand. The volumes of water needed for the different uses could not be provided any longer by systems arising from medieval times such as the Collserola network or the Rec Comtal. It became essential to find new sources and to construct new supply infrastructures (Conillera *et al.* 1986; Voltes, 1967).

19.7 THE WATER SERVICE FROM THE END OF THE 18TH CENTURY TO THE APPROVAL OF THE EIXAMPLE (TOWN EXTENSION) PROJECT

Voltes' (1967) and Conillera *et al.*'s (1986) books as well as the authors mentioned in both describe this chapter in detail. A summary follows of the convulse period described, which started in 1776. From this year on, several situations of water scarcity leaded to troubles in town, from the need to guarantee the safety of the infrastructures to difficulties for financing the new water sources implementation. The same year, the Royal Authorities were forced to send troops to the area were water was diverted from the river Besòs using a weir towards the Comtal canal in 1776 because of bandits attacking the infrastructures.

Also, severe floods in 1777 and 1779 caused damage to the weir and the old canal supply system was abandoned.

A new project consisting of building an aqueduct under the confluence of the Besòs and its main tributary, the Ripoll River, was studied and implemented. The gallery (Figure 19.6) located at the head of the canal was to accumulate water infiltered from the riverbed. This project was financed by several of the stakeholders, started to be built in 1778 and began operating in 1786.

The city used the water obtained for several purposes, like for sending water to the streets of the Raval, and the individuals who wanted to have a supply in their home contributed to finance the works. Consequently, individual consumption gradually increased.

In addition to the unfavourable political and economic conditions and the use of modern production systems, requiring additional and regular water power, a good part of the textile industry located on the

Despite the difficulties and the dispute with the other parties interested in the Montcada water, Barcelona City Council had brought about the first major urban supply initiative of the 19th century.

The political changes succeeded, around 1830, in the breaking-up of the traditional powers with jurisdiction over the water, and the initiative of individual interests and of those of Barcelona City Council, led to a situation of definitive crisis in the management model of the Rec Comtal and the Montcada mine, which had existed since the Middle Ages.

The discontent among the owners of the agricultural land, the impotence of the Batllia General de Catalunya (Civil Government: depending on the regime the authorities took different denominations over time; sometimes changed every other year. Intendent, Head Politician, Civil Governor...) to control the disorders, especially in the summers of 1836 and 1837, due to the negligence and nepotism of the guards, and the infuriating lack of water were the factors which led the Head Politician of Barcelona to take over the control and administration of the water of Montcada and of the Rec Comtal in 1838.

The Head appointed the mayor of Barcelona as provisional administrator and called upon two commissioners for each interested group or institution to try to solve the dual problem of the scarcity of water and the disorder in the distribution. In May 1838 a Board was finally formed for managing the extension of the Montcada gallery, the increase of the volume and the economic arrangement and distribution of the water of the Rec Comtal, established by all the interested parties with the exception of the Royal Assets which maintained a permanent obstructionist attitude, taking steps with the Spanish state administration of the Crown assets and refusing to make its payments.

Throughout 1840 the Board established in 1838 was consolidated and received a definitive impulse thanks to the initiative of the new Head Politician. The first General Meeting of users of the Rec Comtal took place under the chairmanship of the Administrative and Management Board.

The definitive step had been taken towards the transformation of the 1838 Board into a Company of owners interested in the water of the Montcada mine during the February 1842 sessions and new ones between 1844 and 1846. With the definitive approval of the Bylaws and the Articles by the Crown, the Company of owners had taken over the administration and control rights of the water of the Rec Comtal and the Montcada mine which had traditionally been in the hands of the Royal Assets.

The participation of the owners of the Company in the expenses and their degree of influence in the decisions of the Management Board were very uneven: from Barcelona City Council to a small irrigation owner. The divergence of interests, especially between the municipality of Barcelona, the irrigators, the mills and the future steam factories, all of them users of the Montcada water, was important. The central core of the possible divergences between the members of the Company was water management in the summertime, when the low water levels reduced the available flows in an alarming manner.

According to the Articles of the Company, in situations of scarcity the Council could redistribute the canal waters if considered appropriate and the city's urban service was guaranteed by means of the Baix de Montcada conduit. The problem lay in the priority of each use: first irrigation, second urban supply, and finally hydraulic energy for the mills. With this regulatory framework, the conflicts soon appeared. In 1850 the city council suggested to the Board the need to arrange the distribution of water from the outlet of the Montcada gallery in order to guarantee that Barcelona did indeed receive the third of the water which corresponded to it according to an agreement signed in 1822. An agreement was signed in August 1852 and according to it, once the delayed payments of the works corresponding to the municipality had been paid, the town would receive a third of the total water extracted and 1000 m^3 of the two thirds corresponding to the irrigators and the mills. With this agreement the Barcelona City Council and the Management Board of the Company defined the respective rights over the hydrological exploitation of Montcada's waters.

The water achieved for the city with this agreement, 12,000 m^3/d, could not at all confront the truly deplorable health conditions of the city of Barcelona enclosed within the walls in the central decades of the 19th century. The overpopulation, the streets without light or ventilation, the vicinity of the factories and artisans' facilities, the coexistence with domestic and draught animals, were just some of the serious problems. The domestic sanitation procedures mainly used were either latrines discharging to cesspools built without guarantees, or different systems based on the old traditional chamber pots. The scarce and obsolete sewerage network of the city inside the walls and of surrounding neighbourhoods was the result of centuries of works depending on the real circumstances and without any planned procedure. The sewerage ran parallel to or crossed in many places the drinking water transport pipes. The deficient insulation created a permanent health risk.

The supply served the scarce public fountains and the individual homes belonging to the social and institutional elite. In the mid-19th century this supply service had serious deficiencies. It is noted at first the scarce volume of water available, estimated at 53 L/d and inhabitant in 1857. While the volume granted to an individual concession was equivalent to the water consumed by 37 Barcelona residents, each of the 24 fountains documented for 1857 statistically had to serve 9510 residents. Moreover, the eastern sector of the old town, where the wealthy citizens and business core were concentrated, had a denser network of pipes and, therefore, a greater access to water than the Raval neighbourhood.

The droughts, which could reduce the volume by up to 50%, and the bad state of preservation of the distribution network created additional problems. The Baix de Montcada aqueduct suffered from filtrations and continuous water losses. Many sections of the secondary pipes in the old city dated before the third decade of the 19th century or from the 18th century and even earlier had deteriorated. The private pipes accumulated filtrations and scaling.

The legal situation of the city council within the Company of owners hindered its action in two ways: It made it impossible to increase the volume for urban uses, as the agreements with the rest of the owners limited the municipal endowment, and the hierarchy of uses of the gallery waters prioritised irrigation over the resources for the city use and the initiatives of the council to increase the available volumes from Montcada must have the approval of the rest of the Company whose interests did not always coincide with the urban needs.

The impossibility of the municipal water service to entirely meet the needs of the population of Barcelona explains the persistence on the use of wells. In the 18th century this was the most commonly used private supply system and in the mid-19th century it must have been very frequent.

19.8 THE MODERN SUPPLY SYSTEMS OF THE EIXAMPLE AND OF THE PLATEAU OF BARCELONA IN THE SECOND HALF OF THE 19TH CENTURY

At the time of the approval of Ildefons Cerdà's Eixample project (enlargement of the city outside the former walls) for Barcelona in 1859, the different challenges included providing a modern water supply system and a continuous circulation sewerage system. This double network, with characteristics comparable to the ones of the main European cities such as London and Paris, had already been proposed by Cerdà himself among his proposals for the development of Barcelona.

When the Eixample implementation began it was expected to generate higher water demand. The municipality tried to undertake works to extend and arrange the Montcada catchment galleries or to search for new springs together with the Company of irrigators. The City Council also tried to acquire the rights of other users of the Montcada water, without much success.

Other supplies were sought in order to supply the Baix aqueduct. In 1871 there was a water supply tender in order to increase the volume available to the city by 20,000 m^3/d. The initiatives offered presented insurmountable technical, financial and legal difficulties.

An attempt was made to modernise the infrastructure of the Baix de Montcada aqueduct in order to supply water with enough pressure to serve the new buildings of the Eixample. In 1859 the municipal architect Eduard Fontseré proposed installing a steam pumping station in the main distributor to increase, by 20 metres, the pressure of the water from Montcada, with a total of some 40 metres.

In June 1879 the construction of three wells on the right bank of the Besòs, very close to the Montcada gallery, was finished. In August the extraction of water began in order to feed the Baix de Montcada aqueduct. The wells were equipped with Alexander steam machines for pumping with a maximum extraction capacity of 23,000 m^3/d. The wells could supply twice the amount of water that the Montcada galleries had been sending to the city. Moreover, the volume extracted was not subject to the strict legal regime imposed by the Company of owners of the water from the Montcada gallery. This achievement was the result from an order given by the civil governor on February 1878, on the occasion of the exceptional drought being suffered by the town.

The void left by Barcelona City Council in the water service for the Eixample was filled by individual initiatives. Small companies intended to exploit groundwater from the subsoil of the blocks that were being developed or in the closest places possible. Separately, initiatives of companies to supply the whole city and the surrounding towns, with a broader commercial and geographic sense, appeared as there were different and contradictory conditions in the private water market. The groundwater accumulated in the subsoil of Barcelona was a resource that could be better exploited. The other traditional area of exploitation, the Besòs basin, was the most frequented in the search for the desired resources. The lower Llobregat Basin could offer higher volumes and successful prospecting was carried out in this area. There were also projects to transfer water resources from more distant places such as the streams of Argentona and Dosrius the Maresme (northern coastline) or Garraf region (southern coastline) and the Ter basin (60 km northwards, inland).

The new Spanish hydrological regulations, Water Laws passed in 1866 and 1879, supported private initiative in water extraction, transport and distribution. Despite the legal facilities, the supposed demand increase and the proximity of the catchments, the new companies encountered obstacles in the operation of their businesses and the majority underwent bankruptcy in the two next decades, sometimes due to the excessive bureaucracy.

The construction of the collection or pumping systems and the distribution networks entailed huge expenses for the companies. Each of them had to install its own pipes along the same streets as the other, as the most attractive areas for business were precisely those which corresponded to the wealthy neighbourhoods of the Eixample and the central areas of the towns of the Barcelona Plain. The companies capable of delivering water at the lowest price and with the highest pressure, regularity and flexibility in relation to demand would have a much better chance of surviving.

The majority of the companies which entered the water business did not have sufficient capital and technical capacity. The companies with a sound financial base to rely on were exceptional. This was the case of the Empresa del Baix Vallès (Low Vallès Company: Vallès is a region neighbouring Barcelona to the north), linked to the Girona brothers (great bankers), or the Compagnie des Eaux de Barcelone (CAB/Barcelona's Water Company) with Belgian capital.

The evolution of the companies followed the economic and political situations of the time. The expectations generated by the approval of the Eixample project were soon frustrated by the phase of depression which began with the stock market failure of the railway shares in 1866. The political and social circumstances which surrounded the revolutionary six-year period did not favour the essential climate of social peace, political stability and economic confidence necessary for the middle class to risk investments in the water business. At the end of the 1870s the projects to establish supply companies and the interest in obtaining new concessions experienced a substantial revival, in response to the situation of prosperity of Catalan capital known as "gold fever". This phase of expansion did not, however, go beyond the early 1880s.

The municipal urban development was not very inclined towards major sanitation works and projects. Additionally, a high percentage of the industrial and agricultural consumption of water out of the town

19.9 THE FIGHT TO DEFINE THE WATER SUPPLY MODEL IN BARCELONA

At the same time as the private water market was taking shape in Barcelona in the 1880s and 1890s, the city was experiencing a whole series of circumstances with a direct influence on the future of its water supply. First, a growing social and civic concern for urban hygiene and health and for water supply appeared. The hygienic concerns and the new microbiological discoveries describing the role of water in urban epidemics were factors which boosted this interest. In this respect, the 1885 cholera epidemic represented a dramatic lesson for the people of Barcelona. The bourgeois minority became aware of the need to improve the city's hygiene.

As a result, the City Council became more involved in the struggle to improve the city's health conditions. Several facts helped to support this initiative including the 1888 Universal Exposition, the wish of modernising a pro-European town and the planning which translated the new ideas into practice. All of it offered a favourable framework to implement the demands for new municipal initiatives relating to public services responsible for urban hygiene.

In 1891 some specific actions appeared as a result of this process undertaken in the 1880s: the new municipal bylaws, with explicit references to water distribution conditions, the well known engineer García Faria sewerage project and the creation of the Municipal Institute for Hygiene which had to supervise the drinking quality of the water consumed in Barcelona.

Also in 1891, the City Council approved the "*Project for the unification and pumping of Montcada water*". It was intended to go beyond the limitations of the Baix de Montcada aqueduct, the main municipal supply system since 1826, trying to take advantage of all of the municipally owned water resources of Montcada. This included building a safe network which would prevent filtrations, improve the health conditions and bring water to the city with sufficient pressure to distribute it. An aqueduct 6018 m long following the coastal slope of Collserola was planned, which would start from a new pumping station, up to a hill where a tank would be built in a 100 m high location. It was also planned to install 132 km of new distribution pipes. With this project, the municipality could have a distribution service capable of meeting the demand and double the total water reaching Barcelona at the time and could compete in advantageous conditions with the private initiative.

Once the execution of the Alt de Montcada aqueduct project had begun in 1891, the City Council encountered all sorts of obstacles. First, the construction of the conduit gallery was very expensive, technically complex and contested by different stakeholders, including two municipalities of the area and the SGAB. Second, the exploitation of the Montcada water that the aqueduct had to transport was accused of being illegal, as the land on which the wells, built in 1879, were located was not owned by the municipality of Barcelona. The legal claims prevented the aqueduct from being built and affected the availability of water to be transported. The result was that the works came to a standstill in 1897.

The failure of that project, the growing urgency to provide urban services which demanded increasingly high quantities of water and in 1897 the extension of the municipal jurisdiction of Barcelona to the surrounding towns which were annexed, led to a modification of the water policy plans of the city's government.

It was necessary to rely on the private initiative to have new resources to complete the deficient municipal service, based on the water from the Montcada gallery, shared with the Rec Comtal, and that obtained from the three Montcada wells, provisionally declared illegal. An initial supply tender was held for this purpose in 1896. The lack of fully satisfactory offers led to the extension of the tender in 1899, the ruling of which was drafted and approved in the years 1902–03, with a majority presence in the city hall of regionalist and republican councillors. No practical resolution was, however, reached.

In 1905 there were initial discussions between the SGAB and Barcelona City Council. This allowed the negotiations concerning the long dispute on the municipal wells of Montcada to be resumed. Also, contacts began in order to agree on a future municipal acquisition of the service offered by the company. The political

taxes. Moreover, as the water supply management model had to be changed and intervention was needed in the domestic hygiene systems, the profits that the urban owners could obtain on renting their properties were threatened. Part of the associations of industrialists and of traders of Barcelona including especially the Unió Gremial (trade union), were also implied. Some of the small water supply companies still operating in the city other than the SGAB also acted. The owners of some of the projects rejected in the water supply tenders held by the City Council in 1910 and 1911 also participated. Finally, certain scientific institutions such as the Association of Doctors of Barcelona, the Hygiene Academy of Catalonia and the Society for Economic Studies acted.

This heterogeneous opposition group had the support of the conservative political groups which were taking shape as a result of the restructuring of the two main dynastic parties, the Conservatives and the Liberals, directly linked to the interests defending urban property. From 1912 onwards, a good part of this interest group formed the so-called Pro-Barcelona Commission in order to join forces and to organise the strategies to respond to the arguments and actions of the Commission for Water Supply of Barcelona.

The group of forces opposing municipalisation recurrently used two main arguments. First, a frontal rejection of the purchase of the SGAB, arguing the lack of drinking water quality of the water exploited by the Societat General (SGAB) in the Llobregat aquifer was stated. The lack of definition of scientific criteria of drinking water quality applied to the Cornellà water, in the context of tension being experienced, caused serious statements against the City Council and the water supply Commission. The accusation consisted of stating that the Cornellà operation was the cause of the endemic situation of typhoid fever suffered by the city. Second, and as an alternative, the complete use of the water from the Montcada mine and wells available to the City Council was defended, together with the acquisition of the rights of the group which used the volume distributed along the Comtal canal.

The beginning of the First World War, August 1914, considerably hindered the acquisition of the SGAB, a company belonging to a belligerent state, France. Moreover, the lethal and dramatic outbreak of typhoid fever in the autumn of 1914 upset the terms of the social, economic and political struggle that was taking place in Barcelona around the possible government approval of the water supply municipalisation project. The typhus epidemic was caused by filtrations of faecal waters in the Baix de Montcada aqueduct. This supply was discontinued and the alternative service was provided by the Cornellà operation of the SGAB. The rapid end of the epidemic with these measures demonstrated the weakness of the arguments criticising the SGAB. The City Council had to proceed to overhaul the already hundred-year-old municipal service with a complete renewal of the piping infrastructures to the city and with the replacement of the obsolete urban distribution network. These measures had the support of the forces opposing municipalisation, as the modernisation of the Montcada service was included in their programme of demands.

The situation of economic growth, as a result of Spanish neutrality in the First World War, and the intense context of nationalisation of the economy prevailing in the early post-war years, together with the increase in Spanish banking profits and the devaluation of the franc in its exchange with the peseta, aided the naturalisation of the SGAB. In June 1920 the ownership of the company passed to a Spanish banking consortium with the presence of major financial institutions. In August 1920 the new owners of the water company offered it to Barcelona City Council, but this second municipalisation project also failed. It was stopped by the internal political and financial difficulties of the city council, by the context of social conflict experienced by Barcelona from 1920 to 1923, by the business prospects that the banking group soon saw in the water service and by the end of the republican and nationalist city council, imposed by the change of Spanish political regime following the coup d'état of Primo de Rivera on September 13, 1923.

19.10 THE PRIVATE SUPPLY SERVICE: SOCIETAT GENERAL D'AIGÜES DE BARCELONA

The new social and political circumstances starting from 1923, the evolution of the accelerated urban growth experienced by Barcelona and the stimulus of the 1929 International Exposition generated the new context in which the definitive consolidation of the private water supply service was developed.

A dissolution decree of the town council had been executed on October 1, 1923 as a result of the coup d'état. Afterwards, the municipality did not promote any supply project which would increase the pressure on the already huge municipal debt or increase the scarce domestic water consumption.

In this context, from 1923 onwards, the only surviving company in town was SGAB. The company modernised its infrastructures and considerably increased the resources exploited, in order to adapt to the growing demand in the years of highest demographic expansion of Barcelona. Nevertheless, its control over the private service was contested by other parties interested in participating in the water market. The criticism on the SGAB's service was based on the supposed inability of the company to face the urban growth of Barcelona and on the excessive salinity of the Llobregat River water.

The new political panorama that opened up in Barcelona in April 1931 and the effects of the world depression, which began in October 1929, did not have an important repercussion on the city's water service management model. Although the city councils at that time had to consider the municipalisation of supply, the financial weakness of the municipality, the attention paid to other priorities and the general political and economic instability did not make it possible to fulfil any initiative which threatened the control of the SGAB over the supply. The legal management of the water resources of the Catalan basins, which corresponded from 1929 to the Hydrographic Trade Union Confederation of the Eastern Pyrenees (CHPO) did not modify the urban supply models in Catalonia and, in particular, in Barcelona.

Therefore, when the City Council decided not to municipalise the service, the Societat General's (SGAB; SGAB and AGBAR are the same company) service was consolidated and the first steps were taken toward the universalisation of the service. After several problems raised by the poor quality of the Llobregat river water due to excess salinity as mentioned, several solutions were studied, but the ideal project; to build a collector for saline water was postponed until the Civil War broke out in July 1936.

After the Civil War, the main existing company (SGAB) consolidated its owned assets and established the basis for a future development.

19.11 CONCLUSIONS

Big cities experience along their history difficulties and changes in their relation with water. Barcelona is not an exception, and due to the different patterns of development during its long history, different infrastructures were built which reflect differences in engineering capacity and wealth.

Roman aqueducts and medieval conduits, several of them still operating at present, reflect the prosperity times, while the obscure periods did not supply Barcelona with new water related hydrological structures. Over the last two centuries, Barcelona's water supplies experienced huge changes to adapt to changing water demand.

At the end of the 18th century, the start of the industrial revolution led to a much greater demand for water in order to meet the needs of textile manufacturing and the irrigation of crops to feed a constantly growing population.

The water supply needs were met initially by the Rec Comtal Irrigation Canal originating at the river Besós. Subsequently, the first modern drinking water supply system was promoted by Barcelona City Council with the Baix de Moncada Aqueduct.

After decades of competition and of failed municipal projects, in 1929–30, water supply was nearly fully ensured by the Sociedad General de Aguas de Barcelona. A number of the names of streets and squares in town reflect the relationships of the town with water. Like in any town with millenary history, the subsurface of Barcelona keeps secret many hidden infrastructures which will allow better knowledge of the water-related history of the town.

Acknowledgements

The authors greatly appreciate the information provided by Ajuntament de Barcelona (Barcelona Town Hall), Cicle de l'Aigua.

REFERENCES

Anuari estadístic de Catalunya (Yearly Statistics of Catalonia). http://www.idescat.cat/pub/?id=aec (accessed 5 June 2001)

Arandes R. (1998). *Hidrogeologia del Plà de Barcelona* (Hydrogeology of the plain of Barcelona). Ajuntament de Barcelona. Barcelona, Spain (in Catalan).

Conillera P., Llabrés A. and Parés M. (1986). *Descobrir el medi urbà. 4. L'aigua a Barcelona* (Discovering the urban environment. 4. Water in Barcelona). Ajuntament de Barcelona. Barcelona, Spain (in Catalan).

Conillera P. (1991). *L'aigua de Montcada. L'abastament municipal d'aigua a Barcelona* (The Montcada Water. Barcelona's municipal water supply). Barcelona, Ajuntament de Barcelona, Spain (in Catalan).

Fundació Pere Garcia Faria (2002). *Els transvasaments en la història de Catalunya* (Water transferences between basins in the history of Catalonia). Universitat de Girona. Departament d'Arquitectura i Enginyeria de la Construcció. Girona, Spain (in Catalan).

García G. (1990). *L'abastament d'aigua a la Barcelona del segle XVIII (1714-1808)* (Water supply to the 18th century Barcelona (1714–1808)). Fundació Salvador Vives i Casajuana, Barcelona, Spain (in Catalan).

Gran Enciclopèdia Catalana (1971). Ed. Enciclopedia Catalana, S.A. Barcelona, vol. **3**: Barcelona, Spain (in Catalan).

Gran Geografia Comarcal de Catalunya: El Barcelonès i el Baix Llobregat (1982). Ed. Fundació Enciclopèdia Catalana, vol. **8**, Barcelona, Spain (in Catalan).

http://www.bcn.cat/parcsijardins/fonts/sidney.html (accessed 15 January 2012).

http://www.dreamstime.com/search.php?srh_field=Water+tower+Tibidabo+Barcelona+Spain&s_sm=any (accessed 10 January 2012).

IDESCAT. Institut d'Estadística de Catalunya (Catalonian Agency for Statistics). http://www.idescat.cat (accessed 5 June 2011).

INE. Instituto Nacional de Estadística (National Agency for Statistics). http://www.ine.es (accessed 5 June 2011).

Latorre X. (1995). *Història de l'aigua a Catalunya* (Catalonia's Water History). Barcelona, Abecedari, Xavier Latorre i Piedrafita, Spain (in Catalan).

Martín M. (1999). *El Rec Comtal, 1822–1879. La lluita per l'aigua a la Barcelona del segle XIX* (The Count's Canal 1822–1879. The fight for water in the 19th century Barcelona). Fundació Salvador Vives i Casajuana, Barcelona, Spain (in Catalan).

Martín M. (2007a). *Abastament d'aigua, desenvolupament urbà i lluita social a Barcelona (1888–1929)* (Water supply, urban development and social fights in Barcelona (1888–1929)). Revista HMiC, núm. V 2007. Departament d'Història Moderna i Contemporània. UAB, Barcelona, Spain, available at http://webs2002.uab.es/hmic/ (accessed 6 February 2011).

Martín M. (2007b). *Aigua i societat a Barcelona entre les dues exposicions (1888–1929)* (Water and society in Barcelona between the two World Fairs (1888–1929)). Servei de publicacions de la Universitat Autònoma de Barcelona. Theses online of the UAB, Barcelona, Spain, available at www.tesisenxarxa.net (in Catalan) (accessed 6 February 2011).

Martín M. (2009). *Barcelona: aigua i ciutat. L'abastament d'aigua entre les dues Exposicions (1888–1929)* (Barcelona: water and the city. Water supply between the two World Fairs (1888–1929)). Fundació Agbar i Editorial Marcial Pons, Barcelona, Spain (in Catalan).

Miró C. and Orengo H. A. (2010). *El cicle de l'aigua a Barcino. Una reflexió entorn de les noves dades arqueològiques* (The water cycle in Barcino. A reflexion over the new archaeological data). Quaderns d'arqueologia i història de la ciutat de Barcelona – QUARHIS, 6, Spain (in Catalan), pp. 108–133.

Riba O. and Colombo F. (2009). *Barcelona: la Ciutat Vella i el Poblenou. Assaig de geologia urbana* (Barcelona: the Old Town and the New Village). Eds. Institut d'Estudis Catalans i Reial Acadèmia de Ciències i Arts de Barcelona, Barcelona (in Catalan).

Soldevila F. and Coll M. (1979). *Resum d'història dels països catalans* (Summary of the History of Catalan Countries). Edició ampliada a cura de Miquel Coll i Alentorn. Ed. Barcino, Barcelona (in Catalan).

Voltes P. (1967). *Historia del abastecimiento de agua a Barcelona* (History of the water supply to Barcelona). SGAB, Barcelona, Spain (in Spanish).

Chapter 20

Water services in Mexico City: The need to return to the IWRM principles of Tenochtitlán (700 years of water history)

B. Jiménez and D. Birrichaga

20.1 INTRODUCTION

Mexico City is the capital of Mexico, a country with 112 million inhabitants and a GNP of 5200 USD/per capita. The city is responsible for 21% of the national GDP (Gross Domestic Product) due to its commercial, industrial and political activity. This intense activity combined with a huge population living within a closed basin high above sea level has created a peculiar and complex water problem. Mexico City is located on a high plateau in an area where the only natural sources of water are rainfall and groundwater. Because of the high water demand to supply 21 million inhabitants, the city is facing considerable technological and management challenges not only to continue supplying water but also to properly handle the wastewater produced. This chapter describes the management of water in the city from the time of the Aztecs to the present date. It attempts to describe the origin of the present situation and to justify the need to move to a different paradigm that goes beyond Integrated Water Resources Management (IWRM) to consider the planned development of the country. This applies to all 11 cities with greater than 1 million inhabitants that are already threatened by lack of water, in some cases suffering conditions even worse than those of Mexico City.

20.2 MEXICO VALLEY DESCRIPTION

Mexico City lies in the Mexico Valley. It covers an area of around 9600 km^2 and is located in the central part of the country at 2240 m above sea level, between 98°31′58″ and 99°30′52″ west longitude and 19°01′18″ and 20°09′12″ north latitude. The mean annual temperature is 15°C and mean precipitation is 700 mm, varying from 600 mm in the north to 1500 mm in the south. The rainy season is well defined, lasting from May to October (when 80% of the pluvial precipitation falls) and is characterised by intense showers lasting for short periods. One single storm may produce 10–15% of the mean annual pluvial precipitation. With reference to pluvial precipitation the amount of renewable water should be 6646 million cubic metres per year (Jiménez, 2008).

The Mexico Valley was an endorreic (closed) basin with an extensive system of shallow lakes, lagoons and marshes, formed by pluvial precipitation. There were few rivers, most of them perennial. These originated south and west of the valley from the surrounding mountains and springs. A combination of infiltration to the subsoil, and the intense evaporation characteristic of the region, limited the growth of the lacustrine zone.

20.3 THE TIME OF THE AZTECS (1325–1521)

Tenochtitlán (the place of the cactus leaves) was established around 1325 AD by a Nahua Tribe named the Aztecs. Following the instructions of the Sun god (Huitzilopochtli), they travelled from a semi-arid area in the north of Mexico to the south until they found a lake in the middle of which, on a small island, an eagle sitting on a cactus was devouring a snake. According to the legend, this place was where the Aztecs built Tenochtitlán, the capital of their empire. Tenochtitlán was one of the last cities of pre-Hispanic Mexico and lasted only 200 years. It was constructed on an artificially built island that connected to the rest of the valley via four streets which were also used as dykes. The dykes and lakes were operated in such a way so as to control floods, separate the fresh water for supply from the saline water, to manage pluvial water, to dispose of the wastewater and to irrigate crops. The largest dyke, known as *Nezahualcóatyotl*, was 12.6 km long and 6.7 m wide (León Portilla, 1961).

When the Spanish arrived, in 1519 AD, Tenochtitlán was the most important city in Mexico with an area of 15 km^2 and 200,000 inhabitants (Offner, 1983). The streets were geometrically aligned and comprised small houses, parks, gardens, fountains and religious centres, all set around two sites: the Great Temple (Templo Mayor) and the great seat of Tenochtitlán. Roads and bridges were constructed to connect the diverse neighbourhoods and small islands (Becerril & Jiménez, 2007).

The economic, demographic and military power of Tenochtitlán was closely related to the hydraulic organisation. Water was a strategic element for development and domination of the other tribes that lived in the valley (Birrichaga, 2004a).

20.3.1 Water supply

At first, domestic water was conveyed using "cántaros" (mud vessels) from the springs located on the islands (Becerril & Jiménez, 2007). The spring located in the ceremonial centre, called Tozpálatl, was sacred, as according to tradition it was where the city was originally founded. In each household, water was stored in tanks.

The emperor Chimalpopoca realised that on the one hand the growth of the city had almost exhausted the springs and, on the other, the Aztecs' activities were contaminating the lagoon; therefore, he decided to construct an aqueduct to bring water from Chapultepec, a forest located on a hill with a lake fed from springs (León Portilla, 2001). The aqueduct, 12 km in length and 7 m wide, was built up with stakes, reeds and mud, and, due to the low resistance of these materials, several operational problems were experienced (Birrichaga, 2004b).

20.3.2 Water uses

The city gave great importance to sanitary and environmental control. Water was used mainly for irrigation. It was also used in much smaller volumes, though no less importantly, for municipal purposes.

20.3.2.1 Municipal use

To supply water to buildings, an underground distribution system was used that fed pools and gardens. The city enjoyed a great number of green areas with flowers and even a botanical garden. In the home, water was used for housekeeping, drinking, cooking and bathing (Birrichaga, 2004a). Wealthier people received water directly to their homes. It arrived at fountains in the middle of central courtyards from which other underground channels distributed it to the rooms (Cervantes de Salazar, 1985). Some houses had their own wells and even pools. The main sites of Tenochtitlán City and the market received it directly and water was stored in constantly fed tanks. The excess water was discharged to drains that were used to wash clothes and to sewers (León Portilla, 2001). In addition, there were public steam baths called

20.3.3 Sewer system

20.3.3.1 Sanitation

Although the Aztecs had no city-wide sewers, and much of the wastewater ended up in the lake surrounding the city, the city was kept very clean. Streets were swept and washed daily by numerous public employees. The city streets should have been clean enough to walk barefoot without getting dirty. Personal cleanliness and hygiene were highly valued by the Aztecs. They were fond of soaps, deodorants, dentifrices, and breath fresheners. Rubbish was incinerated due to a lack of disposal sites but was also used as fuel to provide lighting. Organic garbage was buried in courtyards, used to feed animals or shipped out of the city to be disposed of on agricultural land (Becerril & Jiménez, 2007).

The upper class had toilets at home, the common people used privies that were in all public places and from which the faecal sludge was collected in canoes. This, along with household sweepings, was applied as fertiliser to chinampas or sold in the market to be used for tanning animal hides. Urine was collected in pottery vessels to be used later as a mordant for dyeing cloth along with coloured mud. Tenochtitlán was a clean environment in comparison to European cities (Becerril & Jiménez, 2007).

20.3.3.2 Flood control

Since Tenochtitlán was located at a lower level than most of the lakes, excess water caused risks. For this reason the Aztecs mastered the art of flood control (Ortiz de Montellano, 1990) by controlling the water level in lakes through a complex set of sluices. Between 1440 and 1450 Nezahualcóyotl, the King of Texcoco, an engineer, architect, urbanist and poet, designed the Albarradón (a dyke 16 km long and 4 m wide) to protect Tenochtitlán from floods, and to independently manage the freshwater from the Xochimilco and Chalco lakes from the saline water of the Texcoco Lake. The Albarradón was a stone and wood canal. If the water level of the Texcoco Lake was high during the rainy season the sluices were closed; during the dry season water from the Chalco and Xochimilco Lakes was allowed to enter. For 103 years the Albarradón prevented floods in Tenochtitlán, a period that, considering the frequency of floods after the Aztec period, is no doubt a record.

In addition the hydraulic system was used for (Birrichaga, 2004a): (a) transport, using canoes to navigate the canals; (b) energy production, using dams to create hydraulic falls; and (c) controlling the salinisation of the chinampas.

20.3.4 Fluvial transport

Mesoamerican cultures did not have animals to carry heavy burdens and did not make practical use of the wheel. The lacustrine system of the valley was criss-crossed by a network of channels and deep drains, most constructed artificially, around which an enormous number of canoes circulated (Becerril and Jiménez, 2007). These channels were the main streets of the city. One went from north to south linking Tepeyac, Tlatelolco, Tenochtitlán and Coyoacán; the other went from east to west along Tlacopan and through the centre of Tenochtitlán.

20.3.5 Water management

The springs were the property of the town and were administered by the government. The political and religious upper classes collected taxes to manage water. All people had an obligation to cooperate, following the instructions of the authorities in order to accomplish hydraulic works.

Urban water was sold by volume or number of vessels. Water was sold in small quantities for domestic use by water vendors who used tanks to transport it on canoes (*acallis).* To pay for water, cocoa and textiles were used (Birrichaga, 2004a).

Curiously – as at that time there were several tribes living in the valley – the Aztecs sought collaborative agreements with other tribes living in the valley in order to integrally manage water; a situation that presently does not occur for political reasons. This organisation was a kind of basin committee and was responsible for building, operating and maintaining the hydraulic infrastructure. As an example, a short time before the Spanish arrived; the Texcoco lord granted the Teotihuacan people the right to use water from his springs for agricultural irrigation. To transport the water a 20 km long canal was built. The water conveyed by the canal could be used by other villages provided they paid for it. To pay for the water the Teotihuacans were also requested to build and maintain the hydraulic infrastructure (Birrichaga, 2004a).

20.4 COLONIAL PERIOD

Tenochtitlán was colonised by the Spanish during the 16th to 17th centuries. When the Spanish arrived in 1521, the Valley of Mexico was ruled by the Aztec Empire, formed by the alliance of three civilizations: Tenochtitlán, Texcoco and Tlacopan. In terms of water use, the conquest resulted in the confrontation of two worlds with very different social, technological and economic views of this resource (Birrichaga, 2004a). To force a city of 200,000 inhabitants to surrender, Cortés, with only 800 soldiers, ordered the destruction of the aqueducts that supplied city. In contrast to the Aztecs who worshipped the lakes, for the Spanish the lakes were a source of problems and a place to dispose of used water. Therefore as a first step they completely isolated the five lakes and used the perennial rivers as sewers (Musset, 1992). The dried areas were used to build new and elegant houses. Since the Spanish were not used to living in a city surrounded by water, they lived in constant fear of floods. As a result they decided to completely dry the lakes, a solution that soon proved to be flawed, and since 1555 the city has been experiencing floods.

20.4.1 Water sources

Once Cortés ruled the city, the aqueducts were repaired, and were actually enlarged to provide additional water. At this time the aqueduct was redirected from Chapultepec to the Salto del Agua public water fountain, instead of the Templo Mayor (where the Cathedral is located today). A second aqueduct was built from the town of Santa Fe, where nowadays an exclusive neighbourhood of Mexico City exists suffering from lack of water, to the Mariscala Bridge (Musset, 1992). The supply system was composed of three sections: (a) the aqueducts transporting water from the sources to the city; (b) water tanks in which water was stored and from there distributed to channels, and (c) the clay or wooden channels or pipelines that distributed water to public fountains, cisterns, private houses and some public buildings (Birrichaga, 1998).

As the Spanish dried the lakes, water vendors using canoes disappeared but the extraction of water from private and public fountains continued. In 1625 water vendors used the Chapultepec aqueduct as supply and transported water in wooden barrels carried on donkeys. In this way, at the end of the 16th century, two aqueducts were feeding 60 public fountains and several private taps. The populace either took water from the public fountains or bought it from water vendors. Box 20.1 presents the terminology used to describe the different water sources at that time. In the 16th century, the "good water" used for human consumption was frequently water taken from springs. This was preferred to water from wells because of its higher quality. In addition, this "discovered water" was cheaper and easier to extract than hidden water (groundwater) (Birrichaga, 2004a).

20.4.1.1 Municipal supply

The Spanish brought in regulations that meant water for domestic use was to be publicly available and had preference over industrial and agricultural use (Margadant, 1989; Lanz Cárdenas, 1984). Water was provided at no cost to public fountains, households, hospitals, and religious buildings (Birrichaga, 1998).

Furthermore, in the Spanish colony there was considerable bureaucracy. Any hydraulic project had to be approved by the City Council, a public hearing and the Spanish Crown, in a process that often took several years (Margadant, 1989).

During Spanish rule, water resources were designated separately to water for public and private use; water for public use was for the inhabitants of the city who could obtain water from public sources. In contrast, those that needed water for industrial or agricultural purposes had to obtain a royal concession called a "*merced*". Those privileged enough to obtain such a concession were called *mercedados*. The granting of a merced did not create permanent rights as water was always considered the property of the Crown. The allocation was granted only on a temporary basis, and could be withdrawn at any time. Users could receive fines. As an example, in 1635, the Viceroy of Albuquerque imposed fines on water vendors who were charging high prices for public water (Birrichaga, 2004a).

As the city grew, the number of water users increased and some began to use water without paying for it. The City Council then created the Water Council to verify water permits and to collect water taxes, but also to operate the hydraulic infrastructure. In addition, the City Council appointed a water judge, "to oversee the use of water", who was in charge of updating the use of the mercedes. This judge was also to advise on the need for additional water (Birrichaga, 1998).

Industries paid for their water annually (González De La Vara, 1991). The money was used to improve the hydraulic infrastructure. To facilitate the construction of public works, in some cases the authorities granted the use of water to private persons if they took charge of the construction of public fountains (Birrichaga, 1998).

In order both to build and maintain the hydraulic works and also to survey and collect water taxes, private companies or independent people were hired. Water workers were Spanish or Indian labourers who were trained as carpenters, building workers, or quarrymen (Birrichaga, 2004).

During the colonial years, water was quantified using a mixture of the Indian and Spanish measurements. The Indian quantities were based on the human body. Water was measured unidimensionally, using a leg or finger as units. If the quantity was larger, then the term "ox-size" was used. This was a useful system as each person was their own reference for these measures (Kula, 1980); the disadvantage was that there were no multiples. The Spanish, in contrast, measured water according to their activities, so, for example, their measures related perhaps to the amount of water that was sufficient to move a mill or to injure a miller. Over time, more precise measurements were used that were understandable to everybody. In 1567 it was agreed that the unit amount of water that a mill needed to operate was the quantity sufficient to grind three rows of seeds of sugar cane (Solano, 1991). In the 16th century, a two dimensional system was introduced. For example, the ox water unit was equivalent to the amount of water that flowed on a rectangular surface of 0.702 m^2 or circular one of 0.038 m^2 (a "vara" of diameter). To standardise this, in 1567 Gastón de Peralta, created a unit system (Box 20.2).

Box 20.2 Hydraulic measurements used during colonial times, from Birrichaga 2004a.

1 ox = 48 furrow = 144 oranges
1 furrow = 3 oranges = 24 reales
1 orange = 8 reales = 8 lemons
1 orange = 2 fingers = 144 straws
1 real or 1 lemon = 18 straws

Furrows and oranges were units used for agricultural or industrial practices, while the straw unit was used to measure the water rights for domestic use (Solano, 1991). As shown above, these units considered the area covered but not the flow. This system was used until the end of the 19th century, when the decimal metric system was imposed (Palerm & Chairez, 2001).

20.4.4 Legal framework

Initially, the Spanish respected the water rights of the influential Indian people. Over time the "Mercedes" changed this. Water began to be allocated to town leaderships and Indian chiefs, conferring upon them the right to commercialise the use of water. This provoked the commercialisation of water resources that were originally considered as a public right, and led to further social water problems. The Spanish government was forced, as a prerequisite to allow settlements to develop, to ask the owners to demonstrate the availability of water (Birrichaga, 2004b).

Competition for water increased among farmers, miners, tanners and domestic users. This obliged the authorities to take better control of water resources but also to set rules to respect public and private ownership of water. Public water was that owned by citizens of villages or towns who had free access to it at the city fountains.

Later, in 1560, the Spanish Crown issued "ordenanzas". In 1563 one ordenanza set out the way in which water should be distributed among users for domestic, forage irrigation, private land and country houses. To solve water conflicts the "oidores" or "hearers" were established as judges who decided how to proceed once complaints were presented (Solano, 1991). The idea was to force Spanish people to respect the rights of Indians to water and land. However, in practice, this was not achieved and several conflicts arose. In parallel, there were also Indian courts that solved water rights issues among the Indian population (Birrichaga, 2004a).

To regulate the use of water, periodically there was a public redistribution that took into account the oldest rights of water, and was advantageous for the Indian population. The distribution was made following the principles of (Birrichaga, 2004a):

(a) *Proportionality* – To consider the land available for cultivation;
(b) *Alternative uses* – To serve different users;
(c) *Economy* – To make efficient use of water by distributing it through a single canal to avoid waste and leaks in minor irrigation conduits;
(d) *Stability* – To ensure maintenance, by requiring users to maintain hydraulic infrastructure at their own expense;
(e) *Cleanliness* – To ensure that users kept the rivers clean.

Besides regulating water rights, the ordenanzas set regulations for public hygiene. Users were required to keep the water clean, to avoid pluvial water discharges to the canals and to prevent cattle from entering water courses. In addition, other ordenanzas banned the washing of clothes or the discharge of waste materials to drains. At the same time ordenanzas obliged users to clean the river bed and basins, springs and water canals (Birrichaga, 2004a).

20.5 THE 18TH CENTURY

At the beginning of the 18th century Mexico City had around 130,000 inhabitants; a figure that included the Indian population. The city was the centre of the colonial economy, as traders and farmers established it as the centre of their financial transactions. At that time the lakes had been dried in order to provide new urban spaces. Palaces and temples had been built and a city grew up with straight streets.

20.5.1 Water sources

During the 18th century water scarcity began to be a problem for the city, due mainly to a water rights system that allowed people to use more water than that allocated, or even to use it without a permit. As part of the measures to improve the supply of water, several ordenanzas were issued requesting the protection of water sources from pollution caused by pluvial water or cattle (Birrichaga 2004b).

20.5.2 Municipal supply

The lack of water in the city forced the authorities to modify the hydraulic system. At the end of the 18th century, Viceroy Revillagigedo tried to reorganise the city based on mechanistic theories, according to which water should circulate in the city to protect public health. Cemeteries, jails, hospitals and slaughterhouses were located based on this theory. This also meant that clean water should be constantly provided and that used water should be eliminated from the city. Despite the reorganisation of the city, in 1776 the lack of water became critical. In order to combat this, the City Council decided to use the water from the springs located in the Desierto de los Leones area, south of the valley, which caused distress to the people using this water in the nearby towns (Reyna, 1988).

Water was constantly extracted without permission from aqueducts and canals. Therefore the surveillance of the permits was reinforced and a program to repair pipelines using lead was instituted. In 1731 the water authorities planned to introduce lead pipelines all over the city but a report from the university concerning the effects of lead on public health resulted in the use of clay pipelines instead (Cooper, 1980). Mechanical methods using flotation devices were used to close leaks from water supplies (Birrichaga, 2006).

At the beginning of the 18th century the quality of domestic water became a source of concern. In 1733, a book written by Joseph Ortiz Barroso, doctor to the Royal Family, stated that "sweet" or freshwater should be similar to the air, that is, transparent, light colourless, odourless and tasteless. It suggested preferential selection of water sources in the following order: springs, pluvial water, well water, rivers, ponds and snow. Barroso also recommended that springs should originate in places where the soil was of good quality, and was not a mineral or sandy type, but of humic constitution. Springs should face west because the sun could then clean the impurities from the air. In addition, their water should be cold during the dry season but hot during winter because this meant that water came from greater depths and was far from the sources of impurities such as the soil or the vegetation growing above the groundwater. If no springs were available and river water had therefore to be used, water should be extracted prior to entering the city to avoid pollution. In Mexico City the water judges were in charge of instituting these recommendations (Birrichaga, 2006).

20.5.3 Sewerage system

At the end of the 18th century the city suffered another significant flood. To alleviate floods, Cosme de Mier y Terán ordered the opening of two canals to drain the Tajo de Nochistongo in 1795.

20.5.4 Water administration

The main characteristic of the 18th century was the attempt to reorder the management of water. Initially, its use was poorly controlled. This is especially true of public baths which used more water than they were entitled to. In 1709, the City Council revoked all of the concessions made to the owners of public baths in order to make a review. Following this, in 1710 the City Council issued laws setting out rules to protect the quality of water sources. In addition to avoiding water pollution, the laws stated the need for aqueducts to be cleaned and repaired (Birrichaga, 2004b).

In 1715, Fernando de Alencastre Noroña y Silva, Duke of Linares and president of the Royal Audience of the New Spain, requested that the Viceroy revoke the law that prohibited the granting of additional water titles in Mexico City. The request was granted and more water was allocated. This aggravated the water availability problem and the authorities were forced not only to periodically review the mercedes but also to repair the water network (Birrichaga, 2004b). In 1740 the City Council appointed the first water judge for the Santa Fe water source.

20.6 19TH CENTURY

The beginning of the 19th century saw a concerted effort to modernise the country, in which rational planning of urban spaces was needed. This required a scientific basis in which water use was considered as part of the planning process. As part of this, politicians promoted the concept of the provision of water to each household to enhance public hygiene. This, along with new technological solutions and demographic changes, modified the way in which water use was managed and conceived. The introduction of water networks in cities dramatically differentiated the urban environment from the rural one. During the first part of the 19th century (in 1821), Mexico obtained its independence from Spain but also suffered from a series of unstable governments rendering the provision of public services difficult.

20.6.1 Water sources

As the city grew, water began to become more scarce and the local springs were insufficient to cope with demand. The City Council began to buy water rights from the nearby towns and even local users. As an example, in 1894 the City bought water from the Hacienda de los Morales, owned by Eduardo Cuevas (Birrichaga, 1998). Later, in 1847 when the city had around 0.5 million inhabitants, in order to respond to the requests of its citizens, the government began to extract water from the subsoil, using wells of a depth of 105 m (Santoyo *et al.* 2005). By 1857, there were 168 artesian wells producing water (Jiménez, 2008). To supply the constantly increasing demand, between 1847 and 1886 the number of shallow wells increased to around 1000 for the whole of the Mexico Valley. This water was complemented by that supplied by three aqueducts (Chapultepec, Santa Fe and Desierto de los Leones). These continued to function until the end of the 19th century.

20.6.2 Municipal supply

20.6.2.1 Infrastructure

At the beginning of the 19th century the average inhabitant of Mexico City had access to 5 to 10 L of water per day. The lack of water resulted in a crisis in 1846. At this time, at the origin of the water problem, there was on the one hand the persistence of a lack of order in the granting of water concessions, and, on the other, the deterioration of the system due to lack of maintenance and damage caused by the constant earthquakes that are characteristic of the Mexico Valley. To restore order, on December 11 of that year, the water police was created. Besides being responsible for repairing the infrastructure, this body also had to inspect the water supply system daily to prevent illegal extraction or pollution of water (Birrichaga, 1998).

Several weeks were required to repair a single water leak as workers did not have access to the proper equipment and there were insufficient valves to isolate short sections of the pipelines. To overcome this problem, taps to control the flow through individual sections of the system, known as saving taps, were installed, but in limited numbers.

In 1866 the City Council decided that the colonial hydraulic infrastructure was inappropriate for the protection of public health and there was a need to build a new system. This was required not only to ensure the provision of clean water, but also to control leaks, supply water with sufficient pressure to buildings on higher ground and to ensure the sufficient flow of water for any kind of use (Birrichaga, 2009). This new concept of the water system required two things: (a) the introduction of a new water network and, (b) the proper operation of the sewerage system. With regard to drinking water, the target was to provide 150 L per day per person. In fact several other American cities had adopted these same objectives. The system was conceived to work in sections in such a way that each sector could be handled independently (Birrichaga, 2006) and use iron pipelines with lead based joints. The idea was to provide water services constantly and universally. As this goal was not fully met, many users began to directly extract water from water tanks and pipelines once again. This was first performed manually by labourers or servants who were paid around 3 pesos per month. Later, at the end of the 19th century, electric pumps costing 5–9 pesos were installed (Marroquín, 1914). This resulted in downstream locations running out of water.

In 1852, the City Council improved the Desierto de los Leones aqueduct, and in 1866 installed efficient valves that improved the function of water and pumps. In 1881 the total cost of operating the steam pumps from the Chapultepec tanks to the city was 700 pesos (Birrichaga, 2004b).

The provision of 150 litres per capita per day also dictated an increase the production of wastewater, and as consequence, the need to handle it properly.

20.6.2.2 Water quality

Public health concerns motivated the interest in controlling water quality. At that time smallpox, typhoid fever, yellow fever and cholera were common diseases. In 1833, a cholera pandemic, which began in 1817, arrived in Mexico causing a 10% mortality rate (Márquez, 1994). To combat cholera, the governor of Mexico City, Joaquín de Herrera, imposed new rules for waste disposal. These applied to cleaning the streets, disposing of wastes from tanneries and butcher's shops via the sewers, and avoiding disposal of used water or washing clothes in public fountains (Carrillo, 1992). In parallel, the aqueducts and pipelines conveying water were repaired and the legal water framework was reviewed to avoid pollution.

During the first half of the 19th century the local government experimented with lead pipelines, as they were cheaper and more easily repaired. Sometime around 1840 the chemist Leopoldo Río de la Loza was requested to prepare an opinion on the safety of this material. He concluded not only that lead had no detrimental health effects, but on the contrary, was beneficial for health as it avoided the infiltrations frequently observed in clay pipelines. Later it was reported that lead poisoned blood. It wasn't until 1884 that lead pipelines were replaced with iron ones constructed by Fábrica de Cañería de Plomo using a patent issued to Iglesias and Valezzi, because they were much more resistant to leaks. This practice lasted until 1915. In 1866, the Central Health Council through the Interior Affairs Ministry, set a standard defining different conditions to provide safe water, these were (Birrichaga, 1998):

(a) *Art. 3* Water should be transported in covered channels, or preferably using iron, clay or lead pipelines.
(b) *Art. 11* It is forbidden to wash clothes in open channels.
(c) *Art. 12* It is forbidden to dispose of industrial wastewater in the channels of the water supply system unless this is water used to provide mechanical energy.
(d) *Art. 15* It is forbidden to tether any animals or to dispose of solid wastes near to public fountains or springs.
(e) *Art. 16* It is forbidden to wash vegetables, linen or any other material in public fountains, or allow animals to drink from them.

In 1876 the entrepreneur Rafael Martínez de la Torre[2] and Doctor Eduardo Liceaga organised a medical conference to promote the provision of proper public services to town houses. New methods to manage wastes and the substitution of clay and lead pipelines with galvanised iron ones were proposed. These proposals were taken up by the government and Mexico City began to be modernised with new wider streets and avenues, public transport, taller buildings and a new water network (Birrichaga, 1998). This was a remarkable time in the 19th century, and during this period the City Council gained a clear conscience with regard to the provision of safe water as an indispensable element for public health: "Water is a vital element for life, hygiene, and the appearance of the city. A city without water is a city with frequent epidemics, spoiled industry which its inhabitants tend to abandon."

20.6.3 The sewerage system

20.6.3.1 Flood control

Floods became a critical problem once again in the 19th century. Between 1803 and 1804 Humboldt analysed the system developed by Enrico Martínez and proposed its completion by the building of the first tunnel (Tequixquiac) and a drain (Gran Canal del Desague). In addition, in 1823 Lucas Alamán undertook the construction of several public works to alleviate the floods in the western part of the city, provoked by the invasion of the United States.

In 1856, Francisco de Garay proposed the realisation of Humboldt's ideas and constructed an open canal 50 km in length to dispose of the wastewater into the Tula River. This canal was to "alleviate forever the city and the nearby towns from floods", (Birrichaga, 1998). To reach the canal a 9 km tunnel, the Tequixquiac tunnel, was needed as suggested by Humboldt. The construction of this hydraulic system began in 1865 when the Emperor Maximilian was ruling Mexico. When the Empire fell, construction stopped. In 1868, Benito Juárez ordered that it should be continued, but it was Porfirio Díaz who inaugurated the "General Drain System of the Valley" in 1900. This system comprised the Tequixquiac tunnel and the Gran Canal, proposed by Franciso de Garay, plus several other regulating tanks. The Gran Canal went from San Lázaro (near the centre of Mexico City) to Zumpango (at the end of the valley) and had a suitable slope to drain storm water from the city without using energy. This system managed both run storm and wastewater. During the inauguration, Porfirio Díaz announced that the city was completely free from floods, but only four months later a new flood hit the city, reaching the height of tramway platforms.

20.6.3.2 Sanitation

Sanitation became an issue during the mandate of the Emperor Maximilian. He ordered new legislation to include new sanitation concepts, as he considered wastewater was 'dark, faulty and inadequate'. This was carried out in 1864 (Birrichaga, 2009).

20.6.3.3 Water reuse

Another milestone of the 19th century related to sanitation is the reclamation of wastewater. As a result of the disposal of Mexico City's wastewater to the Tula Valley, it began to be reused on site. In 1889 reuse was officially performed to produce energy in two hydropower plants, Juandhó and La Cañada, and in 1896 for agricultural irrigation at the Tula Valley when farmers used the non-treated wastewater conveyed by the Salado River on crop fields in Tlaxcoapan, Tlalhuelilpan and Mizquiahuala (Jiménez, 2008). An

[2]Rafael Martínez de la Torre (1828–1876) owned many types of business such as textile industries, sugar cane mills and mines and was the pioneer of condominiums in Mexico City.

irrigation district was officially operated in 1912. This was the origin of the largest irrigation district using non-treated wastewater in the world.

20.6.4 Water administration

During the 19th century, with regard to water services, the City Council was obliged to clean public areas, oversee food quality control, administer the cemeteries, dry marshlands, prevent and control floods and to send the statistical data of the city to the federal government. In 1812 the Cádiz Constitution reorganised the structure of government in the country, creating municipalities with elected governments. For Mexico City, in practice, this meant the need to negotiate with the several new municipalities formed nearby which owned the water sources that were in use. In November 1824 the Congress created the Federal District, an area of two *leguas* (nearly 8.4 km^2) which included Mexico City. During the following years the Federal District was ruled by a governor, Miguel María Azcárate, who maintained a strong relationship with the City Council Mayor, Miguel Lerdo de Tejada. The governor and the mayors of the municipalities around Mexico City constantly informed the mayor of Mexico City of the state of water sources in their territories (Birrichaga, 2004b).

20.6.4.1 Participation of the private sector

In 1855 Mexico rose up against the president, Antonio López de Santa Anna, for respecting the republican principles of a balance between the political strengths of the country. In October, General Juan Alvarez became the interim president. Juan Alvarez ruled Mexico as a democratic republic and promoted private participation, the separation of the government and the Catholic Church, access to non-religious education, and the nationalisation of Mexican resources. To build the new water network the government tried twice to concede the system to the private sector, and so received a proposal from Francisco de Garay, a Mexican trained as an engineer in Paris, and an Englishman, Archibaldo Hope. They proposed granting the service for 35 years during which they agreed to increase the number of pipelines or their diameters, to replace the aqueducts with iron pipes, to maintain all private and public water fountains, to replace taps and manholes, to operate the infrastructure and to install water tanks and filtration systems to ensure the quality of water. In exchange they would provide each household with two and half paja[3] at a tariff of 2 pesos and 4 reales per month. However this proposal was not accepted. Several other attempts to privatise part or all of the service were performed. In 1865 Eusebio Soler wanted to run a water company that would sell water house by house using water tanks. In 1884, the City Council granted the concession of the water system to Carlos Medina for 50 years but the water users opposed this and the contract was revoked. A similar process happened in 1913 (Birrichaga, 1998).

20.6.5 Legal framework

As described above, a significant milestone of the 19th century was the raising of concerns about water quality and its relation to public health. Under this framework, on January 15, 1834, the City Council issued a norm which, under article 17, stated the obligation of water vendors to clean, on the first day of each month, the sludge that accumulated in public and private fountains. Two years later, on May 5, 1836, the government ordered the installation of efficient taps to avoid water spillage when leaks were

[3]A paja was a measurement used for water volume. It is equivalent to $1/20736$ of an ox $= 1/432$ row $= 1/144$ orange $= 1/18$ real. If is expressed per minute, then it is $= 1$ cuartillo $= 1$ water pound $= 45\ cm^3$.

repaired. However, in practice only a limited number of these were fitted. As a result, new norms to better control water rights had to be issued (Birrichaga, 1998).

In 1866, under the mandate of the Emperor Maximilian, the legal framework was completed by a health norm issued by the Central Health Council that announced rules to protect the safety of water.

20.7 20TH CENTURY

At the end of the 19th century and beginning of the 20th, Mexico – with a total population of 13.5 million – entered a prosperous period. The city continued to grow and the water supply problem became an increasing and consistent one. To oust Porfirio Diaz, the president who had held power for several decades, a social revolution was organised. Mexico had several leaders. The armed movement headed by Francisco I. Madero had little effect on the economy and organisation of the country, but once Madero was president several rebel groups appeared. Those headed by Emiliano Zapata and Francisco Villa were very popular and attacked Mexico City. After the Revolution of 1910 the assumption that the municipalities should administer natural resources was questioned. In article 27 the constitution of February 5, 1917, introduced changes to the way in which water was administered. Water was considered part of the national resource and the federal government was designated as the entity to allocate its use. The management of water in a centralised way had begun, with Mexico City as its seat.

20.7.1 Water sources

At the beginning of the 20th century the springs of Xochimilco were used to supply Mexico City (1913), initiating the deterioration of this important recreational and historical area of the city.

During the first four decades, the use of the local aquifer constantly increased. The first wells were sparsely distributed all over the valley. They were shallow and each extracted a relatively small amount of water. However, by the end of the 1940s, exploitation had become more intense. Soon the overexploitation problem became evident, as the city began to subside. To partially control this, in 1942 when the City had grown to almost 2 million, water had to be imported from the Lerma River, located in a basin in the State of Mexico. In 1951, groundwater from the same basin began to be exploited (Jiménez, 2008). The water imported was injected into a network that was deteriorating due to increasing water leaks and most of these new resources were lost. As a result, new wells were dug. At the end of the 1950s the common depth for wells was 100–150 m. In those places where wells from alluvial aquifers were not producing enough water, they were drilled into the consolidated basement rock. Finally, in 1975, when the population had reached 7 million, surface water had to be imported from the Cutzamala region, 130 km away and 1100 m below the level of Mexico City (Jiménez, 2008). The use of Cutzamala water did not stop either the problem of overexploitation or subsidence. From 1964 onwards, new wells were added to the system each year to respond to an increasing water demand from a population that was partly – and unwisely – growing to the south and east of the city, above the natural recharging area. As a result, soil subsidence reached rates of up to 40 cm/year, and in 1954 it was decided to close several wells located in the centre of the City and open new ones in the south, where the soil is volcanic, and in the north (Santoyo *et al.* 2005). However, by that time the centre had already sunk by 7 m (Figure 20.7). The redistribution of wells reduced the sinking rate in the centre, but the western part of the city is still sinking at a rate of 25 cm/yr (Jiménez, 2008).

Iodine, silver and chlorine droplets began to be used to disinfect water. Small drinking water plants were installed in houses, ranging from simple sand filters to a combination of ozone, UV-light, activated carbon and silver colloid filters. This increased the price of water by up to five times.

To assess drinking water quality, faecal coliforms and free residual chlorine are the only parameters systematically measured and globally reported. As a result, official data on drinking water quality is scarce. Furthermore, the information reported on the quality of drinking water by the local and the federal government is contradictory.

20.7.3 The sewerage system

20.7.3.1 Flood control and sewers

Floods continued to be an issue. In order to control them, in 1900 the General Department of Water reviewed all of the projects relating to drainage before beginning a massive program of paving of the streets. Following the construction of the *Gran Canal* (1898) and the Western Sewer (1896) Mexico City's sewerage system was completed with the Central Interceptor in 1975, with all three forming what is now called the Deep Drainage System (Figure 20.1). This system handles, on average, a wastewater volume of around 60 m^3/s, but it is capable of handling between 52 and 300 m^3/s (Jiménez, 2008).

At the end of the 20th century, sewerage covered 94% of the Federal District population and 85% of the municipalities of the State of Mexico (Merino, 2000). It was composed of 10,400 km of pipelines of 0.3–0.6 m in diameter. There were 2369 km of 76–3.05 m pipelines, 96 pumping stations with a total capacity of 670 m^3/s, 91 elevated canals carrying 14.3 m^3/s, 106 marginal collectors, 12 storm tanks with a total capacity of 130,000 m^3, several inverted siphons to overpass the metro, 3 rivers, 29 dams, the Gran Canal, 47 km long, and the Deep Drainage System, 155 km long, with diameters of 3–6.5 m and a depth varying from 20–217 m.

20.7.3.2 Wastewater management

The composition of Mexico City's wastewater is domestic. It has been observed that some pollutants are diluted during the rainy season but others, such as solids and helminth eggs contents, are not. In addition, and contrary to what sanitary engineers hitherto believed, the metal content of the untreated wastewater is not very high, thus meeting thresholds for its use in irrigation without any treatment. This is due to the large amount of household wastewater produced that dilutes the relatively small volumes of industrial discharges (Jiménez, 2008). Mexico City produces, on a year round basis, 67.7 m^3/s of wastewater. There are several wastewater treatment plants in the city: 27 are operated by the Federal District Government, 44 by federal institutions (Ex-Texcoco Lake Commission, Federal Electricity Commission and the army) or private owners and 20 by the municipalities of the State of Mexico (Merino, 2000). The public wastewater treatment plants' total capacity is 15m^3/s, but only 7.7 m^3/s of the wastewater (11% of the total produced) is treated (Jiménez, 2008).

20.7.3.3 Water reuse

Wastewater reuse began in Mexico City when the first wastewater treatment plant was installed in 1956 (Merino, 2000). The effluent was used to irrigate Chapultepec Park, from where the first springs to supply the city originated. Reuse practices continued to grow, and secondary effluents were used to fill recreational lakes and canals (54%, Figure 20.8a), irrigate 6500 ha of agricultural land and green areas (31%), for cooling in industry (8%), diverse purposes in commercial activities (5%) and to recharge the aquifer (2%, Jiménez, 2008). There is no data on the total amount of wastewater treated privately, but it

20.8 DESCRIPTION OF MEXICO CITY IN THE 21ST CENTURY

Mexico City has expanded to 8084 km^2 (i.e. 540 times larger than the city of Tenochtitlán) and the population has increased to nearly 21 million; a factor of 105 compared to that of the Aztec capital. Mexico City, initially located only in the Federal District, grew to cover 37 municipalities of the State of Mexico. There are actually now more people living in the State of Mexico (nearly 60% of the population) than in the Federal District. Of the five lakes that were initially within the Valley, due to urban expansion and the artificial drying of the Valley, there currently only remain small portions of the Texcoco and the Zumpango lakes. Both are fed by excess pluvial water plus treated wastewater in the case of the Texcoco Lake (Jiménez, 2008).

At the present time, only one river in the valley flows throughout the year, the River Magdalena. The others, that were once perennial, carry only wastewater. Most of these river beds have been covered since the 1950s to operate as channels; streets built around or over these channels have acquired the name of the rivers (Jiménez, 2008).

20.8.1 Water sources

At the present time, the city uses 85.7 m^3/s of water (Figure 20.2), 48% (62 m^3/s) of which is supplied through the network system. 19% (16m^3/s) is directly pumped from local aquifers by farmers and industries. The remaining 9% (7.7 m^3/s), corresponds to treated wastewater that is being reused (Jiménez *et al.* 2004). First use water (78 m^3/s) comes from the following sources (Jiménez, 2008):

(a) A total of 1965 registered wells that extract 57 m^3/s from the local aquifer, located mainly to the south and west of the city.
(b) A local river (Río Magdalena) and springs – such as Fuentes Brotantes – (1 m^3/s), located in the southern part of the city.
(c) An aquifer located in the Lerma region, 100 km away and 300 m above the level of Mexico City (3 m^3/s). Initially, 9 m^3/s were extracted, but as the groundwater level in the region decreased and the demand for water in the State of Mexico decreased the volume used by the city decreased.
(d) The Cutzamala system located 130 km from and 1100 m below the level of Mexico City (15 m^3/s).

20.8.1.1 Groundwater overexploitation

Of the local groundwater sources, double the amount of water is extracted than that which is being naturally recharged. Extraction is performed via more than 1965 wells, some privately operated, others run by federal or local water works, operating mainly to the south and west of Mexico City (Figure 20.9). The main barrier to resolving this situation is the independent management of groundwater by the government of the Federal District and the 37 municipalities. While restrictions on the use groundwater have been set in the Federal District, in the municipalities there are simply no limits. From a wider perspective, the local aquifer is still the main source of water. However, the city is expanding to the south and east where the main areas of natural recharge are located.

Impacts of the use of water are not only observed in the Valley of Mexico. As a result of importing water from the Lerma region, which used to be dominated by a lake, fishing disappeared and people were forced to live off the land. Furthermore, local people have gone from using surface water as a supply source to using water from wells. The transfer of water from the Cutzamala region reduced the amount of water available for power generation and caused the loss of a large area of irrigation (Jiménez, 2008).

operate by gravity, transporting up to 200 m^3/s, but its capacity is now of the order of dozens of cubic metres. Due to a loss of drainage capacity, there are around 20–30 large floods (of more than 30 cm) per year causing non-quantified damage of varying magnitudes and nature (SACM, 2006) throughout the city.

(e) Deterioration of groundwater quality. Due to overexploitation, groundwater quality is deteriorating as explained above.

Unfortunately, there is no hard data available on the total cost of the effects of soil subsidence.

Box 20.3 Impacts of soil subsiding.

On July 7, 2007 in the Lomas de San Lorenzo neighbourhood of the Iztapalapa Delegation, the car of Jorge Alejandro Ramírez Arredondo, a 19 year old man, fell into a 14 m deep crack in the subsoil while he was driving through what appeared to be a normal street. Only some days later was his body recovered. The hole was formed below the pavement as result of a water leak. The government is injecting a mixture of bentonite, rocks, volcanic stones and cement to close the hole.

20.8.2 Water quality

There is little available information concerning the quality of water sources, and most of it relates to local groundwater. Various studies reveal that groundwater is being increasingly polluted. In the western part of the Valley, the suspended solids content has increased from 1000 mg/L to 20,000 mg/L, the sodium content from 50–100 mg/L to 600–800 mg/L, the ammoniacal nitrogen content from 0–0.03 mg/L to 6–9 mg/L, and the iron content from <0.1 mg/L to 3–6 mg/L. These increases are due mainly to overexploitation, but also to domestic and industrial pollution. In a specific area in the western part of the city, where uncontrolled dumping sites used to exist, overexploitation is causing leachates to enter the aquifer. High levels of a wide variety of organic compounds have been reported along with increasing concentrations of nitrates, ammoniacal nitrogen and faecal coliforms. Groundwater pollution is also caused by seepage from septic tanks and sewerage. In the southern part of Mexico City, the lack of sewerage and the presence of at least half a million septic tanks discharging directly into a volcanic soil is recharging the aquifer with 1 m^3/s of sewage (Jiménez, 2008).

20.8.3 Water uses

20.8.3.1 Municipal supply

Water services cover 89% of the population of the Federal District and coverage is slightly lower for the municipalities of the State of Mexico. There is no data for the population receiving a regular service in the State of Mexico, but in the Federal District alone there are nearly 1.15 million people receiving a limited amount of water via water tankers (SACM, 2006). These people receive water 1–2 times per week at no cost.

Considering the Mexico City population, the 62 m^3/s that are distributed through the network, represent a water supply of 255 L/inhab · d. However, due to water network leaks that amount to 40% of the total, each person receives on average 153 L/d, a value that falls within the range of 150–170 L/inhab · d recommended by WHO. In actual fact, water use varies along lines of social class. The upper class, representing less that 5% of the total population, uses nearly 600 L/inhab · d while the lower classes, comprising 76.5% of the population uses 129 L/inhab · d. This should motivate the government to invest more money to control water leaks rather than invest large amounts of money on publicity to promote water saving campaigns (Jiménez, 2008).

20.8.4 Water quality

20.8.4.1 Potabilisation

To potabilise water, processes initially employed between the 1940s and 1970s are still in use. Chlorination is used to treat groundwater sources while for surface sources alum coagulation, sedimentation and chlorination are applied. All of these processes were selected based on characterisations performed several decades ago. To treat the groundwater from a small part of the south–western part of Mexico, in the area where the aquifer has the worst water quality, treatment plants installed near the extraction wells are currently using complex processes, including biofiltration, ozonation, activated carbon filtration and reverse osmosis (Félix, 2010). These plants constantly suffer from both low efficiency and poor functioning due to the low quality of the water supply.

20.8.4.2 Drinking water quality

Detailed data concerning drinking water quality produced from the local aquifer exist, but unfortunately, only in a few isolated academic studies. Mazari-Hiriart *et al.* 2000, for instance, found nitrates, chloroform, bromo-dichloro-methane and total organic carbon, in chlorinated groundwater water, with a higher concentration during the dry season. The total trihalomethane content, however, did not surpass the Mexican drinking water norm of 200 μg/L but exceeded the maximum allowable value of 80 μg/L established in the United States. With regard to microbiological quality, these same authors reported the presence of total coliforms, faecal coliforms, faecal *Streptococcus* and other pathogenic bacteria before and also after chlorination. They isolated 84 micro-organisms of 9 genera associated with human faecal pollution. Among these were *Helicobacter pylori* (associated with gastric ulcers and cancer) and MS-2 coliphages (Mazari-Hiriart *et al.* 1999 and 2000).

Almost no information exists with respect to the quality of the drinking water produced from surface sources (Lerma, Cutzamala and Mexico Valley rivers) although it is evident that water sources are becoming more polluted (Jiménez, 2008).

In addition, it has been shown that water quality deteriorates during distribution (Jiménez *et al.* 2004). The water network operates at low pressure and, due to a lack of water; supply across the city is intermittent. To have access to water throughout the day, people must use individual storage tanks and as a result tap water is of a lower quality. As explained previously, domestic potabilisation systems are often used. A family of four earning four times the minimum wage spends 6–10% of its income on bottled water or potabilisation systems (Jiménez, 2008).

The diarrheic disease mortality rate in Mexico City is higher than the mean value for the country, although this small area concentrates 21% of the GDP. Cifuentes *et al.* (2002) found that diarrheic diseases were not linked to the conventional parameters used to evaluate water as many people potabilise water at home. However, for the poorest people who cannot afford their own disinfection systems, diarrheal diseases are high, particularly in children under five years of age, and especially at sites where the service is provided intermittently or via water tankers.

20.8.5 Sewerage system

20.8.5.1 Sewer and flood control

To partially alleviate the city from floods in the low-lying parts of the urban area with respect to the sewers, a pumping station (casa Colorada) was installed in 2010 in the Gran Canal to elevate 40 m^3/s of sewage and pluvial wastewater by 40 metres at a cost of 4 million USD. This station had a limited effect on the problem. It was proposed as an emergency solution in 1995 to control floods until 2002, and wasn't inaugurated until

2007 while other infrastructure was put in place. Unfortunately it took so long for the government to launch this project that in the meantime the city suffered several dramatic floods.

In addition, while once again stating that flooding would be controlled forever, in 2009 the government announced the construction of another deep drainage system (the Western Drainage System). This sewer will convey 150 m^3/s, will be 62 km long and 7 m in diameter and will take 4 years to build. The total cost of construction will be 1 million USD, including a wastewater treatment plant (CONAGUA, 2010).

20.8.5.2 Wastewater treatment

In 1993 the Federal District, the State of Mexico and the Federal Government initiated a project to treat all of the wastewater produced by Mexico City. Several treatment options were explored. At first stabilisation ponds were considered, but there was insufficient area to treat the wastewater of nearly 20 million people. Following this, an activated sludge process was considered. Physicochemical treatment (advanced primary treatment or APT) at a third of the cost of activated sludge was contemplated. This could control diarrheic diseases while maintaining a high level of nutrients in the treated wastewater, which is valuable when the water is used to irrigate. This APT system was analysed as part of a system of four wastewater treatment plants that were supposed to be built to treat 70 m^3/s by the year 2000. However, since 1996, functionaries of the Federal Government, the Federal District and the Mexico State have halted and changed the project on numerous occasions to review it. Finally in 2010 the project was reinitiated, this time only by the Federal Government, and a project based on a combination of activated sludge wastewater treatment plants coupled with physicochemical treatment was selected at a higher cost. This plant is to be operational by the year 2012 although to date, in 2011, construction has not begun.

20.8.5.3 Disposal of non-treated wastewater

As discussed previously, the Valley of Mexico was originally a closed basin that needed to be artificially opened to drain pluvial water and untreated wastewater. Three exit tunnels were built for this purpose: the Gran Canal (1898), the Western Interceptor (1896) and the Central Interceptor (1975) all draining to the Tula Valley. This valley is located 100 km north of Mexico City in the State of Hidalgo, at an altitude of 2100 m in the southern part and 1700 m in the northern part. Pluvial precipitation is 525 mm and occurs during only five months of the year. In contrast, the annual evaporation rate is 1750 mm.

Initially, wastewater was disposed of in the poorest and furthest areas of the Tula Valley ("El Mezquital Valley"). The original vegetation was *Xerophila* scrub. The local population disliked this decision, but in 1920, when an increase in agricultural production became evident, farmers requested that the government send more wastewater. They later requested the president to grant them the concession of Mexico City's wastewater, which was duly granted. The construction of a complex irrigation system was initiated. Today, the irrigated area is around 85,000 ha and has been considered for many years to be the largest irrigated area using wastewater in the world. Thanks to the wastewater and its fertilising properties, it is also considered to be one of the most productive irrigation districts in Mexico (Figure 20.10). The effects on agriculture, soil, ecosystems and crops due to irrigation with wastewater is reported elsewhere (Jiménez, 2005).

Given that in the Tula Valley wastewater is transported in hundreds of unlined channels, stored in dams, and applied to a permeable soil using high irrigation rates (1.5–2.2 $m/ha \cdot yr$) to wash out salts, the aquifer undergoes a significant recharge, estimated to be at least 25 m^3/s, that is, 13 times the natural recharge (Jiménez & Chávez, 2004). As a result, the Tula River flow (partially fed from the aquifer) increased from 1.6 m^3/s to more than 12.7 m^3/s from 1945 to 1995 and the water table rose from 50 m below

20.8.7 Future water demand

At the present time, in order to continuously supply water to Mexico City's entire population there is a need for an extra 1–2 m^3/s of water. To supply the demand for the estimated growth in population over the next five years, an additional 5 m^3/s are needed. Additionally, to prevent overexploitation and inject water to control soil subsidence, at least 15 m^3/s of water is required. Thus, it is estimated that by the year 2010, a total of 38 m^3/s of water will be needed. Part of this flow could certainly come from a leakage control program. Over a five year program, 10–15 m^3/s of water could be saved by investing 1.5 million USD to sectorise and control pressure in the water network, plus 500 million USD per year to repair and change deteriorated pipelines[5] (Capella, 2006). Considering the costs, the time needed to control leaks and recover water and the total volume required, another source of water is needed. This could consist of the transfer of surplus water from other basins and/or the implementation of water reuse programs (Jiménez, 2008).

20.8.8 Water reuse options

A water reuse program for Mexico City needs to consider three sectors: agriculture, industry and local government (Jiménez, 2008).

20.8.8.1 Agricultural reuse

It is commonly believed that a megacity like Mexico City contains only urban areas; however, there are large agricultural areas within the City, located north, north–west, west and south. In total, these areas use 14 m^3/s of water pumped directly from the Valley of Mexico aquifer (25% of the total amount of the water extracted from groundwater and 18% of the fresh water demand). Without significantly reducing the amount of wastewater used in the Tula Valley it would be feasible to treat wastewater on site to reuse it in efficient irrigation systems. For such a project, care must be taken to exchange groundwater for reclaimed wastewater without reducing farmers' incomes. Such a program would certainly need stakeholder involvement (Jiménez, 2008).

20.8.8.2 Industrial reuse

Even though 40% of national industrial activity is registered in Mexico City, this in fact refers to the corporate offices of industries located in other parts of the country, industries with low water consumption (for instance clothes manufacturing) or industries already reusing wastewater. The reason is simple. Before the government initiated industrial reuse programs, a lack of water forced industries to recycle and reuse water and to move their production facilities to other parts of the country. In addition to 44 private facilities reclaiming water, there are two municipal wastewater treatment plants (Lecherían and Aragon) that are privately operated and are mandated to sell reclaimed water to industries. These plants sell 0.99 m^3/s of treated wastewater[6] at a cost of 0.7 USD/m^3 (Jiménez, 2008).

20.8.8.3 Municipal reuse

Because water professionals in the government of Mexico City have always been conscious of the lack of water, municipal reuse is already a common practice. All options for non-potable reuse of wastewater have been applied, and the need now is to reuse wastewater for human consumption. This can be performed in two

[5]The estimated total time needed to repair the whole network is 50 years.

[6]Total capacity of reclamation plants is 1.8 m^3/s.

ways: (a) by locally treating wastewater to a high standard, and (b) by importing the excess water from the Tula Valley aquifer formed by infiltration of Mexico City's wastewater:

(a) *Wastewater treatment on site*. Considering the experience gathered from Windhoek, Namibia this would be a feasible option. However, for Mexico City it has the inconvenience of reducing the amount of wastewater sent to the Tula Valley to around 14.5 m^3/s to render only 10 m^3/s of supply[7]. This is less than the amount needed and would entail sophisticated and expensive treatment. In addition, it appears that this option is less easily accepted by the public especially because so far potable water is not readily available on tap even from conventional water sources.

(b) *Recovering the water sent to the Tula Valley*. As previously mentioned, the rate of wastewater recharge is at least 25 m^3/s in the Tula Valley and wastewater is being treated via its transportation in open channels, use for irrigation and infiltration through the soil. The new aquifer formed by wastewater in the Tula Valley is only 100 km away from and 150 m below the level of Mexico City[8]. Its use is being considered as a new source for Mexico City, and it is actually the cheapest option. A series of studies is currently being undertaken to evaluate the feasibility of the new aquifer as a water supply source. Considering the yields of the groundwater, it is thought that the Tula Valley aquifer can be used to supply 6 to 10 m^3/s. Besides the importance of supplying water to Mexico City, the exploitation of the Tula Valley aquifer, if done properly, could also alleviate the flooding problems in the region, contribute to the control of salinisation of soils, and avoid water losses through evaporation. However, the previous comment can again be made for this option: the local government should first prove its capacity to supply safe water for human consumption from first use water sources.

Table 20.1 Cost comparison of different new water source options for Mexico City (Jiménez & Chávez, 2004).

Source	Flow (m^3/s)	Pumping cost (USD/m^3)	Investment (USD/m^3)	Treatment cost (USD/m^3)	Total cost (USD/m^3)
Temascaltepec	5.0	0.30	0.27	0.016	0.58
Amacuzac	13.5	0.40	0.44	0.016	0.86
Tecolutla 1	9.8	0.48	0.36	0.016	0.85
Potabilisation of raw wastewater on site	3.0	0	0	0.95	0.95
Tula Valley aquifer	6.0	0.09	0.035	0.50	0.73

A cost comparison between the alternative water sources for Mexico City involving reuse and freshwater sources is shown in Table 20.1. The cheapest option, importing water from Temascaltepec, is not viable due to the resistance of the local community to sending its water to Mexico City. In 2003, the plans were to bring more water from Temascaltepec, a site 200 km away from and 1000 m below the level of Mexico City, and from part of the Cutzamala system. When this was discovered by the local population, several public protests took place to challenge the government's plans. Considering the costs and the distances involved in

[7]Considering a 70% recovery of the injected reclaimed wastewater.

[8]One of the current water supplies for Mexico City is the Lerma aquifer, located at more than 1200 m below Mexico City.

implementing the other possible options, the government decided to analyse different and innovative water reuse schemes (Jiménez, 2008). Independently of the option selected, local governments (Federal District and the municipalities of the State of Mexico) must be aware that:

(a) Prior to implementing any of the reuse programs described it is necessary to segregate or treat industrial discharges, even though they are not significant volumes.
(b) The direct reuse of wastewater for human consumption is a controversial option. Any decision taken will need to be based on technical, social, economic and scientific studies that increase costs and take several years to perform. It is estimated that at least 4–6 million USD are needed to obtain all the scientific, technical and practical information needed over a period of 5–7 years.

For all of the possible options, it is striking that although they are frequently cited as a priority and a next step for the city by politicians, functionaries and even academics, these "solutions" have never been submitted for public consultation, despite the fact that rejection by the public could stop the project. In preparation for its start, the Federal District government published a norm (2004) for the reuse of wastewater for human consumption. However, the Federal Government blocked its application for political reasons (Jiménez, 2008).

20.8.9 Integrated management of water

Mexico City has suffered from both lack of water and flooding. At the present time, in 2010, there are nearly one million people without access to water thorough the network. Currently, water is not safe to drink, and only recently a 1.7 m flood formed from a mixture of wastewater and pluvial rain destroyed 7000 houses. The governments of all the political entities forming Mexico City face a number of challenges related to water.

Due to the complexity of the water problem in Mexico City, a Metropolitan Water Authority, bringing together different political regions, sectors and levels of government (federal, regional and local) should be created to manage water in an integrated way. To be effective, this Water Authority needs to have the capacity to take all decisions related to the management of water and have a sufficient budget to ensure its continuing operation. The main task of the Water Authority would be to conceive a short and long-term Integrated Water Resources Management Program, which should consider not only the technical aspects but also those from a social and economic viewpoint. Some of the activities to consider in this program are (Jiménez, 2008):

(a) *Control of soil subsidence.* Action needs to be taken to control soil subsidence. This includes reorganisation of the use of soil in the Valley of Mexico, the increase of natural recharge, the reduction of groundwater overexploitation and the reinjection of highly treated wastewater to the subsoil.
(b) *Protection of groundwater quality.* Mexico City relies on its aquifer for water supply. With this dependence in mind, it is inexplicable why very little effort is being made to stop the alarming increase in groundwater pollution. A program to control non-point source pollution needs to be put in place as well as the construction of sewerage infrastructure in the southern part of the city.
(c) *Leakage control.* Simply by reducing leaks, less water may be extracted from the aquifer. A reduction in losses from the current level of 40% to 20% by the year 2050 would recover 20 m^3/s. This volume could be used both to reduce groundwater overexploitation and to meet the present water deficit.
(d) *Implementation of an aggressive and innovative wastewater reuse program.* Mexico City finds itself in a situation that has no precedent worldwide; therefore an integrated water management

program should consider aggressive and innovative wastewater reuse to fulfill needs not covered thus far.

(e) *Application of innovative and comprehensive educational programs.* In order to obtain the support of the public in implementing activities considered part of the Integrated Water Management Program, it is important to explain Mexico City's water problem. Due to its complexity, this is not an easy task. Educational programs need to address not only society in general, but also targeted sections, such as politicians, bureaucrats, industrialists, academic researchers and NGOs, and so on.

(f) *Improvement of economic tools.* Water commercialisation is performed inefficiently by both the Federal District and the municipalities of the State of Mexico. Mexico City has commercialised water with partial private participation while the municipalities of the State of Mexico provide water services independently of each other and through public participation. The estimated total number of users in the Federal District is 2.1 million, 88% of whom are registered and 60% of whom have a meter. Around 27% (0.57 million) pay a fixed water tariff while 60% pay for the amount of water consumed. The total commercial efficiency of the system (considering billing efficiency and the number of people who actually pay) is 68%. The price of water in Mexico City is heavily subsidised. Water costs at least 2 USD/m^3 to produce while the price charged is 0.2 USD/m^3. Unfortunately, for political reasons, water prices have not been raised. Changes to these public investments should be made directly.

(g) *Rainwater harvesting.* Mexico City receives significant pluvial precipitation representing a volume of 12 m^3/s. Therefore, several NGOs are promoting rainwater harvesting as a water supply option. Families living in specific catchment areas and with the economic resources to build the storage and water treatment facilities required, can obtain 30–50% of their annual water demand this way. Unfortunately, for most of the people in Mexico City this is not a viable solution because the cost of water is very high (10 USD/m^3) and there is no possibility to build large facilities to reduce the cost because of the population density. Rainwater harvesting as a municipal service is considered as a feasible option for collecting less than 1 m^3/s, a volume that is far from the actual water need.

(h) *Implementation of professional public participation programs.* The Mexican and Mexico City governments need to learn how to develop projects with public participation. To date this has not been achieved and initial efforts have proved unsuccessful. To successfully implement the Integrated Water Resources Management Program, public support is needed to sustain projects independently of political parties and governmental changes. For that, the government needs to understand that there are professional tools available to allow public participation.

(i) *Emigration.* As an alternative to the reuse of water for human consumption the government could consider promoting the emigration of people from Mexico City to other regions of Mexico.

20.9 CONCLUSIONS

Since the time of the Aztecs, Mexico City has been considered a megalopolis. In 1521 it had nearly 200,000 inhabitants and was one of the most important cities of Mesoamerica. Today, with 21 million people, and responsible for 21% of the GNP of Mexico it is one of the largest cities in the world, located on a high plateau far from the sea, where desalination may have been an option. This chapter describes how Mexico City grew from a city located on an island (Tenochtitlán), and how, despite the fact that the residents of Tenochtitlán had access to sufficient water and planned their whole social and economic development around this resource, Mexico City became a mega-city, with urban areas suffering both from lack of water and flooding. To redress this situation, options to integrally manage water are proposed, including better use of water, planned reuse and the reduction of the population of the city.

Mexico City has long been accepted as a mega-city. It was conceived in the Aztec era directly due to the availability of water. This changed with the Spanish colonisation and it has not been redressed since. Today, the city is experiencing a very challenging situation concerning its water supply and wastewater disposal system that, as a whole, appears to have no precedents in other parts of the world. From the practical perspective there is a need to forcefully push for water reuse in parallel with the other options presented here. In doing so, it would be interesting to try to apply the same principles used to manage water during the Tenochtitlán period. An additional solution, which may be applied in combination with the other options, is to begin to control the growth of the city and even promote a reduction in its size. Although this solution has been explored since the 1970s it has never received strong political support. It is, perhaps, time to revisit the idea to limit cities' growth based on the amount of water available. Planning growth based on the amount of water resources is important not only for citizenship of Mexico City but also for the inhabitants of the many other cities in the country that are facing the same challenges, surpassing the capacity of their local natural resources, especially with regard to water. The 10 largest Mexican cities, all with populations greater than 1 million, are currently experiencing dramatic problems with water supply.

REFERENCES

Acuña R. (1986). Geographical Relations from the XVI Century: México. 2nd vol. National Autonomous University of Mexico, México (in Spanish).

Becerril E. and Jiménez B. (2007). Potable water and sanitation in Tenochtitlán Aztec culture. *Water Science and Technology*, **7**(1), 147–154.

Birrichaga D. (1998). Water supply to private industries in México (1887–1930). In: The history of the use of water in Mexico. Oligarchy, private industries and municipalities (1840–1940), B. Suarez (ed.), National Water Commission, Mexican Institute of Water Technology and the Research Center for Advanced Studies on Social Anthropology, México, pp. 181–225 (in Spanish).

Birrichaga D. (2004a). The ownership of the hidden and undiscovered waters: The state of water during the Colony in the Center of Mexico during the XVI–XVII centuries. In: Technological racial mix and cultural changes in Mexico. E. Florescano and V. García Acosta (ed.), Miguel Angel Porrúa, Mexico, pp. 91–128 (in Spanish).

Birrichaga D. (2004b). Historical archive for the Federal District: The freshwater of Mexico City during the XVII–XX centuries. Historical Archine of Mexico City (Distrito Federal), Mexico (in Spanish).

Birrichaga D. (2006). The private enterprises of freshwater: private projects for the water supply in Peru and México during 1846–1856. In: Allpanchis, XXXVII (67)k, Mexico, pp. 37–57 (in Spanish).

Birrichaga D. (2009). Water Legislation in the XIX and XX Centuries. In: Brief history of water in Mexico, M. Sánchez, Semarnat, Mexico, pp. 43–59 (in Spanish).

Capella A. (2006). Personal communication.

Carrillo A. (1992). The Cholerae epidemy of 1833 in the State of México. In: The cholera in 1833: a new disease in Mexico, INAH, México, pp. 65–88 (in Spanish).

Cervantes de Salazar F. (1985). Chronicles of the New Spain. Porrúa, Mexico City, México (in Spanish).

Codex of Tepetlaoxtoc (1994). Codex Kingsborough, British Museum Trustees.

Codex Magliabechiano (1970). CL. XIII. 3 (B.R. 232), Biblioteca nazionale centrale di Firenze, Faksimile, Akademische Druck- u. Verlagsanstalt.

CONAGUA (2010). The Eastern Tunnel of the Valley of Mexico. http://www.atl.org.mx/aguadf/images/docs/Folleto_El_Tunel_Emisor_Oriente.pdf (accessed September 2010).

Chávez M. (1994). Water in Ancient Mexico. Salvat Science and Latin-American Culture. México City, Mexico (in Spanish).

Félix Cañedo T. (2010). Removal of emerging pollutants from the water treatment plants of Mexico City. Master thesis, Environmental Engineering, UNAM, Mexico (in Spanish).

Guillermo A. 2000. Geographical localization of the Mexico Basin in Garza, G. In: Mexico City towards the end of the Second Millennium. Distrito Federal Government and El Colegio de México (ed.), México [in Spanish].

González De La Vara M. (1991). The place of snow during 1596–1855. In: Novohispanic Studies. vol. 11. (http://www.ejournal.unam.mx/ehn/ehn11/EHN01103.pdf) [in Spanish] (accessed September 2010).

Jiménez B. (2005). Treatment technology and standards for agricultural wastewater reuse: A case study in Mexico. *Irrigation and Drainage*, **54**(1), 23–35.

Jiménez B. (2008). Water and wastewater management in Mexico City in integrated urban water management. In: Arid and semi-arid regions around the world, L. Mays (ed.), Taylor Francis Ltd., Leiden, The Netherlands.

Jiménez B. and Asano T. (2008). Water reuse: An international survey of current practice, issues and needs. IWA Publishing, London, UK.

Jiménez B., Mazari M., Cifuentes E. and Domínguez R. (2004). The water in the Mexico Valley. In: Water from Academia's perspective. Mexican Academy of Science Mexico, Mexico. pp. 15–32 (in Spanish).

Jiménez B. and Chávez A. (2004). Quality assessment of an aquifer recharged with wastewater for its potential use as drinking source: "El Mezquital Valley" case. *Water Science and Technology*, **50**(2), 269–273.

Kula W. (1980). Measures and Men, Editorial Siglo XXI, México (in Spanish).

Lanz Cárdenas J. (1984). Water legislation in México (1521–1981). Editorial Council of the Tabasco State, Villahermosa (in Spanish).

León Portilla M. (1961). The Ancient Mexicans. Economical Culture Trustee. Mexico City, México (in Spanish).

León Portilla M. (2001). Anahuac's Face and Heart. National Association for the Book. A.C., Mexico City, México (in Spanish).

Márquez Morfín L. (1994). Inequity phase to death: Epidemic, population and society in Mexico City (1800–1850). Siglo XXI, México (in Spanish).

Margadant G. (1989). The water face to the novohispanic rights: Triumph of the realisms and the flexibility. In: History of Mexican Law yearbook, I(113–146), 123–124, 126 (in Spanish).

Marroquin Rivera M. (1914). Descriptive memory of the water works to supply water to Mexico City. Müller Brothers Press and Lithography, México City (in Spanish).

Máynez P. (2002). Sahagú's notebook: a quick glance. 1st edn. Economical Culture Trustee, México City (in Spanish).

Mazari-Hiriart M., Torres Beristain B., Velázquez E., Calva J. and Pillai S. (1999). Bacterial and viral indicators of fecal pollution in Mexico City's southern aquifer. *Journal of Environmental Science Health*, **A34**(9), 1715–1735.

Mazari-Hiriart M., Cifuentes E., Velázquez E. and Calva J. J. (2000). Microbiological groundwater quality and health indicators in Mexico City. *Urban Ecosystems*, **4**(2), 91–103.

Merino H. (2000). Hydraulic system in Mexico City at the end of the millennium. In: G. Garza (ed.), Gobierno del Distrito Federal and El Colegio de México, Mexico (in Spanish).

Musset A. (1992). The Water in the Mexico Valle during the XVI – XVIII Centuries. Center for Mexican and Central America Studies, Mexico City, Mexico (in Spanish).

Offner J. (1983). Law and Politics in Aztec Texcoco. Cambridge University Press, Cambridge, UK.

Ortiz de Montellano B. R. (1990). Aztec Medicine, Health and Nutrition. Rutgers University Press, USA.

Palerm J. and Chairez Araiza C. (2001). Hydrometrics, ancient water metrics and the adventures to convert them to actual metrics, XXIst International Congress of History of Science International Union of History and Philosophy of Science/Division of History of Science, México, D.F., July 8–14 (in Spanish).

Reyna M. (1988). The Lions Desert: its hydraulic resources and the allocation water to its monastery in the XIX Century. In: Stories, INAH, Mexico City, Mexico (in Spanish), pp. 97–107.

SACM (2006). Water in Mexico City, International Tlalocan Festival for Water. Mexico City, March 10–15 (in Spanish).

Solano F. and De (1991). Soil registers. Compilation of the national agronomic legislation during the Colony *(1470–1820)*, National Autonomous University of Mexico, México, D.F. (in Spanish).

Sahagún Fray Bernardino (1985), Historia general de las cosas de la Nueva España, Editorial Porrúa, Mexico, City, Mexico (in Spanish).

Santoyo E., Ovando E., Mooser F. and León E. (2005). Geotechnical synthesis of the Mexico Valley basin, TGC, S.A. de C.V. México (in Spanish).

Santoyo E. and Ovando E. (2002). Underexcavation at the Tower of Pisa and at Mexico City's Metropolitan Cathedral. Proc. International Workshop, ISSMGE-Technical Committee TC36 Foundation Engineering in Difficult, Soft Soil Conditions, CD edition, Mexico City.

Chapter 21

The evolution of water supply throughout the millennia: A short overview

Demetris Koutsoyiannis, Nikos Zarkadoulas, Nikos Manassis, Andreas N. Angelakis and Larry W. Mays

21.1 INTRODUCTION

The importance of water for life, human prosperity and culture development was understood very early in all civilizations. In the initial stages of the civilizations this understanding had been depicted in mythologies and the natural forcings producing the water cycle had been represented by gods and goddesses. The introductory four chapters of this book deal with the evolution of the understanding of the water as a substance, its circulation on earth, and its quantity and quality.

All ancient civilizations had water gods and goddesses showing the significance of water to all of them, as described in *Chapter 1*. The variety among different civilizations of the forms and the features of the gods of water is impressive, illustrating the different perceptions among civilizations of the mechanisms of the water cycle. What is common though in almost all civilizations is the popularity and strength that the gods of water enjoyed.

In *Chapter 2*, the evolution of urban water management in ancient civilization is discussed. During the Neolithic age (*ca*. 10,000–3000 BC), the first successful efforts to control the flow of water were driven by agricultural needs (irrigation). During the Bronze Age, the first water treatment attempts to improve water quality were made. An ancient Hindu source presents, probably, the first water standard, dating 4000 years ago: It dictates that the dirty water should be exposed to the sun and then a hot copper bar to be inserted seven times in it, followed by filtration, cooling and storage in a clay jar. Gradually, as the small settlements grew to cities, the sources of fresh water within the boundaries of the city were not sufficient for the needs of the larger population. Thus, additional water quantities had to be brought from sources outside the boundaries of the city. In Greece, the Hellenistic period marks significant developments in hydraulics, which allowed the construction and operation of aqueducts, cisterns, wells, harbours, water supply systems, baths, toilets, and sewerage and drainage systems. During the Roman period, the technological developments shifted to longer and bigger water transportation systems, namely aqueducts. The basic development in the Byzantine times was the construction of large scale cisterns.

Chapter 3 deals with the historical evidence on waterborne diseases and their prevention. The necessity of water treatment was well understood by ancient peoples. It can be inferred from the writings of

Hippocrates, the father of medicine, that water was considered an important issue for public health. "Plagues" were described and in particular associated with the decimation of the Greek Army near the end of the Trojan War and with massive epidemics in Roman history. The understanding of the ability of pathogens to be transmitted from person to person was said to arise during the Plague of Athens, which occurred in Athens in 430 BC, killing about 30,000 people. Hypotheses have been developed on the aetiology of diseases including influenza, smallpox, bubonic plague, typhus and Staphylococcus. However, even in 1860, while clean water was known to be important, the role of contaminated water in disease transmission was not understood and there were few options for treating water. Sanitation and hygiene were the order of the day due to aesthetic reasons (bad odours) which were still associated with ill health. In recent years, advancements in water treatment technologies and engineering have made outbreaks of water-related diseases more of a rarity for those people fortunate to have sewage treatment and access to drinking water that is consistently clean and safe. However due to a combination of factors (including conflicts, natural disasters, poverty and treatment failure), illness and death due to diseases such as cholera, typhoid and cryptosporidiosis continue to affect millions of people throughout the world.

In *Chapter 4,* the availability of water, especially in Near and Middle East, is examined, along with characteristic examples of water supply practices from ancient Greece. At the earliest stage, people performed all activities near rivers and streams and the surface water was the main source of water supply. Groundwater was also exploited, particularly for water supply. The springs are the natural emergence of groundwater, but drilling wells was also common. Egyptians developed drilling systems in rocks as early as 3000 BC. In Minoan Crete, wells (with a depth of 10–20 m and diameter less than 5 m) and springs were used for water supply. The groundwater exploitation in Cyprus started 7000 years ago and it is connected with the oldest dug wells of the world. The ancient Chinese also developed a drilling tool for water wells, which in principle, is similar to modern machines.

21.2 WATER TECHNOLOGIES IN DIFFERENT CIVILIZATIONS

Civilization prosperity and collapse are strongly related to the availability of water and infrastructures for its use. The Indus Valley civilization (3000–1500 BC) is thought to have collapsed because of the course shift of the Indus River and the continued salinisation. Mohenjo Daro in the Indus Valley (Pakistan) declined after 2000 BC possibly due to climate change, river shifts, and water resources management problems. Droughts possibly caused the collapse of the Akkadian Empire in Mesopotamia in *ca.* 2170 BC. The Minoan civilization declined possibly due to an earthquake damaging the terracotta pipes and other network devices and disrupting the water supply (Gorokhovick, 2005). In addition, the seismic activity could have influenced the hydrological conditions and probably disrupted the groundwater aquifers. Thus, the water supply from all aqueducts (Knossos, Malia, and Tylissos) and wells (Knossos, Palaikastro, and Zakros) were affected. Petra, the Nabataean civilization capital in south-western Jordan, collapsed as a result of the disruption of the elaborate water supply systems caused by an earthquake in May of 363 AD.

In China, as rivers and canals usually pass through the cities, flooding was, historically, the main cause of collapse. A total of 27 floods occurred in Luoyang during the period of the Sui (581–618 AD) and Tang (618–907 AC) Dynasties (*Chapter 8*). After the fall of the Northern Song Dynasty, Kaifeng was severely damaged during the Yuan and Jin Dynasties. Flood disasters were frequent after 1194 AD, when the Yellow River changed its course and flowed towards the south–east. From the Jin to the Qing Dynasty, the Yellow River flooded Kaifeng six times and the nearby areas 40 times. Finally, the once prosperous Kaifeng fell into oblivion (*Chapter 8*). However, the extreme behaviour of nature as a key element on the history of mankind, is best understood by studying the ancient civilizations in the

American continent: crisis overtook all the classic civilizations of Mesoamerica (including the Maya), forcing the abandonment of most of the cities.

In most cases, no conclusive evidence as to why civilizations collapse exists. However, of all natural disasters, the role of drought must have been, historically, the most underestimated. Drought differs from other natural disasters in the following points:

(a) It affects greater populations than any other natural disaster.
(b) It develops slowly and in concealment. It is thus difficult to determine its beginning and its end, and its effects accumulate over a long time and can persist for several years after its expiration.
(c) Its social effects are less visible and extend over much larger geographic areas than other natural disasters (e.g. floods, fires, and earthquakes).
(d) It is difficult to quantify its effects since drought rarely results in destruction of infrastructure.

In contrast, prosperity of civilizations is ever related to successful addressing of water-related challenges. Eleven chapters of this book provide characteristic examples from different civilizations.

Chapter 5 deals with water and water supply technologies in Ancient Iran. With the exception of areas close to rivers, the sole available resource for water is ground water. Therefore, ancient Iranians invented the qanats, through which, they were able to bring water from underground up to the surface without spending any energy thereby solving the problem of agricultural water supply. The history of the ancient dams in Iraq dates back 3000 years and shows that ancient Iranians were among the pioneers in dam construction. In Iran, one of the greatest deeds of Darius the Great was the creation of a "Water Organisation", whose head, called "Ao-Tar" or Water Master, controlled the qanats, dams, rivers and so on. Other developments include subterranean water reservoirs, ice chambers and water mills. Water supply and drainage of the Persepolis complex were first built during the reign of Darius the Great (521–486 BC).

In Iran, over millennia, 22,000 qanat units were constructed, comprising more than 270,000 km of underground channels. In the 1960s, 75% of all the water used in that country for irrigation and domestic consumption was provided by such systems. Qanats were the symbol of sustainable development and operation of groundwater resources, whereas after the more recent expansion of wells as well as the intense use of pumps, these resources, which should be preserved for future generations, are unfortunately facing serious threats.

Over the centuries, the qanat technology was transferred to all civilizations and became known with different names such as "karez" Afghanistan and Pakistan), "kanerjing" (China), "falaj" (United Arab Emirates) and "foggara/fughara" (North Africa). Despite this diffusion of qanat technology throughout the ancient world, the construction of similar structures called puquios in America (*Chapter 11*) indicates a parallel evolution of a similar technology from civilizations that did not communicate.

Chapter 6 deals with the development of water supply in Egypt, and the role of the Nile River in the civilization, life and history of the Egyptian nation. In 3000 BC, Pharaoh Menes began with the construction of basins to contain the flood water, digging canals and irrigation ditches to reclaim the marshy land. He is also credited for diverting the course of the Nile to build the city of Memphis. The Nilometers, which measured the water level in the Nile, were the first hydrological instruments in the world. As a result of their continuous operation through the centuries, the Nile has offered the largest hydrological time series.

Egyptians suffered many losses due to yearly flooding of the Nile. As an effect of that, the first form of government appeared when the Egyptians organised their efforts under leadership of a single authority to avoid these disasters. As time passed, the leader (pharaoh) became more important with more power and influence on Egyptians.

Chapter 7 deals with the impact of climate changes on the evolution of the water supply works in the region of Jerusalem. The oldest water works date to the middle Bronze Age (*ca.* 1800–1500 BC). These water works were altered and extended in later periods. It is assumed that three main reasons dictated the nature of water works in Jerusalem from its earliest history until today: (a) the short-term climatic factor, namely the dry summer months and the occasional years of drought; (b) the long-term climatic factor of the past periods of wet and dry climates; and (c) the need for security which dictated the building of the often fortified settlements on the peak of hills and thus creating the need for secure access to the nearest water resource in the valley. In periods of warming, as happened throughout history, the move of the desert line north has caused aridisation and desertion of cities. During such extreme periods of droughts, springs, particularly those fed by a local perched water table as in the region of Jerusalem, may have dried up and lead to the abandonment of the site. The main conclusion of Chapter 5 is that warm climates spelled dryness and the first to be influenced were the small springs with a limited recharge area and small storage capacity. The discharge of bigger springs was also reduced once the warm and dry period extended over a longer period. Thus the bigger the storage capacity the better are the chances that the storage will mitigate the negative impact of dry periods.

In Jerusalem, the low topographical extension of the Eastern Hill was chosen for settlement instead of the higher level of the Temple Mount (or any of the other neighbouring hills) mainly because it is nearer to the perennial Gihon spring. The oldest part of the city was built on the rather permeable Bina limestone. Thus the construction of cisterns had to wait until the invention of impermeable plaster sealing the bottom and the walls of these structures. The appearance of iron tools for digging water cisterns in bedrock and the invention of impermeable plaster were essential for the operational function of cisterns and aqueducts that enabled the early Israelis to settle in the hills.

In *Chapter 8*, the history of water supply of Chinese dynasties is extensively covered. Despite the unparallel for the ancient world scale of the cities, water infrastructures and their harmonisation with the environment are remarkable. Site selection was of great importance as well. According to the book Guan Zi "the capital should be built either at foot of a great mountain or near a grand river; stay away from dry lands at high elevation in order to get enough water, and stay away from water at low elevation to prevent flood and save drainage and embankment".

Several ancient water supply schemes are still in use, or were in use until the 19th century, evidence of their remarkably advanced design and implementation. Among them, the famous Dujiang irrigation system, built in the third century BC is still in use. In Nanjing, the 16 m high East Water Gate built in the 14th century is still well preserved today. Starting from the 13th century AD, the Old Town of Lijiang built up a water supply system which met the needs of the inhabitants until the end of the 20th century. They are good examples of true sustainability over hundreds of years and are worth considering and using as a reference in modern city planning.

The Karez, a system comparable to the Persian qanat, was developed in Turpan city. With the modern fast agricultural development and overuse of electrically pumped groundwater, the importance of karezes declined and many ran dry because of lowered groundwater levels. Since the end of the 1950s, modern water conservancy projects such as storage reservoirs, diversion works, and pumping wells increased the amount of available water and began to slowly replace the ancient, labour-intensive karez systems. However, sustainable water use and management has become an increasingly critical issue for the further development of the Turpan Region.

The abandonment of small-scale and traditional infrastructures that has occurred very recently is a big step backwards. Apart from qanats, practices and technologies (examined in other Chapters) that should be revisited include: (a) Water cisterns used to collect and storage rainwater; (b) Small-scale decentralised water supply systems; and (c) Water recycling and reuse of water.

Chapter 9 deals with the evolution of water supply in the island of Crete, presenting a number of characteristic examples in selected sites extending from the early Minoan era to present times. During the Minoan era the principles of water supply and sanitation were developed and primitive distribution systems were designed and implemented in the palaces and other settlements on a rather small-scale, but in a cost-effective, decentralised and sustainable manner. The Minoan achievements were not totally forgotten during the Dark Ages. Possible interconnections with Mycenaean and Achaean Greeks had interacted on water supply developments both in Crete and in the mainland of Greece. The prehistoric technologies were further developed during the Hellenistic, Roman and Venetian periods, mainly at an enlarged scale of implementation.

Thus, Crete displays a diachronic continuity on development of water supply technologies. As climatic and geohydrological conditions in the island highly varied through history, water availability was spatially and temporal uneven. It is clear that water technologies, developed through the centuries, such as aqueducts, cisterns, wells and other systems were created and adapted to the local conditions and present an ecological sustainable system with a considerable potential in water supply.

Chapter 10 explores urban water management practices of ancient Greece, from the Archaic to the Hellenistic years. In the Archaic Greek antiquity cities tended to be located at dry places, at a distance from rivers or lakes. Under tyranny, cities grew significantly and the first large-scale urban water infrastructures were developed. The period of democracy that followed, with its small-scale structures and its non-structural measures constitutes a lesson of sustainable management and marks the importance of the institutional progress in water management. During the Hellenistic period, the evolution of the "designed city" is mainly reflected in the scale of the projects, which resulted in water adequacy and more widespread hygienic water use. Some lessons learned are:

- City planning has to include urban water criteria; protection from floods should be a major consideration.
- The use of small-scale infrastructures, in parallel to the large-scale ones, is a big step towards sustainability and resilience. The principles and practices of sustainable water use should not be forgotten even in periods of water adequacy.
- Safety and security of water supply in emergency situations, including turbulent and war periods, should be kept in mind in our designs of urban water systems.

Chapter 11 deals with water supply in Pre-Columbian civilizations in Ancient Peru and South America. Among the environmental conditions that affect water affairs in these areas is the El Niño phenomenon. The institutional and organisational framework have seen great developments as testified in the case of the Inca Empire "Tahuantinsuyo". Among the technological heritage of the Pre-Columbian civilizations, a very interesting case is the so-called puquios, a solution similar to Iranian qanats that indicates a parallel evolution of similar technologies in different places of the world.

Chapter 12 deals with the evolution of water supply in Cyprus and presents a number of characteristic examples in selected sites chronologically extending from the Neolithic age (*ca.* 8500–3900 BC) to present times. During the Hellenistic period, Cyprus was hit by strong earthquakes and several cities, towns and infrastructures were destroyed. Only ancient Amathus was less affected. After the destructive earthquakes, the cities of Cyprus were rebuilt. The Romans built their own structures on the foundations of structures from the Hellenistic period. To satisfy the water demand, water supply technologies such as aqueducts, cisterns, water conduits, chain-of-wells and other systems were developed. The trade with Egypt, Crete, and other Aegean islands had positive influences in the development of the water supply technologies.

In *Chapter 13*, the overview of the different elements of the Roman water supply systems that operated in Middle East from the first century BC to the seventh century AD is presented. Romans mostly settled in

Hellenistic cities with functioning water systems. In most cases Romans reworked and further developed these systems. What really differentiates the Roman period is the scale of the projects, due to the ever increasing water demand. Romans, contrary to Greeks, did not have sufficient documentation for their technologies: only a few Roman writings on water engineering practices have been preserved. Thus, the progress of that period is best demonstrated by the well preserved water supply systems throughout Middle East. Examples include water transportation by aqueducts, water distribution by masonry channels and pipelines and water facilities, such as latrines, baths, toilets, fountains, and cisterns. The inverted siphons, practised in Aspendos, represent a more sophisticated design solution. In Byzantine times, a decline in the quality of the water systems is commonly admitted. However, some counter examples exist.

Chapter 14 compares ancient Greek and Roman water systems and their elements, such as aqueducts, cisterns, reservoirs, water distribution systems, fountains, baths and thermae, and water use for recreational and/or environmental purposes. A continuation of water supply technologies between the two civilizations is observed. During the Roman period sound engineering principles, which were used by Greeks for centuries, were further extended basically by changing the scale of the hydraulic works. Comparison shows that water technologies in Minoan, Greek, and Roman civilizations are not too different from the modern practice, and that hydraulic works of the antiquity are characterised by simplicity, robustness of operation, and the absence of complex controls. Today, many of the water resources exploited in the Roman age are still in use.

Chapter 15 addresses water supply technologies of the pre-Columbian societies in the modern-day south–western United States and in Mesoamerica followed by a discussion of the technology advancement during the post-Columbian era. Three examples from this chapter are very informative:

(a) The Hohokam built the most complex irrigation system in the desert lowlands of the Salt-Gila River Basin, Arizona in and around the present day Phoenix, Arizona. They built more than 483 km of major canals and over 1126 km of distribution canals in the Salt River Valley, which have been identified. Hohokam irrigation systems consisted of three basic types of canals: main canals, distribution canals, and field laterals.

(b) Chaco Canyon is situated in the San Juan Basin in north–western New Mexico. The basin has limited surface water, most of which is discharged from ephemeral washes and arroyos. The water, collected from the side canyon that drained from the top of the upper mesa, was diverted into canals by either an earthen or a masonry dam near the mouth of the side canyon. These canals averaged 4.5 m in width and 1.4 m in depth; some were lined with stone slabs and others were bordered by masonry walls. The canals ended at a masonry head gate, where water was then diverted to the fields in small ditches or to overflow ponds and small reservoirs.

(c) Xochicalco (in the place of the house of flowers), was located on a hill top approximately 38 km from modern day Cuernavaca, Mexico. There were no rivers or streams or wells to obtain water, so rainwater harvesting was the source of water. Rainwater was collected in the large plaza area and conveyed using drainage structures and drainage ditches into cisterns. From the cisterns water was conveyed to other areas of the city using pipes.

21.3 SOME MAJOR CITIES

The next five chapters are devoted to urban infrastructure at some important sites, most of them in the Mediterranean. As most Mediterranean cities were located at a distance from large water systems, climatic variability played a significant role in their evolution. However, during the Classical years and afterwards, no major collapse occurred. The consequences of climatic variability in a certain time period were less intense and were generally connected to the decline of a specific city in a certain time period. The main reasons for the sustainability of the Mediterranean cultures are:

- the existence of distributed urban systems;
- the combination of small-scale and large-scale infrastructures practised in most cases;
- the wise site selection with respect to water resources (not close to large water systems in order to avoid floods, but with water adequacy); and
- the continuous adaptation to new needs.

As ancient Greco-Roman civilization has left us with more evidence (written and archaeological) than any other civilization, the non-stop struggle to cover the demand for water is reflected in the history of cities like Athens (*Chapter 16*) and Rome (*Chapter 17*). The history of cities like Iraklion (*Chapter 18*) and Barcelona (*Chapter 19*) reflects the dependence of water supply infrastructures on the historical and cultural changes.

Chapter 16 investigates the water supply of Athens, Greece under rather adverse conditions, from sixth century BC up to the sixth century AD. Because of the inadequate capacity of springs and rivers in the vicinity of ancient Athens, the development of the underground water resources, became crucial. The water supply infrastructure was based on good geological knowledge, as testified by Solon's rules for wise management of ground water, as well as Plato's discussions on the geological past of limestone hills surrounding the ancient city. Thus, a systematic exploitation of the limited local water resources (springs, galleries and wells) in the city was complemented by transfer of surface and ground water from the surrounding mountainous Attica. The Athenian water system was characterised by (i) combination of small- and large-scale hydraulic works, (ii) sustainability, and (iii) continuous adaptation to new needs (construction of several aqueducts to the city). Drainage and sewerage systems were developed in parallel to the water supply network. It is impressive that some elements of the water supply and drainage designed and constructed in antiquity in a practical and sustainable manner are still operational today.

As described in *Chapter 17*, Rome is another example of water technologies' evolution, whose origin dates back to the Etruscan times. During the Roman Empire, several aqueducts were constructed (11 of which are described). The historical development of water supply in Rome is not a local affair but it served as an example of universal dimension. The developments in Rome, including those in water technology and management, constitute an important part of the history of human civilization and were spread and assimilated in many parts of the world during many centuries.

In the case of the historical city of Iraklion, Crete, as described in *Chapter 18*, the historical developments of water supply in the antiquity were minor. Even during the Roman and Byzantine periods very little progress was made. On the other hand during the Venetian presence on the island the *Morosini* aqueduct of a total length 15.5 km, was designed and constructed within only 15 months. This was one of the major water projects for the city. At the time of the Ottoman occupation of Crete, with the exception of the short period of the Egyptian rule, there was no significant water supply work for Iraklion. After the Union of the island with Greece serious progress in new water supply projects was made in Iraklion and the whole of Crete.

The history of the city of Barcelona (*Chapter 19*) seems to have begun in the Roman period in the time of the Roman Emperor Augustus, when the city was named as "Colonia Iulia Augusta Paterna Faventia Barcino". During that period several aqueducts were constructed. The aqueduct, which started at the Besòs River, was still used during the Late Empire and the Visigoth era. It was most probably abandoned in the eighth or ninth century. The evolution of Barcelona's water supply was highly influenced throughout the 18th century due to: (a) the establishment of the Spanish Bourbon monarchy; and (b) the economic and demographic impulse experienced by Catalonia as the century advanced, with a clear repercussion on the urban expansion of Barcelona. Over the last two centuries, Barcelona's water supplies experienced several achievements to meet the increased water demand.

Chapter 20 relates a different story, as it examines the evolution of water services in Mexico City since the Aztecs' time. It was conceived in the Aztec period directly due to the availability of water. The primeval city located on an island (Tenochtitlán), whose residents had access to sufficient water and planned their whole social and economic growth around this source, became Mexico City, a mega city suffering both from lack of water and flooding. During the Spanish rule, the Mexico City council created the water council to verify water permits and to collect water taxes but also to operate the hydraulic infrastructure. In addition, the city council appointed a water judge, 'to oversee the use of water', who was in charge of updating the use of Mercedes (e.g. the royal concession for industrial or agricultural use of water).

IWA Publishing's authorised EU representative for General Product Safety Regulations is Diane D'Arras, 15 rue Duret, 75116 Paris, France, e-mail: safety@iwap.co.uk.

Printed and bound by CPI Group (UK) Ltd, Croydon, CR0 4YY
12/05/2026
02108863-0014